Quantum Reality

As probably the most successful scientific theory ever created, quantum theory has profoundly changed our view of the world and extended the limits of our knowledge, impacting both the theoretical interpretation of a tremendous range of phenomena and the practical development of a host of technological breakthroughs. Yet for all its success, quantum theory remains utterly baffling.

Quantum Reality: Theory and Philosophy, Second Edition cuts through much of the confusion to provide readers with an exploration of quantum theory that is as authoritatively comprehensive as it is intriguingly comprehensible. The book has been fully updated throughout to include the latest results in quantum entanglement, the theory and practical applications of quantum computing, quantum cosmology and quantum gravity. Needing little more than a school level physics and mathematics background, this volume requires only an interest in understanding how quantum theory came to be and the myriad ways it both explains how our universe functions and extends the reach of human knowledge.

Written by well-known physics author and teacher Dr. Jonathan Allday, this highly engaging work:

- Presents a thorough grounding in the theoretical machinery of quantum physics
- Offers a whistle-stop tour through the early part of the 20th century when the founding fathers of quantum theory forever altered the frontiers of human thought
- Provides an example-filled interpretation of the theory, its applications, and its pinnacle in quantum field theory (QFT), so crucial in shaping ideas about the nature of reality
- Separates fact from speculation regarding quantum physics' ability to provide a starting point for philosophical queries into ultimate understanding and the limits of science

The world beneath the one that we experience with our senses is profoundly mysterious, and while we may never completely unravel that mystery, quantum theory allows us to come closer than ever to understanding where the science leaves off and the mystery begins. **Quantum Reality: Theory and Philosophy, Second Edition** makes that understanding accessible to anyone possessing a quest for knowledge and a sense of awe.

Quantum Reality

Theory and Philosophy

Second Edition

Jonathan Allday

CRC Press is an imprint of the
Taylor & Francis Group, an **informa** business

Cover Image: Harris & Bush, MIT

Second edition published 2023
by CRC Press
6000 Broken Sound Parkway NW, Suite 300, Boca Raton, FL 33487-2742

and by CRC Press
4 Park Square, Milton Park, Abingdon, Oxon, OX14 4RN

First edition published by CRC Press 2009

CRC Press is an imprint of Taylor & Francis Group, LLC

Library of Congress Cataloging-in-Publication Data
Names: Allday, Jonathan, author.
Title: Quantum reality : theory and philosophy / Jonathan Allday.
Description: Second edition. | Abingdon, Oxon ; Boca Raton, FL : CRC
Press, 2022. | Includes bibliographical references and index. |
Identifiers: LCCN 2022014186 | ISBN 9781032127347 (hbk) |
ISBN 9781032122380 (pbk) | ISBN 9781003225997 (ebk)
Subjects: LCSH: Quantum theory. | Quantum theory—Philosophy.
Classification: LCC QC174.12 .A45 2022 | DDC 530.12—dc23/eng20220820
LC record available at https://lccn.loc.gov/2022014186

ISBN: 978-1-032-12734-7 (hbk)
ISBN: 978-1-032-12238-0 (pbk)
ISBN: 978-1-003-22599-7 (ebk)

DOI: 10.1201/9781003225997

Typeset in Times
by codeMantra

To my parents

My point, which you'll hear me rant about again, is that at both the conceptual and the mathematical level, quantum mechanics is not just a funny-looking reformulation of classical physics. The two physical theories are fundamentally, physically different.

Michael A. Morrison

The average quantum mechanic is no more philosophical than the average motor mechanic.

Rev. Dr. John Polkinghorne KBE FRS

Of two alternative futures which we conceive, both may now be really possible; and the one become impossible only at the very moment when the other excludes it by becoming real itself. Indeterminism thus denies the world to be one unbending unit of fact. It says there is a certain ultimate pluralism in it; and, so saying, it corroborates our ordinary unsophisticated view of things. To that view, actualities seem to float in a wider sea of possibilities from out of which they are chosen; and, somewhere, indeterminism says, such possibilities exist, and form a part of truth.

William James

Physics may reveal the mind of God, but only if he happens to be thinking about dirt.

Ken Wilber

Science… proceeds by elucidation, so that feats of genius can become ordinary learning for beginners.

Roland Omnès

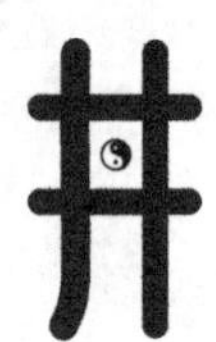

Pictogram Credit: Emrys

Contents

PART 2

Forward

Quantum mechanics is "at first glance and at least in part, a mathematical machine for predicting the behaviors of microscopic particles — or, at least, of the measuring instruments we use to explore those behaviors [Ismael, 2020]." Why does the machine work? This is the question of how to 'interpret' quantum mechanics. It turns out to be intensely controversial. Carroll recalls,

> At a workshop attended by expert researchers in quantum mechanics... Max Tegmark took an...unscientific poll of the participants' favored interpretation ....The Copenhagen interpretation came in first with thirteen votes, while the many-worlds interpretation came in second with eight. Another nine votes were scattered among other alternatives. Most interesting, eighteen votes were cast for "none of the above/undecided." And these are the experts [2010b, 402, n. 199].

There are two reasons why this is not really disagreement over the interpretation of quantum mechanics, in any ordinary sense of 'interpretation'. First, at stake is not what people happen to mean by technical terms, like 'state vector', 'collapse', and so on. This would be a question of (presumably empirical) natural language semantics, and would tell us nothing about the physical world. Second, the 'interpretations' do not even all agree on the machine. For example, Bohmian mechanics (Chapter 31) amends the equations, and makes subtly different predictions.

Physics has made impressive progress without addressing the interpretational question. But there is a growing sense that progress on the deepest mysteries, like how to reconcile quantum theory and General Relativity, may require its resolution [Hossenfelder & Palmer, 2020]. Philosophy is becoming harder to avoid. The situation resembles the one in the early 20th century, when philosophical reflection inspired some of the most penetrating arguments in the history of physics, such as the EPR argument and the Schrödinger Cat thought experiment.

This book is unique in the physics landscape. It is not a textbook, a guide to solving the Schrödinger equation. It is also not a philosophy text, assuming familiarity with metaphysics and epistemology. It is a serious survey for non-specialists of what the mathematics could mean. It offers, all in one place, an accessible introduction to the theory (up through a sketch of quantum field theory), an overview of the 'no-go' results, and a careful discussion of some important interpretations. Philosophers will appreciate the self-contained introduction to the theory, while physicists will learn from the philosophical analysis. Newcomers will delight in all of it.

We need more books like this. Understanding the nature of value, consciousness, mathematical truth, and possibility and necessity, will also require insights from both philosophy and science. Deep interaction between philosophy and mathematics has already born plentiful fruits outside of philosophy proper, like proof theory, model theory, and theoretical computer science. We may hope that a meaningful exchange between philosophy and physics will be comparably fecund.

Justin Clarke-Doane
Columbia University
IAS, Princeton

Preface

The world is not what it seems. Behind the apparent solidity of everyday objects lies a seething shadow world of potentiality which defies easy description, as it is so different from our everyday experience. In some manner, familiar objects such as solid tables, cricket balls, stars, and galaxies arise from what transpires underneath. We do not know precisely how this comes about.

There is a theory that describes the underlying world: *quantum theory*. It is one of the most successful scientific theories of all time and it has profoundly changed our view of the world.

Quantum understanding is vital to our current science and technology; its application is not restricted to esoteric experiments in high-energy physics. The theory certainly helps us understand the inner mechanisms of neutron stars, superconducting materials, and possibly even the early moments of the Big Bang, but without it we would have no appreciation of why the table on which this laptop sits is solid. The LED bulb on the table next to me is generating light (which is a quantum phenomenon) as electrical charge in the form of tiny particles called *electrons* (which we need quantum theory to understand) are passing through a material and transferring energy. The material in the wires leading to the bulb has a property called *resistance*, which can only be fully understood by applying quantum laws.

Yet for all its success, aspects of quantum theory remain utterly baffling. While the mathematics is clear (albeit occasionally hard to deal with), interpreting what it is saying about the world remains a profound challenge.

In the 100-odd years since quantum theory was born, there have been many books written that attempt to explain quantum physics to the interested amateur. This is an important endeavor. The world beneath the one that we experience with our senses is profoundly mysterious, and there are some important philosophical messages about the nature of reality and the limits of science that need to be put across. I hope that this book can contribute to that effort.

THANKS TO THE FOLLOWING (FIRST EDITION)

The Master and Fellows of Gonville and Caius College, Cambridge, and especially Dr. Jimmy Altham for access to the college library and the time and space to work during the summer of 2002 while I was writing the first edition.

Rev. Dr. John Polkinghorne KBE FRS (1930–2021) for an inspiring lunchtime discussion and his encouraging support.

Dr. Lewis Ryder (1941–2018) for reading parts of the manuscript.

Dr. David Wallace for reading the section on the Many Worlds interpretation.

Dr. Grahem Farmello for reading the section on Dirac.

Greg Manson, Md, for help with uncertainty.

THANKS TO THE FOLLOWING (SECOND EDITION)

Many thanks to Mrs Rebecca Hodges-Davies, Dr Kirsten Barr and latterly Dr Danny Kielty all of CRC Press, Taylor & Francis Group for guiding me and this manuscript through the production process. Also the team at codemantra.

Special thanks to Scott Hayek, Bruce Newhall, John Sweeney, Neal Brower (all of the Johns Hopkins University Applied Physics Laboratory, retired) and John R. Moore (retired) who reached out to me with regard to General Relativity and then (undaunted) ventured into quantum reality. They read and commented on sections of the revision with sympathy and in detail. Somehow, they managed to embody my ideal reader and still be willing to talk to me. I'm looking forward to sometime meeting you guys in person!

I am grateful to Dr Philip Davies of Bournemouth University for also contacting me, offering the very flattering, and daunting, prospect of being interviewed on YouTube and then managing to edit me into sounding reasonably coherent. His willingness to then read and comment on some of this second edition was a welcome bonus.

Many thanks to my school friend Professor Simon Hands, University of Liverpool, who read many of these chapters and challenged various bits of wonky physics. Of course, any mistakes remain my responsibility. Here's to the classes of 79/80.

None of this could be done without the continual support of my family and friends. Carolyn has had to carry a lot over the last 18 months never mind the additional burden of a whiney author. My love and thanks to her. Unfortunately, I have already started again…

And finally, many thanks to Emrys who has shown me that the world is exactly as strange as I suspected it to be.

Jonathan Allday
2nd Edition
16th January 2020, Yorkshire
13th February 2022, Worcestershire
Jallday40r@me.com

About the Author

For 30 years, Dr. Jonathan Allday taught physics at a range of schools in the UK. After taking his first degree in Natural Sciences at Cambridge, he moved to Liverpool University where he gained a PhD in particle physics in 1989. While carrying out his research, Dr. Allday joined a group of academics and teachers working on an optional syllabus to be incorporated into A-level physics. This new option was designed to bring students up to date with advances in particle physics and cosmology. An examining board accepted the syllabus in 1993, and now similar components appear on many advanced courses.

Shortly after this, Dr. Allday started work on *Quarks Leptons and the Big Bang*, now published by Taylor & Francis and available in its third edition, which was intended as a rigorous but accessible introduction to these topics. Since then, he has also written *Apollo in Perspective, Quantum Reality* and *Space-time*, co-authored a successful textbook, and contributed to an encyclopedia for young scientists.

Dr. Allday's interest in the physics and philosophy of the quantum world dates back to his school days, where he remembers reading an autobiography of Einstein. As an undergraduate, he specialized in relativistic quantum mechanics and field theory, writing his third-year project on Bell's inequality, as well as taking a minor course in the history and philosophy of science. The idea for this book occurred during a summer placement at Cambridge, hosted by Gonville and Caius College.

Other than physics, Dr. Allday has a keen interest in cricket and Formula 1.

Introduction

> Suppose for example that quantum mechanics were found to resist precise formulation. Suppose that when formulation beyond FAPP [for all practical purposes] is attempted, we find an unmovable finger pointing outside the subject, to the mind of the observer, to the Hindu scriptures, to God, or even only Gravitation? Would that not be very, very interesting?
>
> *J S Bell*[1]
> [my addition]

I.1 PHYSICS

Quantum theory has passed through three distinct stages during its evolution. The first is generally referred to as 'old quantum theory'. During a period roughly between 1900 and 1925 Planck, Bohr and Einstein, among others, grappled with various experimental and theoretical crises by patching up classical (Newtonian) mechanics with some new quantum ideas applied as band-aids. This is not to disparage the work; it was vitally important, and you can only do what is possible at the time. Gradually, however, Heisenberg, Jordan, Schrödinger, Born, Bose, de Broglie and Dirac developed a coherent structure that would be recognized as quantum theory as it exists now. This broadly took place in the mid-1920, but progress did not stop there. Born, Heisenberg, Jordan, Dirac and Pauli pieced together quantum field theory between 1926 and 1929. Other physicists, such as Fermi and Fock later added refinements to the basic structure. In some ways, this evolution can be seen as a progressive acceptance of quantum concepts that were increasingly distant from classical thinking.

This book is divided into three parts to somewhat mirror this evolution. Part 1 covers the basic conceptual and mathematical machinery of mid-1920s quantum theory. In Part 2 we take a whistle-stop tour through old quantum theory and the development of its successor by focussing on the work and views of some of the central figures involved[2]. Finally, in Part 3 we survey some important interpretations and tackle an outline of quantum field theory. At various points, we will draw attention to the key interpretive issues and discuss what is at stake in our view of reality.

The quantum world is very different to the picture painted by classical physics. That being the case and given the extreme nature of some of these differences, we might question the wisdom of accepting the quantum view. The answer is that quantum theory is unsurpassed in its record of explaining different aspects of nature, some of which we will touch upon later. A few significant areas include:

- The otherwise mysterious aspects of radioactivity which found a natural explanation within quantum theory.
- The quantum theory of electrons in atoms allowed us to understand and categorize the spectral lines of atoms and molecules.
- As a parallel development, chemistry, the nature of the chemical bond, the periodic table and the structure of molecules all gained a secure theoretical basis.
- Applying quantum theory to matter allowed an understanding of some anomalous aspects of heat capacities, the conductive (thermal and electrical) properties of materials, and the development of the band theory of solids. Without this key theoretical advance, vital aspects of modern technology, including the development of semi-conductors, would not have been possible.
- Equally, the magnetic properties of materials are a quantum issue. This aspect of theory has led to the development of important magnetic technologies, such as MRI imaging.

DOI: 10.1201/9781003225997-1

- Quantum theory has also proven to be flexible enough to tackle some surprising discoveries, such as superconductivity, the nature of stellar interiors, nuclear fission and fusion and aspects of Big Bang theory.

Given this stellar record from a broad range of different aspects of physical theory, we are bound to take the concepts of quantum theory seriously despite their counter-intuitive aspects.

1.2 PHILOSOPHY

> Scientists are very ambitious. They're very competitive. If they really thought philosophy would help them, they'd learn it and use it. They don't.
>
> *L. Wolpert*[3]

Now I need to say something about philosophy. Don't put the book down, it will all be over in a minute.

Scientists can be quite disparaging about philosophy, sometimes while in the process of espousing, quite stridently, a philosophical position of their own. The fact is all scientific theories point beyond themselves to some degree. At the very least, they open up questions such as: 'what does this theory tell us about reality?', 'what aspects of this theory, if any, directly relate to real elements of the universe?', 'how do we know that any scientific theory is true, and what does truth mean in this context?', 'how does science work as a reliable tool for scrutinizing the world?'. Such issues cannot be fully addressed from within science.

Scientific advances come from the joint application of *experiment* and *theory*. While it is difficult to maintain that these are separate disciplines, when you analyse the situation in detail, the world is too strange and surprising to be understood in all its aspects purely by philosophical reflection. Who would have thought that quantum theory was likely? We need the constant nudge and corrective of experiment to direct the focus of theory.

However, not all the important questions about the world are directly addressable by science. Nor is it sufficient to simply define important questions as being those that are open to scientific techniques, not least because *that is in itself a philosophical position*. An argument along the lines of 'our subject makes progress and yours doesn't, so sucks boo to your subject' is hardly a mature analysis. Perhaps the apparent lack of progress in philosophy is not a reflection of any lack of rigour, but more to do with the difficulty of the problems it seeks to address.

Take the example of quantum theory. The mathematical structure of the theory has been in place for over 100 years, albeit subject to the occasional and useful refinement in that time[4]. However, the interpretation of the theory is still an open question. The fact that there are several competing views is a signal point. Establishing the correct interpretation is not itself a scientific matter. If each approach agrees on the physical content, they have the same predictive power hence they are experimentally indiscernible[5]. Nevertheless, they project radically different outlooks on the nature of reality (contrast the Many Worlds interpretation with Many Minds, just for one example). The choice that each individual makes is made, consciously, or not, on its consonance with their overall worldview. Philosophy naturally enters the discussion, even if tacitly.

Philosophers have a variety of views on the nature of knowledge. At issue are matters relating to how we know something, how reliable our knowledge is, whether all our knowledge comes from the world via our senses or are there some things that we just 'know' etc. Discussions of this kind are covered by a branch of philosophy called *epistemology*. A closely connected, but distinct, area is *ontology*. This is the inquiry into what is actually out there for us to know. As a rough example, the existence of electrons is a matter for ontology; how we know about them and their properties falls to epistemology.

Epistemologically there are two approaches to how science works, or rather what it is that science sets out to do.

If you are a *realist*, then you believe that science is an accurate map of what is really out there. The various ideas and pictures that we come up with (such as electrons, black holes, the Big Bang, and DNA) are elements of reality and we are discovering true information about the world. From this perspective, the purpose of science is clear: to find out as much as possible about what is going on in the world. To a realist, a good theory is one that convinces us that the things it speaks about are not just figments of our scientific imaginations.

However, you might be an *instrumentalist*, in which case you are not too bothered about the accuracy or reality of your ideas, as long as they fit the data and allow us to make accurate predictions. An instrumentalist may not believe that electrons are real. They will agree that various experiments produce clumps of data that can be gathered under the heading "that's an electron" and will use this data to predict another set of experimental readings under slightly different circumstances. However, they will draw short of committing to the objective existence of electrons. You do not have to believe that *Colonel Mustard* is a real person to have fun finding out if he is a murderer in the game *Cluedo*. To an instrumentalist, a good theory is one that allows us to play the game well.

Various scientists have embraced and promoted one approach or another over the years:

Realists

> Physicists believe that there exist real material things independent of our minds and our theories. We construct theories and invent words (such as electron, positron etc.) in an attempt to explain to ourselves what we know about our external world … we expect a satisfactory theory, as a good image of objective reality, to contain a counterpoint for every element of the physical world.
>
> *B. Podolsky*

> A complete, consistent, unified theory is only the first step: our goal is a complete understanding of the events around us, and of our own existence.
>
> *S. Hawking*

> The great wonder in the progress of science is that it has revealed to us a certain agreement between our thoughts and things …
>
> *L. de Broglie*

Instrumentalists

> I don't demand that a theory correspond to reality because I don't know what it is. Reality is not a quality you can test with litmus paper. All I'm concerned with is that the theory should predict the results of measurements
>
> *S. Hawking*

> In science we study the linkage of pointer readings with pointer readings.
>
> *A. Eddington*

There are arguments on both sides. A realist would say that the only satisfactory way of explaining the success of science is by believing that are talking about reality. An instrumentalist would counter by saying that in Newton's age we believed that time was the same for everyone, then Einstein comes along and declares that time is different depending on our state of motion, or if we happen to be in a gravity field. What next? Often our ideas of what is 'out there' change radically, so why believe any of it? If our ideas let us fly to the moon, cure diseases, and make good plastics, who cares?

Many scientists[6] go about earning their daily bread without being bothered about the philosophical niceties; "shut up and calculate" would be their motto. Unfortunately, tackling quantum physics raises questions that are difficult to put aside. It is all to do with the *state* of a quantum system. A realist has some trouble believing that a quantum state is an ontologically real thing, as it seems, at least in part, to depend on our knowledge about a system. An instrumentalist would have no problem believing that states are nothing more than a concise expression of our information about a system. More of a challenge would be explaining why the objects that we study behave in radically different ways if their state changes, which suggests that they have some ontological relevance.

Throughout this book, I am going to try and remain as neutral as possible and point out where realism and instrumentalism have their strengths when applied to quantum theory. You may find that exposure to these ideas forces you to refine your own thinking.

NOTES

1 J. S. Bell, *Speakable and Unspeakable in Quantum Mechanics*, Cambridge University Press; 2nd edition, 2004.
2 As I am not aware of a collective noun for these individuals, I have coined the term *Founding Fathers* to use in this book. I am amused by the whimsical implied connection to the framers. The unfortunate gender specificity in this phrase is a matter of historical fact. Mme Curie, for example, did very important work that should not be ignored, but it is not directly related to our story.
3 L. Wolpert (1929–2021), University College London, Round Table Debate: Science versus Philosophy? https://philosophynow.org/issues/27/Round_Table_Debate_Science_versus_Philosophy.
4 When we get to the chapter on Consistent Histories, we will see something of that nature.
5 Objective collapse is a clear exception to this. As we will see later, this entails some modification to the theory as it stands and hence has different predictions which are, just about, accessible by experiment. However, objective collapse addresses one, critically important, aspect of interpretation, but not the whole philosophical ballpark.
6 Perhaps the majority.

Part 1

1 Our First Encounter with the Quantum World
Light

1.1 SOME OPENING THOUGHTS

The first draft of this chapter was written while sitting in a college garden under a cloudless sky, with the bright sunlight flooding over some particularly well-manicured lawns (Figure 1.1).[1] I clearly remember struggling to see what I was typing over the reflected glare from my laptop screen.

I still find it hard to reconcile the beauty of such scenes with what I know about the nature of light. This is part of the mystery that shrouds quantum reality.

Large-scale (macroscopic) objects, such as trees, bushes, and cricket balls, are made up of small-scale (microscopic) things such as protons, neutrons, and electrons. The laws of physics that describe the large-scale world have been broadly understood since the 1700s. Our first tentative exploration of the physics of the small-scale started in the 1900s. As we rapidly came to realize, the laws governing the small-scale world describe behaviour that, judged by the standards of everyday experience, is utterly bizarre. It is very difficult to see how all the funny business going on at the atomic scale can underpin the regular, reliable world we spend our lives in.

This contrast between the microscopic world ('seen' via experiment) and the macroscopic world (experienced via our senses) is a theme that will recur throughout this book.

1.2 A LITTLE LIGHT READING

When you set out to understand some new phenomenon, it's a good idea to start by looking for similarities with something that you have already figured out. In the case of light, there appears to be two possible comparisons. Light might be a *wave* (a spread out, rather flappy thing that varies in both space and time, like the ripples on a pond) or a stream of *particles* (localized hard lumps, like cricket balls, that simply change their position with time). Up until the 1800s, there was no experimental way of settling the issue, so most people took sides on philosophical or theoretical grounds. Two big-name physicists squared up in opposite corners: Thomas Young favoured a wave view of light with Isaac Newton (see Figure 1.2) championing a particle interpretation.

In 1665 Newton made some fundamental discoveries about light while taking leave from his studies at Cambridge, which was under threat from the plague. In one of his classic experiments, he allowed a thin shaft of sunlight to fall on a glass prism, producing a spectrum of colours. Newton explained this by assuming that the colours of light corresponded to distinct types of particle. In this view, white light was a stream of all the different particles mixed together, rather than a distinct colour of itself. As they passed through the prism, the various types of particle interacted with the glass differently, which caused them to emerge along separate paths, separating to give the observed spectrum.

With Newton as a public supporter, [2] the particle view was bound to hold a certain sway, but the existence of sharp shadows near opaque objects did not hurt the argument.

To the casual glance, shadows have well-defined edges. This is tricky to explain if light is a type of wave. Other examples, such as water waves, clearly bend around objects that get in their way. So, if light is a wave, shadows should have rather fuzzy outlines, or so it was thought in Newton's time.

DOI: 10.1201/9781003225997-3

FIGURE 1.1 The opening of this chapter was written while sitting outside one of the windows of Harvey Court, part of Gonville and Caius College in Cambridge.

FIGURE 1.2 An engraving of Isaac Newton (1642–1727)—his pioneering experiments with light led him to propose that light was composed of a stream of particles.

However, not all were convinced, and in 1801, Young carried out an experiment that was sensitive enough to reveal the wave aspects of light. The key to Young's discovery was the use of two linked sources of light to produce an *interference pattern.* We will go into the details of how interference works in Section 1.5.1, but for a simple illustration imagine dropping two pebbles into an otherwise smooth surface of water. Ripples spread out from each impact point and inevitably overlap somewhere. The result is a complex pattern of motion on the surface of water: an interference

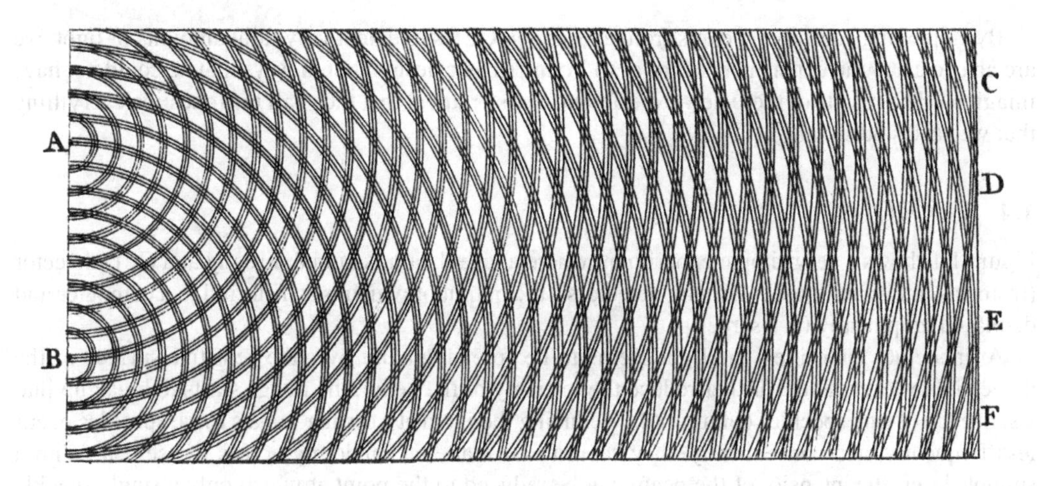

FIGURE 1.3 Thomas Young's original diagram explaining his interference experiment. A and B represent light sources that send out waves of light that spread out in circular patterns centred on each source. These waves look rather like the ripples that would spread out on a lake if pebbles were dropped into the water at A and B. Complex patterns are formed where the waves overlap. C, D, E, and F are places where light and dark bands would appear on a screen.

pattern. In Young's version, specially prepared light from two sources was directed to overlap on a screen. In the region of overlap, instead of a patch of illumination, a series of bands were seen. The natural explanation was that the waves from the two sources were combining, like the ripples on water, causing bright patches where they reinforced each other and dark regions where they got in each other's way (Figure 1.3).[3]

Young was able to use these observations to estimate the wavelength of light. For water waves the wavelength would be the distance between two neighbouring peaks (high points) on the surface. The wavelength of light is a little harder to interpret, as it is related to the electric and magnetic fields that comprise the light wave (Section 1.5.1). However, if we take a wave view of light, the colour is related to the wavelength, with red light being long wavelength compared to blue. Light's wavelength is incredibly tiny, in the region of one-tenth of a millionth of a meter. This explains why we observe sharp shadows. Waves will only bend round objects that are about the same size as their wavelength. The objects that we see casting shadows are much bigger than the wavelength of light, hence the light does not leak around them to blur the edges of the shows.

1.3 LASERS AND VIDEO CAMERAS

In the twentieth century, the study of light was revolutionized by the ability to produce precisely controlled beams that are tightly collimated and made of light that is all the same colour: *laser beams*. These days, lasers are among the most ubiquitous of devices. They are found Blue-ray players, laser pointers, measuring devices and other common pieces of equipment. Lasers are used in schools to carry out experiments similar to that done by Young, but with considerably more ease. They have transformed the teaching of optics.

Another remarkable technological development has been that of the *CCD camera*. CCD stands for Charge-Coupled Device. They are very sensitive detectors of light. Relatively cheap ones are at the centre of digital cameras, converting the light falling on them into electrical signals that can be processed and stored. CCDs are also used in infrared detectors such as those in spy cameras and security alarms. Even more interestingly, they have helped to transform astronomy by making it possible to detect very faint objects in the night sky. Such highly sensitive CCDs have to be cooled to low temperatures so that thermal noise does not mask the image.

By employing controlled beams from a laser and CCDs to detect very faint amounts of light, we are able to carry out experiments similar to Young's basic design, but in ways that he could not have imagined. The results of these new experiments are so radical that they call into question everything that we have said so far.

1.4 PHOTONS

Figure 1.4 shows a very simple experiment where a laser beam is aimed directly at a CCD detector (from now on we will just call them 'detectors') the output of which is transferred to a computer and displayed graphically on a screen.

At moderate intensities, the light seems to be spread equally over the sensitive surface of the detector. However, as we further reduce the intensity of the beam, the image starts to break up into a sequence of tiny speckles (Figure 1.4). Reducing the intensity further makes these speckles occur less frequently, and consequently, they seem to be scattered randomly across the screen. With a suitable laser, the intensity of the beam can be reduced to the point at which only a single speckle occurs at any one time with a notable interval between it and the next one.

A natural way of interpreting these results, aside from thinking that the detector is broken, would be to suggest that the light is a stream of particles. When a particle strikes the detector, it off-loads its energy and produces a single speckle on the screen. At high intensities, there are millions of particles arriving within tiny intervals of time, and the detector records a uniform illumination. Nowadays, we refer to these particles as *photons*.

Now we have two contradictory experiments: one suggests that light is a wave (Young's interference) whereas the other points to the existence of photons. One simple fix would be to suppose that lasers produce photons whereas other sources of light produce waves. Unfortunately, this is not the case. As mentioned earlier, the use of CCDs in astronomy has enabled us to study objects that are so faint that the light is recorded (with the aid of a telescope) a single photon at a time. Clearly, stars and galaxies also produce photons.

Although modern day laser/CCD combinations enable us to perform a simple demonstration that reveals the existence of photons, historically they were detected well before the invention of the laser. Arthur Holly Compton carried out a crucial experiment[4] in 1923 while investigating the scattering of X-rays[5] by atoms.

By 1923 physicists had already successfully produced interference patterns from X-rays, so their wave nature seemed settled. Given this, Compton expected to find that a beam of X-rays would be scattered by electrons inside atoms. The electrons would absorb the energy in an X-ray and then rebroadcast it as a new X-ray sent out in a random direction, but with the same wavelength.

He actually discovered that the X-rays coming off the electrons were of a lower wavelength than those in the incoming beam. Furthermore, the electron struck by the X-ray recoiled as if hit by a physical lump of matter. A detailed examination of Compton's results showed that the energy of the

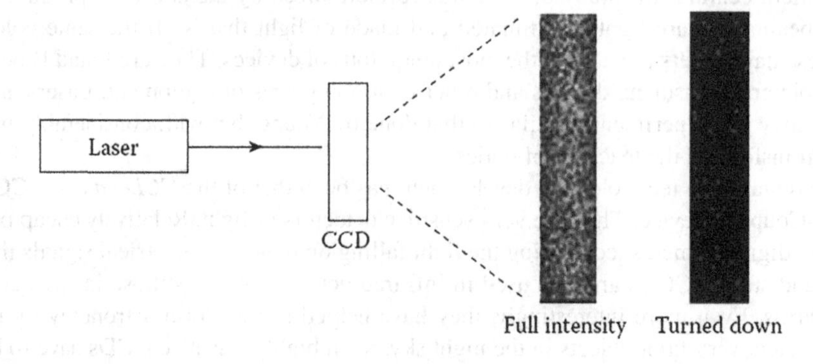

FIGURE 1.4 Using a laser beam to experiment with light.

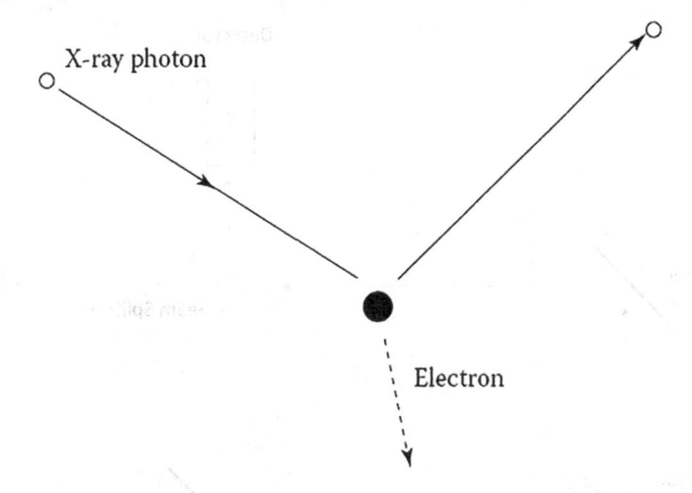

FIGURE 1.5 In Compton's experiment it seemed as if X-ray photons were colliding with electrons like physical lumps of matter. In the process, they transferred some of their energy to the electron, which recoiled from the collision.

incoming X-ray had been passed on to the electron in exactly the same fashion that one snooker ball passes energy onto another when they strike (Figure 1.5). This was completely contrary to the wave picture of light. Compton could explain these results by replacing the wave picture by one that had the X-rays as a stream of photons, but nobody could reconcile this with the interference results.

So, in the mid-1920s physicists found themselves in a bit of a mess. The issue of the wave/particle nature of light, which seemed settled a hundred years before, was now opened up again. However, this time it was worse. Earlier there had been two competing views of the situation waiting for a decisive experiment to declare which one was right. Now there were two contrary experiments, revealing light as a wave in one instance and as a particle in another.

In theoretical terms, a complete resolution to this problem was not to come until the development of quantum field theory and its daughter quantum electrodynamics (a continuous development between the 1930s and the 1950s), both of which are subjects for later. For the moment, we will 'ride the paradox'—thinking on the one hand that light is a particle (photon) and on the other hand that it is a wave - and move on to explore some experiments that demonstrate the split personality of light even more effectively.

1.5 AN INTERFERENCE EXPERIMENT

A key aspect to the next series of experiments is the use of *beam splitters*. An ordinary mirror consists of a shiny or silvered surface designed to reflect all incident light, although in practice, no mirror is ever 100% efficient and some energy is lost. With a beam splitter, 50% of the light is reflected while 50% is transmitted. There are many different varieties of beam splitter, which affect the light as it is transmitted and reflected in slightly different ways. In this context, I have in mind a *dielectric 50:50 beam splitter*, although the details of how it works are not necessary for our analysis.

If you direct a high-intensity laser at such a beam splitter, two beams emerge, each one having half the intensity of the incoming beam.

The next step is to place two ordinary mirrors, one in each path from the first beam splitter, arranged to divert the beams toward a second beam splitter (see Figure 1.6), where the same thing happens to each beam: half of the light arriving passes straight through and the other half is reflected. Finally. two detectors, X and Y, are placed in the beam paths.

Detector X picks up light that has been *reflected* by the first beam splitter (and so travelled the 'top' path), then *transmitted* by the second beam splitter. It will also collect light that was

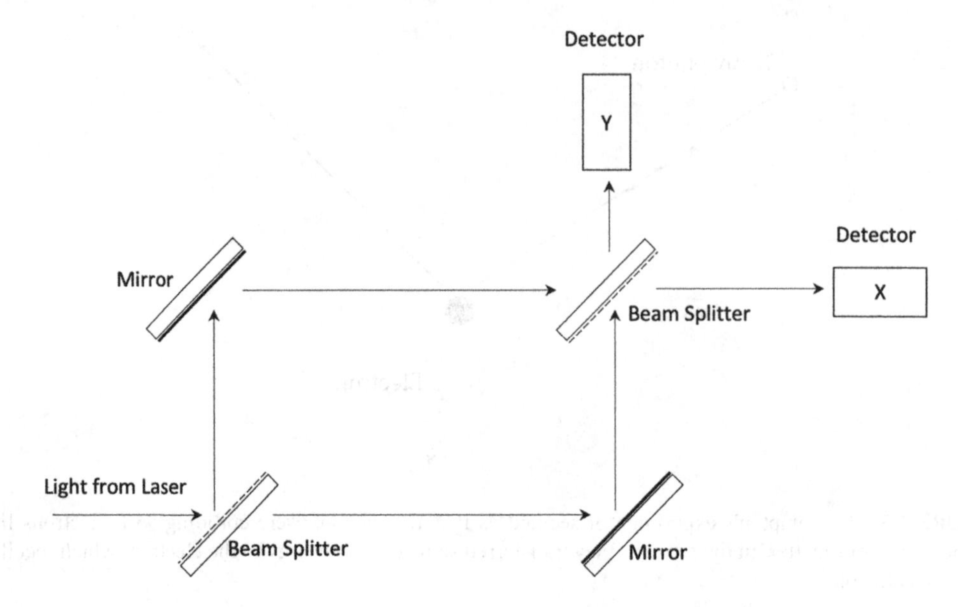

FIGURE 1.6 Using beam splitters to divide a laser beam that is then recombined at a detector.

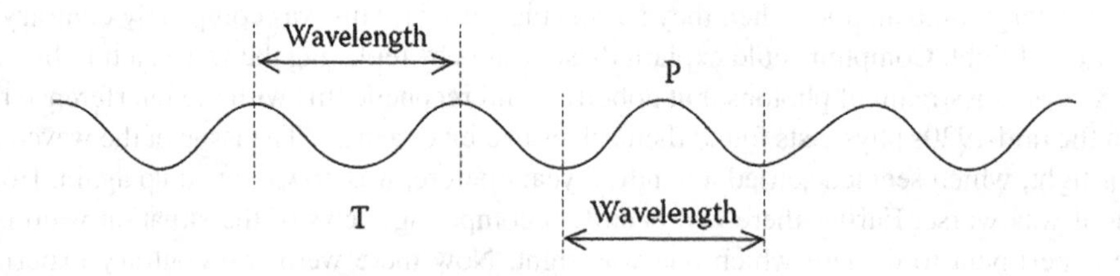

FIGURE 1.7 Light waves, like water waves, have peaks (P) and troughs (T). The wavelength of the wave is the distance between two successive peaks or troughs.

transmitted by the first beam splitter (along the 'bottom' path) and *reflected* by the second one. Any light reaching detector Y must have been either reflected by both beam splitters (top path) or transmitted by both (bottom path).

This arrangement of beam splitters, mirrors and detectors is called a *Mach–Zehnder interferometer* and similar instruments (without modern electronics) have been used for sensitive optical experiments since 1891. Once a Mach–Zehnder interferometer is set up using a standard light source or laser, it is easy to confirm that the intensity reaching each detector depends on the relative distances along the top and bottom paths. If the equipment is very finely adjusted so that these two paths are of *exactly* the same length, detector Y records no light at all, whereas detector X gets all of the intensity entering the experiment. Without this very critical adjustment, X and Y collect light in varying relative amounts: if more light arrives at X, then less will reach Y (and vice versa).

In classical (pre-quantum theory) physics this effect is explained by calling on the idea that light is a wave.

1.5.1 Interference as a Wave Effect

Consider some ripples[6] crossing the surface of a lake. In some places, the water level is higher than normal (these are *peaks*) and in others, it has dropped below normal (*troughs*). The *wavelength* of the ripple is the distance between two successive peaks, which is the same as the distance between successive troughs. The *frequency* of the wave is the rate at which complete cycles (from peak to trough to peak again) pass a fixed point, and the *period* is the time taken for one cycle (Figure 1.7).

The motion of any particle in the surface of the water will track through a repetitive cycle from its starting point to the local peak, then back down to a local trough and up again to the peak. At any moment, the particle is at a certain *phase* of its motion. (This is a term borrowed from lunar observations, where the phase of the moon at any stage in its monthly cycle is an indication of the specific section of that cycle that the moon is currently displaying.) We can quantify the stage of the cycle by comparing the motion of particle or point on the wave with a similar point rotating around a circle, as in Figure 1.8.

The phase is then represented by an angle, φ, measured in *radians*. On this measure, two points on the wave separated by a wavelength will have a *phase difference* of 2π. The phase difference between a peak and a trough is π and that between a peak (or trough) and the central undisplaced line of the wave is $\pi/2$.

As light is composed of electric and magnetic fields,[7] its wave nature is rather more complicated than a simple ripple. The peaks and troughs in a light wave are not physical distances, as in the height of water, instead they are variations in the *strength* of the field. As this is quite a tricky concept to imagine, we can continue to think of a light as being somewhat like a ripple, provided we don't take the analogy too seriously.

Typically ripples on a lake have wavelengths that are comfortably measured in centimetres. Light waves, on the other hand, have wavelengths better measured in nanometres (10^{-9}m), which makes them very sensitive measures of distance. Thinking back to the interference experiment in Figure 1.6, imagine dividing the distance travelled by a light wave into chunks that are equal to the wavelength of the wave. Almost certainly, the distances involved will not be a whole number of such chunks. Equally, the device would have to be very finely calibrated for the two path lengths to be exactly the same number of chunks.

If the distances are not precisely the same, the light travelling along each route will have gone through a different number of complete waves by the time it gets to the detector. As the light has a common source at the first beam splitter, the two beams will set off on their different routes *in phase* (i.e., in step) with each other (see Figure 1.9). If we could see their peaks and troughs directly, they

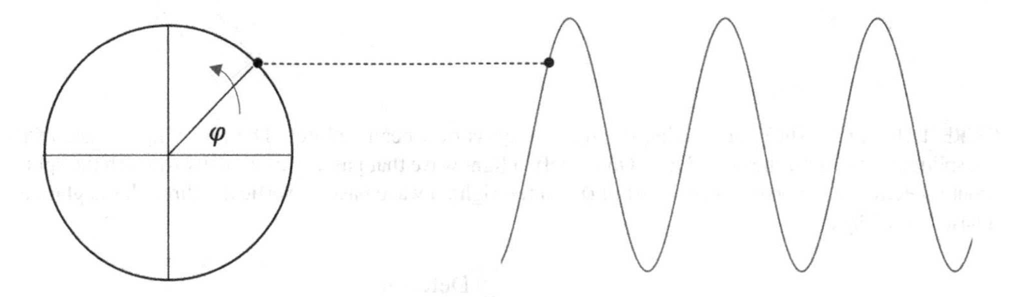

FIGURE 1.8 Comparing the phase in a wave motion with a point rotating around a circle. The phase can then be characterized by an angle, φ, in radians.

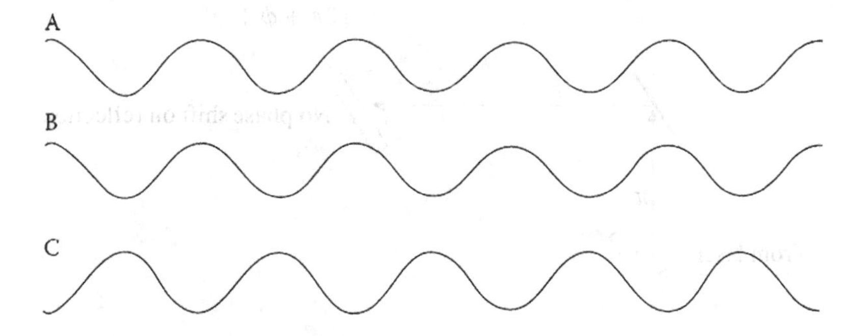

FIGURE 1.9 The waves labelled A and B are in phase with each other (peak to peak and trough to trough); waves B and C are exactly out of phase with each other (peak to trough).

would be marching along peak for peak and trough for trough. However, by the time they get to the detector, the two beams may no longer be in phase, due to the different distance travelled. One could be reaching a peak in its cycle as it arrives and the other a trough (like B and C in Figure 1.9). If this happens, the waves will cancel each other out and there will be little energy entering the detector. Exact cancellation would only happen if the waves met *precisely* peak to trough (π phase difference), which is not possible for any length of time due to small variations in distance (the mirrors will be shaking slightly) and fluctuations in the laser.

To complete a detailed analysis of our experiment, we also need to take into account any phase changes that happen to the light at the various mirrors. Generally, when light bounces off a mirror, the reflected wave is out of phase with the incoming wave by half a wavelength (π phase difference). Things are slightly different with a dielectric beam splitter. Its specially prepared surface, which is bonded to a glass block, can reflect light from either side (the dashed line in Figure 1.6 indicates the reflecting surface). If the reflection takes place from the surface without the light having to pass through the glass block, then the ordinary π phase shift takes place. However, any light that has to pass through the block before reaching the reflecting surface is not phase shifted on reflection. However, there is some phase shift as the wave passes through glass, even if it is not reflected, as illustrated in Figure 1.10.

If we now track the progress of a light wave through the upper arm of the interferometer, we can see that the cumulative phase shift of the wave by the time that it arrives at detector Y is $2\pi + \phi + \phi'$ (Figure 1.11).

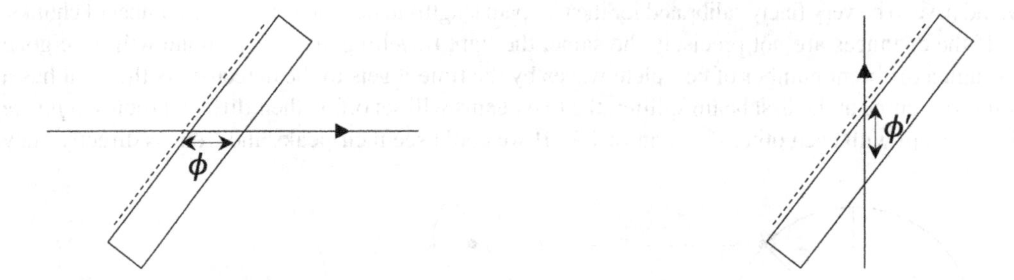

FIGURE 1.10 Phase shifts on passing through the glass of a beam splitter. The reflecting surface of the beam splitter is shown by the dotted line. On the left, a light wave that passes horizontally through the splitter without reflecting, undergoes a phase shift of ϕ. On the right, a wave passing vertically through the glass has its phase shifted by ϕ'.

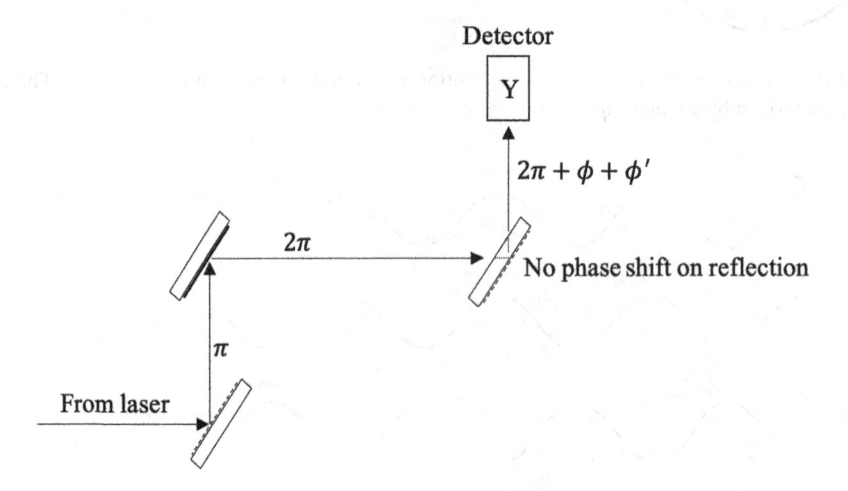

FIGURE 1.11 A wave passing through the top arm of a Mach-Zehnder interferometer on the way to detector Y will undergo a series of phase shifts on reflection and passing through the glass of a beam splitter.

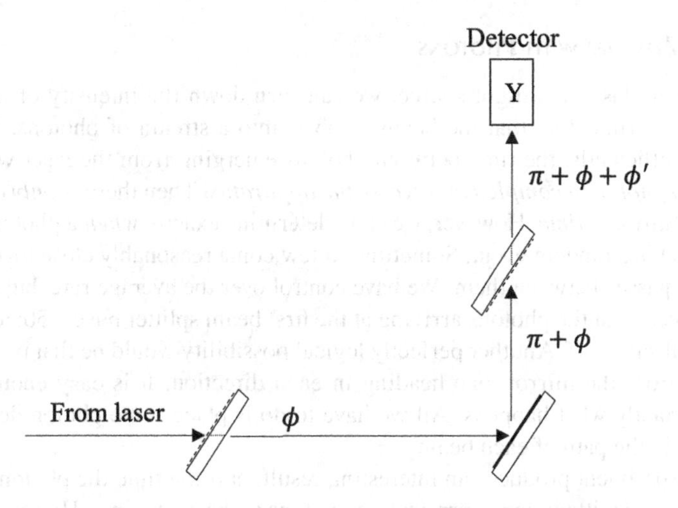

FIGURE 1.12 A wave passing through the lower arm of a Mach-Zehnder interferometer on the way to detector Y will also undergo a series of phase shifts on reflection and passing through the glass of a beam splitter. In this case the cumulative phase shift is different by π compared with that of the top arm.

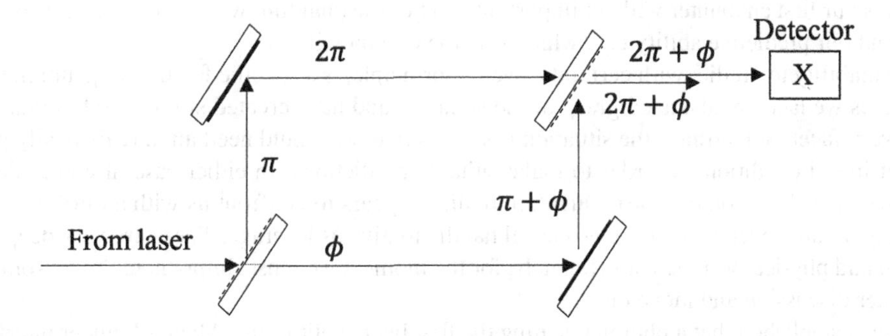

FIGURE 1.13 Light waves arriving at detector X via the top and bottom arms of the instrument arrive exactly in phase with each other, and so constructively interfere.

Applying the same logic to the lower arm of the produces a cumulative phase shift of $\pi + \phi + \phi'$ (Figure 1.12).

In this analysis, we are assuming that the path lengths through the top and bottom arms are identical, so the only phase shifts are due to the reflections and the passage through the glass. Given this fine adjustment of the instrument, the waves arriving at detector Y have a phase difference of $(2\pi + \phi + \phi') - (\pi + \phi + \phi') = \pi$, so they destructively interfere, and no light is seen by the detector.

On the other hand, detector X will see some illumination as the waves arriving there are exactly in phase and so constructively interfere (Figure 1.13).

In most experimental setups, the paths through the interferometer are not equal in length. As we shall see in more detail in Chapters 6 & 7, this also has an impact on the relative phases of the beams, something that has not been incorporated into the argument thus far, on the assumption that both paths were exactly the same length. Given the ability to move one of the fully silvered mirrors, so that the relative path lengths were changed, the experiment could be developed to study the variation of brightness in X and Y as the relative path length varied. In essence, this would be an interference pattern.

Young did not have access to a Mach–Zehnder interferometer (on the very reasonable grounds that they hadn't been invented), but he combined light from two sources to produce an interference pattern on a screen. The results of his experiments could also only be explained by using a wave theory of light.

1.5.2 Mach–Zehnder with Photons

With an appropriate laser as a light source, we can turn down the intensity of light entering the Mach–Zehnder experiment so that the beam resolves into a stream of photons. If we reduce the laser's intensity sufficiently, the time between photons emerging from the laser will be *more than the time it takes a photon to completely traverse the apparatus*. Then there is *only one photon in the interferometer at any one time*. However, we can't determine exactly *when* a photon will be emitted by the laser as that is a random event. Sometimes a few come reasonably close together, sometimes quite a long time passes between them. We have control over the average rate, but nothing more.

One might expect that the photons arriving at the first beam splitter have a 50:50 chance of passing through or reflecting off. Another perfectly logical possibility would be that two reduced energy photons emerge from the mirror, one heading in each direction. It is easy enough to determine experimentally exactly what happens. All we have to do is place some photon detectors just after the beam splitter in the path of each beam.

This simple experiment produces an interesting result: half the time the photon is reflected, and half the time it is transmitted; you never get two photons at the same time. However, there seems to be no inherent difference between the photons that get through and those that reflect. For example, the sequence is not so regular as one reflects and the next goes through, and then the next reflects etc. In fact, there is no pattern to the sequence, except that overall half reflect and half get through.

This is our first encounter with an important facet of the quantum world. Some aspects of nature lie beyond our predictive ability e.g., which way the photon will go.

This inability to predict with certainty, even in principle, is a specific feature of quantum theory. Sometimes we lack an advanced grasp of the situation and have created simple models that do not catch every facet. Sometimes, the situation is so sensitive, we would need an unrealistically precise grasp of initial conditions in order to make suitable predictions. In either case, it comes down to ignorance or inability on our part. Quantum reality appears to confront us with a third possibility: genuinely random events that have no causal handle to give us leverage. These are very deep philosophical and physical waters. Unfortunately, for the moment we must simply note this as something for further discussion and move on.

Having established that a photon reaching the first beam splitter in a Mach–Zehnder interferometer will either *reflect* and travel the *top* path through the device or *transmit* and follow the *bottom* path, our interest must now turn to what happens at the detector end.

For ease, we assume that the detectors are 100% efficient and that no photons get 'lost' while crossing the apparatus.[8] If we add together the number of detections at X and Y over a time period, the answer will correspond to the number of photons emitted by the laser during the same period. However, the relative number of photons arriving at the two detectors depends on the path lengths:

- If they are exactly equal, then no photons *ever* arrive at Y.
- If the paths are not exactly equal, then we find that the detection rate at each detector reflects the intensity of the interference pattern observed when we had the intensity turned up.

To clarify, let's imagine that during a high-intensity experiment, I had arranged for the path lengths to be adjusted until 70% of the total light intensity entering the experiment arrived at X and 30% at Y. Once we turned the intensity down so that we could resolve individual photons, we would find that 70% of the time a photon is detected at X and 30% of the time at Y. There is never a 'double firing' with photons arriving at X and Y together (as long as we have the laser turned down so that there is only one photon in the system at any time). This experiment has been done under extremely well-controlled conditions, and there is no doubt that the photon arrival rate directly reflects the interference pattern in the way described.

Stated rather quickly in this manner, it doesn't sound like there is much of a problem here. Yet there is.

If a photon is a small particle of light, then *how can the different paths have any effect on one single photon*?

We confirmed that photons randomly 'pick' reflection or transmission at a beam splitter. After that they proceed along one path or the other to a detector. It is hard to imagine a single photon *going along both paths at the same time*. Even if we could sustain that idea for a particle, it is not supported by the experimental evidence[9]. Recall that when we put two detectors directly after the beam splitter, they only picked up one photon at a time down one or the other path. There was no sign that the photon traversed both paths simultaneously…

Now a wave can do this. It can spread throughout the experiment (think of the ripples formed when you toss a pebble into a lake) so that parts of the wave travel along each path at the same time. When the two parts of the wave combine at the far side of the experiment, the information about both paths is being compared, which leads to the interference pattern.

A single photon must surely have information about only one path, so how can single photon experiments produce interference patterns?

It transpires that there is a flaw in our argument. It is extremely subtle and cuts to another of the primary issues that physicists have to face when dealing with the quantum world.

We confirmed that the photons randomly divert at the beam splitter by placing detectors in the two paths. However, this eliminates any chance of picking up the interference pattern. If the detectors have stopped the photons, then they have not travelled the paths. In principle, *this does not tell us anything about what might happen when no detectors are present*.

Of course, it is simply 'common sense' to assume that the photons do the same thing with or without these detectors in the experiment, but we have already seen that the interference pattern for photons hardly seems to be a matter of common sense.

There is a way to investigate this further. All one has to do is place just one photon detector, Z, after the beam splitter, say in the path of the reflected beam. If we detect a photon there, then we certainly won't get one at the far side of the experiment, at X or Y. On the other hand, if we don't pick a photon at Z, we can assume that it has passed through the splitter, rather than reflecting, and so we can expect to see it at the far end. The experiment is easily done, given the equipment, and confirms that for every photon leaving the laser we pick one up either at the far end (X or Y) or in the reflected beam (at Z) (Figure 1.14).

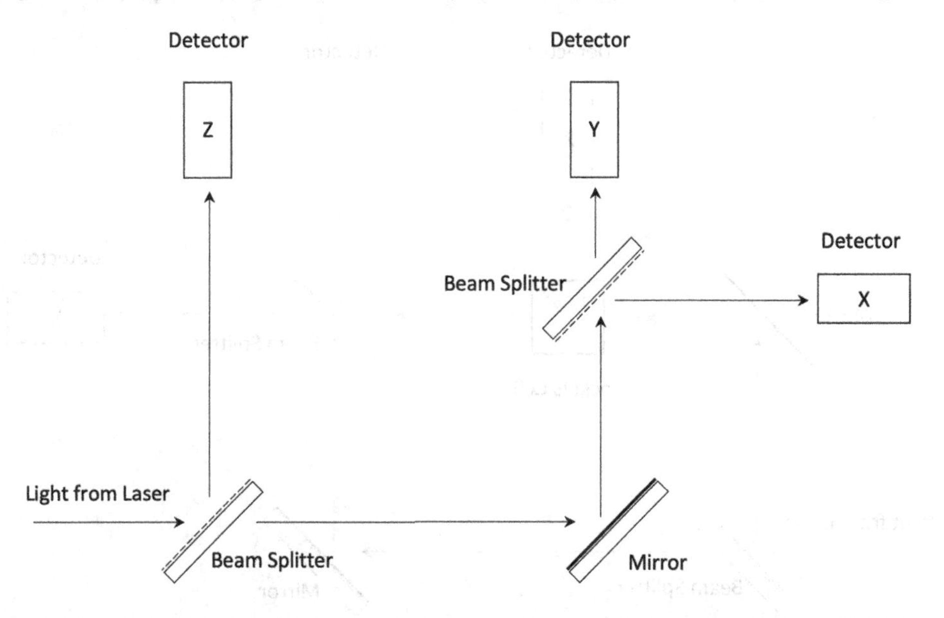

FIGURE 1.14 Mach–Zehnder with photons. A photon arriving at the first beam splitter has a 50:50 chance of being reflected and picked up at the detector. In which case, nothing is seen at *X* or *Y*. However, if the photon is transmitted then there is a 50:50 chance of it arriving at *X* or *Y*, no matter what the length of the path is.

What we find for the transmitted photons is that half of them arrive at Y and the other half at X, *no matter what the length of the path is.* In other words, there is no interference pattern. Removing detector Z opens up the top route to the far side of the experiment. At the same time, it removes any direct knowledge that we might have about the behaviour of the photons at the beam splitter.

It does, however, restore the interference pattern…

Gathering our conclusions:

- with a standard Mach-Zehnder experiment, adjusting the path lengths by moving one or more of the standard mirrors, produces an interference pattern of light at the detectors;
- when we turn the intensity of the laser down, the light beam resolves into a stream of photons;
- the rate of photons emitted by the laser is related to the light intensity;
- reducing the intensity of the beam does not affect the interference pattern—now it's the arrival rate of the photons that depends on relative path lengths[10];
- if we adjust the experiment, so that we can tell which path was taken by the photon (directly or indirectly) at the first beam splitter, then the interference pattern is destroyed;
- if we are unable to tell the path of the photon, then there is an interference pattern, which seems to imply that the photons arriving have information about both routes through the experiment;
- opening up the top path (by removing detector Z) can actually *reduce* the number of photons arriving at Y;
- in the extreme case, if the paths' lengths are the same, opening up the top path means that you *never* get any photons at Y.

1.5.3 Delayed Choice

It is possible to develop the experiment so that the results are even more puzzling.

To do this we introduce a device called a *Pockels cell* (PC) into one of the routes (in Figure 1.15 it can be seen in the reflected route). PCs are crystals that change their optical properties when an

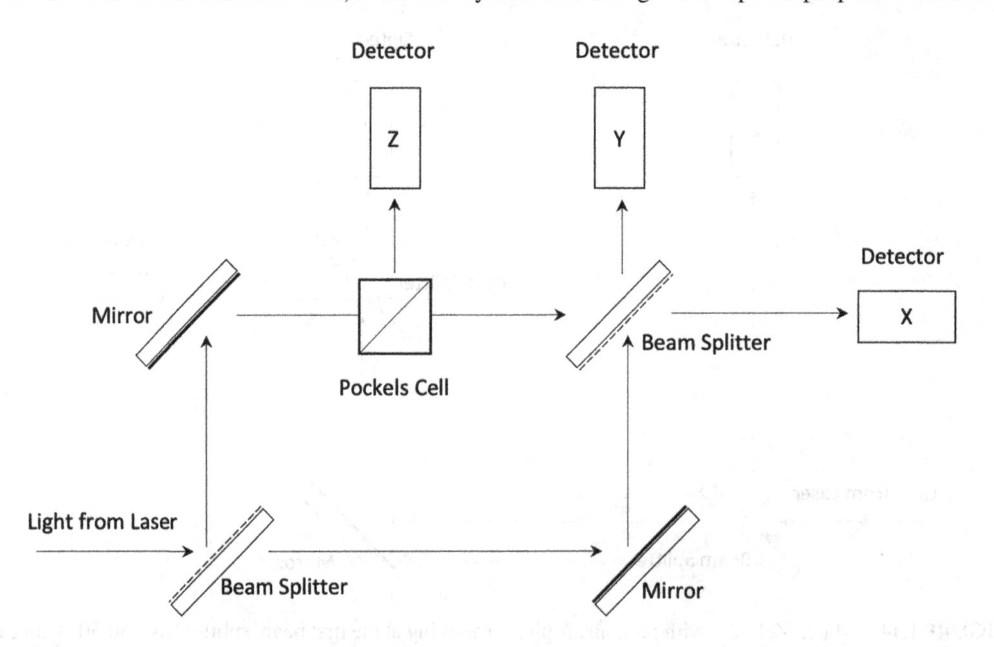

FIGURE 1.15 In this experiment a PC is used. Such a device is capable of passing photons or diverting them to a detector. Passing an electrical current through the cell rapidly changes its setting.

electrical current is applied to them. Without a current, the cell allows photons to pass. Applying a current changes the cell so that it diverts photons, which can then be picked up by another detector.

Consider the scenario shown in Figure 1.15. The PC is initially set to *divert* photons. A photon leaves the laser and arrives at the first beam splitter. If it is reflected, then the setting of the PC will divert it to Z, and we don't see it at X or Y. However, if the photon is transmitted by the first beam splitter, it misses the PC, and it turns up at either X or Y (50:50). In either case there is no interference pattern.

If instead we set the PC to *pass* photons, we get an interference pattern. We will stipulate that the experiment has been finely calibrated so that the paths lengths are an exact match, so that there is no detection rate at Y[11].

So:

- if the PC is set to *transmit,* we get no photons at Y and all of them at X;
- if the PC is set to *divert,* then only half of the photons reach the far side of the apparatus, but they then have an equal chance of being picked up at either X or Y.

This result alone is enough to give us pause. If the photon takes the *lower route* with the PC set to *divert,* then it *can get to X or Y.* If it takes the *lower route* with the PC set to *pass,* then the photon *never arrives at Y.* But if it takes the lower route it doesn't go anywhere near the PC, *so how can the setting of that device affect things*? Is this a further hint that somehow or other the photon travels both routes at the same time?

Now we get devious: we initially set the PC to divert photons, but while the photon is in flight, switch the cell over. As the cell responds quickly to a current, we can make the change *after* the photon has interacted with the beam splitter. One way to do this would be to leave the PC on divert and establish the timing rhythm from a run of photons at a certain laser intensity. That tells us the rate at which photons are being emitted at that intensity. We can easily measure the distance from the laser to the beam splitter, so we know how long it takes a photon to reach that first point in the apparatus. Provided we switch the PC over *after* this time, but *before* the photon has had time to reach detectors X and Y, we will achieve the desired effect. We can then set a switching frequency for the PC.

After the experiment has run for a while, we can use a computer to wade through the data. It will find some photons arriving at Z, which always happens when the detector is set to divert, and some at X and Y. We program the computer to ignore the photons at Z and sort the X & Y group into those that arrived when the cell was set to divert, and those that made it through when it was set to pass (Table 1.1). Remarkably, when the data are separated out in this manner, the photons that arrived at the far side with the PC set to pass show an interference pattern. The other photons that arrived with the PC set to divert (but obviously were committed to the other path and so missed it) show no interference pattern at all.

Recall that in every case the PC was initially set to divert photons and was only switched over *after* they left the beam splitter. With the PC set to divert, we have seen that the photons follow one

TABLE 1.1
The Pattern of Detections for Different Settings of the Pockels Cell

	Detectors		
Cell Setting	**Z**	**X**	**Y**
Divert	Yes, if reflected at first beam splitter (50%)	Yes, if transmitted at first beam splitter (25%)	Yes, if transmitted at first beam splitter (25%)
Pass	Never	Yes, all photons arrive here due to path lengths being equal (100%)	Never, due to the interference pattern

path or another (top route, via the Pockels cell to Z, or bottom route via the beam splitter to X or Y). Whilst they were in flight, we sometimes switched the PC, removing our ability to know which path the photons travelled, and producing an interference pattern. The presence of an interference pattern suggests that the photon travelled both paths.

It's hard to believe that changing the setting of the PC can have an influence that travels backward in time to affect the photon at the first beam splitter. What we can say is that *the ability to deduce the path of the photon (PC set to divert) results in no interference.* If we can't directly or indirectly determine the path of a photon (PC set to transmit), then we do get an interference pattern.

The mathematical machinery of quantum theory describes the photon leaving the mirror as being in a combination (superposition) of two distinct sates, one for travelling each path. In a standard interpretation, when the photon arrives at the PC, this combined state randomly collapses into one or other of the distinct states (corresponding to the photon being on one path or the other). An alternative interpretation talks of parallel worlds, with a photon always travelling along one path in each world, but the two worlds being able to influence each other to some small degree, resulting in an interference pattern.

1.6 SUMMARY

Although this chapter has only been a starting point, we have already come across some fundamental issues. We have seen that a description of light must somehow encompass both wave-like and particle-like natures, depending on the circumstances. The underlying randomness that can appear in the quantum world has made itself known via our inability to tell which way a photon will travel at a beam splitter (in an experiment set up to detect its path). Finally, and in my view most importantly, we have indications that quantum mechanics is going to be a *contextual theory*: an adequate description of the behaviour of a quantum object (light in this case) will require an understanding of the *whole experimental setup*; the behaviour depends on the context.

NOTES

1 My thanks to the Master and fellows of Gonville and Caius College, Cambridge, for the opportunity to spend some time in College writing the first edition of this book.
2 Newton's book, *Optiks,* was published in 1704 and put forward a strong case for the particle view.
3 When ripples are supporting each other, this can cause a patch that is either deeper or higher than normal. In light, bright bands can be caused by *deep* and *high patches*. Dark bands are formed when the light waves oppose one another.
4 For which he earned the 1927 Nobel Prize in physics.
5 X-rays had been discovered in 1895 by Röntgen (for which he was awarded the first Nobel Prize in physics). By this time, their properties had been well established and their nature, as part of the *electromagnetic* spectrum, confirmed.
6 What people normally think of as water waves (the things you see on the beach) are not really waves in the strict sense of physics. Beach waves are a mixture of ripples and tidal movement of water.
7 In a non-photon model, that is.
8 This is an exaggeration, I'm afraid. No detector has 100% efficiency. What I am trying to suggest here is that nothing "odd" happens to the photons in flight. Each one gets through the experiment and is, in principle, detectable at the far end.
9 We will discover that the experimental evidence from one setup is not always directly related to a different setup. Hence it is not quite true to say at this stage that the evidence does not support a photon travelling both paths. Here we have an example of the contextuality of quantum mechanics.
10 Some readers may be worrying about what I mean by the arrival rate of the photons. One picture in your mind might be that, somehow, we are slowing the photons down in the experiment, so it takes longer time for them to get to the far side. In fact, it is a rather more complicated situation. First, we cannot be exactly sure of the moment that a photon leaves the laser (this is due to the uncertainly principle that we will discuss later). Second, the different positions of the detector mean that there will be different travel times, but light is quite quick, so this is not a major factor. You have to take a more overall view of the

experiment. If the detector is at a position that corresponds to a dim part of the pattern, then when we reduce the intensity to the single photon level we have to consider the experiment as a whole—in which case the position of the detector is influencing the probability that a photon leaves the laser. The whole thing from leaving the laser to travelling through the experiment to arriving at the detector at the far side is an interlocking process and each stage has an influence on the others.

11 Note that this is still an interference pattern.

2 Particles

2.1 PARTICLES AND WAVES

> Newton thought that light was made up of particles, but then it was discovered, as we have seen here, that it behaves like a wave. Later, however (in the beginning of the twentieth century) it was found that light did indeed sometimes behave like a particle. Historically, the electron, for example, was thought to behave like a particle, and then it was found that in many respects it behaved like a wave. So it really behaves like neither. Now we have given up. We say: "it is like neither." There is one lucky break, however—electrons behave like light. The quantum behaviour of atomic objects (electrons, protons, neutron, photons, and so on) is the same for all, they are all 'particle waves' or whatever you want to call them.
>
> *Richard Feynman*[1]

In the last chapter, we discussed the paradoxical nature of light: it can appear to be a wave in one situation and a stream of particles in another. To make any progress towards resolving this paradox, we must 'give up', as Feynman suggests. We need to put away the classical descriptions of both waves and particles and try to develop a new set of rules and concepts, consistent with the old ideas but extendible into the reality revealed by quantum experiments. The ultimate answer comes in the form of quantum field theory (Chapter 32), which is rather too complex to jump straight into describing. We need to build our understanding piece by piece. However, Nature has been kind to us in one respect at least (Feynman's lucky break): just as something we thought was a wave (light) can behave like a particle we find that something we thought was a particle (the electron) can behave like a wave. It seems that there is only one paradox to resolve!

Although the wave aspect of electrons is very striking, we are first going to explore their quantum nature by looking at a series of experiments in which electrons interact with a particular form of magnetic field. As we encounter the results of these experiments, we will attempt to build a theoretical description, leading to predictions about further experimental results. Initially, this description will be 'semi-classical', but as we dig deeper, we will come to see that a radically new approach is required. In a sense, we will keep whacking at the pillars holding up a semi-classical description, until we are forced to acknowledge that the edifice has collapsed under the weight of experimental data.

2.1.1 ELECTRONS AND ELECTRON GUNS

At our current levels of experimental sensitivity, there is no evidence for any constituent objects within electrons, hence they are regarded as *fundamental particles*. They have an electrical charge which is negative (specifically $\sim -1.6\times10^{-19}$ C) and a very tiny mass ($\sim 1/2000$th of a proton's mass). Electrons can be found on their own or inside atoms, where they are held to the positively charged nucleus by electrical attraction. However, they can be removed from atoms by using a strong electrical field, such as that produced by a Van De Graaff generator. Beams of electrons can be accelerated and focused using electric and magnetic fields. Indeed, the basic technology that goes into an electron 'gun', a device capable of producing a focused and directed beam of electrons at an arbitrary intensity, has existed since 1869.

In essence, electron guns play a very similar role to the laser for experiments involving light. In this chapter, we imagine the electron gun as part of a hypothetical experiment based on the 1922 research of Otto Stern and Walther Gerlach.

DOI: 10.1201/9781003225997-4

2.2 THE STERN-GERLACH EXPERIMENT

While thinking about experiments involving beams of atoms and magnetic fields, Otto Stern[2] realized that passing a beam of specially selected atoms (those with a net internal magnetic field) through a nonuniform external magnetic field would apply a force to the atomic magnets. This force would achieve two things:

1. it would deflect the path of the atom passing through the external field
2. and cause the atom's internal magnet to align itself to the external field (see Figure 2.1).

Observing the deflection of the atoms proved to be an ideal way of checking their magnetic properties, which was the original purpose of the Stern-Gerlach experiment. The non-uniform field was created by a Stern-Gerlach (SG) magnet with specifically shaped poles as shown in Figure 2.2.

For our purposes, we suppose that an SG magnet will exert a similar magnetic force on any electrons passing between its poles and that this force will deflect their paths as shown in Figure 2.2. In truth, such an experiment is impossible to perform, as the wave nature of the electron gets in the way and obscures the results, However, every result deduced from our hypothetical experiment can be confirmed by slightly more subtle work with neutral atoms[3]. So, in service of a more straightforward and less technical explanation, we will proceed as if a simple electron beam was all that is necessary for these experiments.

The purpose of our hypothetical experiment is to measure how much deflection takes place when you pass an electron beam between the poles of an SG magnet. The best way of obtaining a numerical value for the deflection is to use a simple form of electron detector. Traditionally, this was done by placing a photographic film on the far side of the magnet. In essence, a strip of thin plastic coated with a light-sensitive emulsion will 'fog' wherever electrons strike its surface. An individual electron could not be detected in this way, but after a while, enough electrons will have passed through

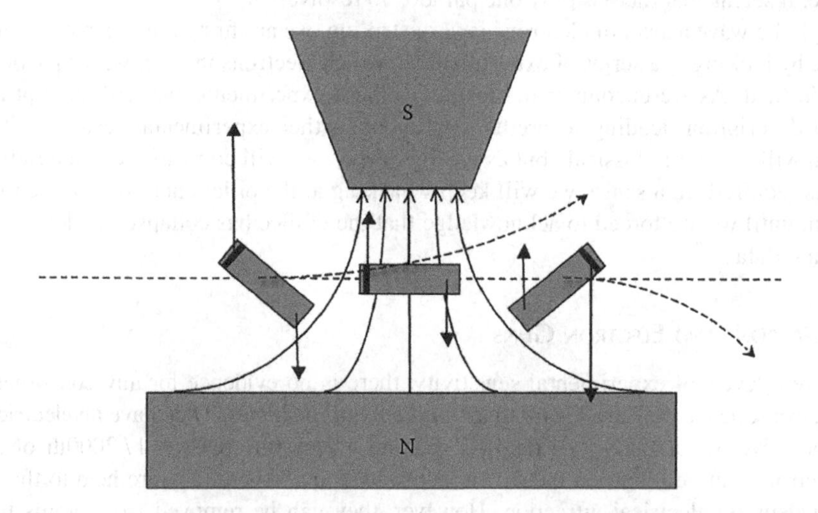

FIGURE 2.1 In this diagram, three small magnets are shown passing in a horizontal direction between the poles of a larger magnet designed to produce a nonuniform field. On each magnet, the north pole is indicated by the black line. The left-most small magnet (leading with its south pole) experiences a combination of forces that tend to turn it in a clockwise direction, so that it aligns with the field. However, due to the nonuniformity of the field, the force on the north pole of the small magnet is stronger than that on its south, so the small magnet will move upward as well as rotate. The right-hand small magnet (leading with its north pole) will experience the same rotation and deflection but the opposite deflection. The small magnet at the centre will simply rotate until it is aligned. Although the small magnets are drawn in different horizontal positions, the effect will occur depending on their orientation as they first enter the field from the left and it will cause them to separate.

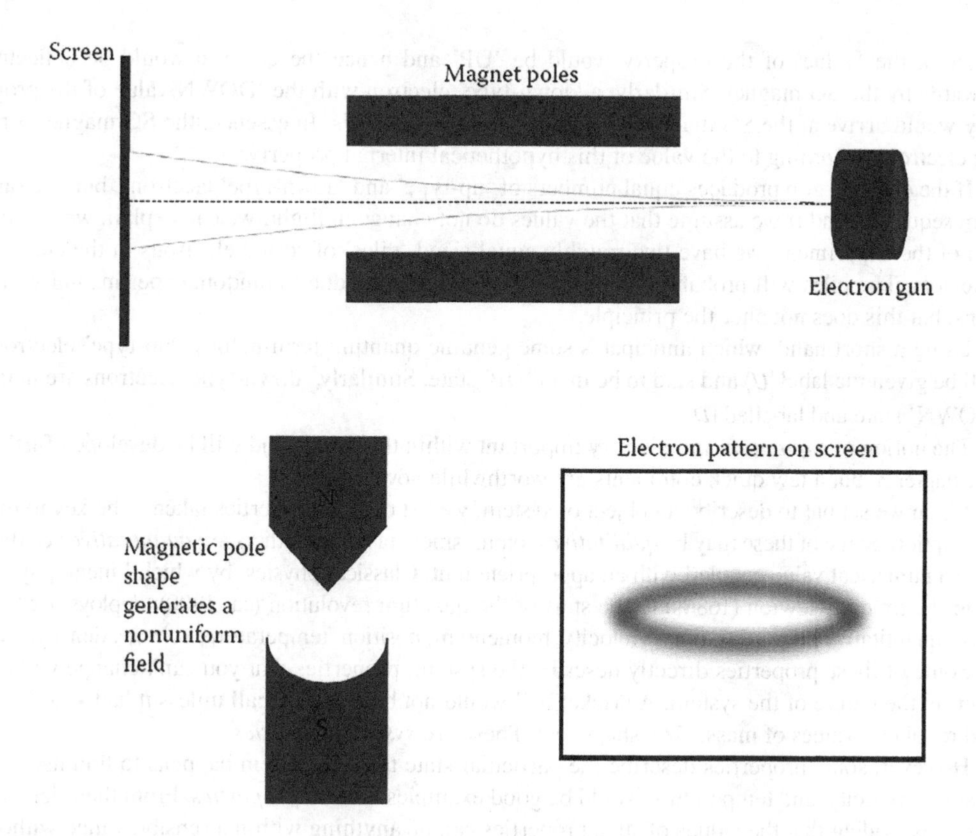

FIGURE 2.2 Deflecting electron paths. With the magnetic field turned on, individual electrons are deflected in one of two directions. Classically, one would expect a smooth range of deflections producing an extended splodge on the photographic plate.

the experiment to adequately expose the emulsion. The film is then treated with chemicals, which makes the fogging permanent and visible to the naked eye.

In this case, two patches would be seen equally spaced above and below the point on the film where an undeflected beam would have impacted (Figure 2.2).

This initial observation is surprising, as it suggests that the amount of deflection, if not the direction, is *the same for each and every electron*. If the electrons were acting like tiny magnets, we would expect their poles to be randomly oriented as they enter the SG field. Consequently, the amount of deflection, which depends on the initial orientation of the electron's magnet, would be slightly different for each one. The result would be an extended smear on the photographic film, not two patches separated by a blank space.

If we want to run a more sophisticated version of this experiment, we can replace the photographic film with a series of electron detectors. There are many advantages in doing this, but the most important is that we can now turn the intensity of the beam right down so that we are only dealing with one electron passing through the experiment at a time[4].

If we were to run an experiment along these lines, we would observe two things:

1. There seems to be no pattern to the deflections: individual electrons either go up or down, apparently at random.
2. Each electron is deflected, upward or downward, by a fixed amount.[5]

A semi-classical explanation for these results would start by assuming the electrons to have some internal property that determines which way they are deflected by the magnet[6]. With an 'up-type' electron, the 'value' of the property would be 'UP' and hence the electron would be deflected upwards by the SG magnet. Similarly, a 'down-type' electron with the 'DOWN value of the property would arrive at the SG magnet and be deflected downwards. In essence, the SG magnet sorts the electrons according to the value of this hypothetical internal property.

If the electron gun produces equal numbers of 'up-type' and 'down-type' electrons, but in a random sequence, and if we assume that the values do not change in flight, we can explain why, at the end of the experiment, we have two roughly equal-sized 'piles' of sorted electrons in the electron detectors. The piles will probably not be exactly the same size due to random experimental variations, but this does not alter the principle.

Using a short-hand, which anticipates some genuine quantum terminology, 'up-type' electrons will be given the label $|U\rangle$ and said to be in an 'UP' state. Similarly, 'down-type' electrons are in the 'DOWN' state and labelled $|D\rangle$.

The notion of a *quantum state* is very important within the theory and will be developed further in Chapter 3, but a few quick comments are worthwhile now.

When we set out to describe an object or system, we list certain properties taken to be key to that description. Some of these may be *qualitative* (colour, smell, taste etc.) others are *quantitative* i.e., they have a numerical value coupled with an appropriate unit. Classical physics, by which I mean physics from the time of Newton (1680s) to the start of the quantum revolution (ca. 1900), deploys a whole set of quantitative properties: mass, velocity, momentum, position, temperature, volume, density, etc.

Some of these properties directly describe the system; properties that you can't change without altering the nature of the system. A cricket ball would not be a cricket ball unless it had strictly set and regulated values of mass, size, shape, etc. These are *system properties.*

However, some properties describe the particular state that the system happens to find itself in; position, velocity, and temperature would be good examples of *state properties.* From their description, it is evident that the values of state properties can be anything within a sensible range without altering the nature of the system[7].

Now, when it comes to experimental work there is a largely unstated assumption, at least in classical physics: *the process of measurement reveals the quantitative values of these properties (in suitable units) to some degree of accuracy.* This seems so obvious, it hardly merits the status of an assumption. Equally, one presumes that any interaction between a system and an apparatus either does very little to alter the values of the state properties or it produces a change which can be compensated for when analysing the results. Furthermore, it was always taken as read that the system had quantitative values of the state properties, *whether they were measured or not.* All these assumptions are called into question when we try to use classical states to describe quantum systems, such as electrons.

In setting out to describe the results of the SG experiment, at least in a semi-classical way, we started from some (presumed) new state property which can have one of two values, placing an individual electron in the $|U\rangle$ or $|D\rangle$ state. The closest we can come to a direct measurement of this property is to see which way the electron emerges from the magnet. On this basis, we conclude that the experiment is sorting the random sequence of $|U\rangle$ / $|D\rangle$ electrons emerging from the gun into two groups, as suggested before. This is shown in Figure 2.3.

Admittedly, it is quite a leap to infer the existence of some new property just from the results of this experiment. Surely it would be perfectly reasonable to conclude instead that the SG magnet is simply randomly booting electrons one way or the other. We are going to need a little more convincing before we accept the reality of this property. We need to test a prediction.

If the electrons exist in some pre-set state of $|U\rangle$ / $|D\rangle$ and are sorted by an SG magnet, then separating out the, say, $|D\rangle$ electrons and passing them through a further SG magnet should reveal that they all exit via the UP channel. This is relatively straightforward to test, as illustrated in Figure 2.4.

The results of this modified experiment are quite conclusive. Electrons that emerge along the UP channel of the first magnet pass through the top second magnet and all of them emerge from that magnet's UP channel. If the SG magnets are randomly booting the electrons, then surely some should come out of the second magnet's DOWN channel.

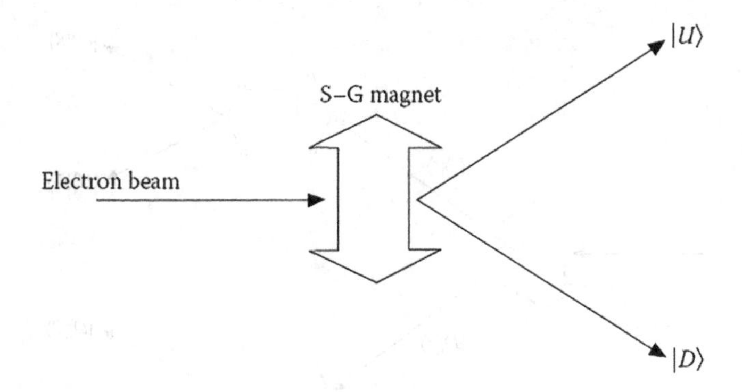

FIGURE 2.3 A summary of the SG experiment with electrons. The double-headed arrow is supposed to symbolize a magnet that has been placed with a vertical orientation.

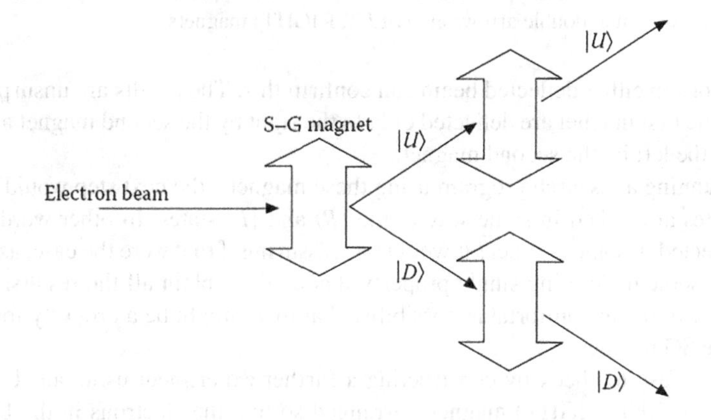

FIGURE 2.4 An experimental arrangement with more than one SG magnet.

Similarly, the $|D\rangle$ electrons emerging from the first magnet are all subsequently deflected downwards by their second magnet.

The second magnets, and any others that we wish to add to the chain, confirm the sorting carried out by the first magnet. These results give the impression that the SG magnets are 'measuring' some state property belonging to the electrons, and reassure us that the interaction between the electrons and the magnets is not altering the value of that state property.

2.2.1 Turning Things Around

So far, we have been using the SG magnets vertically, so that they are deflecting electrons upwards or downwards. If we turn the magnet through 90°, it ought to deflect the electrons to the right or left. There is no reason why the SG magnets shouldn't be placed at any angle we fancy, but for our purposes (UP, DOWN) and (LEFT, RIGHT) will be enough.

The results of running an experiment with the SG magnet turned horizontally are exactly as you might expect. Half of the electrons passing through the poles of the magnet are deflected to the right, and half to the left. Once again there is no obvious pattern to help predict which electron will go which way.

Follow the same line of argument as before, we end up suggesting two possible states for an electron, $|R\rangle$ and $|L\rangle$, which the magnet is sorting. Presumably, the electrons have a second state property that governs their horizontal deflection. Adding two further magnets, also arranged horizontally, to

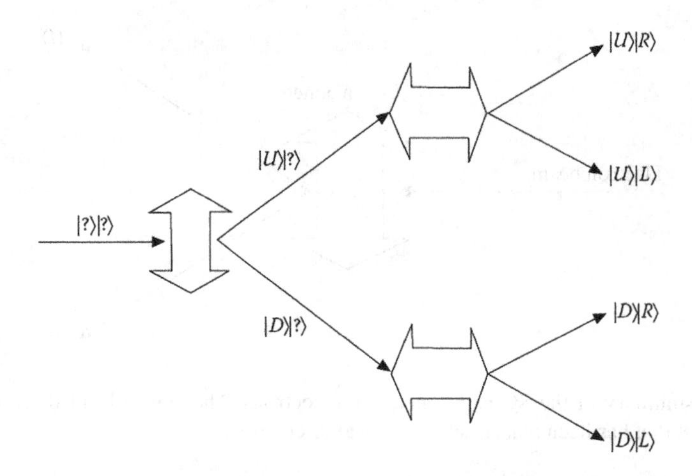

FIGURE 2.5 Sorting out different types of electrons. The vertical double arrow represents an (UP, DOWN) SG magnet and the horizontal double arrows are (LEFT, RIGHT) magnets.

check the electrons in either deflected beam can confirm this. The results are unsurprising. The $|R\rangle$ electrons from the first magnet are deflected only to the right by the second magnet and the $|L\rangle$ ones are deflected to the left by the second magnet.

If we were running a research program using these magnets, the next step would be to see if the $|U\rangle$ and $|D\rangle$ states are linked in some way to the $|R\rangle$ and $|L\rangle$ states. In other words, are the state properties connected in some manner? It would be reassuring if that were the case, as it might reveal the existence of some underlying single property that could explain all the results. Otherwise, we would be faced with the uncomfortable possibility that there might be a property for each conceivable angle of the SG magnet.

This would be easy to check by constructing a further experiment using an (UP, DOWN) SG magnet with two (LEFT, RIGHT) magnets, arranged so that the electrons in the UP and DOWN channels of the first magnet are tested for their $|L\rangle$ / $|R\rangle$ state (see Figure 2.5).

It turns out that a $|D\rangle$ electron passing into a (LEFT, RIGHT) magnet *can come out of either channel*, as can an $|R\rangle$ electron.

It would appear that we are dealing with *four* different combinations of electron states determined by two independent state properties. For example, an electron in state $|U\rangle$ could also be in state $|L\rangle$ or in state $|R\rangle$. The possible combinations are:

$$|U\rangle|R\rangle \quad |U\rangle|L\rangle \quad |D\rangle|R\rangle \quad |D\rangle|L\rangle$$

which the electron gun is evidently producing in equal numbers. The combination of two differently pointing magnets serves to sort the electrons out into their respective states.

In Figure 2.5, I have used the symbol $|?\rangle$ to indicate an undetermined state. When the electrons in the beam arrive at the first magnet, we have no way of knowing either their (UP, DOWN) or (LEFT, RIGHT) state, hence they are $|?\rangle|?\rangle$. The first magnet sorts them into $|U\rangle$ or $|D\rangle$, but tells us nothing about $|L\rangle$ or $|R\rangle$, hence we have $|U\rangle|?\rangle$ and $|D\rangle|?\rangle$ The final pair of magnets completes the sorting so that we now have four piles of distinct state combinations, with roughly equal numbers of electrons in each.

2.2.2 Things Get More Puzzling

The experiment illustrated in Figure 2.5 shows how a combination of three magnets could be used to sort electrons into the four suggested states. As an extension to this experiment, we might consider adding another two magnets to double-check the results, as shown in Figure 2.6.

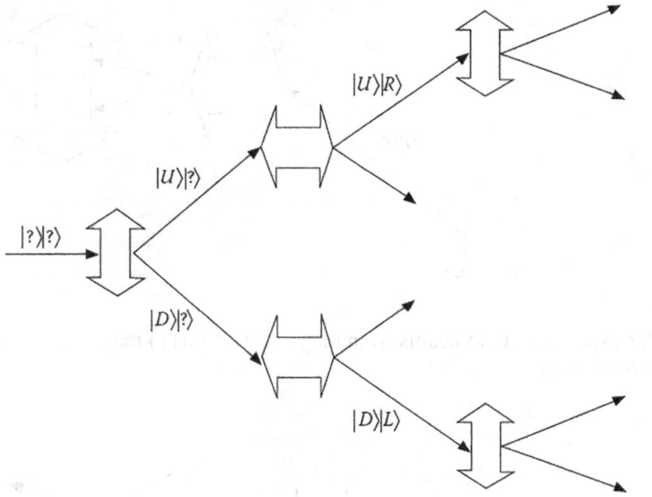

FIGURE 2.6 A further experiment to check our conclusions.

The extra magnets have been placed in the path of the $|U\rangle|R\rangle$ and $|D\rangle|L\rangle$ beams. Any of the four beams could have been chosen, or all four of them using two more magnets, but that would clutter up the diagram and not add anything to the discussion.

The results of this experiment are remarkable.

Electrons from the beam labelled $|U\rangle|R\rangle$ (i.e. containing electrons in the $|U\rangle$ state) pass through this last magnet and emerge from *either* the UP or the DOWN channel. It's as if some $|D\rangle$ state electrons got mixed up with the beam that we thought was pure $|U\rangle$. Unfortunately, this can't be a credible explanation as there are no extra electrons in the beam. In any case, the results show that each of the emerging beams contains roughly half of the electrons.

A more plausible explanation would be that the (LEFT, RIGHT) magnet has somehow changed the state of some of the electrons passing through it. All the electrons arriving at this magnet are in the $|U\rangle$ state, but perhaps after passing through the (LEFT, RIGHT) magnet, a few of them have been flipped into $|D\rangle$.

This immediately suggests a whole new set of questions. What is it about the angle that changes the state? Does the magnet have to be at 90° to do this or will any other angle do? What if the magnet is only *slightly* turned away from being vertical? All of these could be tested by appropriate experiments. None, though, would be as illuminating (and puzzling) as the experiment in the next section.

2.2.3 So, Where Did It Go?

Our next experiment (Figure 2.7) starts with a pure beam of $|U\rangle$ state electrons, extracted from the UP channel of an (UP, DOWN) SG magnet. After passing the $|U\rangle$s through a (LEFT, RIGHT) magnet, there are now two beams, with roughly half of the electrons in each. Finally, we introduce another (UP, DOWN) magnet, but in contrast to the previous experiment, we arrange for it to be very close to the (LEFT, RIGHT) so that *both* $|L\rangle$ and $|R\rangle$ electrons interact with the same magnet. This is not difficult to achieve as the deflections produced by an SG magnet are actually quite small; it is only by allowing the electrons to travel some distance after the magnet that the beams separate by measurable amounts.

Before I reveal what happens, it is worth summarizing our thinking up to this point:

- passing through an (UP, DOWN) magnet splits an electron beam into two samples;
- this implies that electrons can be in one of two states, which the magnet is sorting;

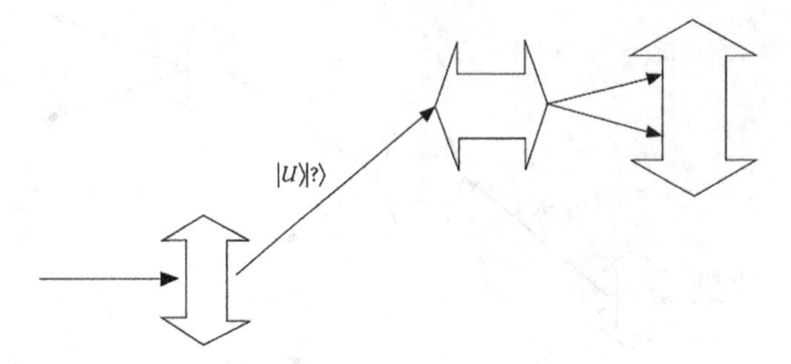

FIGURE 2.7 In this experiment both beams from the (LEFT, RIGHT) magnet are allowed to pass through the same (UP, DOWN) magnet.

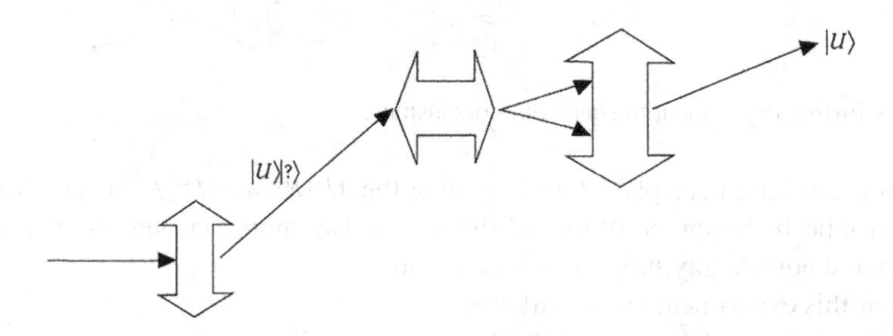

FIGURE 2.8 Passing $|R\rangle$ and $|L\rangle$ beams through the same (UP, DOWN) magnet produces a single beam of the same state as the original beam that entered the (LEFT, RIGHT) magnet.

- we called these states $|U\rangle$ and $|D\rangle$.
- A horizontally arranged magnet sorts electrons into $|R\rangle$ and $|L\rangle$ states.
- This suggests that there are *four* different combinations of electron states: those that are $|U\rangle$ *and* $|R\rangle$, those that are $|U\rangle$ *and* $|L\rangle$, etc.
- However, passing a beam of electrons that should be $|U\rangle|R\rangle$ (having come from an UP channel and a RIGHT channel in that order) into another (UP, DOWN) magnet divides the beam in two again. It would appear that somehow passing through a (LEFT, RIGHT) magnet flips the (UP, DOWN) state of some of the electrons.

Based on this line of thought, allowing both beams from the (LEFT, RIGHT) magnet to pass through a single (UP, DOWN) should produce the same result as having an (UP, DOWN) on each beam. We should get two beams emerging from the single (UP, DOWN) magnet, if we are right in thinking that the magnet has flipped the state of some of the electrons.

Using one magnet to catch both emerging beams *produces just a single beam of pure $|U\rangle$ electrons as in Figure 2.8.*

To be certain that this is correct, you would have to test the beam of $|D\rangle$ electrons from the first magnet as well. Figure 2.9 shows such an experiment in operation and the results match those from the experiment in Figure 2.8.

The conclusion is clear. If the beams from the (LEFT, RIGHT) magnet are passed into *separate* (UP, DOWN) magnets then the $|U\rangle / |D\rangle$ state of the electrons is modified. However, if both beams from the (LEFT, RIGHT) magnet pass through the *same* (UP, DOWN) magnet, then there is no state flip. The original state of the electrons that entered the (LEFT, RIGHT) magnet is preserved.

This is very puzzling. Up to now, everything that we have said about electron states and the way in which electrons are deflected (sorted) by SG magnets could be a simple extension of classical

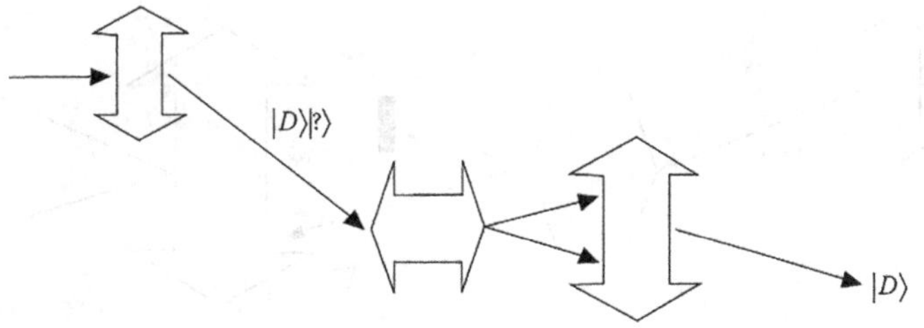

FIGURE 2.9 An experiment that shows that the $|D\rangle$ property of the electrons can be preserved as well.

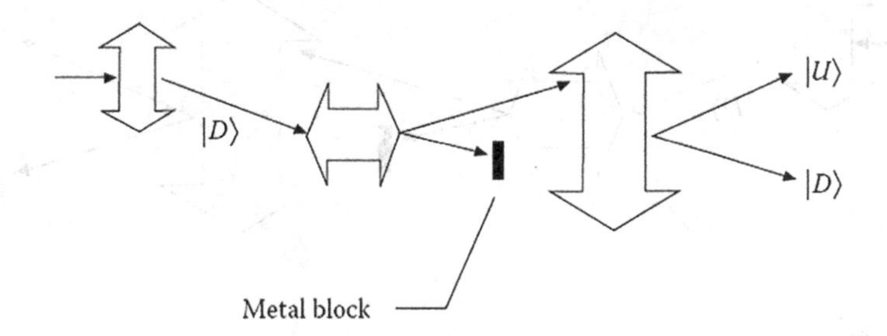

FIGURE 2.10 Blocking the LEFT channel of the (LEFT, RIGHT) magnet scrambles the (UP, DOWN) property of the electrons.

ideas about electrons. Now with this experiment, we're starting to see that these states have a quantum nature, which makes them behave in a rather different way.

One way of trying to retain some common sense would be to speculate that the 'flipping' of the electron's state is a process that needs a certain distance over which to happen. Hence by moving the (UP, DOWN) SG magnet nearer we have not given enough opportunity for the flip to happen.

This might be a reasonable speculation, but we can kill it off, and any similar lines of thought with it, by making a simple modification to the experiment.

A small plate of metal is sufficient to block either of the channels in a (LEFT, RIGHT) magnet and prevent any electrons from getting through. You would just have to make sure that it was not wide enough to stop both channels at the same time. It should be possible to set things up so that we can choose to block either channel, but let's say we pick the LEFT channel for the moment (Figure 2.10).

We have not moved the magnet any further away, so all the $|D\rangle|L\rangle$ electrons will presumably, if our guess about distance being needed is correct, stay in the $|D\rangle$ state and come out of the second magnet along the bottom channel.

Wrong again.

Making this modification just throws another puzzle in our face. Blocking the LEFT channel restores flipping the (UP, DOWN) state. As this experiment doesn't alter the distance travelled by the electrons in the RIGHT channel, we have eliminated any argument based on the flipping needing a certain distance to work. We can turn the flipping on or off *by simply blocking one of the paths and doing nothing to the distance.*

As a matter of routine, you would check that similar results are obtained if the RIGHT channel is blocked instead, as indeed they are. I have summarized the results of all of these experiments in Figure 2.11.

2.2.4 What Does It All Mean?

We started with the notion that electrons possess a certain state property that determines their path through an SG magnet. Presumably, some electrons start in the $|U\rangle$ state and some in the $|D\rangle$ state,

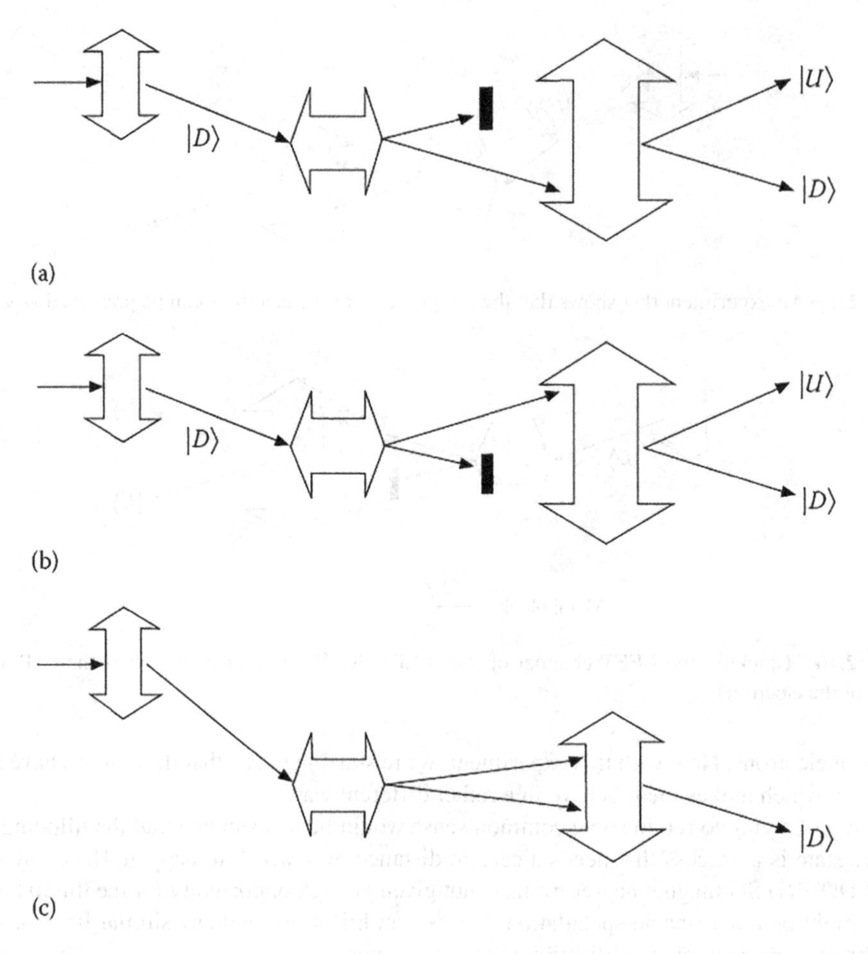

FIGURE 2.11 (a) Blocking the RIGHT channel produces a mixture of $|U\rangle$ and $|D\rangle$ states in the electrons that pass through the LEFT channel. (b) Blocking the LEFT channel produces a mixture of $|U\rangle$ and $|D\rangle$ states in the electrons that pass through the RIGHT channel. (c) Having both LEFT and RIGHT channels open produces only $|D\rangle$ state electrons.

and when the electrons are gathered into a beam, the $|U\rangle$ and $|D\rangle$ ones will be randomly distributed in the beam, so that we can't tell which type is coming next. Roughly half of them are $|U\rangle$ and half are $|D\rangle$, so on average, the beam divides equally when passing through the magnet. This idea had to be extended to include the results of a (LEFT, RIGHT) magnet, suggesting that every electron has two states properties and that a beam contains equal numbers of all four different combinations of states.

Crucially we are assuming that the state of the electron is fully determined *before* it enters any magnet that we might throw in its path. This is the assumption behind the classical idea of a state (that a measurement reveals what is already there).

The results of the experiments shown in Figure 2.11 severely undermine this idea.

1. Passing a collection (beam) of $|D\rangle$ state electrons through a (LEFT, RIGHT) magnet separates them into $|D\rangle|L\rangle$ and $|D\rangle|R\rangle$, each with equal numbers of each.
2. Passing the $|D\rangle|L\rangle$ and $|D\rangle|R\rangle$, electrons into separate (UP, DOWN) magnets produces both $|D\rangle$ and $|U\rangle$ electrons at each magnet, suggesting that the $|D\rangle$ state does not always survive passing through a (LEFT, RIGHT) magnet.

3. Passing the $|D\rangle|L\rangle$ and $|D\rangle|R\rangle$, electrons into the *same* (UP, DOWN) magnet produces a pure $|D\rangle$ beam, suggesting that the $|D\rangle$ state is now preserved.
4. This undermines the thought expressed in point 1 that we can specify the (UP, DOWN) and (LEFT, RIGHT) states at the same time. Perhaps there are no such things as $|D\rangle|L\rangle$ and $|D\rangle|R\rangle$, states after all, just $|U\rangle$ / $|D\rangle$ or $|R\rangle$ / $|L\rangle$ states.
5. Any suggestion that it is the distance travelled by the electrons or the passage through the magnet that causes these effects is contradicted by the experimental results produced by blocking one of the beams.
6. Blocking the left- or right-hand beam through the (LEFT, RIGHT) magnet separately before they reach the same single (UP, DOWN) magnet as used in point 3 results in some of the electrons going up and some going down.
7. The nature of the electron's state seems to depend on the context of the experiment.

There is one point that I haven't mentioned yet, and it makes things even stranger.

If, for example, we block the LEFT channel, then the electrons passing along the RIGHT channel into the (UP, DOWN) magnet emerge in either $|U\rangle$ or $|D\rangle$ states. However, if they passed along the RIGHT channel, *how can they have known that the LEFT channel was closed*? Put it another way, if we suddenly open up the LEFT channel, we *add* more electrons passing into the (UP, DOWN) magnet—those that would have gone through the RIGHT channel anyway and those that were previously blocked in the LEFT channel. Suddenly *all* electrons are now in the $|D\rangle$ state.

None of the results that we have been discussing depend in the slightest way on the intensity of the beam. If we choose, the beam can be turned down so low that only one electron is present in the apparatus at any one time. All our experiments work perfectly well under these conditions and produce the same results. This disposes of any notion that the electrons are 'interacting' or 'getting in each other's way' or any other theory of this sort. As I pointed out before, there is no obvious way that an electron passing through one channel, could be influenced by the other channel being blocked, unless there was another electron in that channel at the same time to mediate the influence. As the experiment gives the same result with a low-intensity beam, that idea can't work either.

One way in which all of these results can be gathered into a coherent whole is to consider what *information* we can obtain from each experiment in turn.

When one of the channels through the (LEFT, RIGHT) magnet is blocked, it's obvious that any electron emerging from the experiment must have passed through the other open channel.

However, with both channels open *we can't tell which path the electrons followed through the (LEFT, RIGHT) magnet.*

Remember that electrons are subatomic particles, so it's not simply a case of leaning close over the experiment and watching them go past. Any equipment that we might add into the experiment to figure out which path the electrons take has the same result as blocking a path.

Here there are obvious similarities with the photon experiments in Chapter 1. Once again, the context of the whole experiment is proving crucial.

Evidently *knowing that an electron is in either a $|L\rangle$ or $|R\rangle$ state prevents us from saying for certain that it is in a $|U\rangle$ or $|D\rangle$ state.*

Look at Figure 2.11: having one path blocked after the second magnet tells us that an electron entering into the (UP, DOWN) magnet is clearly in either the $|L\rangle$ or $|R\rangle$ states, in which case we have lost any notion of it being $|U\rangle$ or $|D\rangle$. With both paths open, there is no information that tells us the $|L\rangle$ / $|R\rangle$ state of the electrons. In this case, it seems that we can retain some information about the $|U\rangle$ / $|D\rangle$ state.

This interpretation is not forced on us by the results of the experiment discussed so far, but as we delve deeper into quantum theory and discuss other experiments, we will see the consistency of this approach. It certainly ties in with our discussion in Chapter 1. Those experiments showed us that interference depended on not being able to tell which path the photons were using. Here we can tell if it is $|U\rangle$ / $|D\rangle$ as long as we can't tell if it is $|L\rangle$ / $|R\rangle$. This is showing us something important about the nature of a quantum state.

2.3 SUMMARY

Probably the most important message to take away from this chapter revolves around the nature of quantum states, in particular how obtaining information about one pair of states can destroy information about another pair of states for the same object. If we couple that with the findings from Chapter 1, we must conclude that it is the whole experimental arrangement that allows us to extract that information, either directly or by inference, which is the crucial aspect.

Clearly quantum states can behave rather differently from classical states. The classical idea that a measurement reveals what is already there has been radically undermined, at least in certain situations[8]. Furthermore, the nature of a quantum state seems to depend on the way in which an experiment has been set up. Some aspects of a quantum state, which are evident in one context, are not evident in another. As we progress further into our development of quantum theory, especially the theory of measurement in Chapter 5, we will discuss how instrumentalist and realist views about quantum states differ. There is, however, a common theme: in some sense, quantum states describe the propensity for a system to behave in certain ways, rather than encapsulating definitive information about how the system will behave.

NOTES

1 The Feynman Lectures on Physics, Addison-Wesley Publishing Company 1963, section 37–1.
2 Otto Stern, 1888–1969, was awarded the Noble Prize in Physics in 1943, "for his contribution to the development of the molecular ray method and his discovery of the magnetic moment of the proton".
3 In practice, the spreading of the electron 'wave packet' and the action of the Lorentz force would introduce an uncertainty into the path of the electrons which would be at least as big as the beam separation (Pauli, 1932). Traditionally these experiments are done using neutral atoms, and information about electrons is inferred from the atomic deflections.
4 A photographic film would not be sensitive enough to detect individual electrons in this manner.
5 The amount of deflection depends on the speed (energy) of the electrons and on how far the detector is away from the magnet.
6 I say semi-classical as the mere existence of two patterns on the film, rather than a continuous stream, has already pushed us outside the bounds of classical physics.
7 The qualification accounts for e.g., the temperature being above the melting point, for example, but even then, we could argue that the atomic nature of the material has not changed.
8 Let's not forget that exposing an $|U\rangle$ beam to an (UP, DOWN) magnet will reveal that all the electrons are in the UP state…

3 Quantum States

3.1 WHERE ARE WE NOW?

During the last couple of chapters, we have considered various experimental results that force us to rethink how matter behaves, at least at atomic scales. For example, an individual photon can apparently follow both paths from a beam splitter, as long as the experimental setup is unable to distinguish the photon's direction of flight. It's as if reality is slightly out of focus until a specific experiment forces the picture to sharpen into one possibility or another.

Of course, we can always accuse our theory of being 'out of focus', or at least incomplete, prior to meekly accepting a radical change in the nature of reality. Maybe in the future, new experiments will reveal some physical variables that we haven't spotted up to now (*hidden variables*), and these will allow us to construct a more advanced theory that resolves what currently appears to be paradoxical. There is no current indication that such a new layer of information exists. Tin fact, there are good grounds for suspecting that if there was a hidden variable theory that was capable of being just as accurate and effective as quantum theory, it would be even more weird in other fundamental ways (see further in Chapter 23).

Our day-to-day experience is based on the large-scale world that we routinely occupy. For example, we are used to throwing cricket balls in from the outfield and observing that they follow a single path. This 'common-sense' understanding colours our view of *all* reality, irrespective of scale, but in truth, there is no guarantee that it will apply outside of our everyday world. Indeed, as we have seen, when we conduct experiments with photons and electrons, we find that they don't act like tiny cricket balls. Consequently, over the 100 or so years since the founding fathers first addressed the problems of the quantum world, the conversation amongst the later generations has shifted from trying to understand the nature of the quantum layer of reality to explaining how a cricket ball (which after all is made of particles such as electrons) can manage to behave in the 'common sense' way, given the underlying strangeness of the quantum world.

However, before we can discuss such puzzles in any greater detail, we need to develop a mathematical scheme for describing our experimental results in a consistent way and which allows some measure of predictability. The path to that end initially passes through a fresh consideration of Newtonian physics.

3.2 DESCRIBING CLASSICAL SYSTEMS

In Chapter 2, I briefly mentioned some ideas about classical and quantum states. Now it's time to put things on a somewhat more formal basis, not in the service of pedantic slavery to precision, but so we can be clear about the contrast between the classical and quantum situations going forward.

USEFUL DEFINITIONS

1. A **system** is a section of the universe that we have earmarked for investigation. Systems can be very large (stars and galaxies) or very small (electrons and photons). They can have lots of pieces within them (a living body) or only a few (a hydrogen atom). We try to isolate any system that we are interested in from the rest of the universe, so we can study its behaviour more easily. This is not always completely possible.

DOI: 10.1201/9781003225997-5

2. **Physical properties** are aspects of a system that we can measure or detect in some fashion. Examples include mass, colour, density, speed, and position. Some of these properties can be given quantitative values (a number and a unit). A system might, for example, have a speed of 20 m/s.
3. **Physical variables** are the symbols that stand for physical properties in a mathematical theory. Speed (velocity), for example, is generally given by the symbol v when we want to refer to it without being fussed about a particular quantitative value. If we want to pin a specific value down, we write something like $v = 20$ m/s. If a theory is *complete*, then all the important physical properties of a system are represented by variables in the theory. Einstein always suspected that quantum theory wasn't complete.
4. **System properties/variables** are the physical properties that describe a system and help us to distinguish one system from another. An electron has system properties such as electrical charge (-1.6×10^{-19}C) and mass (9.1×10^{-31} kg). System properties cannot change without altering the nature of the system.
5. **State properties/variables** are the physical properties of a system that can change without changing what the system is. An electron can have a whole variety of different speeds, and still be an electron. The speed is an aspect of the electron's state.
6. A **classical state** specifies the quantitative values of all the state variables relevant to a given system. In some situations, we can know the overall state of the system without being sure about the individual states of the constituent parts. A gas, for example, is a collection of molecules. The state of the gas, in terms of its pressure, volume, and temperature, can be specified perfectly without knowing the exact state, in terms of velocity and position, of each molecule.

It used to be taken as read that within the bounds of classical (or Newtonian) physics we can calculate a projected final state of a system at some future time, given a complete specification of its initial state and all the forces acting. In practice, though, there are a few things that can get in the way of this straightforward sounding program.

By and large, the collection of properties associated with any interesting physical system is dauntingly large. Something as mundane as a cricket ball has mass, velocity (in three different directions), position (with three different coordinates), volume, size (very hard to be precise about that, given that a cricket ball is not a simple sphere), temperature, surface reflectivity, surface smoothness, and other such things. However, we don't generally have to worry about most of these properties, depending on what we are trying to achieve. Some don't affect the motion of the ball (e.g., its temperature), so they can be omitted from any dynamical calculation. Others do influence motion, but to a limited extent, so we can get away with approximate values. Although the exact shape of the ball will modify the details of its path, we can often be sufficiently accurate by treating it as a simple sphere.

The full details of all the forces acting can also be rather complicated. In the case of the cricket ball, gravity and air resistance are the pertinent factors once it is in flight. Fortunately, at typical cricket ball speeds, we can get by if we ignore air resistance.

The process of simplifying things by ignoring the details of physical properties and forces is called making a *model* of the situation, and it is one of the most important practices in science. As techniques and understanding develop, so more details can be incorporated, and the model may well develop into a theory.

On top of the problems inherent in picking out the crucial details of a system, there is another issue with trying to specify an initial state precisely: there is always some *uncertainty* connected with measurement. No matter how accurate, complicated, or expensive our apparatus may be, there will always be a limit to the precision possible. This in turn will affect any calculations that are made using such measurements as part of a system's initial state. In detailed work, this 'slop' is taken into account. For example, when a player throws a cricket ball, we may not know precisely the value of the velocity at which it left the player's hand. So, the best we can do is to calculate the range of possible landing points, together with some estimation of how likely each one is. We accept this limitation on our predictive power as the price we pay for any sensible calculation. Our knowledge is uncertain, and we have to rely on predicting the probabilities associated with the range of outcomes because we cannot include all the details with perfect accuracy.

Although we can't always know the exact values of all the physical quantities in a classical state, or the fine details of all the forces acting, the process of model making, and approximating has generally proven to be a highly successful strategy for dealing with the large-scale world. After all, it enabled us to put humans on the moon. However, the predictability of Newtonian physics has not only been challenged by quantum physics. In recent decades we have become aware of delicately balanced systems that are more intimately linked with the surrounding universe than most. This has brought its own radical transformation in our understanding of the classical world.

3.2.1 Chaos

> Chaos: When the present determines the future, but the approximate present does not approximately determine the future.
>
> *Edward Lorenz*[1]

Some finely tuned systems are exquisitely sensitive to the precise details of their initial states, so in practice, it is impossible to make any sensible predictions about their development. Saturn's rings are an excellent example of such a situation (Figure 3.1). They are composed of variously sized chunks of rock and ice. At first glance, you'd think that calculating the orbit of any rock in a ring would be nearly as straightforward as calculating the path of a cricket ball. However, this is not the case. Although Saturn's gravity is the dominant force, there are also other significant factors. The rocks will exert tiny gravitational forces on each other, and the orbiting moons of Saturn will pull them in different directions as well. Unlike the cricket ball situation, where we could ignore many of the smaller forces, the motion here is much more critically balanced. Within the volume of the rings, there are certain regions where the precise orbit followed by any chunk of rock depends very sensitively on its velocity. Two rocks placed in exactly the same place, but with a tiny difference in their velocity, would end up moving in very different ways. One might fall in toward the planet, and the other fly off into space. Very few end up meekly orbiting the planet, like their colleagues in other less sensitive regions of the ring plane. This is why the rings have visible gaps in regions devoid of stable orbits.

Finely balanced systems, such as this one, are termed *chaotic*, although the name gives the wrong impression of what's going on. Chaotic systems are unpredictable, as we can never specify

FIGURE 3.1 The rings of Saturn as seen by the Cassini space probe. Clearly visible in this image are the ring gaps, including the large Cassini Division. (Image credit: NASA/JPL/Space Science Institute.)

the initial state precisely enough to single out one final state, but they are not lawless. In principle, everything about a chaotic system is exactly determined, but in practice, all we can do is to estimate the probability of various outcomes. The same was true of the cricket ball, but in that case, the region of uncertainty was quite small. With a chaotic ball, a tiny change in the throw could change the landing point by several kilometres.

Another commonly quoted example of a chaotic system is the atmosphere. We can never measure what is going on in the atmosphere to the level of precision that would be needed for a reliable weather prediction extending more than a few days into the future. The global weather system is finely balanced and small disturbances can produce vast changes in the outcome[2].

Chaotic systems derive their unpredictability from the practical limitation on how well we can pick out initial states. This is important: *chaotic systems are completely determined by their initial states and the laws of Newtonian physics.* We can't in practice make sensible predictions about them, as *no level of precision in our measurements can specify the initial conditions with sufficient acuity.* This contrasts sharply with the unpredictability present in quantum theory, where the fundamental indeterminism is thought, at least without the aid of putative hidden variables, to derive from a genuine randomness at the heart of nature.

3.3 DESCRIBING QUANTUM SYSTEMS

In Chapter 2 we saw how using a semi-classical style of state got us into trouble when trying to describe experiments in the quantum world. A classical state is effectively a list of quantitative values of various physical properties, so it is hard to see how this can be applied to a photon in an interference experiment apparently travelling along two classical paths at the same time. Similarly, any physical property that determines the direction of an electron through a Stern–Gerlach (SG) experiment seems to be influenced by the exact details of how the equipment is set up, which runs contrary to the classical idea that experiments reveal what is already there.

So, it would pay to explore ways of constructing a quantum state to replace the normal classical description of a system. If this is going to work, it's important that certain basic characteristics are designed in from the ground up.

1. The inherent randomness found in some situations must be represented. For example, the description of a photon arriving at a beam splitter has to cope with the photon having an equal chance of being transmitted or reflected, without stating that it will definitely do one or the other.
2. The contextuality of quantum behaviour must be incorporated. We have seen how experimental results are influenced by the overall setup of the equipment. If a photon detector is placed beyond the beam splitter, then the particle will either reflect or transmit. If there is no detector present, then the results imply that the photon explores both possibilities.
3. Quantum systems seem to be able to exist in a mixed state that combines classical states in a way that we would have regarded as being impossible (e.g., reflecting and transmitting at a beam splitter).

Figure 3.2 is an illustrative summary of the kind of structure we are trying to achieve with our description of a quantum state.

The left-hand side of Figure 3.2 shows a system, such as an electron, in some quantum state symbolized by $|\varphi\rangle$, although we have not yet said how this state is going to be specified mathematically. The electron then interacts with some measuring apparatus, M, and as a result one of several possible new states, $|A\rangle$, $|B\rangle$, $|C\rangle$, $|D\rangle$ can come about, with a different probability for each one. A specific example might be a $|U\rangle$ state electron interacting with a (LEFT, RIGHT) SG magnet and as a result emerging in a $|L\rangle$ or $|R\rangle$ state.

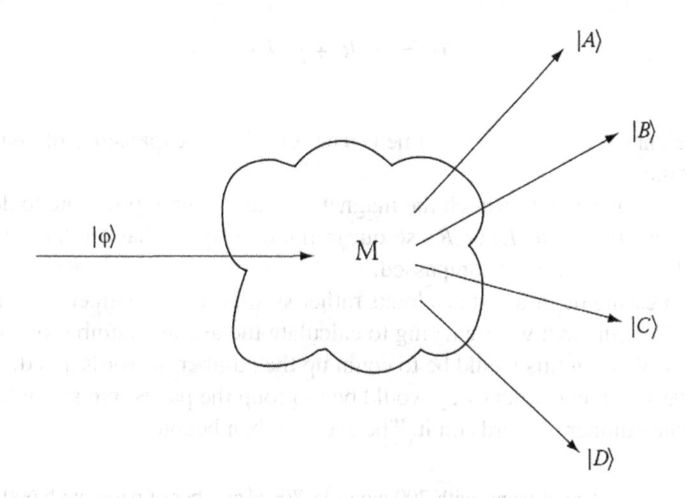

FIGURE 3.2 A quantum description must be able to capture the features shown in the figure. A given state $|\varphi\rangle$ can convert into one of the several other states after an interaction with a measuring apparatus.

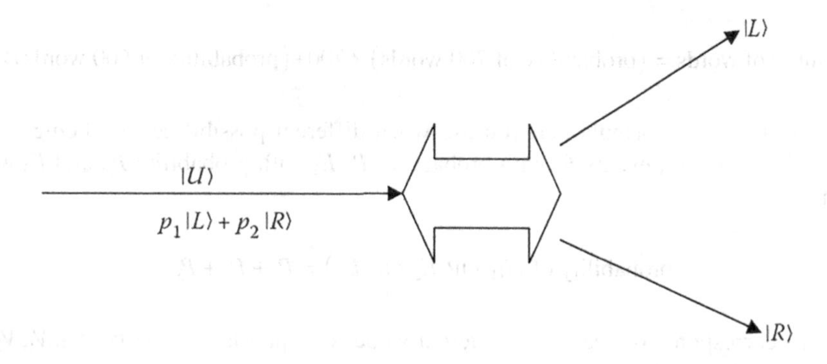

FIGURE 3.3 Writing $|U\rangle$ in terms of $|L\rangle$ and $|R\rangle$.

Taking a direct approach, we could try writing the initial state in the following form:

$$|\varphi\rangle = p_1|A\rangle + p_2|B\rangle + p_3|C\rangle + p_4|D\rangle + \cdots$$

where the numbers $p_1 \ldots p_4$ relate to the probability that the electron would end up in each state $|A\rangle, |B\rangle, \ldots |D\rangle, \ldots$ etc. (I'm not saying that the numbers *are* probabilities, just that they are *connected* to the probabilities in some way yet to be determined.) This is an attractive formulation as it already catches some of the flavours of quantum behaviour. It directly expresses how $|\varphi\rangle$ is, in a sense, 'made up' of all the possibilities $|A\rangle$ through $|D\rangle$ which may subsequently come about. For example, for our SG experiment we would write:

$$|U\rangle = p_1|L\rangle + p_2|R\rangle$$

to represent the initial state of the electron shown in Figure 3.3.

In this representation, the order of the terms is not significant. In other words, the two possible expressions:

$$|U\rangle = p_1|L\rangle + p_2|R\rangle$$

$$|U\rangle = p_2|R\rangle + p_1|L\rangle$$

are exactly equivalent to each other. The same is true for all such expansions of quantum states that we will come across.

After the electron has passed through the magnet, it's no longer appropriate to describe it by the state $|U\rangle$. Now it's in either state $|L\rangle$ or $|R\rangle$, so our initial description has 'collapsed' into one of the two alternatives that it originally encompassed.

This way of expressing quantum states looks rather similar to the manner in which probabilities can be combined. Imagine that we are trying to calculate the average number of words per page in this book. One way of doing this would be to count up the number of words and divide by the number of pages. However, an equivalent way would be to group the pages into sets where each page in the set had the same number of words on it. The average then becomes:

$$\text{average number of words} = \frac{(\text{no of pages with 700 words})\times 700 + (\text{number of pages with 600 words})\times 600 + \cdots}{\text{total number of pages}}$$

or:

$$\text{average number of words} = (\text{probability of 700 words})\times 700 + (\text{probability of 600 words})\times 600 + \cdots$$

which looks just like the formula that you use when different possibilities are being considered. Given, mutually exclusive, events E_1 with probability P_1, E_2 with probability P_2, and E_3 with probability P_3, then:

$$\text{probability of } (E_1 \text{ OR } E_2 \text{ OR } E_3) = P_1 + P_2 + P_3.$$

If these events correspond to measuring different values of a physical property (e.g. V_1, V_2 and V_3), then the average value of that property after many trials is:

$$\text{average value of V} = (V_1 \times P_1) + (V_2 \times P_2) + (V_3 \times P_3)$$

which certainly looks something like our quantum mechanical state. If the two formulations were *exactly* the same, then the terms $p_1 \ldots p_4$, etc., in:

$$|\varphi\rangle = p_1|A\rangle + p_2|B\rangle + p_3|C\rangle + p_4|D\rangle + \cdots$$

would have to be probabilities, but there is a problem with that as we will see in the next section.

3.3.1 Specific Example: Mach–Zehnder Again

Having devised a trial representation of a quantum state, we need to apply it to a specific experimental situation and see how it copes. A useful example to choose is the Mach–Zehnder interferometer from Chapter 1. If you remember, this experiment seems to need both wave and particle descriptions of light, so it will be a good test of our quantum state formulation.

At the first beam splitter, photons can be reflected or transmitted (pass straight through). So, it would be appropriate to write the initial quantum state of a photon before striking the mirror as:

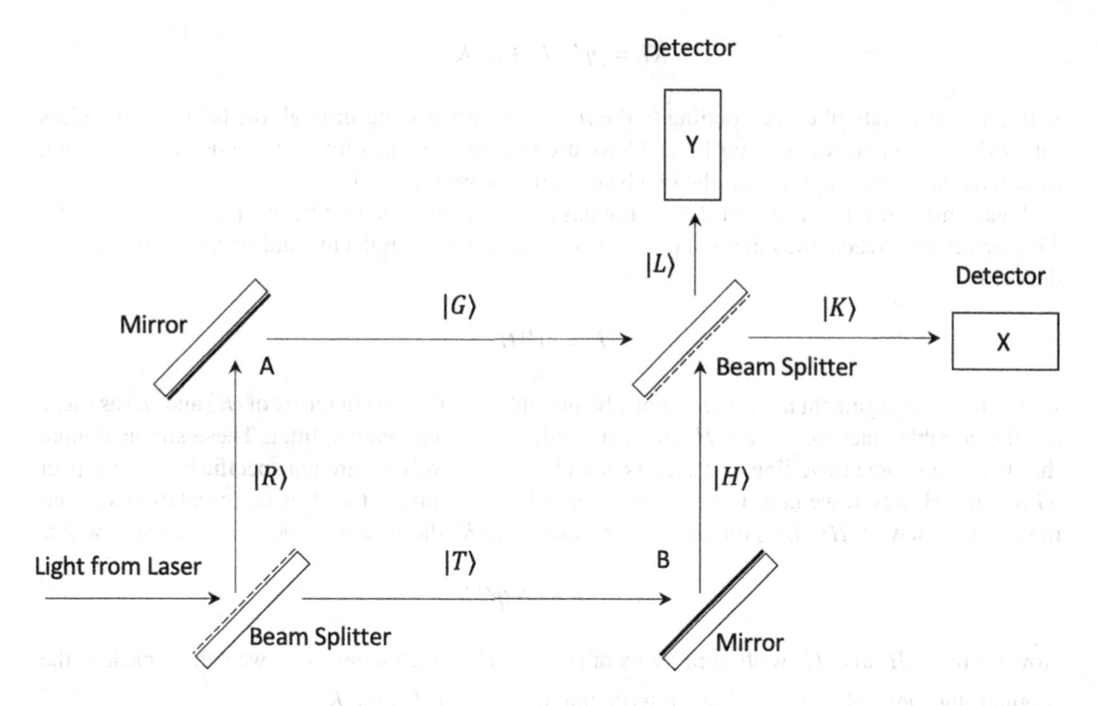

FIGURE 3.4 Another look at the Mach–Zehnder device.

$$|\varphi\rangle = (\text{some factor})|T\rangle + (\text{another factor})|R\rangle$$

where $|T\rangle$ represents the transmitted state and $|R\rangle$ the reflected one (not RIGHT from an SG magnet). The relative probability of transmission or reflection, and hence the values of the factors, will be determined by the construction of the mirror.

The reflected photon *aspect*[3] goes on to meet the fully silvered mirror at A (Figure 3.4). Consequently, its quantum state must change at this mirror to, say, $|G\rangle$. As all photons are reflected (assuming a perfect mirror that does not absorb any photons), there is no relative probability involved. However, as there is a phase shift on reflection, it would be prudent to introduce another factor to correspond to that:

$$|R\rangle = nr'|G\rangle$$

where the factor n is allocated to the π phase shift and r' the reflection. As it is 100% reflection, we might be thinking that $r' = 1$ so why bother including it, but it will make aspects of the argument clearer as we develop this approach.

To be consistent, we must use the same factor to indicate the π phase shift on reflection at the beam spitter as well. Hence, we ought to write:

$$|\varphi\rangle = tp|T\rangle + nr|R\rangle$$

where t is the factor corresponding to transmission, p the factor for the ϕ phase shift through the glass, n the π phase shift on reflection and r the factor for reflection itself, which is not the same as r' as at the beam splitter, we do not have 100% reflection.

At the next half-silvered mirror, the photon aspect $|G\rangle$ can be either reflected up, $|L\rangle$ (without phase shift, but with a shift due to passing through the glass), or transmitted through, $|K\rangle$, so we ought to be able to expand $|G\rangle$ as follows:

$$|G\rangle = pp'r|L\rangle + tp|K\rangle.$$

with the coefficients p' corresponding to the ϕ' phase shift passing through the height of the glass rather than the width (see Section 1.5.1). I have used the same factors for transmission and reflection as before, on the assumption that the two beam splitters are identical.

Meanwhile the other aspect of the photon has emerged from the first beam splitter in a state $|T\rangle$. This aspect will 'reach' the mirror at position B, where it will be reflected and so change into a state denoted $|H\rangle$:

$$|T\rangle = nr'|H\rangle$$

Using the same argument as before, it must be possible to write $|H\rangle$ in terms of $|K\rangle$ and $|L\rangle$ as these are the possible outcomes when $|H\rangle$ interacts with the second beam splitter. These simply denote the state of a photon travelling to detectors X and Y respectively, so are not specific to either aspect $|G\rangle$ or $|H\rangle$. However, we have to be slightly careful when constructing this representation as from the point of view of $|H\rangle$, $|L\rangle$ is the *transmitted state* and $|K\rangle$ the *reflected* one. So, we have to write:

$$|H\rangle = nr|K\rangle + tp'|L\rangle.$$

Now we have $|H\rangle$ and $|G\rangle$ written in terms of $|L\rangle$ and $|K\rangle$, which means that we can go back to the original state $|\varphi\rangle$ and figure out how to write that in terms of $|L\rangle$ and $|K\rangle$:

$$\begin{aligned}
|\varphi\rangle &= tp|T\rangle + nr|R\rangle \\
&= tpnr'|H\rangle + n^2rr'|G\rangle \\
&= tpnr'\left[nr|K\rangle + tp'|L\rangle\right] + n^2rr'\left[pp'r|L\rangle + tp|K\rangle\right] \\
&= tprn^2r'|K\rangle + t^2pp'nr'|L\rangle + r^2n^2pp'r'|L\rangle + tprn^2r'|K\rangle \\
&= 2tprn^2r'|K\rangle + \left(t^2pp'nr' + r^2n^2pp'r'\right)|L\rangle
\end{aligned}$$

We have now produced a representation of the initial photon state $|\varphi\rangle$ in terms of the two possible final outcomes $|K\rangle$ and $|L\rangle$. The factors $2tprn^2r'$ and $\left(t^2pp'nr' + r^2n^2pp'r'\right)$ relate to the probabilities that the photon will be detected at X and Y respectively.

At this point we need to think back to the actual experimental results. If the distances in the detector are balanced out, then all the photons passing through the device are picked up at detector X, and none of them reaches Y. Consequently, *the final state of the photon in these circumstances cannot include* $|L\rangle$, *implying that* $\left(t^2pp'nr' + r^2n^2pp'r'\right) = 0$.

That being the case, we must have:

$$pp'nr'\left(t^2 + nr^2\right) = 0$$

or[4]:

$$t^2 + nr^2 = 0$$

At this stage, it is possible to develop the argument in several different ways, all of which lead to a similar conclusion.

Prima facie, if the beam splitter is genuinely 50:50 and 100% efficient, we would expect the factors t and r to have the same value. In which case, $t^2 + nr^2 = 0$ implies $n = -1$.

Initially, we suggested that the factors in a state expansion such as:

$$|\varphi\rangle = p_1|A\rangle + p_2|B\rangle + p_3|C\rangle + p_4|D\rangle + \cdots$$

were *related* to the probability of each state transition. If $n = -1$ in the expansion $|T\rangle = nr'|H\rangle$, then the factors cannot actually be probabilities, as a negative value would make no sense.

In the instance $|T\rangle = n|H\rangle$ awe are saying that all photons in state $|T\rangle$ transform into $|H\rangle$ (with a phase shift) so n must related to a probability=1. A sensible speculation regarding the relationship between a factor and a probability might then be (as we have established that $n = -1$):

$$\text{probability} \propto (\text{factor})^2$$

as this would always result in a positive number. We will pursue this line of approach further in a few paragraphs.

For the moment, let us turn back to the other factors that have been involved in this calculation. For example, p and p' which have been set up to represent phase shifts that happen while a photon aspect is passing through the glass of a beam splitter. We have not suggested any value for these shifts, as the argument from Section 1.5.1 shows that the specific values do not matter. However, imagine for the moment that we have constructed a beam splitter of sufficient thickness to produce a phase shift of $\pi/2$. In which case the factor p would be indicative of a $\pi/2$ phase shift. If we then arranged for another sheet of glass of the same thickness to be paced directly after the beam splitter, so that all transmitted photon aspects would pass through that as well, then we would have an overall phase shift of π and in terms of a state representation:

$$|U\rangle = pp|T\rangle = p^2|T\rangle$$

However, we already know how to represent a π phase shift, as we obtained $n = -1$ where n was the factor controlling the phase shift at an ordinary mirror. Hence, we see that:

$$|U\rangle = p^2|T\rangle = n|T\rangle = -1|T\rangle$$

so $p^2 = -1$.

This poses a further problem. We have already established that the factors are not in themselves probabilities: having $n = -1$ effectively dashed that hope. We did suggest that the relationship between a probability and a factor might be:

$$\text{probability} \propto (\text{factor})^2$$

but here we have the square of a factor being -1, so that can't be right either.

While we're about it, having $p^2 = -1$ is in itself an extremely significant step. It would appear that some of the factors in an expansion such as:

$$|\varphi\rangle = p_1|A\rangle + p_2|B\rangle + p_3|C\rangle + p_4|D\rangle + \cdots$$

have to be *imaginary numbers*, at least in some contexts.

3.3.2 Probability Amplitudes

An equation such as $p^2 = -1$ has the solution $p = i$, where $i = \sqrt{-1}$. However, if the factors in our state expansions are imaginary (or even possibly complex numbers[5]), what can they *mean?* They certainly can't be probabilities. Instead, they are referred to as *probability amplitudes*, or often just *amplitudes* for short. Hence in a construction such as:

$$|\varphi\rangle = a_1|A\rangle + a_2|B\rangle + a_3|C\rangle + a_4|D\rangle + \cdots$$

a_1, a_2, a_3, etc., are the probability amplitudes for the states $|A\rangle$, $|B\rangle$, $|C\rangle$ etc.

We started with the notion that the numbers we used to multiply 'sub states' were *related* to the probability that a particular sub-state would come about as a result of a measurement. Any hope that the numbers might actually *be* the probability has been dashed by applying the idea to the Mach–Zehnder experiment. *We have to use complex numbers if we are going to represent the factors associated with phase shifts.*

In some instances, the use of complex numbers in a physical theory is a useful mathematical convenience (for example in the analysis of electrical circuits). Their deployment gives access to elegant and powerful techniques that can be a big help. However, there are other (longer and more strenuous) ways of doing the same calculations, so the move to use complex numbers is not forced.

The same is not true in quantum theory.

In some deep and puzzling sense, *the theory will not work without complex numbers.* This is an intriguing and unique result. Generally, complex numbers are considered less 'real' than, say, integers and rational numbers as the latter both 'appear' in nature. You can count 10 swans but would be hard pressed to count $10i$ swans. The need to use complex numbers to represent amplitudes is one of the principal reasons why some physicists avoid a fully realistic view of quantum theory[6]. There is no other physical theory, aside from those developed out of quantum mechanics, that *requires* the use of complex numbers.

3.3.3 Relating Amplitudes to Probabilities

Any probability obtained from an amplitude must have all the factors of i removed from it, as probabilities are definitively real numbers. Mathematicians have already provided a recipe for turning complex numbers into real numbers: multiply the complex number by its *complex conjugate*[7].

We therefore have a possible rule for to converting *probability amplitudes* into *probabilities:* we *multiply the amplitude by its complex conjugate* (which is also called 'taking the complex square''):

IMPORTANT RULE 1: THE BORN RULE

If

$$|\varphi\rangle = a_1|A\rangle + a_2|B\rangle + a_3|C\rangle + a_4|D\rangle + \cdots$$

then:

$$\text{Prob}\left(|\varphi\rangle \rightarrow |A\rangle\right) = a_1^* \times a_1 = |a_1|^2$$

$$\text{Prob}\left(|\varphi\rangle \rightarrow |B\rangle\right) = a_2^* \times a_2 = |a_2|^2$$

etc., named after Max Born[8] who first formulated this approach in 1926.

This is one of the fundamental rules of quantum theory.

Rules like these can't be *proven* in any mathematical sense. Pure mathematics can't tell us what probability amplitudes *mean*. That is the job of physics, and the only way of doing it is to relate mathematics to experimental results. We have to assume the rule, use it to do some calculations, and then check and see if we are right from confirmatory experiments. If all works out, then the rule gets accepted. In this case, the relationship between amplitudes and probabilities is a cornerstone of quantum theory; the success of the whole construction relies on its being correct. Quantum theory has been around for more than 100 years now, so we can regard this rule as being well checked by experiments.

The story on this is not entirely settled though. There is one version of quantum theory that some physicists claim they can use to *prove* Important Rule 1, but we will come to that in Chapter 28.

3.3.4 Amplitudes, Complex Numbers and Phase

Any complex number, $a + ib$, can be identified with a point on the *complex plane*, Figure 3.5. In this diagram, the horizontal arrow labelled x shows that we are plotting the *x coordinate* along that direction. On a normal graph, there is also a vertical arrow, y, denoting the *y coordinate* lies in the vertical direction. In this case, though, I have labelled the vertical line as iy, which I'm not really allowed to do. You can't plot a coordinate *iy as there is no such distance*. The point of a graph is to represent quantities (numbers with units) as lengths according to some scale. Really, I should have just put y on the vertical axis and then said that any complex number is going to be of the form:

$$z = (x \text{ coordinate}) + i(y \text{ coordinate})$$

I haven't done this in order to emphasise, in Figures 3.5 and 3.6, that neither the x-direction nor the y-direction is an *actual physical direction in space*. The figures represent the way of picturing complex numbers. This becomes very important later when we look at the amplitude function for the so-called free particle. It would be easy to (mistakenly) think of this amplitude in terms of a physical 'size' of some form, which is a point that we will come back to in Chapter 7.

Imagine standing at the origin looking along the real line (stretching off into the distance). If you turn in an anticlockwise direction through an angle ϑ and proceed to walk in that direction a distance R, then provided you have figured the right values of ϑ and R, you will end up at the point $a + ib$ (Figure 3.6).

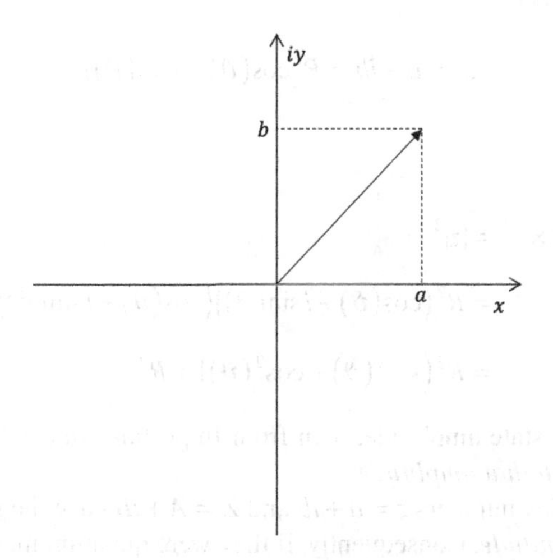

FIGURE 3.5 A complex number can be visualized as a representation of a specific point on a 2-D plane. The horizontal axis is sometimes known as the real line and the vertical axis is the imaginary line.

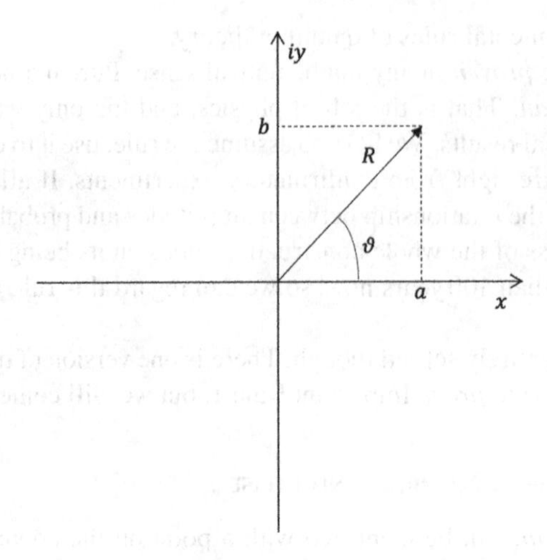

FIGURE 3.6 The same point on the complex plane can be reached by marching a distance R in the direction ϑ or by walking horizontally to a and then turning 90° and walking up to b.

If R and ϑ are to get you to the same place as a and b, then:

$$a^2 + b^2 = R^2$$

$$a = R\cos(\vartheta)$$

$$b = R\sin(\vartheta)$$

So, for any complex number z:

$$z = a + ib = R\big(\cos(\vartheta) + i\sin(\vartheta)\big)$$

The complex conjugate z^* is then:

$$z^* = a - ib = R\big(\cos(\vartheta) - i\sin(\vartheta)\big)$$

Leading to:

$$\begin{aligned} z \times z^* &= |z|^2 \\ &= R^2\big(\cos(\vartheta) - i\sin(\vartheta)\big)\big(\cos(\vartheta) + i\sin(\vartheta)\big) \\ &= R^2\big(\sin^2(\vartheta) + \cos^2(\vartheta)\big) = R^2 \end{aligned}$$

Given that z is a quantum state amplitude, then from Important Rule 1 (The Born Rule) R^2 *is the probability associated with that amplitude.*

In Figure 3.7 the complex numbers $z = a + ib$ and $Z = A + iB$ have the same value of R, which is sometimes called the *magnitude*. Consequently, if they were quantum mechanical amplitudes, they would both lead to the same probability. Indeed, there are an infinite number of complex amplitudes leading to the same probability.

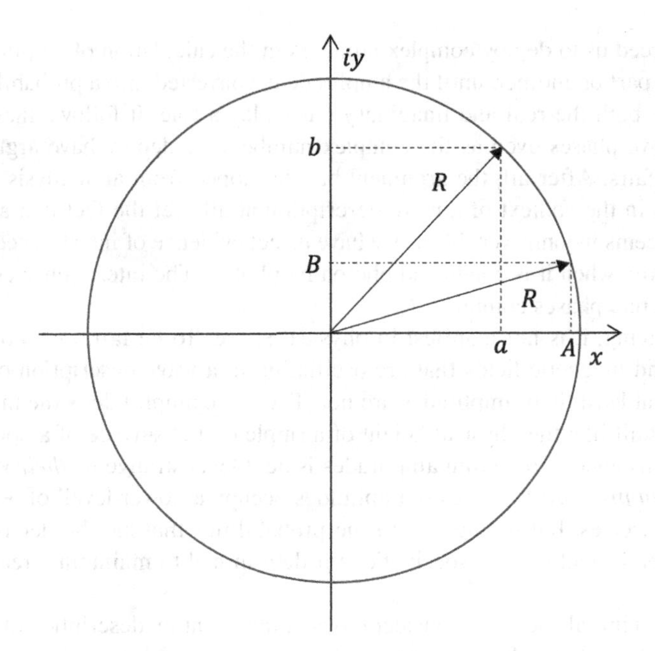

FIGURE 3.7 In this figure, both complex numbers, $a+ib$ and $A+iB$, have the same magnitude R.

All the complex numbers with the same R differ only in the value of ϑ, which is the *argument* of the complex number.

As the magnitude of amplitude is related to the probability, it is natural to enquire after an interpretation of the argument. In one sense, the argument is irrelevant, or at least the absolute value of the argument is. After all, all amplitudes with the same magnitude but different arguments lead to the same probability. This directly mirrors what we have already found. In describing the phase shift of a photon aspect while travelling through the glass of a beam splitter, we found ourselves in need of an amplitude that represented a 100% probability (no photons were harmed in passing through the glass) but a change of phase. Perhaps the argument of the amplitude is the phase.

This thought gains plausibility when we consider the suggestion that $n = -1$ denotes a π phase shift. After all, if we represent n in the form $n = R(\cos(\vartheta) + i\sin(\vartheta))$ with the specific value $\vartheta = \pi$ we get:

$$n = R(\cos(\pi) + i\sin(\pi)) = -R$$

hence $R = 1$. In Section 3.3.1 when I set up the amplitude calculation for the Mach-Zehnder interferometer, I intimated that the requirement for 100% probability in amplitude nr' and yet to have some difference in the phase was a hint to the nature of the amplitude. Now we see this in full detail – having the amplitudes as complex numbers allow us to have a *range of different numbers, all of which have the same magnitude*, specifically 1 in this case.

While we now have a plausible link between the argument of a complex number and the phase of a state, we need to be clear about how very different this situation is from classical mechanics. Often, while analysing the properties of waves, it is convenient to represent that wave using complex numbers, with the magnitude as the amplitude of the wave and the argument as the phase. However, this is a mathematical convenience that allows us to deploy the range of tools that have been developed for work in the complex plane. At the end of the calculation, we reject any part of the resulting formula which is a multiple of i (in other words we 'take the real part') and focus on the bits that remain. We can't do that with the quantum amplitude. Consideration of the Mach-Zehnder experiment has forced us to deploy complex numbers in the calculation of amplitudes. At no point can we isolate one part or another, until the amplitude is converted into a probability via Important Rule 1. Even then, both the real and imaginary parts play a role. It follows inevitably from this, that amplitudes have phases even as the complex numbers we deploy have arguments. This is a curious state of affairs. After all, the argument has developed from an analysis of an experiment using photons, not in the context of a wave description at all. Yet the fact that amplitudes have a wave-like nature seems incontrovertible as we have direct evidence of interference coming about in this experiment, even when it is conducted photon by photon. The interference comes about when amplitudes of different phases combine.

Whatever is waving, it is not manifest in physical space. To be fair, one would rightly argue that the electric and magnetic fields that are oscillating in a wave description of light are not in physical space, or at least their amplitudes are not. The wave amplitude is the maximum strength of the fields, not at all like the physical height of a ripple on the surface of a body of water. Even then, the specific instance of quantum amplitudes is next level strange *as their wave nature gives rise to variations in probability*. It's as if amplitudes occupy a 'lower-level' of reality, not directly observable in themselves, but giving rise to the probabilities that can be determined in experiments. At least, one is lead to such speculations if determined to maintain a realistic view of the physics involved.

The fact that such initial and basic considerations of the quantum description of experiments lead directly to profound questions about the nature of reality is extraordinary.

3.3.5 States in Stern–Gerlach Experiment

Having spent a while discussing the Mach–Zehnder experiment, it's time to see if we can apply our ideas to SG experiments as well.

If we pass a $|U\rangle$ state electron through a (LEFT, RIGHT) magnet, it can emerge from either channel with equal probability. Similarly, if we send a $|D\rangle$ state electron into a (LEFT, RIGHT) magnet, it will also emerge from either channel with equal probability. So, using the ideas from the previous section, we must be able to write the two quantum states $|U\rangle$ and $|D\rangle$ in the form:

$$|U\rangle = a|R\rangle + b|L\rangle$$

$$|D\rangle = c|R\rangle + d|L\rangle$$

where a, b, c, and d are the relevant probability amplitudes. Now we need to figure out what the values of these numbers might be, and we have some clues to help us out.

First, the pair of values (a, b) must be different in some way to the pair of values (c, d) as $|U\rangle$ and $|D\rangle$ are manifestly different states.

Second, if the probability of emerging from either channel is the same, then Important Rule 1 tells us:

$$aa^* = bb^* = cc^* = dd^* = \frac{1}{2}$$

which suggests:

$$a = b = c = d = \frac{1}{\sqrt{2}}$$

but that can't be right as it runs counter to our first clue. We have to be more careful and use the fact that 1/2 can be either $\left(+1/\sqrt{2}\right)^2$ or $\left(-1/\sqrt{2}\right)^2$ to construct:

$$|U\rangle = \frac{1}{\sqrt{2}}|R\rangle + \frac{1}{\sqrt{2}}|L\rangle$$

$$|D\rangle = \frac{1}{\sqrt{2}}|R\rangle - \frac{1}{\sqrt{2}}|L\rangle$$

which is the correct combination, but unfortunately, we can't access a proof at this stage.

To figure out the values of a, b, c, and d, we have applied:

IMPORTANT RULE 2: NORMALIZATION

If:

$$|\varphi\rangle = a_1|A\rangle + a_2|B\rangle + a_3|C\rangle + a_4|D\rangle + \cdots$$

then:

$$|a_1|^2 + |a_2|^2 + |a_3|^2 + |a_4|^2 + \cdots = 1$$

You can see how this works out in the case of our $|U\rangle$ state where:

$$|a_1|^2 + |a_2|^2 = \left(\frac{1}{\sqrt{2}}\right)^2 + \left(\frac{1}{\sqrt{2}}\right)^2 = \frac{1}{2} + \frac{1}{2} = 1.$$

Important Rule 2 is telling us that the total probability that you get by adding the probability for each possibility in our state must come to 1. In other words, something has to happen[9]! If we have got things right and all the probabilities associated with the amplitudes in the state add up to 1, we say that the state has been *normalized*.

3.3.6 General Stern–Gerlach States

When we talked in detail about the SG experiment in Chapter 2, we only discussed magnets that were at 90° to one another. Now consider a beam of $|U\rangle$ state electrons arriving at an SG magnet with its axis tilted at some angle ϑ to the vertical. Electrons will emerge from this magnet along one of two paths as before, but in this case, we will refer to them as $|1\rangle$ and $|2\rangle$ respectively. Direct observation indicates that the split of electrons passing down each channel is not 50:50, indicating that the amplitudes are not the same:

$$|U\rangle = a|1\rangle + b|2\rangle$$

with the proviso that $aa^* + bb^* = 1$, following from Important Rule 2.

A detailed mathematical analysis shows that the correct amplitudes are:

$$|U\rangle = \cos\left(\frac{\vartheta}{2}\right)|1\rangle + \sin\left(\frac{\vartheta}{2}\right)|2\rangle \tag{3.1}$$

$$|D\rangle = \sin\left(\frac{\vartheta}{2}\right)|1\rangle - \cos\left(\frac{\vartheta}{2}\right)|2\rangle \tag{3.2}$$

These states have to be consistent with what we had before. So, if we send our $|G\rangle$ and $|D\rangle$ states into a (LEFT, RIGHT) magnet, $\vartheta = 90°$ and $\vartheta/2 = 45°$ giving:

$$\sin\left(\frac{\vartheta}{2}\right) = \sin(45°) = \frac{1}{\sqrt{2}}$$

$$\cos\left(\frac{\vartheta}{2}\right) = \cos(45°) = \frac{1}{\sqrt{2}}$$

yielding the states that we had earlier. As $\sin^2(\vartheta/2) + \cos^2(\vartheta/2) = 1$ for any ϑ, the result is also consistent with Important Rule 2.

3.3.7 Some Further Thoughts

> An elementary particle is not an independently existing, unanalyzable entity. It is, in essence, a set of relationships that reaches outward to other things.
>
> *Henry Stapp*[10]

Summarizing the story so far:

QUANTUM STATES

The mathematical representation of an initial quantum state uses a symbol such as $|\varphi\rangle$, and an expansion (summed list) over a series of final quantum states as follows:

$|\varphi\rangle = a_1|A\rangle + a_2|B\rangle + a_3|C\rangle + a_4|D\rangle + \cdots$

The amplitudes are a collection of complex numbers related to the probability that the initial state $|\varphi\rangle$ will change into one of the $|n\rangle$ final states as the result of a measurement.

Important Rule 1 gives us the relationship between amplitudes and probabilities. Amplitudes have magnitudes, which determine the probability, and phases which give rise to interference when two or more amplitudes combine as coefficients of the same state.

The list of possible final states is called the *basis* of our expansion.

So, we know how to represent the amplitudes and what they mean; but what about the basis states? How can we write down something such as $|n\rangle$ in mathematical terms? Is there some equation or formula for $|n\rangle$ itself?

Up to now, we have simply written states such as $|\varphi\rangle$ in terms of a set of basis states, and then those in turn have been written as a combination of a further basis. For example, we wrote $|U\rangle$ as a combination of $|L\rangle$ and $|R\rangle$. In turn, $|L\rangle$ can be written as a combination of $|U\rangle$ and $|D\rangle$ (as can $|R\rangle$). We seem to be caught in a regression of writing one thing in terms of another without actually getting anywhere. However, this is not entirely fair.

For one thing, the structure of our quantum state is a reflection of the contextuality of quantum physics, something that we referred to earlier. A state such as $|U\rangle$ can be written as:

$$|U\rangle = \frac{1}{\sqrt{2}}|R\rangle + \frac{1}{\sqrt{2}}|L\rangle$$

in the context of a (LEFT, RIGHT) magnet, or as:

$$|U\rangle = \cos\left(\frac{\vartheta}{2}\right)|1\rangle + \sin\left(\frac{\vartheta}{2}\right)|2\rangle$$

when the magnet is at some other angle, ϑ.

Furthermore, each of the $|n\rangle$ states in our basis represents a possible result of a measurement. As we have been talking about SG magnets and interference experiments, these states have not been anchored to a specific physical property, but as we go further into quantum theory this link will come up. What we are missing at the moment is some way of extracting quantitative information about a physical property from a state such as $|n\rangle$, which is discussed in Chapter 5.

3.4 WHAT ARE QUANTUM STATES?

The average quantum mechanic is about as philosophically sophisticated as the average motor mechanic.

John Polkinghorne

According to the picture we have built up in this chapter, the quantum state of a system contains a series of complex numbers related to the probability that the system will collapse into a new state when a measurement takes place. Each of these new states represents a possible result of the measurement, which might be a path or a specific quantitative value of some physical variable. However, this simple description of a quantum state hides a number of difficulties.

If we make a measurement on a specific system, for example, an electron, then the result will be a distinct value of the physical property being measured. However, such a measurement can't confirm the probability of finding the electron with that value. It can't tell us if we have the amplitudes right. We have to make the same measurement several times and see how often each specific value comes up. This gives us some practical difficulties to deal with. For example, how do we ensure that the electron is prepared in *exactly* the same state every time we wish to make a measurement? We might be better off using a collection of electrons, if we can put all of them in the same initial state and perform one measurement on each. If this is the best way of carrying out our measurement, we are entitled to ask if our quantum state represents an individual electron, or the collection.

This is more than just a debate over terminology: it raises important questions about the nature of probability itself.

Probabilities pop up in science in various ways. Sometimes, there is some physical aspect of the system that reflects this. A good example would be throwing a fair dice, which will result in each face coming up one-sixth of the time, precisely because there are six faces to choose from. The probability is directly related to the shape of the object being thrown. On the other hand, if we have a collection of balls in a bag and half of them are red and the other half are white, then the probability of drawing a red ball out of the bag (without looking) is 1/2. In this case, the probability is not a direct reflection of some property of each ball. The probability only exists when the balls are placed in the collection. Perhaps we might prefer to say that the probability state describes only the collection and not the individual balls within it.

So, if the quantum state can only be taken to refer to a collection of systems, the probability might come from having members within the collection that are not quite identical. Then the amplitudes wouldn't be representing a genuine unpredictability that is inherent to a system. They would be expressing our ignorance of what is going on at a 'deeper level.' Perhaps there are some hidden variables at work and if we only knew what they were and what values they take, exact predictions would be possible. Our collection of systems would actually have various possible values of these hidden variables, we just couldn't tell which one was which. The probability would simply be telling us how many of each type were in the collection.

However, if our quantum state refers to a single system, the probabilities involved might reflect the physical nature of the system. This is a more absorbing possibility, as it may open up a new way of looking at reality. If the quantum state of a system is represented by a set of probability amplitudes, then, in a manner of speaking, we are describing the state in terms of what it can *become* as a result of a measurement or interaction. After the measurement, one of the possibilities has taken place (manifested itself), so the system is in a new state. This state in turn is best described in terms of what it can become after the next measurement. We are therefore continually describing systems in terms of what they become or change into, arguably never what they are. Perhaps there is nothing more to describing what something is than saying what it can become or what it can do. The quantum description is then one of *processes*. David Bohm[11] used to talk about *implicate* and *explicate orders*. His ideas on this were quite complicated, but part of what he was driving at was how the current state of a quantum system has its future implicit within it. Once a measurement has taken place one of the implicate possibilities becomes explicate.

This is a very abstract picture, but we shouldn't worry about that. The classical state of a system is also abstract, in the strict sense of the word, as the state is represented by a series of quantities abstracted from the physical properties of the system. Nevertheless, it seems more real as the speed, position, mass, etc., of an object are familiar concepts, and we're quite used to the idea that they can be captured in quantitative terms.

The meaning we give to the amplitudes in a quantum state is a philosophical question. Provided that the quantum description allows us to make calculations that correctly predict the outcome of an experiment (given the random nature of some quantum outcomes), there is no experimental way in which different ways of thinking about the quantum state can be distinguished.

The majority of physicists take a very pragmatic view of this. As far as they are concerned, quantum theory works: it allows them to make calculations, do experiments, and build careers. The deeper questions about what it all means are thought to be rather fruitless as they are not accessible to experimental resolution. This approach, however, does not suit everyone[12].

There is no denying the distinct element of weirdness about this. The probability amplitudes seem to be detached from any tangible reality. At one level this is simply the normal puzzlement that we feel when learning about an unfamiliar subject. On another level our unease reflects a genuine problem in quantum theory: how can it be that the familiar everyday world that we experience results from the underlying quantum reality described by probability amplitudes which are complex numbers?

When quantum mechanics was first developed, the founding fathers struggled to understand how the bizarre outcomes of their experiments could be described in terms of the familiar classical world. As their understanding and experience deepened, they learned to replace some classical concepts, but the direction of their thinking always tended to be from the familiar classical toward the bizarre quantum. Now, physicists have become more used to dealing with the quantum world and the thinking is more from the other direction: accepting quantum reality and wondering how the everyday world can be distilled from it. Various approaches have been taken to this from the *Many Worlds interpretation* through *many minds* to *consistent quantum histories*. These sound just like slogans at the moment, but we will discuss each of them when we come to Chapter 26 onwards.

NOTES

1 Edward Norton Lorenz (1917–2008) American mathematician and meteorologist best known as the founder of chaos theory, a branch of mathematics dealing with the behaviour of systems that are highly sensitive to initial conditions.

2 This is sometimes known as the *butterfly effect*: after the suggestion that a butterfly flapping its wings in one part of the world could result in the thunderstorm in the opposite corner. The comic writer Terry Pratchett once suggested that all the money spent researching complex chaotic systems could be better spent mounting an expedition to find the butterfly and get it to stop.

3 Terminology is very difficult in this sort of description. Unless there is a measurement device situated on either arm of the experiment, we can't say that the photon has been either transmitted or reflected. Consequently, I shouldn't imply that the photon is going in any particular direction. The alternative would be to say something like "the reflected photon state moves along the upper arm," but states do not move. I have settled on saying things like "the reflected photon aspect."

4 Unless any of p, p', n, r' happens to be equal to zero, which is not likely in this context.

5 Technically, a number which is a straight multiple of i, such as bi, is referred to as an imaginary number. A construction such as $a+ib$ which has both real (a) and imaginary (ib) segments is a complex number.

6 My own view, for what it is worth, is that reality is a little more subtle than that…In fact, I see the forced inclusion of complex numbers as being a pointer to a realistic interpretation. If the only way of fully exploring experimental results dictates a certain form of mathematics, surely that is reflecting something significant about the nature of reality.

7 If $z = a+ib$, then the complex conjugate of z, denoted $z^* = a-ib$.

8 Max Born (1882–1970) Nobel Prize winner in Physics 1954.

9 If our probabilities added up to a number <1, this would mean there was a probability that something would happen that we had not included in the list.

10 Henry Pierce Stapp 1928 – theoretical physicist, Lawrence Berkeley National Laboratory, quote from "S-Matrix Interpretation of Quantum Theory", *Phys. Rev. D* 3, 1303, March 1971.

11 David Bohm (1917–1992) was professor of theoretical physics at Birkbeck College, London.

12 I should put my cards on the table at this point. Personally, I hold onto a realistic interpretation of quantum theory. I do not think that some form of hidden variable theory will replace it in the future. I will try to be fair to all points of view in this book but will also argue to support the realistic view.

4 Amplitudes

4.1 MORE ON AMPLITUDES

Amplitudes play a key role in quantum mechanics, as they are the link between theory and experiment. Theorists use quantum mechanics to calculate the amplitudes for various situations, and then experimenters set up appropriate apparatus and take measurements to check the theoretical conclusions.

Normally in a physics experiment, the same particle/atom/system, or whatever, is passed through a sequence of measuring devices, each of which will try and extract some information about what is going on. Most likely, each device is designed to have an effect on the state of the system, so the set of amplitudes describing what might happen at one will depend on the outcome of the previous measurement. Consequently, our theory requires some techniques for tracing amplitudes through a sequence of situations. We need to know how to combine and convert amplitudes.

Physicists have had to abstract the basic rules for combining amplitudes from the way that nature behaves, as revealed by experimental results. As with Important Rules 1 and 2 from the last chapter, *these principles can't be proven in a strictly mathematical sense.* We have to follow nature's lead and discern the right thing to do.

IMPORTANT RULE 3: COMBINING AMPLITUDES

Changes in states (*transitions*) are governed by amplitudes.

1. When one transition directly follows another, the respective amplitudes are multiplied together.
2. When two or more alternatives are possible:
    a. the probabilities add if the alternatives can be distinguished in the experiment
    b. the amplitudes add if the alternatives cannot be distinguished (and then the probability is the complex square of the total amplitude)

The second clause in Important Rule 3, dealing with the situation of indistinguishable alternatives, is really the hinge point that pivots quantum mechanics away from being a refined version of classical mechanics. Here we glimpse how interference comes about and the curious way that a quantum state can be a combination of states that would not be allowed in the classical domain. However, we need to see how this rule works in practice, before we can explore its more subtle aspects.

Let's go back to our SG type experiments again, with an arbitrary initial state $|\varphi\rangle$ where:

$$|\varphi\rangle = a|U\rangle + b|D\rangle$$

I have chosen $(|U\rangle, |D\rangle)$ as the basis for this expansion as the first part of our experiment is going to be an (UP, DOWN) magnet. After that, each beam from the first magnet passes into one of two other magnets arranged parallel to each other, but at some angle ϑ to the original (vertical) magnet. The emerging states are then $|L'\rangle$ and $|R'\rangle$. Expanding both $|U\rangle$ and $|D\rangle$ over a new basis $(|L'\rangle, |R'\rangle)$ we get:

$$|U\rangle = m|L'\rangle + n|R'\rangle$$

$$|D\rangle = p|L'\rangle + q|R'\rangle$$

DOI: 10.1201/9781003225997-6

At the first magnet, the electrons divide into an UP beam containing a fraction $|a|^2$ of the original beam, and a DOWN beam containing a fraction $|b|^2$ of the original beam. At the next magnet, the $|U\rangle$ states collapse into either $|L'\rangle$ with probability $|m|^2$ or $|R'\rangle$ with probability $|n|^2$. A similar division happens to the $|D\rangle$ electrons, as you can see in Figure 4.1. So, the probability that an electron starting in state $|\varphi\rangle$ will end up in state $|L'\rangle$ having gone through magnet 1 is:

$$\text{Prob}\left(|\varphi\rangle_{\text{via magnet 1}}^{\rightarrow}|L'\rangle\right) = |a|^2 \times |m|^2$$

a result that can be obtained by simply considering the fraction of the original beam that makes it through each stage. However, we get the same result using clause 1 of Important Rule 3:

$$\text{Amplitude}\left(|\varphi\rangle_{\text{via magnet 1}}^{\rightarrow}|L'\rangle\right) = \text{Amplitude }\left(|\varphi\rangle \rightarrow |U\rangle\right) \times \text{Amplitude }\left(|U\rangle \rightarrow |L'\rangle\right) = a \times m$$

and as probability = $|\text{amplitude}|^2$, we get:

$$\begin{aligned}\text{Prob}\left(|\varphi\rangle_{\text{via magnet 1}}^{\rightarrow}|L'\rangle\right) &= |a \times m|^2 \\ &= \left(a^* \times m^*\right)\left(a \times m\right) \\ &= \left(a^* \times a\right)\left(m^* \times m\right) \\ &= |a|^2 \times |m|^2\end{aligned}$$

confirming that the first clause aligns with experimental results.

Now let's ask a slightly different question: what is the probability of an electron ending up in state $|L'\rangle$ irrespective of which (L', R') magnet it went through? It's simple enough to see what this ought to be if we look back at Figure 4.1:

$$\text{Prob}\left(|\varphi\rangle_{\text{via magnet 1}}^{\rightarrow}|L'\rangle\right) = P_1 = |a|^2 \times |m|^2$$

$$\text{Prob}\left(|\varphi\rangle_{\text{via magnet 2}}^{\rightarrow}|L'\rangle\right) = P_2 = |b|^2 \times |p|^2$$

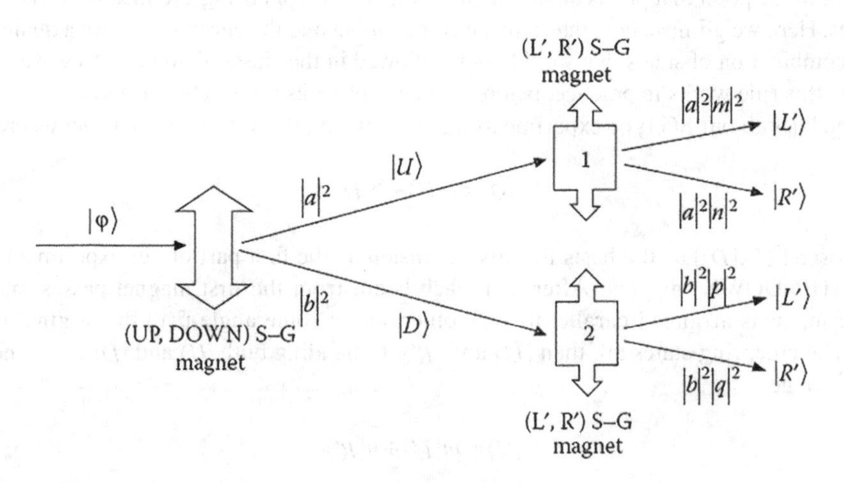

FIGURE 4.1 A set of SG magnets set up to illustrate the rules for combining amplitudes. The (L′, R′) SG magnets have been labelled 1 and 2 so that I can refer to them later.

giving an overall probability:

$$\text{Prob}\left(|\varphi\rangle_{\text{either magnet}}\overrightarrow{\ }|L'\rangle\right) = P_1 + P_2 = |a|^2 \times |m|^2 + |b|^2 \times |p|^2 = |am|^2 + |bp|^2 \quad \text{(R4.1)}$$

This is an application of a standard rule in probability calculations, which we mentioned in Section 3.3: when you have one event OR another, the probabilities add. Clause 2a of Important Rule 3 which states "the probabilities add if the alternatives can be distinguished in the experiment" is effectively a subtle variation on this rule of probability. The crucial part being the phrase "can be distinguished in the experiment". Information about which alternative a particular system follows has to be available, *even if we don't choose to use that information during the experiment.*

That leaves us with the final possibility, when the alternatives can't be distinguished within the experiment. We need an example to see how it works. The most convenient situation is to use the previous experiment with a slight alteration: we get rid of one of the (L′, R′) SG magnets and pull the other one forward, so it gets *both beams* from the (UP, DOWN) magnet, as in Figure 4.2.

Now we ask again: what is the probability of an electron starting in state $|\varphi\rangle$ and ending up in state $|L'\rangle$? With this arrangement, there is no way to tell which channel is used by an electron on its way from the (UP, DOWN) magnet to the (L′, R′) magnet. The possibilities are indistinguishable, in the sense that the experiment does not allow us to distinguish between them. As we have seen in previous chapters, when possibilities can't be distinguished, a quantum system seems to take all the options at once. The final clause of Important Rule 3 deals with such cases.

Accordingly, we need to trace the amplitude through each possible path individually:

$$\text{Amplitude}\left(|\varphi\rangle_{\text{Top path}}\overrightarrow{\ }|L'\rangle\right) = a \times m$$

$$\text{Amplitude}\left(|\varphi\rangle_{\text{Bottom path}}\overrightarrow{\ }|L'\rangle\right) = b \times p$$

and then add the amplitudes together, giving:

$$\text{Amplitude}\left(|\varphi\rangle_{\text{Can't tell which path}}\overrightarrow{\ }|L'\rangle\right) = a \times m + b \times p = am + bp$$

From here it is a straightforward job to calculate the probability by complex squaring the amplitude:

$$\text{Prob}\left(|\varphi\rangle_{\text{Can't tell which path}}\overrightarrow{\ }|L'\rangle\right) = |am + bp|^2$$

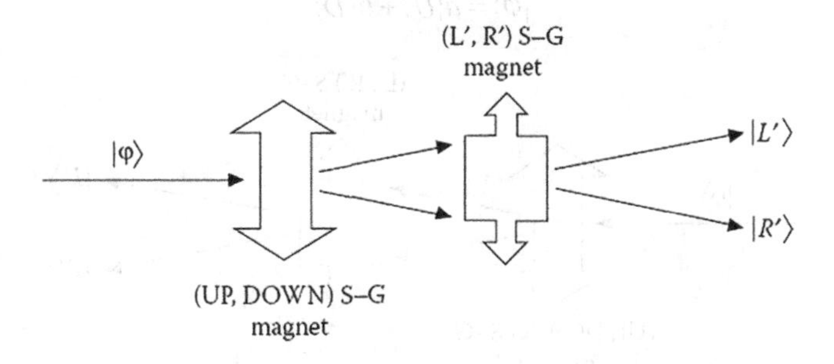

FIGURE 4.2 This is the sort of experimental arrangement in which you have to apply the last clause of Important Rule 3, as there is no way to tell which path an individual electron follows out of the (UP, DOWN) SG magnet.

$$= (am + bp)^* (am + bp)$$

$$= (a^* m^* + b^* p^*)(am + bp)$$

$$= (a^* m^* am + a^* m^* bp + b^* p^* am + b^* p^* bp)$$

$$= |am|^2 + |bp|^2 + (a^* m^* bp + b^* p^* am)$$

Now compare this with result R4.1 from above. The extra term that this probability contains, the term in the brackets $(a^* m^* bp + b^* p^* am)$, represents the 'interference' between possibilities and the phases of the amplitudes will be important.

Basically, this is the theoretical representation of what happens in an interference experiment, as discussed further in Chapter 6, but I want to make one comment now. As we are adding the amplitudes before they are squared (when we can't tell which path is being followed), the two amplitudes can 'interfere' with each other. If $(a^* m^* bp + b^* p^* am)$ is positive, then the combined probability is bigger than that for the distinguishable situation. However, if $(a^* m^* bp + b^* p^* am)$ is negative, then the probability decreases. In a standard interference experiment, the phase of the amplitudes will change depending on the path through the equipment. This change affects the sign of the amplitude (we have seen how a phase shift of π multiplies the amplitude by –1) so we have an overall probability that gets bigger or smaller depending on the path length.

The rule that we add the amplitudes before complex squaring if the paths are indistinguishable has no equivalent in normal probability calculations nor in classical physics.

4.1.1 Change of Basis

The idea that states can be expanded over a suitable basis is a very important aspect of quantum theory. This is how, in part, quantum theory reflects the contextual nature of quantum reality, i.e., that the behaviour of a system is influenced by the experimental context.

Every quantum state can be expanded in various ways, depending on what sort of experiment is involved. Consequently, it is always a good idea to choose a basis that is useful in describing the measurement that's going to take place. To illustrate this idea, let's have another look at the indistinguishable path experiment (Figure 4.3).

As they enter the experiment, electrons in state $|\varphi\rangle$ come across an (UP, DOWN) SG magnet. So, it's sensible to expand $|\varphi\rangle$ over the basis formed by $|U\rangle$ and $|D\rangle$:

$$|\varphi\rangle = a|U\rangle + b|D\rangle \tag{4.1}$$

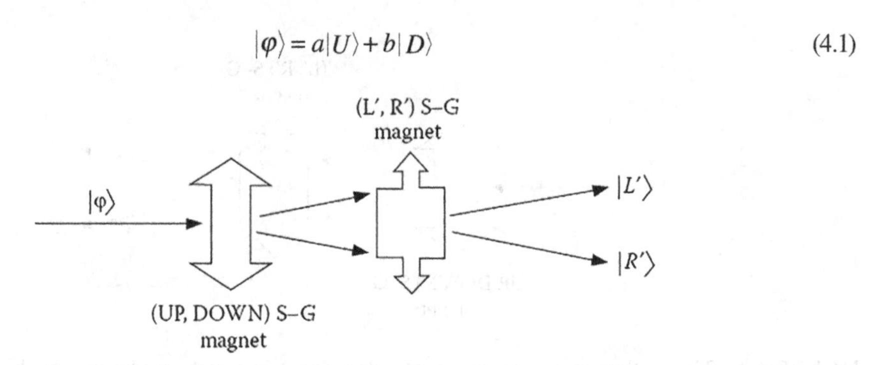

FIGURE 4.3 Another way of thinking about the probability of the $|D\rangle$ state turning into an $|L'\rangle$ state by the end of the experiment is to think about the change of basis involved.

The electrons then hit the (L', R') magnet, which makes it reasonable to consider an expansion of $|U\rangle$ and $|D\rangle$ states in terms of $|L'\rangle$ and $|R'\rangle$:

$$|U\rangle = m|L'\rangle + n|R'\rangle$$

$$|D\rangle = p|L'\rangle + q|R'\rangle \tag{4.2}$$

In terms of the overall situation, our initial state is $|\varphi\rangle$ and the final state we are interested in is $|L'\rangle$, so what we really need is an expansion of $|\varphi\rangle$ in terms of $|L'\rangle$ and $|R'\rangle$. We can get this by plugging Eq. 4.2 into Eq. 4.1:

$$|\varphi\rangle = a|U\rangle + b|D\rangle$$

$$= a\left[m|L'\rangle + n|R'\rangle\right] + b\left[p|L'\rangle + q|R'\rangle\right]$$

$$= am|L'\rangle + an|R'\rangle + bp|L'\rangle + bq|R'\rangle$$

$$= (am + bp)|L'\rangle + (an + bq)|R'\rangle$$

The result is very interesting. Here we have the amplitude for $|\varphi\rangle$ to transition into $|L'\rangle$ and in the same expression the amplitude $|\varphi\rangle \to |R'\rangle$:

$$|\varphi\rangle = (am + bp)|L'\rangle + (an + bq)|R'\rangle$$

$$\text{Amplitude}\left(|\varphi\rangle \to |L'\rangle\right) = (am + bp)$$

$$\text{Amplitude}\left(|\varphi\rangle \to |R'\rangle\right) = (an + bq)$$

The first, $(am + bp)$, being exactly what we calculated in the previous section by using Important Rule 3. In this case, we have worked through by manipulating basis expansions. Clearly, the two approaches are equivalent. You might like to think about why is it that this calculation produces the same answer[1].

4.2 DIRAC NOTATION

The first person to introduce symbols such as $|\varphi\rangle$ for quantum states was the celebrated English physicist, Paul Dirac[2]. In Dirac terminology, objects like $|\ \rangle$ are called *kets*. However, Dirac's creativity did not stop there; he also brought *bras* into the physics vocabulary. A bra, $\langle\ |$, is an alternative way of representing a state by using an expansion over the complex conjugates of the amplitudes, rather than the amplitudes themselves:

$$\langle\varphi| = a_1^*\langle 1| + a_2^*\langle 2| + a_3^*\langle 3| + \cdots + a_n^*\langle n|$$

Bras and kets can be multiplied together, either for the same state:

$$\langle\varphi|\varphi\rangle = \left(a_1^*\langle 1| + a_2^*\langle 2| + a_3^*\langle 3| + \cdots + a_n^*\langle n|\right) \times \left(a_1\langle 1| + a_2\langle 2| + a_3\langle 3| + \cdots + a_n\langle n|\right)$$

or indeed for different states. Dirac suggested that when we build bra×ket combinations such as $\langle\varphi|\varphi\rangle$ or $|\psi\rangle\varphi\rangle$, we should call them a 'bra-ket' or bracket.

This is the sort of thing that passes for whimsical humour among physicists.

Dirac notation has become the standard alphabet for working in quantum theory. Aside from its elegance, which will become more apparent as we move on, it has the virtue of illuminating some important physics.

The first thing that we can do is to use Dirac notation to express another important rule for dealing with amplitudes.

IMPORTANT RULE 4

If a system starts off in state $|\varphi\rangle$ and ends up in state $|\psi\rangle$, then the amplitude for the transition can be calculated by taking the *bra of the final state* and multiplying by the *ket of the initial state*:

$$\text{Amplitude}\big(|\varphi\rangle \to |\psi\rangle\big) = \langle\psi| \times |\varphi\rangle = \langle\psi|\varphi\rangle$$

Let's see how this works in a specific case:

$$|\varphi\rangle = a|U\rangle + b|D\rangle$$

According to Important Rule 4, the amplitude governing the transition $|\varphi\rangle$ into $|U\rangle$ should be:

$$\text{Amplitude}\big(|\varphi\rangle \to |U\rangle\big) = \langle U| \times |\varphi\rangle = \langle U|\big[a|U\rangle + b|D\rangle\big] = a\langle U|U\rangle + b\langle U|D\rangle$$

From the state expansion we see by inspection that the amplitude for $|\varphi\rangle \to |U\rangle$ is a. If our result is going to be consistent with this, it follows that:

$$\langle U|U\rangle = 1$$

$$\langle U|D\rangle = 0$$

which makes sense. According to Important Rule 4, $|U\rangle U\rangle$ should be the amplitude for $|U\rangle$ to change into $|U\rangle$, which if we have normalized the states properly (Important Rule 2) should give 1. However, $|U\rangle$ can't change into $|D\rangle$ (at least not directly), so $\langle U|D\rangle = 0$. Hence:

$$\text{Amplitude}\big(|\varphi\rangle \to |U\rangle\big) = \langle U|\varphi\rangle = a\langle U|U\rangle + b\langle U|D\rangle = a \times 1 + b \times 0 = a$$

4.2.1 Orthonormal Bases

Let's see what happens when we take a bra and multiply it by the ket for the same state:

$$\begin{aligned}\langle\varphi|\varphi\rangle &= \big(a_1^*\langle 1| + a_2^*\langle 2| + a_3^*\langle 3| + \cdots + a_n^*\langle n| + \cdots\big)\\ &\quad \times\big(a_1|1\rangle + a_2|2\rangle + a_3|3\rangle + \cdots + a_m|m\rangle + \cdots\big)\\ &= a_1^*a_1\langle 1|1\rangle + a_1^*a_2\langle 1|2\rangle + a_1^*a_3\langle 1|3\rangle + \cdots\\ &\quad + a_n^*a_m\langle n|m\rangle + \cdots\end{aligned}$$

Here I have used a collection of states $|1\rangle$, $|2\rangle$, $|3\rangle$, etc., as my basis. From now on, when I want to refer to a set of states $|n\rangle$, numbered 1, 2, 3, etc, I will use the abbreviation $\{|n\rangle\}$ for the collection.

If I have made a sensible choice of basis states to expand over, two things follow:

1. terms such as $\langle n|m\rangle$, where n and m are not the same, vanish as $\langle n|m\rangle = \langle 2|1\rangle = 0$ etc.
2. terms of the form $\langle n|m\rangle$, where n and m are the same, $\langle n|m\rangle = \langle 1|1\rangle = \langle 2|2\rangle = 1$ etc.

Such combinations come up so often, so we write $\langle j|i\rangle = \delta_{ij}$, where $\delta_{11} = \delta_{22} = \cdots = \delta_{ij} = 1$ for $i = j$ and $\delta_{10} = \delta_{01} = \delta_{21} = \cdots = \delta_{ij} = 0$, for $i \neq j$.

These rules are not pulled out of thin air; they directly reflect experimental facts. If the states $|1\rangle$, $|2\rangle$, $|3\rangle$, etc., represent different measurement results and the experiment has separated out distinct paths or quantitative values of a physical variable, one state can't 'overlap' with another. Hence the amplitude for a direct transition from $|m\rangle$ to $|n\rangle$ is zero.

Any two states for which $n \neq m$ and $\langle n|m\rangle = 0$ are *orthogonal states.* In a nice, neat basis set, all the states are orthogonal to one another and properly normalised, so the collection is an *orthonormal basis.*

Assuming that my basis, $\{|n\rangle\}$, is orthonormal, the calculation of $\langle\varphi|\varphi\rangle$ reduces nicely:

$$\begin{aligned}\langle\varphi|\varphi\rangle &= a_1^*a_1\langle 1|1\rangle + a_2^*a_2\langle 2|2\rangle + a_3^*a_3\langle 3|3\rangle + \cdots + a_n^*a_n\langle n|n\rangle + \cdots \\ &= a_1^*a_1 + a_2^*a_2 + a_3^*a_3 + \cdots + a_n^*a_n + \cdots \\ &= 1\end{aligned}$$

the last step following from Important Rule 2. Notice that the process of multiplying a bra by a ket has reduced to just dealing with numbers, there are no states left in the expansion. This will always happen, if the same orthonormal basis is used for both the bra and the ket.

4.2.2 New Light Through...

The elegance of Dirac's notation pays dividends if we return to the argument of Section 4.1.1 and recast it (briefly) using bra and ket terminology.

The previous starting point was to write $|\varphi\rangle = a|U\rangle + b|D\rangle$ (Eq. 4.1) which would now become:

$$|\varphi\rangle = \langle U|\varphi\rangle|U\rangle + \langle D|\varphi\rangle|D\rangle \tag{4.3}$$

as $a = \langle U|\varphi\rangle$ and $b = \langle D|\varphi\rangle$.

Similarly, we can recast the expansions of $|U\rangle$ and $|D\rangle$ in terms of $|L'\rangle$ and $|R'\rangle$:

$$|U\rangle = m|L'\rangle + n|R'\rangle = \langle L'|U\rangle L'\rangle + \langle R'|U\rangle|R'\rangle$$

$$|D\rangle = p|L'\rangle + q|R'\rangle = \langle L'|D\rangle L'\rangle + \langle R'|D\rangle|R'\rangle$$

and substitute them into the $|\varphi\rangle$ expansion. This produces an expression for the amplitude governing the transition from $|\varphi\rangle \rightarrow |L'\rangle$:

$$|\varphi\rangle = \langle U|\varphi\rangle|U\rangle + \langle D|\varphi\rangle|D\rangle \tag{4.3}$$

$$= \langle U|\varphi\rangle\{\langle L'|U\rangle|L'\rangle + \langle R'|U\rangle|R'\rangle\} + \langle D|\varphi\rangle\{\langle L'|D\rangle|L'\rangle + \langle R'|D\rangle|R'\rangle\}$$

$$= \{\langle U|\varphi\rangle\langle L'|U\rangle + \langle D|\varphi\rangle\langle L'|D\rangle\}|L'\rangle + \{\langle U|\varphi\rangle\langle R'|U\rangle + \langle D|\varphi\rangle\langle R'|D\rangle\}|R'\rangle$$

So, the required amplitude, $|\varphi\rangle \rightarrow |L'\rangle$, is:

$$\langle U|\varphi\rangle\langle L'|U\rangle + \langle D|\varphi\rangle\langle L'|D\rangle = \langle L'|U\rangle\langle U|\varphi\rangle + \langle L'|D\rangle\langle D|\varphi\rangle$$

where in the last step I have simply tinkered with the order of terms.

Previously we found:

$$\text{Amplitude}\left(|\varphi\rangle_{\text{cannot tell which path}} \overset{\rightarrow}{} |L'\rangle\right) = (am + bp) = (ma + pb)$$

now we have, in Dirac notation:

$$\text{Amplitude}\left(|\varphi\rangle_{\text{cannot tell which path}} \overset{\rightarrow}{} |L'\rangle\right) = \langle L'|U\rangle\langle U|\varphi\rangle + \langle L'|D\rangle\langle D|\varphi\rangle$$

beautifully illustrating how amplitudes combine together. Look at it closely. The first term is the amplitude that takes us from $|\varphi\rangle$ to $|L'\rangle$ via $|U\rangle$. The second term takes us from $|\varphi\rangle$ to $|L'\rangle$ via $|D\rangle$. At the very least, this is an aesthetically pleasing combination….

This expression is an example of a more general rule:

IMPORTANT RULE 5

Any amplitude governing a transition from an initial state to a final state via an intermediate state can be written in the form:

$$\langle \text{final state}|\text{initial state}\rangle = \langle \text{final state}|\text{intermediate state}\rangle\langle \text{intermediate state}|\text{initial state}\rangle$$

If there is more than one indistinguishable intermediate state, we have to sum over all of the transition amplitudes to get the overall amplitude:

$$\langle \text{final state}|\text{initial state}\rangle = \sum_i \left[\langle \text{final state}|i\rangle\langle i|\text{initial state}\rangle\right]$$

The notation for this rule uses an abbreviation that we're going to come across quite a lot from now on: Σ[ ], which means "add up a collection of terms of the type contained in the square bracket":

$$\sum_i \left[\langle \text{final state}|i\rangle\langle i|\text{initial state}\rangle\right]$$

$$= \langle \text{final state}|1\rangle\langle 1|\text{initial state}\rangle + \langle \text{final state}|2\rangle\langle 2|\text{initial state}\rangle$$

$$+ \cdots\langle \text{final state}|n\rangle\langle n|\text{initial state}\rangle + \cdots$$

Important Rule 5 is an extension of Important Rule 3, written in a more formal way.

4.2.3 Going the Other Way

In Dirac notation, the amplitude for a transition between some initial state, $|i\rangle$, and a final state, $|j\rangle$, is $\langle j|i\rangle$ but it is always possible that the process is running in the opposite direction, with $|j\rangle$ as the initial state and $|i\rangle$ as the final state. That amplitude would be $\langle i|j\rangle = \langle j|i\rangle^*$. In summary:

QUITE IMPORTANT RULE 6

the amplitude to go from $|i\rangle$ to $|j\rangle$ is the complex conjugate of the amplitude to go from $|j\rangle$ to $|i\rangle$, in other words: $\langle j|i\rangle = \langle i|j\rangle^*$

which will turn out to be useful later on.

If you're not convinced by this quick argument, try expanding $\langle i|$ and $|j\rangle$ over the same orthogonal basis set and then multiplying the two expansions together. Then do the same thing for $\langle j|$ and $|i\rangle$. You will find that you get the same result.

NOTES

1 The primary reason is as follows. If the paths are distinguishable, Expansion 1 is no longer valid. What happens instead is that $|\varphi\rangle$ turns into one of $|D\rangle$ or $|U\rangle$, whereas Expansion 1 is assuming that both are valid at the same time; in other words, the state has not collapsed so we can't tell which path is happening.

2 Paul Adrien Maurice Dirac (1902–1984), Nobel Prize in Physics 1933.

5 Measurement

5.1 EMBRACING CHANGE

Change is built into the structure of a quantum state.

We have seen how a state can be expanded over a set of basis states, which represent different possible outcomes of an experiment. Once the experiment has been performed, the original state has collapsed into one of the basis states.

State collapse is a very particular and distinctive process, but it is not the only way that quantum states can change. For most of the time, various quantum systems that exist in nature[1] wander about the universe without being involved in experiments, yet clearly, things can and do change. Any self-respecting theory must be capable of describing the ordinary business of a world in which things interact with one another, transform, and develop. Part of our dealings in this chapter will be to cultivate the machinery that quantum theory uses to portray the changes inherent in ordinary development as well as discuss the rather more mysterious changes that take place in a measurement.

5.2 TYPES OF STATES

When thinking about quantum states, it's helpful to divide them into two broad groups.

A state such as $|U\rangle$ represents a particle with a definite fixed value of a certain property (although we have not yet been completely clear about what this particular property is). In the terminology of quantum physics, these are *eigenstates* and measuring that property will result in the appropriate value, and just that value.

On the other hand, a state such as $|\varphi\rangle = \sum_n |n\rangle$ is a combination of many basis states, and so we cannot predict the value when a measurement is carried out. These states are called *mixed states* or *superpositions.*

5.2.1 EIGENSTATES

Eigenstates describe systems as having a definite fixed value of a given physical property. Such states are very important. If we know that an object is in an eigenstate, we can predict with absolute certainty what a measurement is going to produce. We discussed this in Chapter 2 when we fired $|U\rangle$ electrons into an (UP, DOWN) SG magnet; they all emerged from the magnet along the top path. If we followed this up with a second (UP, DOWN) measurement, once again all the electrons emerged from the top path. This is another important property of an eigenstate: if we make a measurement of the physical property associated with some particular eigenstate, then *that measurement does not change the state of the object, no matter how many times we repeat the measurement.*

If we choose to be awkward and measure a property *different* from the one fixed by the eigenstate, we can't predict with certainty what will happen. For example, if we send a $|U\rangle$ state electron into a (LEFT, RIGHT) magnet, we don't know which of the two paths it will emerge from. While it is an eigenstate of (UP, DOWN), $|U\rangle$ is not an eigenstate of (LEFT, RIGHT).

It is possible for a system to be in an eigenstate of more than one physical property at the same time. An example of this will crop up in Chapter 7 where we will find that the free-particle state is an eigenstate of energy *and* momentum. With such a state, you can measure either property, predict with certainty what will happen, and not change the state during the measurement.

DOI: 10.1201/9781003225997-7

5.2.2 Mixed States

A good example of a mixed state would be $|R\rangle = \frac{1}{\sqrt{2}}|U\rangle + \frac{1}{\sqrt{2}}|D\rangle$. If we choose to measure an electron in this state using an (UP, DOWN) SG magnet, we will get either $|U\rangle$ or $|D\rangle$ with a 50:50 probability. After a long run of such experiments, we would expect to get $|U\rangle$ and $|D\rangle$ with about equal frequency, but as the process is random, we would not be concerned if the fractions were not exactly 50:50 in any one experimental run.

Mixed states give rise to a lot of trouble when we try to understand the implications of quantum mechanics regarding the nature of reality. Crucially, the mixture is often of a sort that would not be allowed in a classical situation. When we considered the Mach–Zehnder device in Chapters 1 and 3, we ended up with a mixed state that combined two different paths through the device. Any self-respecting classical particle would follow only one path and not seem to be in two places at the same time. Yet the mixed state was unavoidable. Without it, we couldn't explain the experimental fact that none of the photons ended up at one of the detectors (with two equal path lengths through the device).

In his brilliant textbook on quantum mechanics, Paul Dirac[2] identified the existence of quantum mixed states as the central puzzle of quantum theory. Dirac points out that a quantum mixed state *is not some sort of average.* It doesn't describe an existence that is a blend of the separate states, which a classical 'mixed state' would. When you try to observe a quantum mixed state, the mixture collapses into one of the component states. The mixture simply tells us the relative probability of finding the separate component states via the amplitudes that go into the mixture.

A convenient way of thinking about mixed states is to recognize that they show us a set of tendencies for something to happen. An admittedly rather clichéd way of expressing this would be to say that the state was 'pregnant with possibilities' and that when a measurement is made, one of the possibilities is turned into an actuality (a process sometimes referred to as *instantiation*). Most physicists probably think of mixed states on similar lines. Where they tend to disagree is on the extent to which they believe this propensity or latency in the state is ontologically related to the physical nature of the object.

It is worth noting that the distinction between a mixed state and an eigenstate is not an absolute divide. States $|U\rangle$ and $|D\rangle$ are eigenstates as far as (UP, DOWN) measurements are concerned but mixed states for (LEFT, RIGHT) measurements. Similarly, states $|R\rangle$ and $|L\rangle$ are eigenstates for (LEFT, RIGHT) measurements but mixed states for (UP, DOWN) measurements.

5.3 EXPECTATION VALUES

Let's imagine that we have a whole set of electrons available to us, all of which are in the same state $|U\rangle$. We fire a collection of them at an (UP, DOWN) SG device. We know with absolute certainty that the results will always be UP as $|U\rangle$ is an eigenstate of (UP, DOWN). However, if we choose to fire the remainder of our electrons at a (LEFT, RIGHT) device, our ability to predict what happens becomes rather more limited. Given a specific electron, all we can say is that it might emerge from the LEFT channel or it might come out of RIGHT. We have no way of telling what is going to happen to each electron in turn. This is how randomness finds its expression in the quantum world.

This doesn't stop us from saying what will happen *on average.*

If we give UP the value (+1) and DOWN (−1), then after a run of such measurements, the average value across all the electrons taking part will be close to zero[3].

Now imagine that our collection of electrons is in some other state $|\varphi\rangle$, where:

$$|\varphi\rangle = a|U\rangle + b|D\rangle$$

with $|a|^2 + |b|^2 = 1$ but $a \neq b$.

On subjecting this collection of electrons to a (UP, DOWN) measurement, we will find that a fraction of them equal to $|a|^2$ will come out UP and another equal to $|b|^2$ will come out DOWN. Again, we have no way of knowing what will happen to each individual electron. If $|a|^2$ of them are (+1) and $|b|^2$ of them are (–1), then we can compute the average value as:

$$\text{Average value} = |a|^2 \times (+1) + |b|^2 \times (-1) = \left(|a|^2 - |b|^2\right)$$

In quantum theory, the average value of a series of identical measurements made on a collection of identically prepared systems (in the same state) is called the *expectation value*. Note that the expectation value only applies to a series of measurements. In the case of a collection of electrons, each individual electron in the collection will give either $(+1)$ or (-1) *and so can't be compared with the expectation value, which belongs to the collection as a whole.*

This association between the expectation values and the results of a collection of measurements convinces some people that quantum states are representative of the collective, not the individual. According to this view, when we say that the electrons are in state $|\varphi\rangle$, we are not accusing each individual electron of being in the state $|\varphi\rangle$. It is the collection as a whole that has the state. Indeed, such thinking would discourage the view that *any* state applies to an individual system.

There is clearly some sense to this view. After all, and to some people this is a crucial point, *we can never tell from a single measurement what state a system is in.* Firing an electron at an (UP, DOWN) magnet and seeing it come out UP does not allow us to distinguish between $|U\rangle$ as the initial state, which was bound to give us UP, and an initial state of $|\varphi\rangle$, which means that we got lucky in that one instance. It wouldn't help to repeat the experiment on the same electron. Once it has emerged from the UP channel, we already know that it is *now* in state $|U\rangle$, *no matter what state it started off in.* The only way of telling what was going on would be to have a collection of these electrons and measure each one in turn. Finding any of them coming out in the DOWN channel would enable us to tell the difference between $|U\rangle$ and $|\varphi\rangle$; but it would not allow us to tell the difference between $|U\rangle$, $|R\rangle$ and $|L\rangle$.

So, there is something to be said for the idea that states refer only to collections of systems. Such a view ties in with an instrumentalist interpretation of quantum theory. A realist would prefer to know what was happening to the 'real' individual electrons out there.

5.4 OPERATORS

It's very important to be able to tell if a state is a mixed state or an eigenstate with reference to a given measurement. The mathematical machinery of quantum theory must allow for a way of doing that, *otherwise all we can do is construct a state after the fact.* This is where objects called *operators* play a crucial role.

Mathematically, an operator takes some sort of expression and transforms it into something else. A simple example of an operator would be sin( ) which when applied to an angle, ϑ, returns the value of sine for that angle. In quantum theory, operators take us from one state to another, a process governed by strict rules for each operator.

There are many different operators in quantum theory, which have different jobs to do. Perhaps the most important ones are the operators that represent the process of measuring a physical property.

An immediate example of such an operator would be the (UP, DOWN) SG operator, [4] called $\hat{S}_z$. (I will use the convention that a symbol with a hat, "^", indicates an operator.) The role of this operator is to 'pull from' a state information about how that state will react to an (UP, DOWN) measurement. The rules that govern how it works are very simple. As $|U\rangle$ and $|D\rangle$ are the eigenstates of (UP, DOWN) measurements, we have:

$$\hat{S}_z|U\rangle = +1|U\rangle \text{ and } \hat{S}_z|D\rangle = -1|D\rangle$$

The operator takes an eigenstate and multiplies it by a number equal to the value of the quantity that would be found by a measurement of the particle in that state[5]. This is pretty much a definition of what we mean by an operator in this context. By the way, the value that multiplies the state, which is the value that an experiment would reveal without fail, is called the *eigenvalue* of the eigenstate. The complete set of eigenstates for an operator connected with a physical variable form a basis set.

If the state concerned is not an eigenstate of vertical SG measurements, then applying $\hat{S}_z$ will make rather a mess of it:

$$\hat{S}_z|R\rangle = \hat{S}_z\left(\frac{1}{\sqrt{2}}(|U\rangle + |D\rangle)\right) = \frac{1}{\sqrt{2}}((+1)|U\rangle + (-1)|D\rangle) = \frac{1}{\sqrt{2}}(|U\rangle - |D\rangle) = |L\rangle$$

Although this doesn't look like it's much use to anyone, the action of an operator on states that are *not* one of its eigenstates has an important role to play. Watch what happens when we expand the curious construction $\langle R|\hat{S}_z|R\rangle$:

$$\begin{aligned}
\langle R|\hat{S}_z|R\rangle &= \left[\frac{1}{\sqrt{2}}(\langle U| + \langle D|)\right]\hat{S}_z\left[\frac{1}{\sqrt{2}}(|U\rangle + |D\rangle)\right] \\
&= \left[\frac{1}{\sqrt{2}}(\langle U| + \langle D|)\right]\left[\frac{1}{\sqrt{2}}(\hat{S}_z|U\rangle + \hat{S}_z|D\rangle)\right] \\
&= \left[\frac{1}{\sqrt{2}}(\langle U| + \langle D|)\right]\left[\frac{1}{\sqrt{2}}((+1)|U\rangle + (-1)|D\rangle)\right] \\
&= \left[\frac{1}{\sqrt{2}}(\langle U| + \langle D|)\right]\left[\frac{1}{\sqrt{2}}(|U\rangle - |D\rangle)\right] \\
&= \frac{1}{\sqrt{2}}\frac{1}{\sqrt{2}}(\langle U|U\rangle - \langle U|D\rangle + \langle D|U\rangle - \langle D|D\rangle)
\end{aligned}$$

As $\langle U|D\rangle = \langle D|U\rangle = 0$ and $\langle U|U\rangle = \langle D|D\rangle = 1$, this reduces to:

$$\langle R|\hat{S}_z|R\rangle = \frac{1}{2} - \frac{1}{2} = 0$$

On the face of it, this hardly appears to be an earth-shattering calculation. However, having rehearsed the process with a simple state like $|R\rangle$, let's try again with something slightly meatier, like the mixed state $|\varphi\rangle$:

$$\begin{aligned}
\langle\varphi|\hat{S}_z|\varphi\rangle &= \left[a^*\langle U| + b^*\langle D|\right]\hat{S}_z\left[a|U\rangle + b|D\rangle\right] \\
&= \left[a^*\langle U| + b^*\langle D|\right]\left[a(+1)|U\rangle + b(-1)|D\rangle\right] \\
&= \left[a^*\langle U| + b^*\langle D|\right]\left[a|U\rangle - b|D\rangle\right] \\
&= |a|^2\langle U|U\rangle - |b|^2\langle D|D\rangle \\
&= |a|^2 - |b|^2
\end{aligned}$$

which, you will recall from earlier, is the expectation value of an (UP, DOWN) measurement on this state. For any operator, $\hat{O}$, representing a physical property, the expectation value of a series of measurements made on a collection of systems in state $|\varphi\rangle$ is:

$$\langle \hat{O} \rangle = \langle \varphi | \hat{O} | \varphi \rangle$$

5.4.1 Operators and Physical Quantities

In general, operators act to their right. With the combination $\hat{O}|\psi\rangle$ the operator is applying to the ket. However, it is sometimes necessary to apply the operator to a bra, in which case we write $\langle \psi | \hat{O}^T$ and the operator $\hat{O}^T$ acts to its left. This operator is known as the *transpose* of the original.

So far, we have regarded the action of turning a ket into a bra as simply taking the complex conjugate, $\langle \psi | = (|\psi\rangle)^*$. However, when we see how they can be represented as *vectors* (section 10.3.1), it will be clear that there a little more to it. Equally, the operator $\hat{O}^T$ in *matrix form* (Section 10.3.1 as well) is not the same as $\hat{O}$, but the differences are not *physically relevant* and for the moment can be ignored.

Not every operator can represent a physical variable. Suitable operators must obey a quite stringent mathematical condition. If $\hat{O}|\psi\rangle = (a+ib)|\psi\rangle$, then $\hat{O}^*|\psi\rangle = (a-ib)|\psi\rangle$, equally $\langle \psi | \hat{O}^T = (a+ib)\langle \psi |$ and $\langle \psi | (\hat{O}^*)^T = (a-ib)\langle \psi |$. Then if we impose $(\hat{O}^*)^T = \hat{O}$, we must have:

$$\langle \psi | (\hat{O}^*)^T | \psi \rangle = (a-ib)\langle \psi | \psi \rangle = (a-ib)$$

$$\langle \psi | \hat{O} | \psi \rangle = (a+ib)\langle \psi | \psi \rangle = (a+ib)$$

i.e., $(a-ib) = (a+ib)$, which can only happen if $b = 0$. Hence, the operator has real-valued eigenvalues, *which is the condition for representing a physical variable.* The construction $(\hat{O}^*)^T$ is abbreviated to $\hat{O}^\dagger$, for notational compactness, and an operator where $\hat{O}^\dagger = \hat{O}$ is *Hermitian.*

HERMITIAN OPERATOR:

Any operator, $\hat{O}$, with $\hat{O}^\dagger = \hat{O}$ is Hermitian.
The eigenvalues of such operators are real-valued. Operators representing physical variables must be Hermitian.

5.4.2 Classical and Quantum

The connection between a physical quantity and an operator in quantum theory is very different from the way in which classical physics deals with such things.

In classical physics, as we know, a state is a collection of quantities that describe an object at a given moment. Those quantities can be speed (momentum), mass, energy, etc., and each is given a number that represents the value of this quantity, as measured in suitable units. The classical laws of nature provide rules that connect various quantities together so that when we put actual values in place, we can predict future values.

In quantum mechanics, a state is a collection of amplitudes for the object to have values of physical quantities. The physical quantity is associated with an operator $\hat{O}$, which can be used to find out the expectation value of that physical variable. There is a strong temptation to call this the average value of the *quantity,* but it's not really. It is the average value obtained from a repeated run of measurements on identical systems, but none of them will have a value equal to the expectation value[6]. This is particularly pertinent when we discuss something like position. Classically, an electron flying through space has a position from moment to moment, so over a period of time, it is sensible to calculate an average position. However, as we will come to understand in more detail, in quantum theory, an electron does not have a well-defined position between measurements of position. We can say that it is now here, and was there before, but between times, all bets are off. In that case, *the average position for the electron between measurements does not have a meaning.* If we calculate the expectation value of a position measurement, it tells us the average value of a set of position measurements on a collection of electrons, which once again is *not really the average position of an individual electron.*

Operators tell us nothing by themselves; they need to act on states for any information to be forthcoming.

IMPORTANT RULE 7: OPERATORS & THINGS

Every physical property is associated with a Hermitian operator $\hat{O}$, ($\hat{O}^\dagger = \hat{O}$), with eigenstates defined by:

$$\hat{O}|\psi\rangle = \alpha|\psi\rangle$$

where α is the predictable value obtained if you conduct a measurement of that physical property on a system in the eigenstate $|\psi\rangle$.

The complete set of eigenstates $\{|\psi_i\rangle\}$ for a given operator forms a basis, so for any other state $|\varphi\rangle$:

$$|\varphi\rangle = \sum_i a_i|\psi_i\rangle$$

The operator associated with a physical variable can be used to calculate the expectation value, $\langle\hat{O}\rangle$, of a series of measurements made on a collection of systems in the same state $|\varphi\rangle$: $\langle\hat{O}\rangle = \langle\varphi|\hat{O}|\varphi\rangle$

5.5 HOW STATES EVOLVE

Some operators are not tied to physical variables. For example, the *evolution operator* $\hat{U}(t)$ moves a state forward in time:

$$\hat{U}(t)|\Psi(T)\rangle = |\Psi(T+t)\rangle$$

where $|\Psi(T)\rangle$ is the state of a system at time T and $|\Psi(T+t)\rangle$ the state at a later time $T+t$.

As $\hat{U}$ does not represent a physical variable, it is not required to be Hermitian. However, there is a different, and very important constraint on its mathematical behaviour. As a state evolves through time, its normalization must remain the same. If this were not the case, the probability sum rule would not apply. Imaging we decompose an arbitrary state $|\Psi\rangle$ using a basis $\{|i\rangle\}$:

$$|\psi\rangle = \sum_i a_i|i\rangle$$

The amplitudes are constrained by $\sum_i a_i^2 = 1$. In other words, when we measure the physical variable represented by $\hat{O}$, which is the operator owning the eigenstates $\{|i\rangle\}$, we must get one of the eigenvalues and $|\Psi\rangle$ collapses into $|I\rangle$ with probability $|a_i|^2$. These probabilities must sum to 1, or we have missed something out. So, the time evolution must act on the amplitudes as follows:

$$|\Psi(T)\rangle = \sum_i a_i(T)|i\rangle \qquad \sum_i a_i^2(T) = 1$$

$$\hat{U}(t)|\Psi(T)\rangle = |\Psi(T+t)\rangle = \sum_i a_i(T+t)|i\rangle \qquad \sum_i a_i^2(T+t) = 1$$

Expressing this a different way, $\langle\Psi(T+t)|\Psi(T+t)\rangle = 1$, but $|\Psi(T+t)\rangle = \hat{U}(t)|\Psi(T)\rangle$ which makes $\langle\Psi(T+t)| = \langle\Psi(T)|\hat{U}^\dagger$. Putting this together:

$$\langle\Psi(T+t)|\Psi(T+t)\rangle = \langle\Psi(T)|\hat{U}^\dagger(t)\hat{U}(t)|\Psi(T)\rangle = 1$$

As $\langle\Psi(T)|\Psi(T)\rangle = 1$, we must have $\hat{U}^\dagger\hat{U} = \hat{I}$, or in other words $\hat{U}^\dagger$ is the same as the *inverse operator* $\hat{U}^{-1}$. If $\hat{U}(t)$ takes a state from $T = 0$ to $T = t$, then $\hat{U}^{-1}(t)$ takes it back from $T = t$ to $T = 0$. Hence $\hat{U}^{-1}(t)\hat{U}(t)$ will take the state for a short time-ride, lopping back to the start[7].

UNITARY OPERATOR:

Any operator, $\mathcal{U}$, with $\hat{\mathcal{U}}^\dagger = \hat{\mathcal{U}}^{-1}$ is unitary.
Applying a unitary operator to a state will transform the state but preserve the normalization.

Physicists are constantly on the lookout to make sure that their calculations *preserve unitarity*, i.e., that they have not mucked up the probability sum rule.

We have seen how $\hat{U}(t)$ moves a state through time by adjusting the values of the amplitudes in the state's decomposition. Now, imagine that we take the time interval, t, and divide it into many smaller chunks, δt. Nothing we have seen in the evolution operator so far has suggested that it works any differently for very small intervals of time compared with longer ones. So, rather than applying $\hat{U}(t)$ once to take the state from $|\Psi(T)\rangle$ to $|\Psi(T+t)\rangle$, we ought to be able to get the same answer by applying $\hat{U}(\delta t)$ many times, one after the other. Applying $\hat{U}(\delta t)$ for the first time gives:

$$\hat{U}(\delta t)|\Psi(T)\rangle = \hat{U}(\delta t)\big(a(T)|U\rangle + b(T)|D\rangle\big) = a(T+\delta t)|U\rangle + b(T+\delta t)|D\rangle$$

and it is very unlikely that $a(T+\delta t)$ will be radically different from $a(T)$ (the same is true for $b(T)$ and $b(T+\delta t)$). Applying $\hat{U}(\delta t)$ once again:

$$\hat{U}(\delta t)|\Psi(T+\delta t)\rangle = \hat{U}(\delta t)\big(a(T+\delta t)|U\rangle + b(T+\delta t)|D\rangle\big) = a(T+2\delta t)|U\rangle + b(T+2\delta t)|D\rangle$$

and so forth. Hence, the state's evolution from $|\Psi(T)\rangle$ to $|\Psi(T+\delta t)\rangle$ takes place via *a continuously smooth change from one moment to the next.* Furthermore, and this is very important, *this evolution is perfectly determined by the physics of the system* (expressed in the structure of $\hat{U}(t)$); there is no randomness involved here. To take a simple example, an electron in state $|U\rangle$ moving through space toward an SG magnet is very unlikely to be passing through a complete vacuum with no gravitational, electrical, or magnetic fields lying around. Any such fields will exert a force on the electron and quite possibly disturb its SG orientation. As a result, the state will evolve from $|U\rangle$ into some other state $|\varphi\rangle$. As it turns out, the energy of the electron as it interacts with these fields will determine the way the state evolves. Consequently, the evolution operator is constructed from the energy operator for the electron, as we'll see when we discuss energy and momentum operators in Chapter 12.

Of course, states also evolve in a sharp and unpredictable manner when a measurement takes place. Indeed, a possible definition of a measurement is that it produces a *sharp and unpredictable change in the state of the system.*

When an electron in a state such as $|\Psi\rangle$ reaches an (UP, DOWN) SG magnet, the state will change into $|U\rangle$ with probability $|a|^2$ or $|D\rangle$ with probability $|b|^2$. After the measurement, we can say that the state has 'evolved' into $|U\rangle$ (i.e., $a \to 1, b \to 0$) or $|D\rangle$ (i.e., $a \to 0, b \to 1$). In this instance the amplitudes a and b have not changed in a smooth manner, as produced by the $\hat{U}(t)$ operator. The dramatic and unpredictable change in a quantum state due to a measurement is often referred to as the *collapse of the state*. State collapse *does not preserve unitarity*, nor should it as we are manifesting one of the possibilities within the state.

As state collapse is such a radically different sort of process from $\hat{U}(t)$ evolution, *it is mathematically impossible for the equations of quantum theory, as we currently understand them, to describe state collapse*[8]. It remains a vital 'bolt on' assumption, rather than something that can be predicted from within the theory.

This is a significant point.

Quantum theory can happily describe the evolution of states from one moment to the next; but that is not the world that we live in. We appear to be 'protected' from seeing systems in the strange mixed states that quantum theory allows. We have to add to the theory something that caters for measurement processes and state collapse. The world is not observed to be a set of evolving possibilities. The possibilities encoded into a quantum state have to be linked to actual events in the world. Some physicists[9] have suggested that quantum theory needs to be modified in order to introduce some new equations that can describe measurement in a satisfactory manner. States would not then collapse (snap change) but would rather evolve in a different manner, over a very short timescale.

If this is true, then there must be some physics that we have missed up to now. It would have to remain as a very small-scale effect until the particle (or whatever) is interacting with a measuring device, as the current equations work very well, except under those circumstances. Clearly the physicists who are working to modify the theory along these lines are *striving for a realistic view of state collapse*. We will explore this option in Chapter 29.

The collapse of a state as a result of measurement is an encouragement to some physicists to adopt an *instrumentalist* view of quantum states. They would argue that a snap change in the state of a system shows that the state represents our information about the system. After all, before we throw a die, our information about it comprises the one-sixth probability of each face coming up. After the throw, our informational state collapses, as we now know which face has landed uppermost. The dice has not changed in any physical manner. Correspondingly when a quantum state collapses this does not necessarily signal any physical difference in the system being described, just a change in our knowledge about it.

To counter this, a realist would say that a quantum state can't just be our knowledge of the system, as how can our knowledge have a direct effect on the system's behaviour? We have previously considered various experiments in which our ability/lack of ability to infer information about the state of a photon had a direct result on the outcome of the experiment. Therefore, our knowledge must be a reflection of something real to do with the system.

Quantum states may well be rather weird and spooky compared to the rather cosy classical reality that we have all been brought up in, *but there is no reason to give up believing that science is revealing truths about the world just because we find these truths surprising.*

SUMMARY

The $\hat{U}(t)$ operator evolves a state forward in time in a *smooth* and *predictable* manner:

$$\hat{U}(t)|\Psi(T)\rangle = |\Psi(T+t)\rangle$$

The exact mathematical form of $\hat{U}(t)$ depends on the physics of the system.
State collapse is a *sharp* and *unpredictable* change in state as a result of a measurement (if the state is not an eigenstate of the physical quantity being measured).

5.5.1 Why Is State Collapse Necessary?

It is reasonable to ask ourselves how we ever managed to get into this mess in the first place. Why were we driven to construct a theory that requires such an odd concept as state collapse? The argument, boiled down to its essentials, would run as follows.

Certain events in the microworld, such as the reflection or transmission of photons from a beam splitter, seem to be random (at least at our level of current understanding). We are forced to represent the state of microscopic objects in terms of probabilities. We assign a number to each possibility, and this number tells us the probability that things will turn out this way.

A detailed analysis of the Mach–Zehnder experiment shows us that these numbers (we call them amplitudes) have to be complex numbers, so they can't in themselves be probabilities. The link between amplitudes and probabilities is guessed (that probability = $|\text{amplitude}|^2$) and subsequently checked by experiment. We are therefore led to represent the quantum state of a system by a collection of amplitudes.

In certain situations, we see that the appropriate quantum state is a mixture of states that would have to be separate classically. For example, there are many experiments that can only be understood if the intermediate state of a photon appears to follow two different classical paths at the same time. Without this, interference effects would not take place and the actual experimental results could not be explained.

Although quantum theory allows these mixed states to occur, we seem to be protected from them, as they are never directly observed.

It is clear that the quantum state of a system is prone to change. If we prepare an appropriate beam of electrons for a Stern–Gerlach (SG) experiment, we find that 50% of them exit the device along each channel, indicating that all the electrons were in one of these mixed states to start with. However, if we pass all the electrons from one channel into another identical SG device, then all of them emerge from the same channel. The first experiment has changed their quantum state from one of these unobservable mixed states into a classical-like eigenstate that is definitively $|U\rangle$ or $|D\rangle$.

This is the nub of the problem. Certain experiments can only be explained if classically forbidden mixtures are allowed. However, we live in a classical world so that quantum mixed states must collapse into more classical-like states that we can directly observe.

The realists among us would say that state collapse is really happening out there, so we had better just get used to it.

Instrumentalists would probably confess that they don't like the idea, but after all a theory is only a way of getting experimental predictions out, and we have to do what works.

From Chapter 26 onward we will be looking more closely at the different interpretations of quantum theory that have surfaced over the decades. Although they can be broadly classed as instrumentalist or realist, the subtle differences between them often hinge on the gloss that is given to measurement (collapsing states) and the nature of mixed states.

5.5.2 Behind the Veil

Consider again an electron flying toward an SG measuring apparatus. For a change let's assume that it has been prepared by a previous SG experiment, aligned at some angle ϑ, and that we take only the electrons from the top channel out of this magnet. We will call this state $|\alpha\rangle$. These electrons are then passed on to another experimental team, *but we don't tell them how they were prepared.* Is there a set of measurements that the second team can use to find the state of the electrons, and maybe the angle ϑ?

The answer, unfortunately, is no.

As team 2 doesn't know the state, they are forced to set their SG magnet in some randomly chosen direction. Effectively they are asking the question, "Are you UP in this direction?" They will get the answer "yes" or "no" depending on the (random) collapse of the state $|\alpha\rangle$ once they measure it; *but in the process they destroy the original state.*

The state determines the relative probabilities of the "yes/no" answer to their question but is not in itself determined by these answers. Every state determines the *measurable information* that can be extracted from it. Indeed, a workable definition of a quantum mechanical state would be "the sum total of measurable information that can be determined about a system".

However, there seems to be something *objective* about the state. What if, by chance, team 2 happened to point its SG magnet along the same angle ϑ as the magnet that produced $|\alpha\rangle$ in the first place? They would receive the answer "yes" to the question, "Are you UP in this direction?" with absolute certainty. This information is encoded into the state *even if they don't happen to hit that direction by chance.* Is this enough to claim that there is some objective aspect to the state? If so, then it is a curious form of *veiled* reality. Some aspects are hidden from us, but we can see enough of what is there to gain an overall impression.

There is something really puzzling and profound here. This simple little argument has the effect of pushing us right up against the realist/instrumentalist debate. An arbitrary SG state, $|\alpha\rangle$, contains the seemingly objective information that it will record UP if we happen to get the direction of an SG magnet correct. Being able to predict the outcome of a measurement, with 100% certainty in a repeatable manner, is normally taken as an indication that a particle has, in an objective sense, a given property (yes, that's UP and I can see that it's UP!). However, if we picked the wrong direction for the SG magnet, then state $|\alpha\rangle$ will collapse into UP or DOWN along this direction with some probability What, in that situation, has happened to the real objective property?

5.5.3 Determinism and Free Will

Oracle: I'd ask you to sit down, but, you're not going to anyway. And don't worry about the vase.
Neo: What vase?
[Neo turns to look for a vase, and as he does, he knocks over a vase of flowers, which shatters on the floor.]
Oracle: That vase.
Neo: I'm sorry...
Oracle: I said don't worry about it. I'll get one of my kids to fix it.
Neo: How did you know?
Oracle: Ohh, what's really going to bake your noodle later on is, would you still have broken it if I hadn't said anything?
The Oracle to Neo in the film, ***The Matrix.***

The moment at which a measurement takes place is the moment at which the randomness lying at the heart of quantum reality expresses itself. Up to that point, amplitudes change in a completely predictable, and more importantly, calculable way (if we know $\hat{U}(t)$ for the system).

A measurement produces a snap change in the state and an outcome that can't be predicted. We know the possible outcomes and various amplitudes that govern their probabilities, but we can't know which outcome is going to happen.

Some people, Einstein being a good example, really can't live with this. They feel that the world is a structured and rigid web where effect follows cause and all things should be predictable, given the right information. Einstein was bitterly opposed to quantum theory, at least in its conventional form, but his feelings were not directed at the *physics* as such (he acknowledged that quantum theory *worked*), but at the *philosophy* that it seemed to imply. He was sure that the physics was *incomplete* and that a more detailed understanding would give us the ability to predict the outcome of an experiment with certainty. The basis of his view was a philosophical conviction that the world did not include random events: an objection summed up in Einstein's widely quoted saying, "God does not play dice."

Throughout the ages there has been a philosophical debate between those who believe in free will and those who think everything is determined. Determinists, like Einstein, say that the world is a complex machine with interlocking parts that we can, in principle, understand. If we know how all the parts work and fit together, then we can also predict what will happen when one piece causes another to move, and so on.

As we are part of the machinery of the world, all our actions are predictable as well. Whether you choose to throw this book across the room or not would be completely predictable, if we knew enough about the immediate circumstance that you are currently enveloped in. The word 'choice' should not really apply. You might think you have a 'choice', but actually, your brain is a complex machine with the 'cogs' whirring round to produce a predictable action.

However, there are people who believe in free will. They think that every choice we make is a free one selected from a range of available options according to our own moods and thoughts at the time. Sometimes they point to quantum physics as a way of justifying this belief. If the world is, at root, governed by randomness, then surely we can let a bit of free will in?

Personally, I'm on the side of free will, but I think that trying to use quantum physics to justify this idea is misguided.

Firstly, making a choice to do something is anything but random. I hate the idea that we are just complex mechanisms ticking our way through a predetermined world, but I also hate the thought that there is some quantum dice being thrown in my head and that all my carefully thought-out decisions are, at root, random[10].

Secondly, the randomness of quantum theory comes in with state collapse and that seems to be connected to measurement. Whatever measurements are, they are very specific situations and probably linked to what happens when something very small (a particle) bumps into something very big (a measuring device). There seems to be no obvious way to connect a measurement with the business going on in our heads.

Finally, quantum mechanics is basically a deterministic theory about the world. Amplitudes can be calculated in a precisely governed manner. The ultimate solution to the *measurement problem*, whatever form that may take, may not lie inside our current quantum theory. As we don't know what the form of this ultimate solution will be, and I for one think that the randomness will remain, it seems risky to base a philosophical viewpoint on the part of quantum theory that we understand the least.

NOTES

1 Which is, of course, potentially all systems…
2 PAM Dirac, *The Principles of Quantum Mechanics,* 4th Ed., Oxford Science Publications, New York, 2004.

3 The size of the deflection in an SG magnet is determined by the strength of the field and the spin of the particle. Spin is a complicated subject discussed in Chapter 10. There we will find out the size of an electron's spin, and that it is measured in a basic unit $\hbar/2$ and so it is perfectly valid to say that UP has a value +1 on the $\hbar/2$ unit and DOWN has – 1 on the same scale.
4 The normal convention is to point the z-axis vertically, the y-axis then would be (left, right), and the x-axis along the flight path (forward, backward). Spins can be aligned (forward, backward) as well.
5 Remember that we have assigned the values $(+1)$ to UP and (-1) to DOWN.
6 There will be situations where one of the measurements produces a value that is by coincidence equal to the average value. In a string of numbers 1, 0, – 1, the average is 0, which is equal to the second value, but you would not say that the second value was the average.
7 I am reminded of a scene from the Dr Who 50th anniversary episode where two incarnations are both using their screwdrivers to 'reverse the polarity' at the same time. They conclude that as a result they are 'confusing the polarity'. Well, it amused me…
8 Note for the technically aware: $\hat{U}(t)$ is a *linear operator*, which is why it can produce infinitesimal changes. The evolution of a state in a measurement is a distinctly nonlinear affair, hence the impasse.
9 Notably Roger Penrose.
10 Of course, hating an idea does not make it wrong. I also think that I have got some pretty good rational arguments for believing in free will, one being that we can actually argue about this. People who are determinists also have to accept that they can't argue for it—after all you can't convince me that I'm right unless my cogs happen to line up that way. I am more than happy to debate these ideas with anyone over a pint of beer, provided they *choose* to buy me one.

6 Interference

6.1 HOW SCIENCE WORKS?

Science, as it is generally described at school, starts with experimental data. Scientists then have to find some *hypothesis* which explains these results. Using this hypothesis, a *prediction* is made about what might happen in a slightly different experiment. An experimental check is carried out, and if the prediction turns out to be correct the hypothesis can be upgraded into a *theory*. You then return to 'GO' and start again with expanded experimental data.

You don't have to work in science for very long to realize that this picture is unsophisticated and oversimplified. Sometimes, it can be extraordinarily difficult to find a hypothesis that explains the data. Sometimes there are many hypotheses, and the data are insufficient to decide between them. Often, we must start with a rough idea of what is going on, under rather simplified circumstances, and try to add in more detail as things become more familiar and we gain experience. Occasionally the theories that we use are too mathematically intricate to allow us to extract exact answers. In which case, we have to content ourselves with some sort of partial answer obtained by using ingenious approximations, limited attention to detail, and often a great deal of coffee.

One technique that sometimes helps is to imagine a simplified experiment that, if it could be made to work, would illuminate aspects of the theory. Einstein was an absolute master at this. He called them *Gedankenexperiments*, which translates as 'thought experiments.' It was never the intention that the *Gedankenexperiments* be carried out, and in many cases that would not be possible. Rather the idea was to imagine a possible experiment, stripped of technical detail, and to see how the theory would deal with describing what would happen. During the development of quantum mechanics in the 1920s, Einstein would often argue with Bohr and Heisenberg about the nature of quantum theory and produced some brilliant thought experiments designed to catch the theory out and show that it would produce inconsistent results. In the end, this turned out to be a losing battle and Einstein had to admit that quantum theory was at least consistent[1], although he never accepted that it had the last word to say about the microworld.

Occasionally, simplified experiments of this sort become part of the scientific culture and are used as teaching aids when the ideas are being passed on to the next generation. There is a danger that mistakes, and misinterpretations, can be inherited and accepted as part of the folklore, which is why it can be timely when technology catches up with the creativity of the scientist, and a *Gedanken experiment* becomes possible.

Quantum theory has such an experiment. It is integrated into the folk memory, used as a teaching aid, and widely repeated in books. Richard Feynman believed that all the mysteries of quantum theory could be seen in the results. It is known as the 'double-slit experiment', and in 2002 was elected[2] "the most beautiful experiment of the twentieth century".

6.2 THE DOUBLE-SLIT EXPERIMENT

The double-slit experiment (illustrated in Figure 6.1) is a modified version of the setup first used by Thomas Young in 1801, which was briefly discussed at the start of Chapter 1. Young's version is a classic experiment that, at the time, seemed to have settled the dispute between wave and particle views of light decisively in favour of waves.

In Young's version, light from a single source was allowed to illuminate an opaque screen which had two tiny slits cut into it. Light passing through the slits combined to form a pattern of illumination on a second screen a short distance away. Today this experiment is easily performed in a

DOI: 10.1201/9781003225997-8

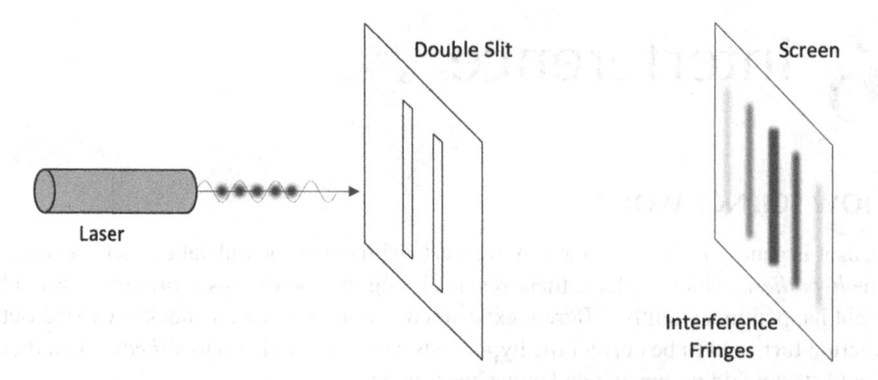

FIGURE 6.1 The double-slit experiment. Waves from a light source illuminate an opaque screen into which two closely spaced, narrow slits have been cut. Light passing through the slits reaches a second screen in the distance, where a pattern of light and dark bands is revealed. The lower image shows the results of a similar experiment conducted with a laser beam. Typically, the slits are separated by about 1 mm and the screen can be a few metres away.

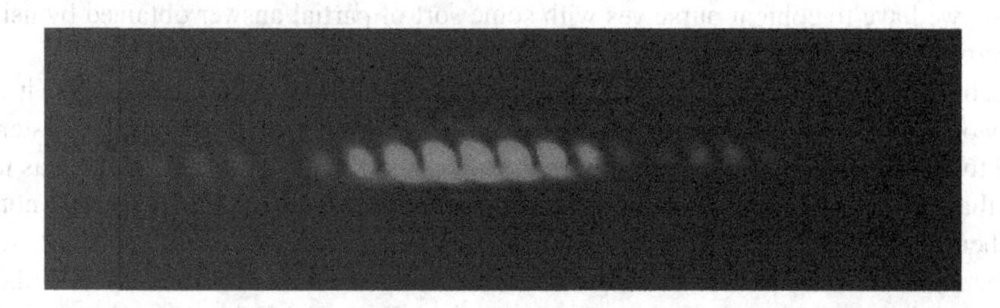

FIGURE 6.2 The pattern of light and dark bands formed by passing green light from a laser beam through a double-slit system. The main interference pattern is the sequence of dots with spaces between them, governed by the slit separation. The broader variation of intensity across the whole pattern is due to the width of the slits. (Image credit: Dr Graham Beards https://en.wikipedia.org/wiki/Double-slit_experiemnt#/media/File:Young's_slits.jpg, CC4 https://creativecommons.org/licenses/by-sa/4.0/deed.en.)

school lab using a laser and 'off-the-shelf' equipment. Students can then clearly see a sequence of uniformly spaced light and dark bands on the far screen (see Figure 6.2).

In the standard explanation for this pattern, waves from the laser arrive at the slits and are absorbed by the opaque material, except where the slits provide a gap for them to pass through[3]. With the correct orientation, a wavefront will pass through both slits together. On the other side, with narrow slits, the waves will appear to be concentric circular wavefronts centred on each slit (see Figure 1.3). Due to their common source, the two sets of waves emerging from the slits will be in phase with each other. However, by the time they have reached the far screen, the waves will have travelled different distances[4], developing a phase difference between them. For a given point on the screen, this phase difference will lead to a degree of constructive or destructive interference. In Figure 6.3 the path lengths from slit A and slit B have produced a phase difference of π at the screen leading to completely destructive interference and a dark band. This is repeated periodically across the screen, whenever the path difference produces a phase shift which is a whole number multiple of π.

You might think that the same effect could be achieved by firing two lasers at the screen, dispensing with the slits altogether. Unfortunately, this doesn't work as the lasers are independent light sources, so there will be no fixed phase relationship between the waves that they produce. The path difference would still produce a phase difference at the screen but given the fluctuations between the lasers, this would vary from constructive to destructive from one moment to the next. All you

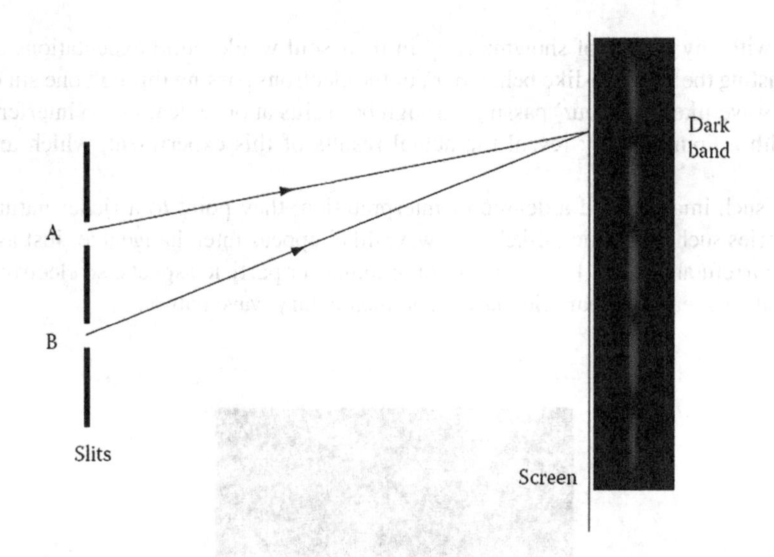

FIGURE 6.3 Light travelling to a point on the screen has to cover a different distance depending on which slit it emerges from. This path difference gives rise to interference as the waves combining at the screen have performed different numbers of complete cycles.

would get is a rapidly and randomly shifting pattern, which averages to a featureless illumination on the screen. A single laser beam covering both slits will split into two parts with the same phase as the light passes through the slits. Consequently, *the effect can only be made to work when waves from a source pass through both slits at the same time.*

6.2.1 The Double Slit with Electrons

> We choose to examine a phenomenon which is impossible, absolutely impossible, to explain in any classical way, and which has in it the heart of quantum mechanics. In reality, it contains the only mystery.... We should say right away that you should not try to set up this experiment. This experiment has never been done in just this way. The trouble is that the apparatus would have to be made on an impossibly small scale to show the effects we are interested in. We are doing a "thought experiment," which we have chosen because it is easy to think about. We know the results that would be obtained because there are many experiments that have been done, in which the scale and the proportions have been chosen to show the effects we shall describe.
>
> *Richard Feynman*[5]

Surprisingly, when Feynman wrote these words, he was not up to date on this point. The lecture course on which his book was based started in 1961, the same year as the first attempt at a genuine double-slit experiment with electrons.

Feynman chose to discuss this modified double slit using electrons, despite believing it to be impractical to carry out, in order to challenge student expectations. They would have been taught to think that light is a wave and probably seen this demonstrated using some variation of Young's experiment. Equally, their understanding of what it is to be an electron would have them thinking in terms of miniature balls of solid matter with an electrical charge. In his account, Feynman likens electrons to tiny bullets fired from a machine gun. With this in mind, students would expect that a sequence of electrons launched one after the other into the apparatus would either be absorbed without trace in the first screen or pass through one of the slits to be detected on the far side. After a period of time, the data revealed by the electron detector would amount to an accumulation of electron strikes opposite each slit, with a few stray hits elsewhere.

Teachers with any degree of showmanship in their soul would build expectations towards this result, contrasting the 'particle-like behaviour', of the electrons passing through one slit or the other, with light's 'wave-like behaviour' passing through both slits at once, leading to interference. Then, doubtless with a flourish, they reveal the actual results of this experiment, which are shown in Figure 6.4.

Although such images need a degree of interpretation, they point to a richer nature of reality where categories such as 'particle-like' and 'wave-like' appear interchangeable. Just as light, once thought to be irrefutably wave-like, can take on granular or particle aspects, so electrons, often the model of what we mean by a particle, have a complementary wave nature.

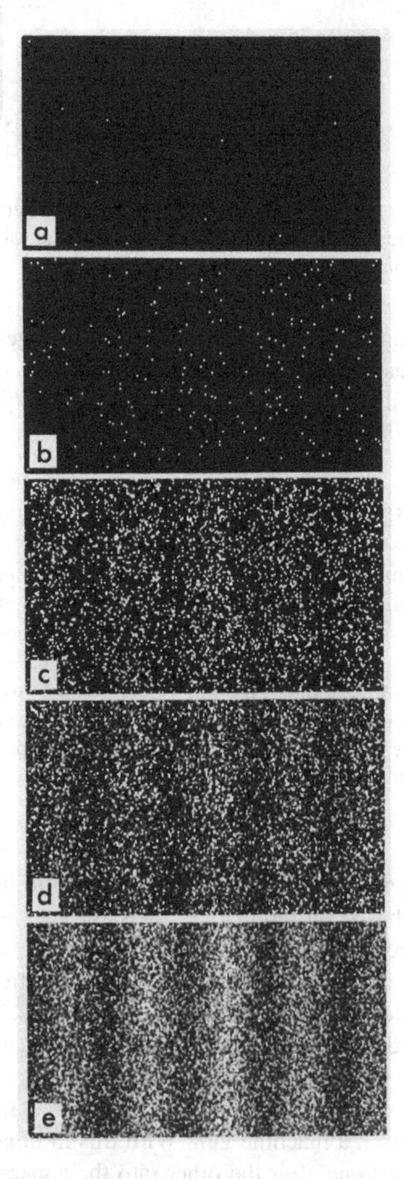

FIGURE 6.4 Images from the electron double-slit experiment performed by the Hitachi team. The images show how the interference pattern built up over 20 minutes. The numbers of electrons involved are (a) 10, (b) 200, (c) 600, (d) 40,000, and (e) 140,000. (Image credit: Provided with kind permission of Dr. Tonomura, CC3 https://creativecommons.org/licenses/by-sa/3.0/deed.en.)

Although it is technically challenging, the basic structure of Young's experiment can be adapted to use electrons rather than light. The greatest single problem derives from the accessible electron wavelengths being very much smaller than the photon wavelengths obtained from a laser. Consequently, the whole apparatus has to be scaled down. Various approaches are deployed, but all of them replace the second screen with some form of electron detector. Electrons will not illuminate a sheet of material the way that light does, so electronic devices have to be used which render an electron 'strike' as a spot of light on a monitor.

Claus Jönsson and his colleagues at Tübingen[6] first performed a double-slit experiment with electrons in 1961. They used special techniques to manufacture very thin single, double and multi-slit arrangements, rather like those used in a Young-type apparatus. Then, in 1974, a team of researchers led by Pier Giorgio Merli at the University of Milan managed to achieve experimental results with only one electron at a time passing through the apparatus (just as we discussed for photons in Chapter 1). Puzzlingly, this work did not achieve widespread recognition. For some time afterwards, people continued to talk about electron double-slit experiments as impractical thought experiments, although in 1976 a short film was made featuring the Milan team and their results, which can now be downloaded[7].

A very convincing and secure electron double-slit experiment was performed in 1989 by Akira Tonomura and his team from the Hitachi Advanced Research Laboratory in Saitama, Japan[8]. Their source produced less than 1000 electrons per second so, given the size of the apparatus and the speed of the electrons, it was extremely unlikely that two or more electrons would be in transit at the same time. As with the earlier Merli experiment, Tonomura et al used an *electron biprism* arrangement, rather than genuine slits, in their apparatus (Figure 6.5).

A very sensitive electron detector allowed the evolution of the resulting pattern to be displayed over a 20 min period, as shown in Figure 6.4. This remarkable and highly significant series of images deserves careful study.

In Figure 6.4a and b, you can see individual dots, each one of which was produced by a single electron arriving at a localized point in the detector. These dots reinforce our impression of electrons as simple lumps of matter travelling through space, like snooker balls rolling across a table. If all we had to look at were pictures such as Figure 6.4a and b, we would have no problem interpreting these results. A single electron produced by the source travels to the screen. If it happens to hit the screen, it will be absorbed and play no further part in matters. If the electron is lucky enough to arrive at one of the slits, it can pass through and enjoy life a bit longer until it smacks into the detector, producing a spot of light on the monitor.

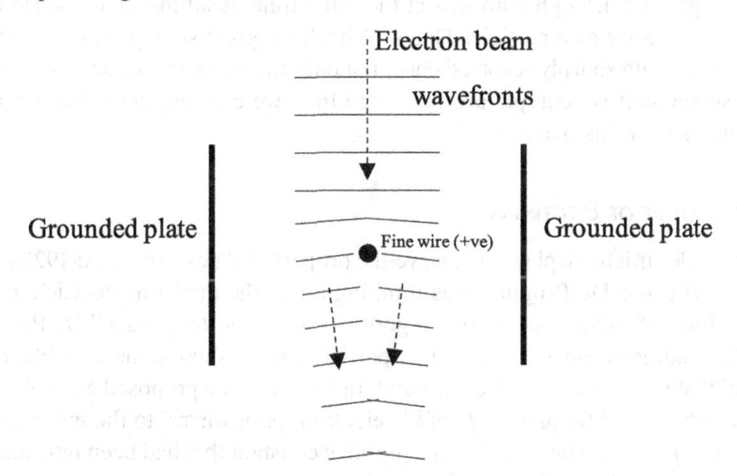

FIGURE 6.5 An electron biprism consists of a fine wire placed between two vertical parallel plates. The wire cuts the electron beam at 90° and is set to a positive potential with respect to the grounded plates on either side of the beam. Electron wavefronts are deflected either side of the wire and overlap to form an interference pattern.

Unfortunately, we can't just ignore the latter photographs from this sequence (Figure 6.4c–e). What initially appeared to be a more-or-less random distribution of spots in Figure 6.4a and b, slowly resolves into a more detailed pattern as more hits accumulate. By the time we get to Figure 6.4e, the image has a striking resemblance to the interference pattern in Figure 6.2. The most paradigm-shifting aspect of the pattern, however, is that it continues to be made up of individual dots on the screen: *the 'snooker-ball' electrons are being shepherded into arriving at the screen one by one according to an interference pattern.*

6.2.2 Wave/Particle Duality

> The elementary quantum phenomenon is a great smoky dragon. The mouth of the dragon is sharp, where it bites the counter. The tail of the dragon is sharp, where the photon starts. But about what the dragon does or looks like in between we have no right to speak, either in this or any delayed-choice experiment. We get a counter reading but we neither know nor have the right to say how it came. The elementary quantum phenomenon is the strangest thing in this strange world.
>
> *John A Wheeler*[9]

If we interpret the presence of individual dots on the screen as evidence for electrons being tiny, localized, lumps of matter, then we also have to face the paradoxical way that the dots are marshalled into an interference pattern, which requires some form of wave-like underpinning.

Blocking off one of the slits or placing a barrier on one side of an electron biprism, confirms the interference effect at work. Under these conditions, the banding inherent in the distribution of dots disappears, to be replaced by a clump of strikes opposite the clear path through the experiment.

The mystery is made all the more profound by the conditions inherent in the experiment: only one electron is in transit at any one time. Even so, each electron, from the first to the last, seems to have 'knowledge' of the final interference pattern. This eliminates any notion that the banding arises from pairs of electrons influencing each other as they pass, more or less at the same time, through different slits.[10] Incidentally, the indication that each individual electron 'understands' the whole interference pattern is the most direct challenge to any interpretation that regards quantum states as referring to *collections* of objects and not *individuals* (see Section 5.3).

It's hard to avoid seeing this as evidence for every single electron moving through the equipment as a wave (hence passing through both slits at the same time, resulting in an interference pattern), but arriving at the detector as a particle. This is Wheeler's 'great smoky dragon', although he was referring to photons, with sharply resolved ends, but only the vaguest outline in the middle.

To make it sound as if we can picture this more than we can, physicists have coined the term *wave/particle duality* for this paradox.

6.2.3 Wave Nature of Electrons

The idea that particles might display some wavelike properties dates to the mid-1920s when a young PhD student called Louis De Broglie[11] was thinking about the dual wave/particle nature of light. Working on the lines of "what's sauce for the goose is sauce for the gander," De Broglie suggested that, in the right conditions, electrons and other particles might show some wavelike properties due to a 'pilot wave' that helped to guide them around. In his thesis, he proposed a couple of formulas to link the wavelength, λ, and frequency, f, of his electron 'pilot waves' to the more familiar energy, E, and momentum, p, of a particle via h, a quantitative constant that had been introduced into physics by Max Planck in 1900 (see Chapter 17):

$$\lambda = \frac{h}{p} \quad \text{and} \quad f = \frac{E}{h}$$

In their experiment, the Hitachi team used 50,000 V to accelerate electrons to ~ 120,000 kms^{-1} (40% of the speed of light), giving them a De Broglie wavelength[12] of ~ 6.1×10^{-12} m. In contrast, the wavelength of green light used in Figure 6.2 ~ 5.5×10^{-7}m. This illustrates the contrast of scale involved in the two experiments.

In the classic Young-type experiment with light, the spacing of the interference bands can be measured and used to calculate the wavelength of light involved. Equally, in an electron double slit, measuring the band spacing allows the electron wavelength to be established, confirming the accuracy of De Broglie's equations.

The De Broglie formulas catch the weirdness of wave/particle duality in a very striking manner. The energy and momentum of the particle relate to the frequency and wavelength of the wave[13]. De Broglie's formulas link these ideas together, but the bare equations do not help us *picture* this. We are still left struggling to see how the same object can appear like a wave at one moment and a particle the next.

De Broglie's notion of a 'pilot wave' was soon abandoned by most of the founding fathers, in favour of *amplitude functions*. However, Erwin Schrödinger notably persevered with a form of pilot waves, and in more modern times it was picked up by David Bohm and his collaborators. I will talk more about De Broglie and Schrödinger in Chapter 20 and David Bohm's version of quantum mechanics will come up in Chapter 31.

6.3 DOUBLE-SLIT AMPLITUDES

A quantum mechanical description of the double-slit experiment must involve a set of amplitudes governing the passage of an electron through the slits to the far detector. We have already seen how to deal with amplitudes of this sort. They are covered by Important Rule 5 from Chapter 4.

We start by writing:

$$\langle y|I\rangle = \langle y|A\rangle\langle A|I\rangle + \langle y|B\rangle\langle B|I\rangle$$

where $|y\rangle$ is the state corresponding to an electron arriving at position y in the detector, $|A\rangle$ and $|B\rangle$ to that passing through slits A and B respectively, and $|I\rangle$ the initial state of an electron emerging from the electron source (Figure 6.6).

We have added together two terms in the overall amplitude, as the two possibilities (travelling through one slit or the other) can't be distinguished in the context of the experiment as it is set up (IR3 clause 2b). Normally when we combine amplitudes in this fashion, some sort of inference results. If we block a slit, then one of these terms will disappear and hence there can be no interference.

The probability that an electron arrives at y is the complex square of the total amplitude:

$$|\langle y|I\rangle|^2 = |\langle y|A\rangle\langle A|I\rangle + \langle y|B\rangle\langle B|I\rangle|^2$$

$$= \left(\langle y|A\rangle\langle A|I\rangle + \langle y|B\rangle\langle B|I\rangle\right)^* \left(\langle y|A\rangle\langle A|I\rangle + \langle y|B\rangle\langle B|I\rangle\right)$$

$$= \langle y|A\rangle^*\langle A|I\rangle^*\langle y|A\rangle\langle A|I\rangle + \langle y|A\rangle^*\langle A|I\rangle^*\langle y|B\rangle\langle B|I\rangle + \langle y|B\rangle^*\langle B|I\rangle^*\langle y|A\rangle\langle A|I\rangle + \langle y|B\rangle^*\langle B|I\rangle^*\langle y|B\rangle\langle B|I\rangle$$

$$= |\langle y|A\rangle\langle A|I\rangle|^2 + \left[\langle y|A\rangle^*\langle A|I\rangle^*\langle y|B\rangle\langle B|I\rangle + \langle y|B\rangle^*\langle B|I\rangle^*\langle y|A\rangle\langle A|I\rangle\right] + |\langle y|B\rangle\langle B|I\rangle|^2$$

The first and last terms give the conventional probabilities that we might expect for an electron that just goes through one slit or the other. The middle term in square brackets is where the interest lies: it is an *interference term.*

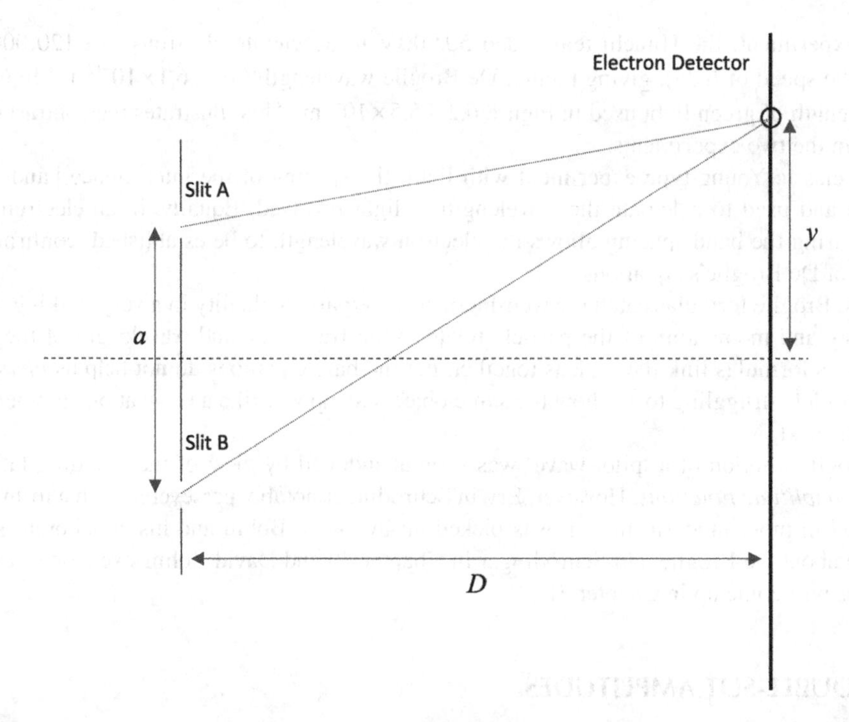

FIGURE 6.6 Terminology used in the description of the double-slit experiment for electrons. The marked point, a distance y from the centre of the electron detector, is at the middle of a dark band. The distance between dark bands is related to the wavelength λ of the electron waves by the formula, spacing=$D\lambda/a$.

6.3.1 Phase and Physics

If there is a dark band at y on the detector, the probability of an electron arriving there must be zero. Hence:

$$|\langle y|I\rangle|^2 = |\langle y|A\rangle\langle A|I\rangle + \langle y|B\rangle\langle B|I\rangle|^2 = 0$$

Given an electron source placed midway between the slits, it seems sensible to set $\langle A|I\rangle = \langle B|I\rangle = \langle \text{slits}|I\rangle$ and consequently:

$$|\langle y|I\rangle|^2 = |\langle y|A\rangle + \langle y|B\rangle|^2 |\langle \text{slits}|I\rangle|^2 = 0$$

So, either $|\langle \text{slits}|I\rangle|^2 = 0$, which isn't very interesting as the electrons aren't even getting to the slits in the first place, or $|\langle y|A\rangle + \langle y|B\rangle|^2 = 0$, which presumably arises due to amplitude interference[14].

Clearly, if $|\langle y|A\rangle + \langle y|B\rangle|^2 = 0$ then $\langle y|A\rangle = -\langle y|B\rangle = (-1)\langle y|B\rangle$, which is reminiscent of the π phase shift factor from section 3.3.1. Of course, this ties in directly, as in the case of wave interference, a π phase shift between two waves arriving at the same point will lead to total destructive interference at that point.

Back in Chapter 3, we learnt to think of amplitudes as complex numbers which can, in turn, be written in the form:

$$z = R[\cos\vartheta + i\sin\vartheta]$$

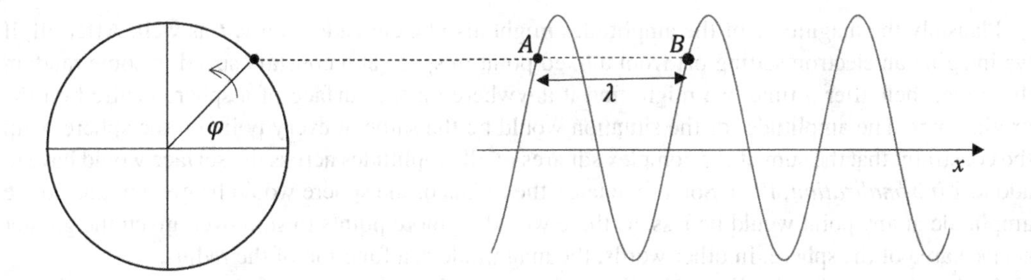

FIGURE 6.7 In the waveform on the right, B is the closest point to A which has the same phase as A. The distance between A and B is equal to the wavelength of the wave.

with R being the magnitude of the amplitude and ϑ its phase. A beautiful theorem from the mathematics of complex numbers allows us a further simplification:

$$z = R[\cos\vartheta + i\sin\vartheta] = Re^{i\vartheta}$$

relating the trigonometric functions $\sin\vartheta$ and $\cos\vartheta$ to the exponential function e^x. It also follows that:

$$z^* = R[\cos\vartheta - i\sin\vartheta] = R[\cos(-\vartheta) + i\sin(-\vartheta)] = Re^{-i\vartheta}$$

Given this new mathematical structure, we can rewrite our two amplitudes as:

$$\langle y|A\rangle = R_A e^{i\vartheta_A(y,\,t)}$$

$$\langle y|B\rangle = R_B e^{i\vartheta_B(y,\,t)}$$

where I have made the phases explicitly dependent on position, y, and time, t. This is a slightly unusual approach that requires a short diversion to explain…

In a conventional wave, the wavelength is the shortest distance between two points of the same phase (Figure 6.7).

Generally, we would be tracking how an amplitude varied along the direction of an electron's flight, which would be x in Figure 6.7 if the wave illustrated is the De Broglie wave of the electron. Consequently, the amplitude would have the form:

$$\langle x|\text{wave}\rangle = Re^{i\vartheta(x,\,t)}$$

with $\vartheta(x+\lambda,\,t) = \vartheta(x,\,t) + 2\pi$, making $\vartheta(x,\,t)$ sensitive to shifts in x of the order of an electron wavelength, $\sim 10^{-12}\,\text{m}$. It is easy to show that:

$$\vartheta(x,\,t) = \frac{2\pi}{\lambda}x - 2\pi ft$$

but this is something that we will discuss in more detail in Chapter 7.

In this case, however, we are more interested in the phase of the amplitude when the electron arrives at position y *on the electron detector* than we are along the line of the electron's path. Some simple geometry would enable us to relate the two together but would clutter up the equations needlessly[15].

I have also made the phase explicitly dependent on time, t, to emphasise that we need the phase of both amplitudes at y *at the same time*. This is *not* suggesting that y is changing with time.

Plausibly the magnitude of the amplitudes might also be dependent on y, t as well. After all, if we imagine an electron setting off from a fixed point in space at a constant speed in some random direction, then after a time, we might find it anywhere on the surface of a sphere centred on the origin point. The amplitude for the situation would be the same at every point on the sphere, with the constraint that the sum of the complex squares of all amplitudes across the surface would have to add to 1 (*normalization*, IR2). Sometime later, the radius of the sphere would be greater, and so the amplitude at any point would be less, as there would be more points to sum over, given the greater surface area of the sphere. In other words, the magnitude is a function of the radius.

In the electron double-slit, each slit would act as the origin point for spherical wavefronts. However, given that the slits are close to each other, the path difference between $A \rightarrow x$ and $B \rightarrow x$ will be small enough to have no discernible impact on the magnitudes and we can take $R_A \approx R_B = R$.

This concludes our slight diversion.

6.3.2 An Experiment with Phase

Having written the two amplitudes in the form:

$$\langle y|A\rangle = R_A e^{i\vartheta_A(y,\,t)}$$

$$\langle y|B\rangle = R_B e^{i\vartheta_B(y,\,t)}$$

leaving the magnitudes different for the moment, we can substitute them into our probability calculation and see what we get. Recall that we had:

$$|\langle y|I\rangle|^2 = |\langle y|A\rangle + \langle y|B\rangle|^2 |\langle \text{slits}|I\rangle|^2$$

with

$$\begin{aligned}
|\langle y|A\rangle + \langle y|B\rangle|^2 &= \left(\langle y|A\rangle^* + \langle y|B\rangle^*\right)\left(\langle y|A\rangle + \langle y|B\rangle\right) \\
&= \left(R_A e^{-i\vartheta_A(y,\,t)} + R_B e^{-i\vartheta_B(y,\,t)}\right)\left(R_A e^{i\vartheta_A(y,\,t)} + R_B e^{i\vartheta_B(y,\,t)}\right) \\
&= R_A^2 + R_A R_B e^{i[\vartheta_B(y,\,t) - \vartheta_A(y,\,t)]} + R_B R_A e^{i[\vartheta_A(y,\,t) - \vartheta_B(y,\,t)]} + R_B^2 \\
&= R_A^2 + R_B^2 + R_A R_B \left[e^{i[\vartheta_B(y,\,t) - \vartheta_A(y,\,t)]} + e^{i[\vartheta_A(y,\,t) - \vartheta_B(y,\,t)]}\right]
\end{aligned}$$

Once again, we have an interference term, in this case $\left(R_A R_B \left[e^{i[\vartheta_B(y,\,t) - \vartheta_A(y,\,t)]} + e^{i[\vartheta_A(y,\,t) - \vartheta_B(y,\,t)]}\right]\right)$, which acts to modulate the overall probability. Unpacking this interference term further:

$$\begin{aligned}
R_A R_B \left[e^{i[\vartheta_B(y,\,t) - \vartheta_A(y,\,t)]} + e^{i[\vartheta_A(y,\,t) - \vartheta_B(y,\,t)]}\right] &= R_A R_B \left[e^{-i[\vartheta_A(y,\,t) - \vartheta_B(y,\,t)]} + e^{i[\vartheta_A(y,\,t) - \vartheta_B(y,\,t)]}\right] \\
&= R_A R_B \big[\cos\left(\vartheta_A(y,\,t) - \vartheta_B(y,\,t)\right) - i\sin\left(\vartheta_A(y,\,t) - \vartheta_B(y,\,t)\right) \\
&\quad + \cos\left(\vartheta_A(y,\,t) - \vartheta_B(y,\,t)\right) + i\sin\left(\vartheta_A(y,\,t) - \vartheta_B(y,\,t)\right)\big] \\
&= 2R_A R_B \cos\left(\vartheta_A(y,\,t) - \vartheta_B(y,\,t)\right)
\end{aligned}$$

making the overall result:

$$|\langle y|I\rangle|^2 = \left[R_A^2 + R_B^2 + 2R_AR_B\cos\left(\vartheta_A(y,\,t) - \vartheta_B(y,\,t)\right)\right]|\langle \text{slits}|I\rangle|^2$$

If slit A was blocked off, $R_A = 0$, and we would just have $R_B^2|\langle \text{slits}|I\rangle|^2$. Equally, if $R_B = 0$ the distribution of electron strikes would just be governed by $R_A^2|\langle \text{slits}|I\rangle|^2$. It's the interference term that provides the pattern of light and dark bands across the detector.

6.3.3 The Interference Term

In a fully detailed mathematical analysis of the double-slit experiment, we would have to account for the different path lengths having an impact on the magnitudes of the amplitudes as well as the phases. However, as mentioned before, provided the two slits are close together, $R_A \approx R_B = R$ giving:

$$\begin{aligned}|\langle x|I\rangle|^2 &= \left[R_A^2 + R_B^2 + 2R_AR_B\cos\left(\vartheta_A(y,\,t) - \vartheta_B(y,\,t)\right)\right]|\langle \text{slits}|I\rangle|^2 \\ &= \left[2R^2 + 2R^2\cos\left(\vartheta_A(y,\,t) - \vartheta_B(y,\,t)\right)\right]|\langle \text{slits}|I\rangle|^2 \\ &= 2R^2\left[1 + \cos\left(\vartheta_A(y,\,t) - \vartheta_B(y,\,t)\right)\right]|\langle \text{slits}|I\rangle|^2\end{aligned}$$

The constraint for a dark band at y is $|\langle y|I\rangle|^2 = 0$, so:

$$\left[1 + \cos\left(\vartheta_A(y,\,t) - \vartheta_B(y,\,t)\right)\right] = 0$$

implying

$$\cos\left(\vartheta_A(y,\,t) - \vartheta_B(y,\,t)\right) = -1$$

or

$$\vartheta_A(y,\,t) - \vartheta_B(y,\,t) = n\pi$$

with n being an odd integer. As a result, if we look across the span of the electron detector along the y axis, every time the path difference $\left[\text{distance}(A \to y) - \text{distance}(B \to y)\right]$ is a whole number of half wavelengths, the phase difference for the amplitudes will be $n\pi$, resulting in destructive interference. Between these regions will be bands of constructive interference.

Having different values for R_A and R_B will just mean that there is never completely destructive interference at the centre of a dark band.

6.3.4 Amplitudes and Electron Strikes

Stepping back from calculating the probability of an electron arriving at y, we ought to consider the state of the electron as it arrives at the detector:

$$|\Psi\rangle = R\sum_n\left[e^{i\vartheta_A(y_n,\,t)} + e^{i\vartheta_B(y_n,\,t)}\right]\langle \text{slits}|I\rangle|y_n\rangle$$

To write down the state in this form, I have divided the impact points into a very large set of small regions y_n and set out a basis $\{|y_n\rangle\}$ to expand over. This enables me to approximate the state, without having the complexity of a continuous value of y to deal with (that comes later...). Extracting the probability for the electron arriving at a chosen y_n, say y_8, would then be a case of multiplying by $\langle y_8|$:

$$\langle y_8|\Psi\rangle = \langle y_8|\left\{R\sum_n\left[e^{i\vartheta_A(y_n,t)} + e^{i\vartheta_B(y_n,t)}\right]\langle \text{slits}|I\rangle|y_n\rangle\right\} = R\sum_n\left[e^{i\vartheta_A(y_n,t)} + e^{i\vartheta_B(y_n,t)}\right]\langle \text{slits}|I\rangle\langle y_8|y_n\rangle$$

$$= R\left[e^{i\vartheta_A(y_8,t)} + e^{i\vartheta_B(y_8,t)}\right]\langle \text{slits}|I\rangle$$

(using $\langle y_8|y_n\rangle = 0$ for $n \neq 8$, an example of the rules we introduced in Section 4.2.1)

and then complex squaring to get the same result as before.

The electron's state as it arrives at the detector is formed from a collection of amplitudes and a position basis spanning the width of the detector. Yet, we observe a collection of dots indicative of a series of localized detections, one per electron. Evidently, the interaction between an electron and the detector brings about the random collapse of the state into one of the $|y_n\rangle$. In Wheeler's rather picturesque terms, this is the sharp bite of the dragon at the detector. Without the state collapsing into one of the $|y_n\rangle$, the amplitude would never resolve into a definitive instantiation.

6.4 LAST THOUGHTS

We have already seen that the quantum world radically challenges our comfortable classical understanding. The results from the Mach-Zehnder interferometer and the Stern-Gerlach experiments have provided ample evidence for that. However, in my view, there is something more immediate and visceral about the electron double-slit experiment. The sequence of images in Figure 6.4 (or better, watching the associated video) shows us the interference pattern, which is so hard to explain without some form of wave at work, building up from individual spots, which appear to be persuasive evidence for particles arriving at the detector.

In their earliest attempts to grapple with this paradox, at least some of the founding fathers flirted with the idea of a physical pilot wave guiding particle trajectories. David Bohm's approach to quantum theory echoes some aspects of this idea by deploying a, so-called, *quantum potential* to act as a guide for particles. Bohm's antological interpretation is worthy of serious consideration (Chapter 31), as is much of his related philosophical thinking but, for better or worse, it is not part of the mainstream approach to these matters.

Circling back, it is evident that seeking some form of physical pilot wave to explain these effects would not be lightly rejected unless some other alternative suggested itself. Here we come to the role of the amplitude.

Our specific analysis of the Mach-Zehnder interferometer demonstrated the need to employ complex numbers to enumerate quantum amplitudes. By their nature, complex numbers can be written as a magnitude and a phase. The strictures of clause 2b of Important Rule 3 require that we add amplitudes if the alternatives cannot be distinguished in the context of the current experiment. Invariably, the amplitudes that we combine will have different phases, which in turn may well vary in time and space. This is exactly the conditions required for interference to take place, not between physical waves, but within the more mysterious 'reality' of the quantum amplitude. The answer to the question "what is waving" is "the quantum amplitude" or, in a more expressive and accurate fashion, "the quantum probability amplitude is waving". Once again, we are driven into a philosophical consideration of the reality of the quantum amplitude and with it the instrumentalist vs realist debate.

Our mental unease is not helped by reflecting on interference in the context of light. Thomas Young's experiment was adequately explained by presuming light to be a wave phenomenon.

However, the data did not shed any light[16] on the nature of this wave. It needed the work of James Clerk Maxwell in the early 1860s to suggest that light was an electromagnetic wave – a periodic variation in the strengths of coupled electric and magnetic fields. These fields are in themselves not easily pictured, but they have a more concrete reality than the quantum amplitude…

However, with the advent of lasers we can carry out Young's experiment photon by photon, with results very similar to those seen for the electron double-slit. The interference pattern builds over time from a series of photon dots. In this context, we are led to the same conclusion: that the photon amplitude is 'waving' and giving rise to the interference effect. The challenge now is to find some explanatory link between the electric and magnetic fields that we thought we understood and the quantum amplitudes that appear to be at work. This is where quantum field theory enters, and we must regrettably postpone any further consideration until Chapter 32.

According to Feynman, the electron double-slit experiment "has in it the heart of quantum mechanics" and "is impossible, absolutely impossible, to explain in any classical way." The quantum explanation is precisely set out:

- the interference pattern arises from mixing amplitudes of different phase
- the localized strikes on the electron detector arise from state collapse at the instant of measurement

but there are deep philosophical waters lying beneath these simple statements. A serious study of the double-slit experiment for electrons is a worldview-altering exercise.

NOTES

1. Einstein was trying to catch out the theory by showing that it was possible to have an experiment measure position and momentum (or energy and time) simultaneously to a degree of accuracy forbidden by the uncertainty principle (see Chapter 13).
2. Physics World Magazine, *May 2002*.
3. We have arranged it so that the diameter of the laser 'spot' is greater than the distance between the slits.
4. Unless we happen to be thinking about the point on the screen directly between the two slits.
5. The Feynman Lectures on Physics, *Vol. 3*.
6. Jönsson, C. *Am. J. Phys.* 42, 4–11, https://doi.org/10.1119/1.1987592 (1974).
7. https://www.bo.imm.cnr.it/users/lulli/downintel/electroninterfea.html.
8. A. Tonomura, J. Endo, T. Matsuda, T. Kawasaki and H. Ezawa 1989 "Demonstration of single-electron build- up of an interference pattern" *Am. J. Phys.* 57 117–120.
9. W. A. Miller and J. A. Wheeler, "Delayed-Choice Experiments and Bohr's Elementary Quantum Phenomenon" S. Kamefuchi et al., eds., Proceedings of the international symposium on foundations of quantum mechanics, Tokyo: Physical Society of Japan, 1984, 140–152.
10. Of course, the distance between slits is bigger than any notion of the size of an electron in its 'lump of matter' form. It's not that the electron 'breaks in two' at the slits.
11. Louis De Broglie (1892–1987) Nobel Prize in Physics 1929. The closest pronunciation of his name in English is "De Broy".
12. Try calculating this for yourself, given that the mass of an electron is $9.10938188 \times 10^{-31}$ kg and Planck's constant is $6.626068 \times 10^{-34}\ \mathrm{m^2kgs^{-1}}$.
13. Whatever these waves are, and I am being deliberately coy about that as things stand, the energy and momentum of the particle are not linked to any energy or momentum present in the wave. That in itself is a surprising feature of wave / particle duality.
14. Of course, both might be true, but that would mean that we've really messed up the experiment.
15. Oh, all right…..using the terminology of Figure 6.6 $x_A = \sqrt{(y-a)^2 + D^2}$, $x_B = \sqrt{y^2 + D^2}$
16. Pun (sort of) not intended…

7 Free Particles

7.1 THE POSITION BASIS

If we wish to calculate the probability of finding a particle at a specified location, x, then we need to expand the particle's state over a basis set $\{|x\rangle\}$, each member of which corresponds to the particle being found at a given point.

Straightaway we are faced with a problem: position is a continuous variable which can take an infinite number of values in the smallest span of distance. We need a specific mathematical technique to deal with this situation and the infinite basis set that naturally accompanies it. However, for the moment, so that we can focus on other aspects of quantum theory without getting too caught up in mathematical machinery, we are going to fudge the issue. Rather than having a continuous set of x values, we imagine dividing the region of interest into a very large, but finite, number of different segments, each of which is centred on a specific value x_n. We then focus on finding the probability of our particle turning up inside one of these segments. This means that we have an approximate state:

$$|\psi(x_n,t)\rangle = \sum_n a(x_n, t)|x_n\rangle$$

which will suffice for our immediate purposes, although we shall return to continuous variables in Chapter 12.

In this state expansion, I have written the amplitude in the form $a(x_n, t)$ rather than as a discrete list $a_n(t)$ one for each location x_n, as this will generalize more easily into the continuous case when we get there later.

7.2 THE AMPLITUDE FOR A FREE PARTICLE

If we're going to get any specific information out of our very general expansion $|\psi(x_n,t)\rangle = \sum_n a(x_n, t)|x_n\rangle$, we need a mathematical expression for $a(x_n, t)$ which is relevant to the physical situation of interest.

In many cases, like the double-slit experiment from the last chapter, it can be quite difficult to find the mathematical form of $a(x_n, t)$. We must solve the equations of quantum mechanics and put in some initial conditions to get the right answer. As we're not yet ready to talk about such things in detail and have not seen the relevant equations in any case, we shall start with a simple example - that of a particle moving in one dimension with no forces acting on it: the *free particle*.

Textbooks on quantum theory generally take one of the following two approaches at this juncture:

1. To write down the equations of quantum theory and then derive the free-particle amplitude from them
2. To write down the free-particle amplitude and then, working backwards, figure out the simplest equation that would yield this amplitude as a solution.

In either case, nothing is *proven*, simply proposed, and then confirmed by experimental data. Arguably, even the confirmation is provisional, as the next experiment we perform could always come up with data that does not fit in with our previous understanding. This is the nature of science. The founding fathers inched their way to a satisfactory version of quantum theory by a series of inspired guesses, half thought-out models and mathematical insight. The chief equation of

DOI: 10.1201/9781003225997-9

conventional quantum theory is the famed *Schrödinger equation*, which will make an appearance in Chapter 14. Its status as an established cornerstone of our understanding rests on:

- the correct predictions obtained from its use
- the deeper understanding it brings.

We are going to take a hybrid approach, presenting the free particle amplitude, and exploring its properties in this chapter, and then in Chapter 12 refining the free particle state into a fully continuous form, extracting plausible operators to play the role of position and momentum measurement, and then showing how they can be combined to form the Schrödinger equation.

So, without further delay, the free particle amplitude function $a(x_n, t)$ is:

FREE PARTICLE AMPLITUDE

$$a(x_n, t) = Ae^{i\left(\frac{2\pi}{\lambda}x_n - 2\pi ft\right)} = A\exp\left(i\left[\frac{2\pi}{\lambda}x_n - 2\pi ft\right]\right)$$

Here I have used the exponential format, using e^x, but also the equivalent functional form, using $\exp(x)$. The latter has the advantage of being more readable when there are a lot of variables to pack into the small-fonted power in e^x. In both cases, A, the magnitude of the complex amplitude, is constant. We will pick up on the significance of A in Section 7.2.3.

7.2.1 Classical Waves

In classical wave theory, a transverse wave (Figure 7.1) would be described by the equation:

$$y = A\sin\left(\frac{2\pi}{\lambda}x - 2\pi ft\right)$$

with A, in this case, being the amplitude of the wave.

Justifying the spatial factor is straightforward. The wavelength, λ, is defined as the shortest distance between two points on the wave that have the same phase *at the same instant in time*. So, if we presume the phase to be linearly proportional to the distance along the wave, $\vartheta(x) = Kx + \xi$, then:

$$\vartheta(x+\lambda) = Kx + K\lambda + \xi = \vartheta(x) + 2\pi = Kx + \xi + 2\pi$$

$$\therefore\ K\lambda = 2\pi \quad \Rightarrow K = \frac{2\pi}{\lambda}$$

The constant factor, ξ, which cancelled out of the calculation, is effectively moving the origin to the left or right. We can see how this works by a slight realignment of definition:

$$\vartheta(x) = K(x + \phi') = Kx + K\phi$$

$$\xi = K\phi$$

$$\vartheta Kx + \xi = Kx + K\phi = K(x + \phi).$$

so that ϕ records how far the origin has moved, and $K\phi$ is the resulting phase shift applied to all points on the wave.

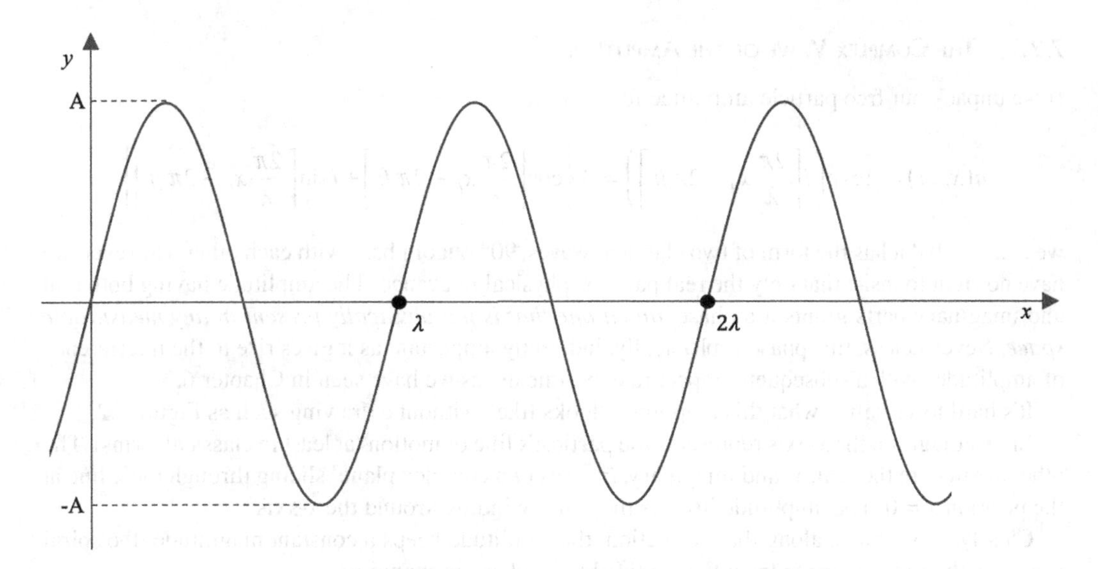

FIGURE 7.1 A transverse wave of amplitude A and wavelength λ. The wave is pictured frozen at an instant in time.

Let us now presume that the wave is moving[1] to the right at a constant speed v. If we view this wave by moving alongside at the same speed,[2] then it appears to us to be stationary, with the origin retreating to the *left* at speed v. In other words:

$$\phi = -vt + \Phi$$

The standard wave equation relates wave speed to frequency and wavelength, $v = f\lambda$, giving $\phi = -f\lambda t + \Phi$, and $K\phi = -Kf\lambda t + K\Phi = 2\pi ft + K\Phi$. As a result:

$$\vartheta(x, t) = \frac{2\pi}{\lambda}x - 2\pi ft + K\Phi$$

All that remains is to ensure that we start our clock when we pass through the original origin, so that $\vartheta = 0$ when $x = 0$ and $t = 0$, which implies that $\Phi = 0$.

The *period* of a wave, T, is the shortest time interval between equivalent phases at the same point:

$$\vartheta(x, t+T) = \vartheta(x, t) - 2\pi$$

$$\vartheta(x, t+T) = \frac{2\pi}{\lambda}x - 2\pi f(t+T) = \frac{2\pi}{\lambda}x - f\lambda t - 2\pi fT = \frac{2\pi}{\lambda}x - 2\pi ft + 2\pi$$

$$\therefore\ 2\pi fT = 2\pi \Rightarrow T = \frac{1}{f}$$

which is exactly what we would expect.

7.2.2 The Complex Wave of the Amplitude

If we unpack our free particle amplitude for a moment:

$$a(x_n, t) = A\exp\left(i\left[\frac{2\pi}{\lambda}x_n - 2\pi ft\right]\right) = A\left\{\cos\left[\frac{2\pi}{\lambda}x_n - 2\pi ft\right] + i\sin\left[\frac{2\pi}{\lambda}x_n - 2\pi ft\right]\right\}$$

we can see that it has the form of two classical waves, 90° out of phase with each other. However, we have no right to insist that only the real part has physical relevance. The amplitude having both real and imaginary parts grants it a phase, *albeit one that is not physically present in any measurable space*. Nevertheless, this phase is physically, indirectly, important as it gives rise to the interference of amplitudes with a subsequent impact in experiments, as we have seen in Chapter 6.

It's hard to visualize what this amplitude 'looks like', without a drawing such as Figure 7.2.

In that diagram, the x-axis represents the particle's line of motion (at least in classical terms). The other arrows are the real, y, and imaginary, iz, axes of a complex plane[3] slicing through the x line at the position $x = 0$. The amplitude itself is the spiral winding around the x-axis.

Clearly, as we move along the x-direction, the amplitude keeps a constant magnitude (the spiral is always the same distance from the x-axis), but its phase is changing.

To expand on this, we take a sequence of vertical slices through the figure at different distances along the x-axis (Figure 7.3).

In each figure, the amplitude at a specific x value is represented by an arrow pointing from that point on the x -axis to the spiral line in Figure 7.2.

Figure 7.3a is for $x = 0$, so the phase of the amplitude is fixed by the $2\pi ft$ part of the formula, as $2\pi x/\lambda$ clearly vanishes. If we arrange to start the clock timing $(t = 0)$ the particle's motion as it crosses $x = 0$, then this contribution to the phase will also be zero, which is why the arrow points along the y-axis. Provided each other slice is taken at the same moment, $2\pi ft$ will vanish in each figure.

In Figure 7.3b we are looking at a point further along the x-axis at the same moment. The phase is consequently greater by $2\pi x/\lambda$ and the arrow has rotated round by a fraction, x/λ, of a complete circle. As we move still further down the x-axis, the phase continues to change, until in Figure 7.3d it has advanced by π radians compared to Figure 7.3a. Evidently, we have moved a distance $\lambda/2$ along in x.

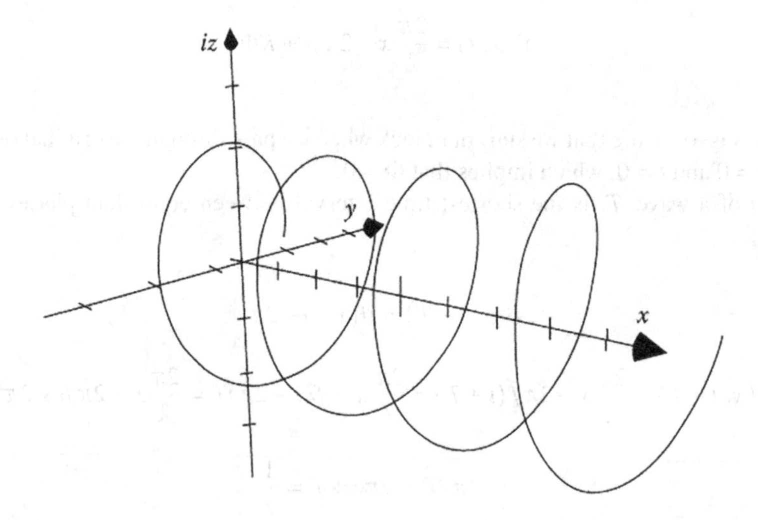

FIGURE 7.2 A representation of the free-particle amplitude. The x-direction points along the 'flight' of the particle, whereas y and iz represent the complex plane. Consequently, x is the only 'physical' direction on this figure.

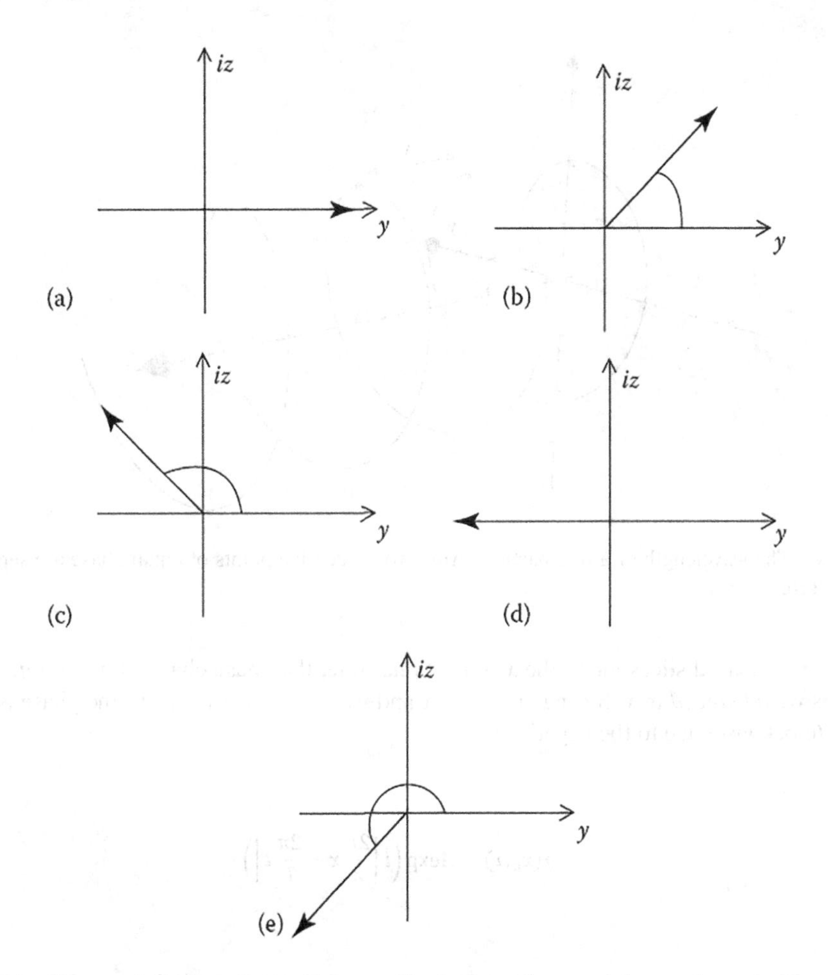

FIGURE 7.3 Five vertical slices through the amplitude for a free particle are shown in Figure 7.2. The amplitude is represented by the black arrow, which is drawn from the x-axis to touch a point on the spiral. These slices represent the situation at five different points, but at the same moment. This does not imply that the particle has moved from a to b to c, etc.

Eventually, the amplitude's phase will reach 2π when $x = \lambda$ a situation shown in Figure 7.4.

Here is one more hint on how to read these figures. Go back to our electron interference experiment in Chapter 6. If one of the paths (say the top one) from slit to screen was a distance x_1 corresponding to the left black blob in Figure 7.3, then we know what the phase would be for that path. A second drawing like that in Figure 7.3 might represent the amplitude from the other slit (the bottom one). If its path to the screen was a distance x_2, corresponding to the right-hand black blob, then the two amplitudes would have the same phase and constructive interference would occur.

7.2.3 Frequency

In the previous section, we set our clock so that $t = 0$ corresponded to the particle crossing the origin. While this is convenient from the numerical point of view, nothing much would change if we did not impose that condition. The phase for every point on the wave, and so the arrows in each of Figures 7.3a–e, would shift around by the same amount $2\pi ft$

Alternatively, we could sit at a specific x value and watch what happens to the phase as time passes. This behaviour is governed by the $2\pi ft$ part of the formula or, using $f = 1/T$, it becomes $2\pi t/T$ so that the factor t/T tells us the fraction of a complete circle that the phase moves over time.

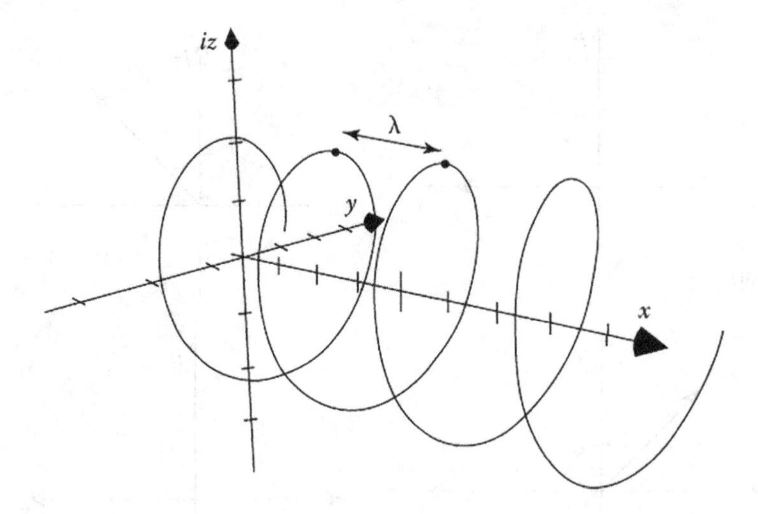

FIGURE 7.4 The wavelength of a free particle. Any two successive points of equal phase are separated by one wavelength.

When we compared slices along the x-axis at one time, the phase changed in an *anticlockwise* direction as we *advanced* in x. Sitting at a fixed x and watching the time pass, the phase is rotating *backward* (clockwise) due to the negative sign.

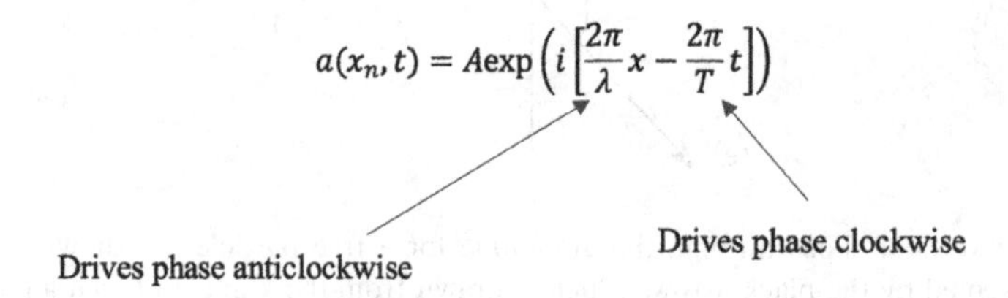

$a(x_n, t) = A\exp\left(i\left[\frac{2\pi}{\lambda}x - \frac{2\pi}{T}t\right]\right)$
Drives phase anticlockwise
Drives phase clockwise

From the vantage point of a fixed x position and looking along the x line, we would see the coil of the amplitude rotating around the axis, like a stretched spring set into rotation about its long axis. This 'movement' of the amplitude is important, as it is a factor in interference for example, but it is *not* directly representative of a particle's movement through space.

7.2.4 What Does the Amplitude Tell Us about the Motion of a Free Particle?

Surprisingly, the free particle amplitude does not represent a particle showing 'visible' motion. The state is:

$$|\psi(x_n, t)\rangle = \sum_n a(x_n, t)|x_n\rangle = A\sum_n e^{i\left(\frac{2\pi}{\lambda}x_n - 2\pi ft\right)}|x_n\rangle$$

so, to calculate the probability of finding the particle in a specified segment, x_3, we proceed as follows. The first step is to obtain the amplitude $\langle x_3|\psi(x_n, t)\rangle$:

$$\langle x_3|\psi(x_n,t)\rangle = A\langle x_3|\sum_n e^{i\left(\frac{2\pi}{\lambda}x_n - 2\pi ft\right)}|x_n\rangle = A\sum_n e^{i\left(\frac{2\pi}{\lambda}x_n - 2\pi ft\right)}\langle x_3|x_n\rangle = Ae^{i\left(\frac{2\pi}{\lambda}x_3 - 2\pi ft\right)}$$

We then take the complex square:

$$\left|\langle x_3|\psi(x_n,t)\rangle\right|^2 = \left(Ae^{i\left(\frac{2\pi}{\lambda}x_3-2\pi ft\right)}\right)^*\left(Ae^{i\left(\frac{2\pi}{\lambda}x_3-2\pi ft\right)}\right)$$

$$= A^2e^{-i\left(\frac{2\pi}{\lambda}x_3-2\pi ft\right)}Ae^{i\left(\frac{2\pi}{\lambda}x_3-2\pi ft\right)} = A^2$$

This is a striking result. The probability of finding the particle in a given segment, x_3, is A^2, *no matter what x_3 I pick. Furthermore, that probability does not change with time.* The particle is equally likely to be found anywhere along the 'direction of flight' at any moment. This doesn't sound very much like a particle moving along at constant speed.

As always with quantum mechanics, we need to stop and think carefully about what we're doing. Our classical instincts are no longer a reliable guide. We need to analyze what we mean by a casual statement such as 'the electron is moving at constant speed in the x direction'. In the macroscopic world, we 'see' things like this all the time, but we need to remember that visual imagery derives from photons bouncing off an object and reaching the eye. That is not going to happen with electrons. The progress of a particle would be inferred from a sequence of measurement reporting its presence *here, now,* and *there, later.*

Carrying out an experiment to measure the position of a particle known to be in the free particle state, $|\psi(x_n,t)\rangle$, will trigger state collapse into one of the position eigenstates, $|x_n\rangle$. After this measurement, the state will evolve according to the $\hat{U}(t)$ operator, albeit in a simple form if the particle continues to be free. However, the state evolving from $|x_n\rangle$ will not be the free particle state $|\psi(x_n,t)\rangle$. There are two connected reasons for this.

Firstly, the act of measurement inevitably involves an interaction with the object being measured. In classical physics, the scale of this interaction is either negligible or controllable, so its impact on the state can be accounted for. In the microworld, no matter how delicate we are, the scale of the interaction is significant and will markedly alter the state. If nothing else, it can bring about state collapse. Furthermore, the scale of the interaction is not known. Imagine bouncing a photon off an electron. Detecting the scattered photon will enable us to measure its energy and direction of travel, hence giving some handle on the electron's position when it scattered the photon. However, it does not tell us how much energy the photon passed onto the electron. To know that we would have to measure the energy of the photon before the impact; a process which would either destroy the photon or change its energy by an amount that was not measured. If nothing else, in this context, the first position measurement would scramble the particle's direction of travel.

Secondly, while $|x_n\rangle$ is a position eigenstate, it is *not* a momentum eigenstate. So, collapsing $|\psi(x_n,t)\rangle$ into a specific $|x_n\rangle$ destroys any information we have about momentum. Position and momentum are said to be *conjugate variables*, indicating that there is no single basis which is formed from eigenstates of both variables. This is an example of an extremely important principle of quantum theory, first described by Heisenberg and now carrying his name: *Heisenberg's uncertainty principle*, which is the subject of Chapter 13.

The free particle state consequently does not represent a particle smoothly moving through space in a classical sense. The quantum equivalent of that situation is more correctly described by a *wave packet*, assembled from a set of free particle states, as discussed in Chapter 15.

7.2.5 Amplitudes, Energy, and Momentum

In the previous section, I indicated that a position eigenstate is not a momentum eigenstate. Given that the free particle state is an expansion over a position basis, it is sensible to ask if it is an eigenstate of some form. The De Broglie equations help us out here.

As:

$$|\psi(x_n,t)\rangle = A\sum_n \exp\left[i\left(\frac{2\pi}{\lambda}x_n - 2\pi ft\right)\right]|x_n\rangle$$

and:

$$\lambda = \frac{h}{p} \quad \text{and} \quad f = \frac{E}{h}$$

we have:

$$|\psi(x_n,t)\rangle = A\sum_n \exp\left[i\left(\frac{2\pi}{h}px_n - \frac{2\pi}{h}Et\right)\right]|x_n\rangle = A\sum_n \exp\left[i\frac{2\pi}{h}(px_n - Et)\right]|x_n\rangle$$

The combination $h/2\pi$ comes up frequently in quantum theory, so it has been given its own abbreviation, $\hbar = h/2\pi$, which tidies things up slightly:

$$|\psi(x_n,t)\rangle = A\sum_n \exp\left[\frac{i}{\hbar}(px_n - Et)\right]|x_n\rangle$$

In this form, we can see that the state contains a single momentum, p, and energy, E. The latter is hardly surprising given the former, as being a free particle, the only energy must be kinetic and K.E. $= p^2/2m$ in non-relativistic physics. The free particle state is an eigenstate of both momentum and energy, hence its importance. In Chapter 12 we will use the free particle momentum eigenstate to extract the mathematical form of the momentum operator.

7.3 WHERE NEXT?

The free-particle amplitude has a much wider range of application than you might think, given that it refers to a particle which is not being acted on by any force. In Chapter 12 we see how it can be generalized to cater for a continuous range of positions. Free particles will come up again when we consider a few applications of quantum theory to more real-life situations in Chapter 15. It will also feature in the context of quantum field theory, where it acts as a building block for more complicated situations.

In Chapter 13 we're going to discuss the uncertainty principle, which is probably the most famous result to come out of quantum theory. There we will see how being a momentum eigenstate prevents the free-particle state from describing a particle localized to a region of space.

NOTES

1. Strictly what we should say is that *energy* is flowing through the wave medium at speed v. It is the nature of a wave that no 'material' medium has a net movement. Consequently, the waves that lap up on a beach are not really waves in this sense. The tidal motion does cause a net flow of matter (water). A better example is the ripples that spread out when a stone is dropped into a still lake.
2. Not that we can do that for a light wave…
3. Potential confusion warning! On previous complex plane figures, I was using x and iy to signify the real and complex axes. Here I want to use x to mean a physical direction in space, as that is a more familiar usage. It's all the same really. I could have used y and iz on the previous complex planes, but I wanted to use z to stand for a specific complex number. These are the sorts of tangles that authors occasionally get into. Have some sympathy for us, and just go with the flow.

8 Identical Particles

8.1 SOME OPENING THOUGHTS

An atom consists of a nucleus, containing protons and neutrons, with various electrons swarming around the outskirts. The electrons are held in the vicinity of the nucleus as they are negatively charged and hence attracted by the positively charged protons.

Although atoms are complicated objects, there are certain situations where they can be treated as simple lumps of matter. Their internal structure has no relevance to the details of what is happening. This sort of situation is exemplified by experiments in which atoms are deliberately smashed into one another at comparatively low energy and as such provide an ideal start in a study of the quantum physics of identical particles.

8.2 PARTICLE DODGEMS

Experimental physics is a very careful and structured occupation. Many fine details need to be sorted if you're going to cajole nature into revealing her secrets. Even after the experiment is completed, there's generally a long period of data processing required to extract a clear understanding.

For our purposes, such details are not relevant. They may be fascinating in themselves, but they distract from the physics that we need to draw from specific experiments. For this reason, we have been deliberately sketchy about how experiments are carried out in practice.

In order to experiment with collisions, you need to produce two beams of atoms which are moving at tightly controlled speeds and then aim them at one another so that a significant proportion of the atoms from one beam impact those in the other. You also need a way of detecting the atoms after they have collided, coupled with a means of identifying which atom has ended up at which detector.

Such an experiment is illustrated, in outline, in Figure 8.1 where A and B represent sources for generating beams of atoms. The two beams collide at X and the atoms scatter off one another to be detected at either 1 or 2 where the type of atom arriving can also be recorded.

While running this experiment, we find that no matter how well we control the parameters involved, it is impossible to predict which of the atoms ends up where. If the atoms are of different types (say A contains hydrogen and B contains helium), then we can record which one went where once they have arrived, but we cannot predict that *this* hydrogen produced at A is *bound* to end up at 1, or similar. This is not simply due to a lack of control over the energy and direction of flight of the atoms. We are, once again, facing the random nature inherent in some quantum events.

We can, however, measure the *probability* that an atom produced at A will end up at 1, and the probability that it will end up at 2 (Figure 8.2).[1] This is simply a case of running the experiment with a large number of atoms and counting how many arrive at 1 and 2. The probabilities are then[2]:

$$PA1 = \frac{\text{number of } A \text{ atoms ariving at 1}}{\text{total number of collisions}}$$

$$PA2 = \frac{\text{number of } A \text{ atoms ariving at 2}}{\text{total number of collisions}}$$

$$PB1 = \frac{\text{number of } B \text{ atoms ariving at 1}}{\text{total number of collisions}}$$

$$PB2 = \frac{\text{number of } B \text{ atoms ariving at 2}}{\text{total number of collisions}}$$

DOI: 10.1201/9781003225997-10

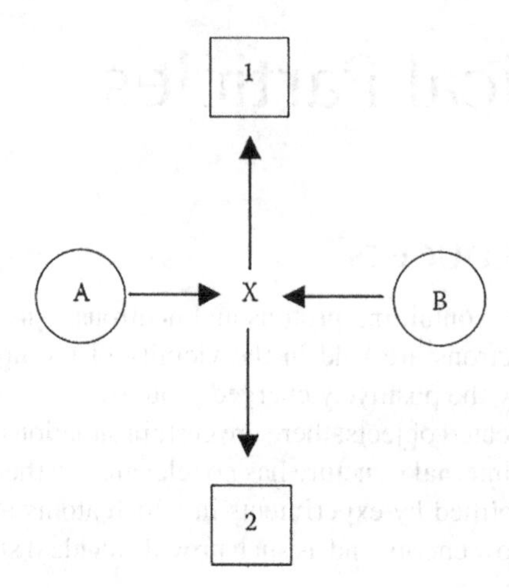

FIGURE 8.1 An experiment to collide atoms. *A* and *B* produce atoms that fly toward X where they scatter off one another to be collected at 1 or 2.

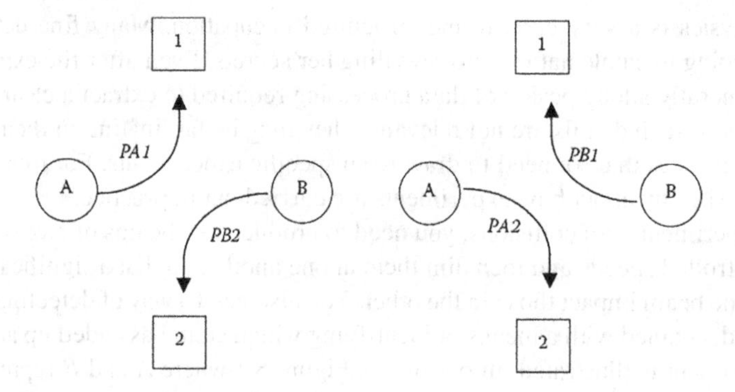

FIGURE 8.2 There are various possibilities for how the atoms will travel from source to collector.

Note that in these calculations, we have not specified that atoms in beam *A* have to be of a different type to those in beam *B*.

Let's say that we run this experiment but choose not to worry about *which* atom ends up at 1 and which one at 2. We just want to measure the probability that one of the atoms arrives at 1 while the other[3] gets to 2.

As a result of this experiment, and after a suitable period of data massage, we have a measure of the probability of getting an atom at 1 and an atom at 2, Prob(1, 2).

The next step would be to check that the experimental measurement of Prob(1, 2), agrees with any calculated probability based on sensible assumptions. In this case, we need a relatively simple application of probability theory.

Frequently we have to combine probabilities to calculate an overall likelihood. We have discussed this in the context of having one distinct event OR another in Section 3.3: you add the probabilities when it is event 1 OR[4] 2. There is also a rule for combining probabilities in the case of having one event AND another if the events are distinct. For example, we may want to know the chance of Cambridge winning the Boat Race AND the horse *Galloping Lump* winning the Grand National.[5] The general rule for independent events is that you *multiply* probabilities when you want events 1 AND 2 to happen. In this experiment, we're interested in one atom arriving at 1 and another at 2. It might be that *A* has ended up at 1 $(A \rightarrow 1)$ while *B* has gone to 2, $(B \rightarrow 2)$. The probability for this would be (*P*A1 AND *P*B2) = $PA1 \times PB2$.

However, it could be that $A \rightarrow 2$ and $B \rightarrow 1$, which we also need to consider. The probability for this would be $PA2 \times PB1$.

As we're not bothered about which atom ends up where, we have to combine these separate results to get the overall probability:

$$\text{Prob}(1, 2) = (A \rightarrow 1 \text{ AND } B \rightarrow 2) \text{ OR } (A \rightarrow 2 \text{ AND } B \rightarrow 1)$$

$$= PA1 \times PB2 + PA2 \times PB1$$

If we are to make any further progress, we need values for $PA1$, $PB2$, etc.

Now A and B are simply labels that we have set up for the atom sources. Provided that we've constructed everything well enough, i.e., that the sources and detectors are functioning equally effectively and that they are symmetrically placed in the experimental configuration, A and B should run identically. Hence the experiment would work just as well *if we changed the labels around*, calling "A" "B" and vice versa. That being the case, $PA1 = PB1$ and $PA2 = PB2$. So, we can write:

$$PA1 \times PB2 = PA2 \times PB1 = \text{P}$$

$$\text{Hence, Prob}(1, 2) = PA1 \times PB2 + PA2 \times PB1 = 2\text{P}$$

This is the result that we would expect as our measured probability, as we have deliberately chosen to ignore any information that tells us what type of atom ends where, and so we don't know from which source a detected atom started.

It would be perfectly possible to run the experiment so that A and B were producing the same type of atom (say, both hydrogen). In this case, the information at 1 and 2 would be of no use. Finding a hydrogen atom at 1 does not tell us if it came from A or B. Would this affect our probabilities?

Classically the answer is clearly "no." If the atoms are different, we can *potentially* tell which source they started from (hydrogen is A, helium is B) whereas if both of them are the same, we don't have this ability. However, if we choose not to record what type of atom arrives at a detector, the two experiments appear to be in the same boat. Surely Prob(1, 2) *must be the same in both cases*?

The answer, of course, is to run the experiment and find out. Setting it so that hydrogen atoms[6] come from both A and B we find, experimentally, that:

$$\text{Prob}(1, 2)_{\text{identical atoms}} = 4\text{P}$$

A measured probability that is *twice the calculated value* and *twice the measured value for an experiment with distinct atoms*, where we chose not to use the data available on which ended up where. Not being able to tell, *even in principle*, which atom is which has made a *fundamental and measurable difference to the result*. The inescapable conclusion of this experimental data is that identical particles (atoms in this case) in the quantum realm bring new physics. Their behaviour does not correspond to classical probability theory, but that does not mean that we can't figure out what is happening.

8.2.1 Scattering Amplitudes

Our discussion so far has been couched entirely in terms of probabilities and experimental data. It will be illuminating to see how our rules for combining quantum amplitudes deal with this situation.

Starting with the case of distinguishable atoms, the possibilities are shown in Figure 8.3, which echoes Figure 8.2.

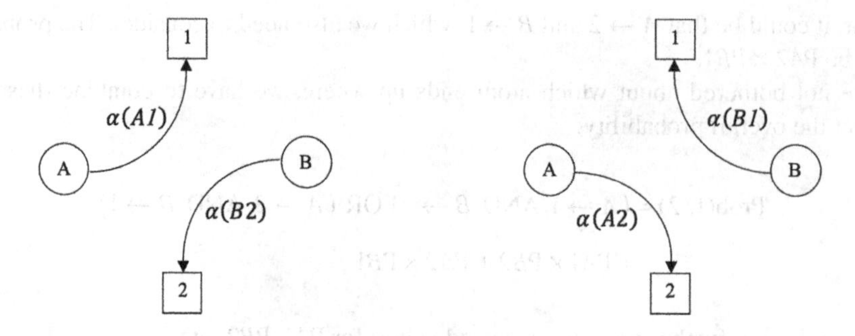

FIGURE 8.3 Scattering amplitudes for atoms moving from sources A and B to detectors 1 and 2. The term $\alpha(A1)$, for example, is the amplitude for an atom from source A to end up at detector 1.

Developing our earlier thinking:

$$\text{Probability } (A \rightarrow 1) = \text{P}A1 = |\text{Amp}(A1)|^2 = |\alpha(A1)|^2$$

$$\text{Probability } (B \rightarrow 2) = \text{P}B2 = |\text{Amp}(B2)|^2 = |\alpha(B2)|^2$$

$$\text{Probability } (A \rightarrow 2) = \text{P}A2 = |\text{Amp}(A2)|^2 = |\alpha(A2)|^2$$

$$\text{Probability } (B \rightarrow 1) = \text{P}B1 = |\text{Amp}(B1)|^2 = |\alpha(B1)|^2$$

where I have introduced α as a symbol for the amplitude, in order to make the subsequent equations a little more readable.

Previously we argued that P$A1$ = P$B1$ and P$A2$ = P$B2$ because A and B were simply labels, hence changing their attribution around could not alter the underlying physics. That being the case, we infer that:

$$\alpha(A1) = \alpha(B1) \Rightarrow \alpha^*(A1) = \alpha^*(B1)$$
$$\alpha(A2) = \alpha(B2) \Rightarrow \alpha^*(A2) = \alpha^*(B2)$$

Given that the cases illustrated by the left-hand and right-hand sides of Figure 8.3 are distinguishable, we deploy clauses 1 and 2a of Important Rule 3:

$$\begin{aligned}\text{Prob}_{\text{distinguishable}}(1, 2) &= \text{Prob}[(A \rightarrow 1 \text{ AND } B \rightarrow 2) \text{ OR } (A \rightarrow 2 \text{ AND } B \rightarrow 1)] \\ &= |\alpha(A1)\alpha(B2)|^2 + |\alpha(A2)\alpha(B1)|^2 \\ &= \alpha(A1)\alpha(B2)\alpha^*(A1)\alpha^*(B2) + \alpha(A2)\alpha(B1)\alpha^*(A2)\alpha^*(B1) \\ &= |\alpha(A1)|^2|\alpha(B2)|^2 + |\alpha(A2)|^2|\alpha(B1)|^2 \\ &= \text{P}A1 \times \text{P}B2 + \text{P}A2 \times \text{P}B1\end{aligned}$$

as expected.

In our earlier calculation we also made the connection:

$$\text{P}A1 \times \text{P}B2 = \text{P}A2 \times \text{P}B1 = P$$

so that $\text{Prob}_{\text{distinguishable}}(1, 2) = 2P$. Working backwards:

$$PA1 \times PB2 = PA2 \times PB1 = P$$

$$\therefore |\alpha(A1)|^2 |\alpha(B2)|^2 = |\alpha(A2)|^2 |\alpha(B1)|^2 = P$$

$$\therefore \alpha(A1)\alpha(B2)\alpha^*(A1)\alpha^*(B2) = \alpha(A2)\alpha(B1)\alpha^*(A2)\alpha^*(B1) = P$$

which we will need shortly.

Switching to the case of identical atoms, the left-hand and right-hand sides of Figure 8.3 are now indistinguishable, which means that we have to add the amplitudes for the different 'paths' first before we complex square to get the probability (clause 2b of IR3):

$$\begin{aligned}\text{Prob}_{\text{identical}}(1, 2) &= |\alpha(A1)\alpha(B2) + \alpha(A2)\alpha(B1)|^2 \\ &= (\alpha(A1)\alpha(B2) + \alpha(A2)\alpha(B1))(\alpha^*(A1)\alpha^*(B2) + \alpha^*(A2)\alpha^*(B1))\end{aligned}$$

The algebra is now a little tedious, but worth the effort:

$$\begin{aligned}\text{Prob}_{\text{identical}}(1, 2) &= \alpha(A1)\alpha(B2)\alpha^*(A1)\alpha^*(B2) + \alpha(A1)\alpha(B2)\alpha^*(A2)\alpha^*(B1) \\ &\quad + \alpha(A2)\alpha(B1)\alpha^*(A1)\alpha^*(B2) + \alpha(A2)\alpha(B1)\alpha^*(A2)\alpha^*(B1) \\ &= P + \alpha(A1)\alpha(B2)\alpha^*(A2)\alpha^*(B1) + \alpha(A2)\alpha(B1)\alpha^*(A1)\alpha^*(B2) + P \\ &= 2P + \alpha(A1)\alpha(B2)\alpha^*(B2)\alpha^*(A1) + \alpha(A2)\alpha(B1)\alpha^*(B1)\alpha^*(A2) \\ &= 4P\end{aligned}$$

where I have used $\alpha(A1) = \alpha(B1)$ and $\alpha(A2) = \alpha(B2)$ in the last but one step.

Clearly, despite the result being something of a surprise, the machinery of quantum theory is well up to the task of dealing with scattering identical atoms. This also illustrates how we should extend the interpretation of 'indistinguishable alternatives' from two or more paths through an experiment for the same quantum object to include *different paths for indistinguishable particles.*

Our atomic measurements are hinting at a very important quantum approach: *if we can't tell the difference between two situations, then they are not really two situations. They are just one event.*

The left-hand side of Figure 8.4 shows the classical version of the experiment with identical atoms. If an atom ends up at 1, this could be due to one of two separate events. Either ($A \rightarrow 1, B \rightarrow 2$) or ($A \rightarrow 2, B \rightarrow 1$) will do the job. The quantum world of identical particles works differently. Atoms from A and B go into the mix and then emerge heading towards 1 and 2 (right-hand side of Figure 8.4). They lose their identity in the middle of the experiment. This has further profound implications for the physics of identical bosons, which we will take up in Chapter 9.

8.2.2 The Moral of the Story

When we are dealing with atoms (and the particles from which they are made), there is something more to being 'identical' than we can appreciate with larger-scale objects. Atoms of the same type are *really* identical, measurably identical if you like, which can make a considerable difference in some physically important situations.

Clearly, quantum theory has to take into account the nature of identical particles. Much of the rest of this chapter contains a discussion of the quantum physics of identical particles, but first, we

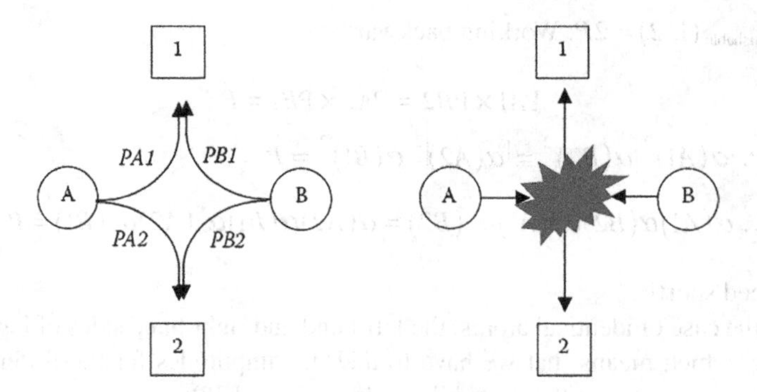

FIGURE 8.4 In a classical world (where snooker balls live) we would think about the chance of getting an atom to 1 as well as another to 2 like the left-hand side figure—even if the atoms were of the same type; atoms, though, do not live in this world. The experimental results tell us that the probability is rather different—as if A going to 1 and B going to 1 were the same event.

need to see how quantum theory deals with states that contain more than one particle, identical or otherwise.

8.3 STATES OF MORE THAN ONE PARTICLE

In most real-world situations, the systems that quantum theory will have to describe contain more than one particle. So far, we have only discussed single-particle states. Adding even one more particle to the mix can produce significant complications.

As a starting point, consider a very simple 'universe' (a toy universe if you like) in which particles, such as electrons, can only exist in one of five different boxes (Figure 8.5). According to quantum theory, there's a separate amplitude for the particle to be in each of the boxes. This is the equivalent of a position amplitude in the real world.

We can write these amplitudes as a sequence of complex numbers: A_1, A_2, A_3, A_4, A_5. The probability of finding an electron in box i is then $|A_i|^2 = A_i^* \times A_i$ according to our standard scheme.

Now we introduce a second particle into the toy universe, for example, a neutron so that there is no issue with distinguishing the particles. In this case, the probability of a particle being in a

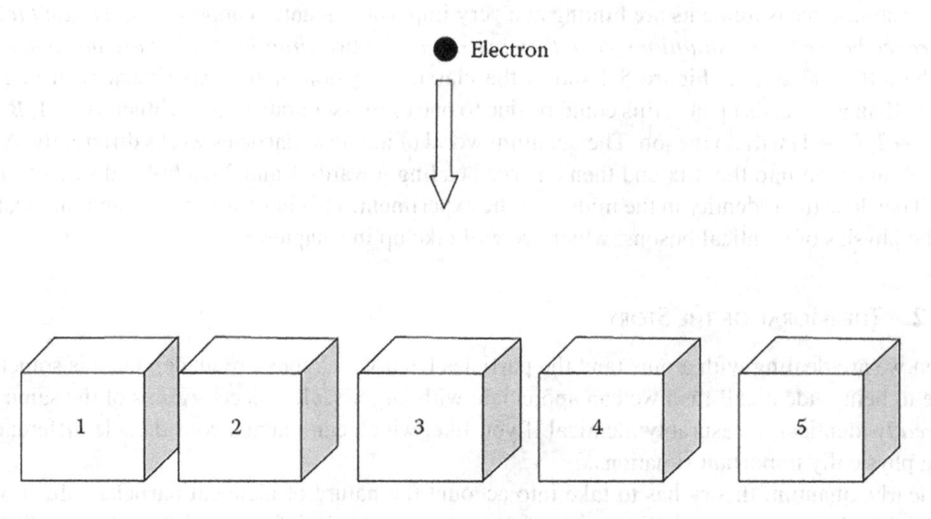

FIGURE 8.5 In our toy universe an electron is constrained to sit in one of five boxes.

box is not affected by the presence of the other particle anywhere in our toy universe. There is no electrostatic repulsion between them, for example. The location of the neutron will be controlled by another set of amplitudes B_1, B_2, B_3, B_4, B_5. Consequently, the amplitude for finding an electron AND a neutron in box 1 would be A_1B_1, that for finding an electron in box 1 AND a neutron in box 3 would be A_1B_3, etc.

The state of the electron can be written as:

$$|\varphi\rangle = A_1|e_1\rangle + A_2|e_2\rangle + \cdots + A_5|e_5\rangle$$

by expanding it over a basis $\{|e_i\rangle\}$ representing states in which the electron is found in box 1, box 2, or box n.

I can use a similar basis to expand the state of the neutron:

$$|\psi\rangle = B_1|n_1\rangle + B|n_2\rangle + \cdots + B_5|n_5\rangle$$

using the basis $\{|n_i\rangle\}$ which is *not* the same as $\{|e_i\rangle\}$. A state such as $|n_i\rangle$ indicates that a neutron has been found at the given position in space, whereas $|e_i\rangle$ is finding an electron. The basis states are *particle states*, not general *positions* in space.

The simplest way to construct a state for the combination of particles is to multiply the separate states:

$$\begin{aligned}|\Psi\rangle = |\psi\rangle|\varphi\rangle &= \left(B_1|n_1\rangle + B_2|n_2\rangle + \cdots + B_5|n_5\rangle\right)\left(A_1|e_1\rangle + A_2|e_2\rangle + \cdots + A_5|e_5\rangle\right) \\ &= B_1A_1|n_1\rangle|e_1\rangle + B_1A_2|n_1\rangle|e_2\rangle + \cdots + B_2A_3|n_2\rangle|e_3\rangle + \cdots + B_5A_5|n_5\rangle|e_5\rangle\end{aligned}$$

with 25 terms in the expansion. Note that the basis here is the product of the two separate bases sets. A term such as $|n_i\rangle|e_j\rangle$ indicates that the neutron is in box i and the electron in box j. All such possibilities together form the basis.

If the particles are allowed/able to interact with one another, we have to be rather more careful. It might be that $|\varphi\rangle$ describes the state of the first particle *in the absence of the second,* and $|\psi\rangle$ describes the state of the second *in the absence of the first,* but we cannot rely on $|\varphi\rangle$ and $|\psi\rangle$ working when they *both* are present. That being the case, $|\psi\rangle|\varphi\rangle$ will not represent the state for both particles. It's far more likely that we will need to employ a new state $|\Phi\rangle$ where:

$$|\Phi\rangle = C_{11}|n_1\rangle|e_1\rangle + C_{21}|n_2\rangle|e_1\rangle + \cdots + C_{13}|n_1\rangle|e_3\rangle + \cdots + C_{55}|n_5\rangle|e_5\rangle$$

with C_{11} being the amplitude for the first particle to be in box 1 and the second also in box 1; C_{13} has the first in box 1 and the second in box 3, etc. Importantly, $C_{13} \neq B_1A_3$, etc., or in other words:

$$|\Phi\rangle \neq |\varphi\rangle \times |\psi\rangle$$

$|\Phi\rangle$ is a genuine multiparticle state that cannot be factored into a product of separate single particle states. States like this produce some interesting physics.

Authentic multiparticle states such as $|\varphi\rangle$ contain more information than single-particle states or even proto-multiparticle states such as $|\Psi\rangle$. To specify state $|\psi\rangle$ for a particle on its own, we needed five amplitudes (as there are five boxes in our toy universe). The same was true for $|\varphi\rangle$. When we constructed the combined state $|\Psi\rangle = |\psi\rangle|\varphi\rangle$ the 25 amplitudes we needed were products of the five from $|\psi\rangle$ and the five from $|\varphi\rangle$, *so there was no new information added*. The 'information count', if you like, is still 10. However, in the interacting multiparticle state, the 25 amplitudes cannot be constructed from the separate particle amplitudes ($C_{13} \neq B_1 A_3$, etc.) We have picked up 15 more 'information counts', which is a clue to the richness of the physics contained in these states.

8.3.1 Identical Particles

In Section 8.2.1, we demonstrated the consequences of experimenting with identical quantum particles. It is completely impossible to tell them apart, which gives rise to some surprising experimental outcomes. Specifically, in that case, the measured probability for finding a particle in each detector was twice the classically expected value. This is an experimentally verified confirmation that identical particles in the quantum realm are far 'more identical' than classical particles. In the classical case, we can always distinguish objects no matter how careful we are to make them identical. No manufacturing process can produce a sequence of objects that survive microscopic examination without finding some distinguishing feature. At the quantum level, two electrons, for example, are identical in every measurable aspect, so classical expectations do not apply. Incidentally, this profound identity of particles arises from their underlying nature as aspects of a quantum field, which we will explore in Chapter 32.

Given what we have seen with experiments on identical atoms, we might expect that our multiparticle state $|\Phi\rangle$ will be different if the two particles that it represents are identical.

Firstly, the basis can no longer be of the form $\{|n_i\rangle|e_j\rangle\}$ as that implies the ability to tell one from the other. Instead, we have to use a basis $\{|i,j\rangle\}$ which is not composed of products of single-particle basis states.

Secondly, the identity of the particles places an extra constraint on the amplitudes. In the expansion:

$$|\Phi\rangle = C_{11}|1,1\rangle + C_{12}|1,2\rangle + C_{13}|1,3\rangle \cdots + C_{21}|2,1\rangle + C_{31}|3,1\rangle + \cdots + C_{55}|5,5\rangle$$

C_{13} must be linked with C_{31}, for example. After all, the particles' identity means that we are unable to distinguish between {particle 1 in box 1 with particle 2 in box 3} and {particle 2 in box 1 with particle 1 in box 3}. The probability can only be {a particle in box 1 AND another in box 3}. As a result (and generalizing):

$$|C_{ij}|^2 = C_{ij}C_{ij}^* = C_{ji}C_{ji}^* = |C_{ji}|^2$$

with the implication that either

$$C_{ij} = C_{ji} \Rightarrow C_{ij}^* = C_{ji}^*$$

or

$$C_{ij} = -C_{ji} \Rightarrow C_{ij}^* = -C_{ji}^*$$

giving us two possible expansions:

$$|\Phi\rangle_S = C_{11}|1,1\rangle + C_{12}|1,2\rangle + C_{13}|1,3\rangle \cdots + C_{12}|2,1\rangle + C_{13}|3,1\rangle + \cdots + C_{55}|5,5\rangle$$

or

$$|\Phi\rangle_A = C_{11}|1,1\rangle + C_{12}|1,2\rangle + C_{13}|1,3\rangle \cdots - C_{12}|2,1\rangle - C_{13}|3,1\rangle + \cdots + C_{55}|5,5\rangle$$

But this is slightly too hasty in the second case. If we're saying that $C_{ij} = -C_{ji}$, then if we apply this rule consistently for *all* i and j including cases where $i = j$, we have to conclude that $C_{ii} = C_{jj} = 0$. After all, how can $C_{44} = -C_{44}$ (for example) unless $C_{44} = 0$. This simple conclusion leads to some amazing physical consequences, which we will come to shortly.

Writing out the expansions for $|\Phi\rangle_S$ and $|\Phi\rangle_A$ in full is a rather lengthy process, but we can compress the notation somewhat by writing the amplitudes as an array. For example, displaying the original state $|\Phi\rangle$ as an array:

$$|\Phi\rangle = \left\{ \begin{matrix} C_{11} & C_{12} & C_{13} & C_{14} & C_{15} \\ C_{21} & C_{22} & C_{23} & C_{24} & C_{25} \\ C_{31} & C_{32} & C_{33} & C_{34} & C_{35} \\ C_{41} & C_{42} & C_{43} & C_{44} & C_{45} \\ C_{51} & C_{52} & C_{53} & C_{54} & C_{55} \end{matrix} \right\}$$

To be clear, I am still implying that the state is expanded over a basis with these amplitudes. I am just not writing down each term with the basis in place, only the amplitudes.

The array for $|\Phi\rangle_S$ is:

$$|\Phi\rangle_S = \left\{ \begin{matrix} C_{11} & C_{12} & C_{13} & C_{14} & C_{15} \\ \underline{C_{12}} & C_{22} & C_{23} & C_{24} & C_{25} \\ \underline{C_{13}} & \underline{C_{23}} & C_{33} & C_{34} & C_{35} \\ \underline{C_{14}} & \underline{C_{24}} & \underline{C_{34}} & C_{44} & C_{45} \\ \underline{C_{15}} & \underline{C_{25}} & \underline{C_{35}} & \underline{C_{45}} & C_{55} \end{matrix} \right\}$$

where I have underlined the terms that I have set equal to others, so the pattern is more visible. That for $|\Phi\rangle_A$ is:

$$|\Phi\rangle_A = \left\{ \begin{matrix} 0 & C_{12} & C_{13} & C_{14} & C_{15} \\ -C_{12} & 0 & C_{23} & C_{24} & C_{25} \\ -C_{13} & -C_{23} & 0 & C_{34} & C_{35} \\ -C_{14} & -C_{24} & -C_{34} & 0 & C_{45} \\ -C_{15} & -C_{25} & -C_{35} & -C_{45} & 0 \end{matrix} \right\}$$

There are no underlines in this case, as the minus signs achieve the same job.

With the amplitudes set out in this form, we can more easily show the key properties of $|\Phi\rangle_S$ and $|\Phi\rangle_A$, and hence explain why the states have the subscripts S and A.

As these states have been built on the premise that we can't distinguish the two particles involved, it is sensible to consider what happens if we exchange the labels on the particles, so that 1 becomes 2 and vice versa. In the case of our array for $|\Phi\rangle_S$ this means:

$$|\Phi\rangle_S' = \text{Exchange}_{1\leftrightarrow 2}\left[|\Phi\rangle_S\right] = \left\{ \begin{matrix} C_{11} & C_{21} & C_{31} & C_{41} & C_{51} \\ \underline{C_{21}} & C_{22} & C_{32} & C_{42} & C_{52} \\ \underline{C_{31}} & \underline{C_{32}} & C_{33} & C_{43} & C_{53} \\ \underline{C_{41}} & \underline{C_{42}} & \underline{C_{43}} & C_{44} & C_{54} \\ \underline{C_{51}} & \underline{C_{52}} & \underline{C_{53}} & \underline{C_{54}} & C_{55} \end{matrix} \right\}$$

but, as $C_{ij} = C_{ji}$:

$$|\Phi\rangle_S' = \text{Exchange}_{1\leftrightarrow 2}\left[|\Phi\rangle_S\right] = \begin{Bmatrix} C_{11} & C_{21} & C_{31} & C_{41} & C_{51} \\ C_{21} & C_{22} & C_{32} & C_{42} & C_{52} \\ C_{31} & C_{32} & C_{33} & C_{43} & C_{53} \\ C_{41} & C_{42} & C_{43} & C_{44} & C_{54} \\ C_{51} & C_{52} & C_{53} & C_{54} & C_{55} \end{Bmatrix}$$

$$= \begin{Bmatrix} C_{11} & C_{12} & C_{13} & C_{14} & C_{15} \\ C_{12} & C_{22} & C_{23} & C_{24} & C_{25} \\ C_{13} & C_{23} & C_{33} & C_{34} & C_{35} \\ C_{14} & C_{24} & C_{34} & C_{44} & C_{45} \\ C_{15} & C_{25} & C_{35} & C_{45} & C_{55} \end{Bmatrix}$$

$$= |\Phi\rangle_S$$

In summary, $|\Phi\rangle_S' = \text{Exchange}_{1\leftrightarrow 2}\left[|\Phi\rangle_S\right] = |\Phi\rangle_S$, which explains the subscript S, which stands for *symmetric*. The state is unchanged (symmetric) by the exchange of the two particles.

On the other hand:

$$|\Phi\rangle_A' = \text{Exchange}_{1\leftrightarrow 2}\left[|\Phi\rangle_A\right] = \begin{Bmatrix} 0 & C_{21} & C_{31} & C_{41} & C_{51} \\ -C_{21} & 0 & C_{32} & C_{42} & C_{52} \\ -C_{31} & -C_{32} & 0 & C_{43} & C_{53} \\ -C_{41} & -C_{42} & -C_{43} & 0 & C_{54} \\ -C_{51} & -C_{52} & -C_{53} & -C_{54} & 0 \end{Bmatrix}$$

Now we apply $C_{ij} = -C_{ji}$ to get:

$$|\Phi\rangle_A' = \begin{Bmatrix} 0 & -C_{12} & -C_{13} & -C_{14} & -C_{15} \\ -(-C_{12}) & 0 & -C_{23} & -C_{24} & -C_{25} \\ -(-C_{31}) & -(-C_{32}) & 0 & -C_{34} & -C_{35} \\ -(-C_{41}) & -(-C_{42}) & -(-C_{43}) & 0 & -C_{45} \\ -(-C_{51}) & -(-C_{52}) & -(-C_{53}) & -(-C_{54}) & 0 \end{Bmatrix}$$

$$= \begin{Bmatrix} 0 & -C_{12} & -C_{13} & -C_{14} & -C_{15} \\ C_{12} & 0 & -C_{23} & -C_{24} & -C_{25} \\ C_{13} & C_{23} & 0 & -C_{34} & -C_{35} \\ C_{14} & C_{24} & C_{34} & 0 & -C_{45} \\ C_{15} & C_{25} & C_{35} & C_{45} & 0 \end{Bmatrix}$$

$$= -|\Phi\rangle_A$$

Or, in summary $|\Phi\rangle_A' = \text{Exchange}_{1\leftrightarrow 2}\left[|\Phi\rangle_A\right] = -|\Phi\rangle_A$ which is indicated by the A subscript for *antisymmetric*.

All of this manipulation has come from a seemingly innocent observation. With a set of two identical particles, the probability for particle 1 being in box i and particle 2 being in box j must be $|C_{ij}|^2 = C_{ij}C_{ij}^* = C_{ji}C_{ji}^* = |C_{ji}|^2$, which can be true if either $C_{ij} = C_{ji}$ (symmetric) or $C_{ij} = -C_{ji}$ (antisymmetric).

Astonishingly this has led to *two physically distinguishable situations*, for if our combination of identical particles is in an antisymmetric state then we can *never find both particles in the same box* $(C_{ii} = C_{jj} = 0)$.

This remarkable observation is found to hold in nature for a whole class of particles called *fermions* (the electron is a prime example of this breed).

Particles whose states are symmetric under a switch in their labels are called *bosons* (the photon is a boson). There will be a lot more to say about the differences between fermions and bosons as we progress.

8.3.2 States in Real World

Moving from our toy universe to the real world will entail switching from a finite number of positions 'in boxes' to a continuous range of possible positions. Consequently, the states $|\varphi\rangle$ and $|\psi\rangle$ will be[7]:

$$|\varphi\rangle_1 = \sum_{x_1} \varphi(x_1)|x_1\rangle$$

$$|\psi_2\rangle = \sum_{x_2} \psi(x_2)|x_2\rangle$$

if we choose to expand them over a position basis.[8] Of course, it's always possible that particle 1 is in state $|\psi\rangle$ and particle 2 in state $|\varphi\rangle$, so we need to keep the other two possibilities in mind:

$$|\varphi_2\rangle = \sum_{x_2} \varphi(x_2)|x_2\rangle$$

$$|\psi_1\rangle = \sum_{x_1} \psi(x_1)|x_1\rangle$$

If we try to construct a multiparticle state from these single-particle states, we have the following six possibilities to consider:

$$|\Phi\rangle = |\varphi_1\rangle|\psi_2\rangle$$

$$|\Psi\rangle = |\varphi_2\rangle|\psi_1\rangle$$

$$|\Phi'\rangle = |\varphi_1\rangle|\varphi_2\rangle$$

$$|\Psi'\rangle = |\psi_1\rangle|\psi_2\rangle$$

$$|\Omega\rangle = \frac{1}{\sqrt{2}}\left(|\varphi_1\rangle|\psi_2\rangle + |\varphi_2\rangle|\psi_1\rangle\right)$$

$$|\mathrm{X}\rangle = \frac{1}{\sqrt{2}}\left(|\varphi_1\rangle|\psi_2\rangle - |\varphi_2\rangle|\psi_1\rangle\right)$$

In the case of the last two combinations, the factors of $1/\sqrt{2}$ ensure that the overall state is properly normalized, e.g.:

$$
\begin{aligned}
\langle X|X\rangle &= \left[\frac{1}{\sqrt{2}}\left(\langle\varphi_1|\langle\psi_2| - \langle\varphi_2|\langle\psi_1|\right)\right]\left[\frac{1}{\sqrt{2}}\left(|\varphi_1\rangle|\psi_2\rangle - |\varphi_2\rangle|\psi_1\rangle\right)\right] \\
&= \frac{1}{2}\left[\langle\varphi_1|\varphi_1\rangle\langle\psi_2|\psi_2\rangle - \cancel{\langle\varphi_1|\varphi_2\rangle\langle\psi_2|\psi_1\rangle} - \cancel{\langle\varphi_2|\varphi_1\rangle\langle\psi_1|\psi_2\rangle} + \langle\varphi_2|\varphi_2\rangle\langle\psi_1|\psi_1\rangle\right] \\
&= \frac{1}{2}\left[\langle\varphi_1|\varphi_1\rangle\langle\psi_2|\psi_2\rangle + \langle\varphi_2|\varphi_2\rangle\langle\psi_1|\psi_1\rangle\right] \\
&= \frac{1}{2}[1+1] = 1
\end{aligned}
$$

as required.

If the two particles involved are distinct (e.g., an electron and a neutron), then each of these multiparticle combinations is in contention, unless it is physically impossible for them to be in the same state, which would rule out $|\Phi'\rangle$ and $|\Psi'\rangle$. However, if they are identical *bosons,* we can't be sure which is in state $|\psi_1\rangle$ and which in $|\varphi\rangle$. In this case, only $|\Phi'\rangle$, $|\Psi'\rangle$ and $|\Omega\rangle$ are viable *as they are the only ones that are symmetrical under a switch between 1 and 2*:

$$|\Phi\rangle_E = \text{Exchange}_{1\leftrightarrow 2}\left[|\varphi_1\rangle|\psi_2\rangle\right] = |\varphi_2\rangle|\psi_1\rangle \neq |\Phi\rangle$$

$$|\Psi\rangle_E = \text{Exchange}_{1\leftrightarrow 2}\left[|\varphi_2\rangle|\psi_1\rangle\right] = |\varphi_1\rangle|\psi_2\rangle \neq |\Psi\rangle$$

$$|\Phi'\rangle_E = \text{Exchange}_{1\leftrightarrow 2}\left[|\varphi_1\rangle|\varphi_2\rangle\right] = |\varphi_2\rangle|\varphi_1\rangle = |\Phi'\rangle$$

$$|\Psi'\rangle_E = \text{Exchange}_{1\leftrightarrow 2}\left[|\psi_1\rangle|\psi_2\rangle\right] = |\psi_2\rangle|\psi_1\rangle = |\psi'\rangle$$

$$|\Omega\rangle_E = \text{Exchange}_{1\leftrightarrow 2}\left[\frac{1}{\sqrt{2}}\left(|\varphi_1\rangle|\psi_2\rangle + |\varphi_2\rangle|\psi_1\rangle\right)\right] = \frac{1}{\sqrt{2}}\left(|\varphi_2\rangle|\psi_1\rangle + |\varphi_1\rangle|\psi_2\rangle\right) = |\Omega\rangle$$

$$|X\rangle_E = \text{Exchange}_{1\leftrightarrow 2}\left[\frac{1}{\sqrt{2}}\left(|\varphi_1\rangle|\psi_2\rangle - |\varphi_2\rangle|\psi_1\rangle\right)\right] = \frac{1}{\sqrt{2}}\left(|\varphi_2\rangle|\psi_1\rangle - |\varphi_1\rangle|\psi_2\rangle\right) = -|X\rangle$$

Note that $|\Phi\rangle$ and $|\Psi\rangle$ are neither symmetric nor antisymmetric, as the state changes completely under $1 \leftrightarrow 2$ exchange.

Having established the possible boson states, it is also clear that $|X\rangle$ is the appropriate state for a pair of identical fermions, as it is antisymmetric under the $1 \leftrightarrow 2$ exchange.

Developing this line of thought further, if the two individual states happen to be the same, $|\varphi\rangle = |\psi\rangle$ then combined fermion state $|X\rangle = 0$. Once again, we see that identical fermions can't be in the same state. However, there is an even stronger constraint than this. If we take the time to expand out $|X\rangle$ using our position bases, we find:

$$
\begin{aligned}
|X\rangle &= \frac{1}{\sqrt{2}}\left(|\varphi_1\rangle|\psi_2\rangle - |\varphi_2\rangle|\psi_1\rangle\right) \\
&= \frac{1}{\sqrt{2}}\left(\sum_{x_1}\varphi(x_1)|x_1\rangle\sum_{x_2}\psi(x_2)|x_2\rangle - \sum_{x_2}\varphi(x_2)|x_2\rangle\sum_{x_1}\psi(x_1)|x_1\rangle\right)
\end{aligned}
$$

which, with all those summations around, looks distinctly messy. However, the key insight comes when we consider the two particles being at the same point in space, so that $x_1 = x_2 = x$:

$$|X\rangle = \frac{1}{\sqrt{2}}\left(\sum_x \varphi(x)|x\rangle \sum_x \psi(x)|x\rangle - \sum_x \varphi(x)|x\rangle \sum_x \psi(x)|x\rangle\right) = 0$$

Clearly the result we found with the model universe, that two identical fermions can't be found in the same box, carries forward into a more realistic scenario. This opens up all sorts of significant questions, not the least of which is how do we picture the mechanism that prevents the two from being in the same place? We will pick this thought up in Section 8.4.

8.3.3 Overall States

There is a way in which the fermions can co-occupy the same location, as the spatial amplitude is not the only aspect of the complete state. Electrons, which are fermions, have *SG* states, which need to be taken into account as well. The imposition of antisymmetry, which we have established using spatial amplitudes, derives from the fundamental nature of the appropriate quantum field (see Chapter 32). In that context, it becomes apparent that the constraint applies to the fully expressed state as a whole. In other words, the product of the spatial and *SG* aspects of the state has to be antisymmetric. So, *it is possible to have a symmetric spatial aspect with an antisymmetric SG state*, and vice versa:

$$|X\rangle_A = \left[\frac{1}{\sqrt{2}}\left(|\varphi_1\rangle|\psi_2\rangle + |\varphi_2\rangle|\psi_1\rangle\right)\right]\left[\frac{1}{\sqrt{2}}\left(|U_1\rangle|D_2\rangle - |U_2\rangle|D_1\rangle\right)\right]$$

or

$$|X\rangle_{A'} = \left[\frac{1}{\sqrt{2}}\left(|\varphi_1\rangle|\psi_2\rangle - |\varphi_2\rangle|\psi_1\rangle\right)\right]\left[\frac{1}{\sqrt{2}}\left(|U_1\rangle|D_2\rangle + |U_2\rangle|D_1\rangle\right)\right]$$

In either case, switching the particles affects the *SG* states as well, so that the state is still antisymmetric overall although the space-dependent part happens to be symmetric in the case of $|X\rangle_A$, and so the particles can be in the same place.

Another point about combined states of this form, whether symmetric antisymmetric or otherwise, is that they are *mixed states* (see Section 5.2.2); we can't definitively say that a given particle is in one or the other of the single states.

8.3.4 More Than Two Particles

Things get rapidly more complicated if we're dealing with more than two particles. Sticking with the fermion case, it is evident that in a three-particle collection there must be at least three states to choose from. After all, if three electrons had only two states to work with, then inevitably two of them would have to be in the same state, so the overall combination would vanish. A three-particle antisymmetric state looks something like:

$$|\Phi\rangle_A = \frac{1}{\sqrt{6}}\left[|\phi_1\rangle|\varphi_2\rangle|\psi_3\rangle + |\psi_1\rangle|\phi_2\rangle|\varphi_3\rangle + |\varphi_1\rangle|\psi_2\rangle|\phi_3\rangle - |\psi_1\rangle|\varphi_2\rangle|\phi_3\rangle - |\varphi_1\rangle|\phi_2\rangle|\psi_3\rangle - |\phi_1\rangle|\psi_2\rangle|\varphi_3\rangle\right].$$

So, you can see how intricate things can become.

8.3.5 More General States

Each of the multiparticle states quoted so far has relied on the separate particle states, applicable for a particle in isolation, still being valid in the multiparticle situation. In general, this won't work, although it's always a good starting place if you're trying to construct an approximation to the true state.

The most general way of constructing a multiparticle state, in this form of quantum theory,[9] is:

$$|\Phi\rangle_A = \sum_{x_1, x_2, \ldots, x_n} \psi_A(x_1, x_2, \ldots, x_n)|x_1\rangle|x_2\rangle \ldots |x_n\rangle.$$

or

$$|\Phi\rangle_S = \sum_{x_1, x_2, \ldots, x_n} \psi_S(x_1, x_2, \ldots, x_n)|x_1\rangle|x_2\rangle \ldots |x_n\rangle$$

here ψ_A and ψ_S are antisymmetrical and symmetrical amplitude functions respectively (incorporating any *SG* states).

A symmetrical amplitude function remains the same (is invariant) under an exchange of any two particles in the list, i.e.:

$$\psi_S' = \text{Exchange}_{\text{any two particles}}\left[\psi_S\right] = \psi_S$$

With an antisymmetrical function, it depends on how many switches you carry out:

$$\psi_A' = \text{Exchange}_{\text{any two particles, m times}}\left[\psi_A\right] = (-1)^m \psi_A$$

Note that if we switch particles 1 and 12 in a n-particle antisymmetric function $\psi_A(x_1, x_2, \ldots, x_{12}, \ldots x_n) \rightarrow \psi_A(x_{12}, x_2, \ldots, x_1, \ldots x_n) = -\psi_A(x_1, x_2, \ldots, x_{12}, \ldots x_n)$ and then try to make $x_1 = x_{12} = x$ we get $\psi_A(x, x_2, \ldots, x, \ldots x_n) = -\psi_A(x, x_2, \ldots, x, \ldots x_n)$ which can only be true if $\psi_A = 0$. Once again, we see that the structure of the overall state prevents two identical fermions having exactly the same state properties.

8.3.6 A More Elegant Approach

Throughout the whole of this chapter so far, we have been engaged in cobbling together multiparticle quantum theory from single-particle states. This is one reason why we end up dealing with intricate combinations when trying to construct states. Perhaps a different approach is required.

Quantum field theory is, at least in part, a genuine attempt to take multiparticle theory in a different direction, especially when dealing with identical particles. Even in the aforementioned states $|\Phi\rangle_A$ and $|\Phi\rangle_S$, we're not taking the identity of these particles completely seriously. After all we have retained the labels 1, 2,…, n in a position basis, as if we could label the particles themselves:

$$|\Phi\rangle_A = \sum_{x_1, x_2, \ldots, x_n} \psi_A(x_1, x_2, \ldots, x_n)|x_1\rangle|x_2\rangle \ldots |x_n\rangle$$

$$|\Phi\rangle_S = \sum_{x_1, x_2, \ldots, x_n} \psi_S(x_1, x_2, \ldots, x_n)|x_1\rangle|x_2\rangle \ldots |x_n\rangle$$

As we've seen from our atomic experiments at the start of this chapter, any procedure to label particles (such as which source they come from) is bound to fail once they can mix with one another. In groups, identical particles lose their individuality. In quantum field theory (Chapter 32), an entirely new way of constructing states (called *Fock space)* is used, which removes any particle labels and gives us a significant insight into the nature of wave/particle duality.

8.4 FINAL THOUGHTS

In no particular order of importance:

- It's in the nature of fermions to always be in an antisymmetric state. Likewise, it's in the nature of bosons to be in a symmetric state. This leads to some very different physical consequences for the two classes of particles.
- The difference between fermions and bosons exploits a mathematical 'loophole' in the relationship between amplitudes and probabilities (Section 8.3.1). We could cite this as an example of a quantum 'rule of thumb' which historical experience lends plausibility to: *anything goes unless it's forbidden.* Nature might have chosen to only recognize $C_{ij} = C_{ji}$, but in the absence of a specific law of nature that forbids the $C_{ij} = -C_{ji}$ possibility, both have been exploited. The consequences are profound as we shall see, but to give something of a hint of what it to come, very roughly fermions give rise to the macroscopic experience of matter and bosons give rise to classical fields.
- Given that the physical differences between fermions and bosons arise from different properties of their amplitudes, one might be tempted to construct an argument that the amplitudes are in some objective sense 'real', especially in light of the next comment.
- One of the most intriguing consequences of fermion state antisymmetry is the impact that it can have on the spatial location of the particles. If they are in a symmetric *SG* state, then the overall antisymmetry means that they can't occupy the same location in space.[10] As we are trained by experience and earlier education to think in classical terms, we try and visualize some force acting to keep the particles apart. Unfortunately, this does not work for various reasons, not the least of which that there is no obvious force that can do the job, but more profoundly, the effect seems to be independent of how spatially spread the particle amplitudes appear to be. The identical fermions simply avoid one another like polite passers-by on the street. In the end, one arrives at the rather Zen-like answer that the fundamental antisymmetry of the state is keeping them apart, something that can only be expressed in its mathematical structure. This is another aspect of the contextuality and non-local nature of quantum theory. Extraordinarily this turns out to have important physical consequences, for example, in white dwarf stars (Chapter 15).

NOTES

1. Strictly speaking, you can't measure a probability – you can count occurrences and then estimate probabilities. However, it's a convenient phrase to use.
2. In the following equations, total number of collisions refers to the number that have been detected, see also note 3 below.
3. This is not as trivial as one might think [e.g., it's not 100%] as the atoms might fly off at some angle and miss 1 or 2, but that's an experimental detail.
4. It is worth noting that in probability calculations "OR" is always taken to mean that either 1 happens, or 2, or both.
5. The Grand National is a famous UK horse race that takes place in April. If you don't know what the Boat Race is, then we have nothing more to say to each other…
6. A note that relates to later in this book. This experiment is using ${}^{1}_{1}\mathrm{H}$ atoms, which are *bosons* and hence not subject to the exclusion principle that applies to *fermions* such as ${}^{2}_{1}\mathrm{H}$.

7 With the range of possible positions being continuous, I should strictly use an integral to express this state. However, I am trying to keep the argument as simple as possible. Think of this as being an expansion over a very great number of discrete positions lined up one after each other: a great number of tiny boxes in a row, if you like.
8 Of course, this is still not the 'real' real world, as I am only expanding the states over the x axis. However, that does not detract from the argument.
9 Quantum field theory uses a different formalism which is actually more suited to dealing with multiple particle states. This will be discussed in Chapter 32.
10 Classically, material objects can't occupy the same position in space no matter what. Here though, we are in principle dealing with point particles – objects that appear to have no material dimension. Furthermore, we are using the shorthand 'cannot occupy the same position in space' to mean 'the amplitude for them to be both in the same spatial location is zero'. An actual location is only manifest when a measurement is carried out.

9 Scattering Identical Bosons

9.1 SCATTERING

Consider a specific experiment[1] in which particles A, B, C are scattered by some force, or collision with another object so that their paths are deflected into a detection apparatus. For example, particle A, with initial state $|A\rangle$, is scattered so that it strikes the sensitive surface of the detector at a point (x_1, y_1), using some convenient coordinate system to demark the surface. We also assume that the detector is arranged so that it presents an area at 90° to the line of flight.

In practice, however, it is very unlikely that we will be able to detect *exactly* where the particle arrives. Equally, as we will discuss further in Chapter 12, the probability of a particle arriving at a *precise* point is effectively zero.[2] More correctly, on both grounds, we should record the impact point as being within some small region of area $\mathcal{A}_1$ centred on the point (x_1, y_1). So, we specify $|1\rangle$ to be the state of a particle arriving at the detector but normalized in such a way as $|\langle 1|A\rangle|^2$ is the probability of the particle arriving *within a unit area of the point* (x_1, y_1). We then multiply by $\mathcal{A}_1$ to get the probability that the particle will arrive in the region of interest:

$$\text{Prob}(\text{single particle}) = |\langle 1|A\rangle|^2 \mathcal{A}_1$$

(the area $\mathcal{A}_1$ is small enough for the probability not to vary significantly across its bounds).

The setup is illustrated in Figure 9.1. Particles A, B, and C have been separately scattered by some force or collision (in Figure 9.1 this is indicated by the small explosions) so that they fly into a detector (the large oval). The particles hit specific regions (1, 2, 3) with areas $(\mathcal{A}_1, \mathcal{A}_2, \mathcal{A}_3)$ respectively.

If the particles in this experiment are distinguishable from one another, and they act independently, the probability of detecting all three particles is the product of their separate probabilities (this is the standard rule for combining probabilities):

$$\begin{aligned}\text{Prob}(\text{three particles}) &= |\langle 1|A\rangle|^2 |\langle 2|B\rangle|^2 |\langle 3|C\rangle|^2 \mathcal{A}_1\mathcal{A}_2\mathcal{A}_3 \\ &= |\langle 1|A\rangle\langle 2|B\rangle\langle 3|C\rangle|^2 \mathcal{A}_1\mathcal{A}_2\mathcal{A}_3\end{aligned}$$

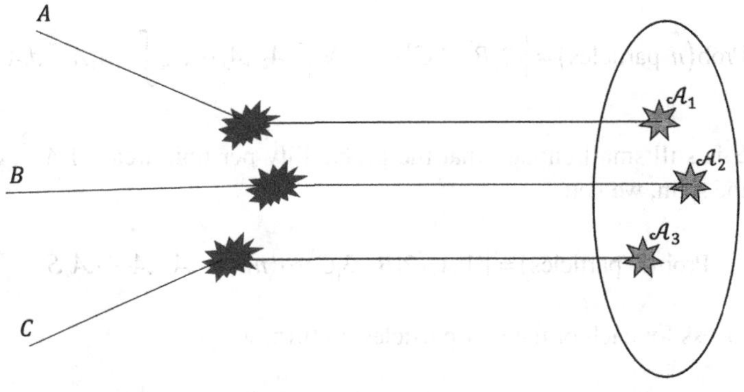

FIGURE 9.1 Scattering particles into a detector. A, B, and C have been deflected from their original paths into a particle detector, illustrated by the oval shape in the diagram. Regions $\mathcal{A}_1, \mathcal{A}_2, \mathcal{A}_3$ represent the areas of the sensitive detector surface where the particles make contact.

DOI: 10.1201/9781003225997-11

Equally, if we had n particles scattering into the detector, the probability would end up as:

$$\text{Prob}(n \text{ particles}) = \left|\langle 1|A\rangle\langle 2|B\rangle\langle 3|C\rangle\cdots\langle n|N\rangle\right|^2 \mathcal{A}_1\mathcal{A}_2\ \mathcal{A}_3\cdots\mathcal{A}_n$$

However, this is not quite the end of the calculation. In truth, we are not that interested in which region of the detector the particles hit, as long as they impact *somewhere* on its surface. The best way to adapt our equation is to generalize it so that the points 1, 2, 3, ..., N can be anywhere on the surface of the detector.

If we allow impacts on various regions on the detection surface $(X, Y, Z, P, Q, \text{etc.})$, then particle A can be scattered to $|1\rangle$ at X or Y or Z, etc. As these are *indistinguishable* events, we *add* the probabilities together[3] (Important Rule 3).

Using $\mathcal{A}_1(X)$, $\mathcal{A}_1(Y)$, $\mathcal{A}_1(Z)$ etc., as the different areas that point 1 could occupy on the whole surface, we get:

$$\begin{aligned}\text{Prob}(n \text{ particles}) &= \left|\langle 1|A\rangle\langle 2|B\rangle\langle 3|C\rangle\ldots\langle n|N\rangle\right|^2 \mathcal{A}_1(X)\mathcal{A}_2\ \mathcal{A}_3\cdots\mathcal{A}_n\\ &+\left|\langle 1|A\rangle\langle 2|B\rangle\langle 3|C\rangle\cdots\langle n|N\rangle\right|^2 \mathcal{A}_1(Y)\mathcal{A}_2\ \mathcal{A}_3\cdots\mathcal{A}_n\\ &+\left|\langle 1|A\rangle\langle 2|B\rangle\langle 3|C\rangle\cdots\langle n|N\rangle\right|^2 \mathcal{A}_1(Z)\mathcal{A}_2\ \mathcal{A}_3\cdots\mathcal{A}_n+\cdots\end{aligned}$$

We can pull out of the sum all the parts that do not change from term to term, which gives us:

$$\begin{aligned}\text{Prob}(n \text{ particles}) &= \left|\langle 2|B\rangle\langle 3|C\rangle\cdots\langle n|N\rangle\right|^2 \mathcal{A}_2\ \mathcal{A}_3\cdots\mathcal{A}_n\\ &\left[\left|\langle 1|A\rangle\right|^2 \mathcal{A}_1(X)+\left|\langle 1|A\rangle\right|^2 \mathcal{A}_1(Y)+\left|\langle 1|A\rangle\right|^2 \mathcal{A}_1(Z)+\cdots\right]\end{aligned}$$

Allowing each of the $\mathcal{A}_i$ to reduce in size to $\delta\mathcal{A}_i$, and increasing the number of them to compensate, this becomes:

$$\text{Prob}(n \text{ particles}) = \left|\langle 2|B\rangle\langle 3|C\rangle\cdots\langle n|N\rangle\right|^2 \mathcal{A}_2\ \mathcal{A}_3\cdots\mathcal{A}_n\sum_i\left|\langle 1|A\rangle\right|^2 \delta\mathcal{A}_i$$

In essence, we are integrating over the whole surface, $\mathcal{S}$, of the detector:

$$\text{Prob}(n \text{ particles}) = \left|\langle 2|B\rangle\langle 3|C\rangle\cdots\langle n|N\rangle\right|^2 \mathcal{A}_2\ \mathcal{A}_3\cdots\mathcal{A}_n\int_{\mathcal{S}}\left|\langle 1|A\rangle\right|^2 d\mathcal{A}$$

Assuming that $\mathcal{S}$ is still small enough that the probability per unit area, $\left|\langle 1|A\rangle\right|^2$, can be taken as constant across the span, we get:

$$\text{Prob}(n \text{ particles}) = \left|\langle 1|A\rangle\langle 2|B\rangle\langle 3|C\rangle\cdots\langle n|N\rangle\right|^2 \mathcal{A}_2\ \mathcal{A}_3\cdots\mathcal{A}_n\mathcal{S}$$

Repeating the process for each of the ther particles, in turn, we get:

$$\text{Prob}(n \text{ particles}) = \left|\langle 1|A\rangle\langle 2|B\rangle\langle 3|C\rangle\cdots\langle n|N\rangle\right|^2 \mathcal{S}^n \tag{9.1}$$

as each integration produces a factor of $\mathcal{S}$, and there will be n integrations in total.

This is the final result that we need.

9.2 THE SAME, BUT DIFFERENT: IDENTICAL PARTICLES

So far, so good. We've calculated the probability that n particles will be scattered into the same detector, provided we can distinguish between the particles. Now we have to see what happens to the overall probability when we are dealing with *identical* particles. We'll take the boson case as that leads to some interesting physics, and we don't have to worry about the antisymmetry of states.

We have already seen an experiment similar to this at the start of Chapter 8. Taking up the key point from that example and applying it to this case, we must realize that the events, particle A scattering to 1 and particle B scattering to 1, can't be distinguished from one another if the particles are identical. Consequently, the more complicated events $(A \rightarrow 1, B \rightarrow 2, C \rightarrow 3)$ and $(B \rightarrow 1, A \rightarrow 2, C \rightarrow 3)$ can't be distinguished either. Hence, following Important Rule 3, we have to add the *amplitudes* first before we complex square to get the probability.

Starting with just three particles, so that we can get the hang of things, we end up with:

$$\text{Prob}(3 \text{ identical bosons}) = \left| \langle 1|A\rangle\langle 2|B\rangle\langle 3|C\rangle + \langle 1|B\rangle\langle 2|A\rangle\langle 3|C\rangle + \langle 1|C\rangle\langle 2|B\rangle\langle 3|A\rangle \right.$$
$$\left. + \langle 1|A\rangle\langle 2|C\rangle\langle 3|B\rangle + \langle 1|C\rangle\langle 2|A\rangle\langle 3|B\rangle + \langle 1|B\rangle\langle 2|C\rangle\langle 3|A\rangle \right|^2 \mathcal{A}_1\mathcal{A}_2\,\mathcal{A}_3$$

If you carefully pick your way through this expression, you will see that I have added all the different combinations of scattering amplitudes built by switching A, B, C around in the six different possible ways.

Note that I have also gone back to specifying precise points 1, 2, and 3 where the particles arrive: one thing at a time.

Clearly if I have more than three identical particles, say n of them, life gets considerably more complicated.

In the three-particle case, there are six possibilities. To see this, imagine three holes into which you can drop three marbles (maximum one marble per hole). The first marble can go in any one of the three holes, so there are three possibilities. The second marble has only two holes to choose from, and the third marble has only one hole. So, the total number of different ways in which the marbles can be dropped into the holes is $3 \times 2 \times 1$, which is 6.

The argument extends to n marbles and holes quite easily. The first marble has n holes to choose from, the second has $(n-1)$ holes, the third $(n-2)$ holes, etc. In the end, the $(n-1)$th marble has got only 2 holes to choose from, and the nth marble doesn't have a choice as there's only 1 hole left. The total number of possibilities is:

$$n \times (n-1) \times (n-2) \times (n-3) \times \cdots \times 3 \times 2 \times 1 = n!$$

The symbol $n!$, which is pronounced *n factorial,* is the mathematical shorthand for the product of all numbers up to and including n.

Going back to my scattering calculation, the jump to n identical particles would mean constructing a horrendous expression with $n!$ different terms.

Fortunately, we can simplify things somewhat if we make another, quite reasonable, assumption: the directions that a particle can be scattered into are so similar that the amplitudes are the same, that is, $\langle 1|A\rangle = \langle 2|A\rangle = \langle 3|A\rangle = \cdots \langle n|A\rangle$. This has to be the case if the detection area, S, is relatively small, as otherwise, the particles would miss the detector.

By the same argument, $\langle 1|B\rangle = \langle 2|B\rangle = \langle 3|B\rangle = \cdots \langle n|B\rangle$, etc.

Applying a similar assumption to our three-boson case transforms the probability from:

$$\text{Prob}(3 \text{ identical bosons}) = \left| \langle 1|A\rangle\langle 2|B\rangle\langle 3|C\rangle + \langle 1|B\rangle\langle 2|A\rangle\langle 3|C\rangle + \langle 1|C\rangle\langle 2|B\rangle\langle 3|A\rangle \right.$$
$$\left. + \langle 1|A\rangle\langle 2|C\rangle\langle 3|B\rangle + \langle 1|C\rangle\langle 2|A\rangle\langle 3|B\rangle + \langle 1|B\rangle\langle 2|C\rangle\langle 3|A\rangle \right|^2 \mathcal{A}_1\mathcal{A}_2\mathcal{A}_3$$

to:

$$\text{Prob}(3 \text{ identical bosons}) = \left|\langle 1|A\rangle\langle 2|B\rangle\langle 3|C\rangle + \langle 2|B\rangle\langle 1|A\rangle\langle 3|C\rangle + \langle 3|C\rangle\langle 2|B\rangle\langle 1|A\rangle + \langle 1|A\rangle\langle 3|C\rangle\langle 3|B\rangle + \langle 3|C\rangle\langle 1|A\rangle\langle 2|B\rangle + \langle 2|B\rangle\langle 3|C\rangle\langle 1|A\rangle\right|^2 \mathcal{A}_1\mathcal{A}_2\mathcal{A}_3$$

$$= \left|6\langle 1|A\rangle\langle 2|B\rangle\langle 3|C\rangle\right|^2 \mathcal{A}_1\mathcal{A}_2\mathcal{A}_3$$

To achieve this simplification, I have used the fact that $\langle 1|A\rangle = \langle 2|A\rangle = \langle 3|A\rangle$ to replace every $\langle 2|A\rangle$ and $\langle 3|A\rangle$ with $\langle 1|A\rangle$, every $\langle 1|B\rangle$ and $\langle 3|B\rangle$ with $\langle 2|B\rangle$ and finally every $\langle 1|C\rangle$ and $\langle 2|C\rangle$ with $\langle 3|C\rangle$. In the process, I have made each of the six terms identical.

Applying the same idea to the n -boson case (and without writing out all the terms to start with and then making them equal…) produces:

$$\text{Prob}(3 \text{ identical bosons}) = \left|n!\langle 1|A\rangle\langle 2|B\rangle\langle 3|C\rangle\cdots\langle n|N\rangle\right|^2 \mathcal{A}_1\mathcal{A}_2\ \mathcal{A}_3$$

9.2.1 Using the Whole Detector

At this stage, when we were dealing with *distinguishable* particles, we integrated the various detection regions over the area of the detector, to end up with our $\mathcal{S}^n$ term.

We can do the same thing again here, but now we have to be quite careful. We have broken the link between regions and particles, as we can't tell if A ends up at $\mathcal{A}_1$ or $\mathcal{A}_2$ or at any other spot. If we allow each region to range over the total area separately, *we overestimate things*, as we will be covering the whole area with indistinguishable events, *but more than once*. Bluntly it does not matter if $\mathcal{A}_1$ is sitting on a given part of the detector or $\mathcal{A}_2$ is there instead, it amounts to the same thing with indistinguishable particles.

The way to account for this is to let every spot go wherever it wants, but then divide the result by $n!$. This compensates for the number of equivalent ways that the spots can be arranged.

If we do this, we end up with a formula for the probability that n indistinguishable bosons will be scattered and detected:

$$\text{Prob}(n \text{ identical bosons}) = \frac{\left|n!\langle 1|A\rangle\langle 2|B\rangle\langle 3|C\rangle\cdots\langle n|N\rangle\right|^2 \mathcal{S}^n}{n!}$$

$$= \frac{(n!)^2\left|\langle 1|A\rangle\langle 2|B\rangle\langle 3|C\rangle\cdots\langle n|N\rangle\right|^2 \mathcal{S}^n}{n!} = n!\left|\langle 1|A\rangle\langle 2|B\rangle\langle 3|C\rangle\cdots\langle n|N\rangle\right|^2 \mathcal{S}^n$$

$$= n! \times \text{Prob}(n \text{ particles})$$

comparing with Eq. (9.1) from earlier. So, the probability of scattering n identical bosons is $n!$ times greater than what you would expect for distinguishable particles.

This is a fascinating and very important result, well worth the effort it took to get to this point.

SCATTERING IDENTICAL PARTICLES

$$\text{Prob}(n \text{ identical bosons}) = n! \times\ \text{Prob}(n \text{ non-identical particles})$$

By the way, this is entirely consistent with the calculation from the start of Chapter 8. There we ended up with a probability of $2P$ for the distinguishable case, and $4P$ when the experiment used identical particles. Our new rule would lead us to:

$$\text{Prob}(2 \text{ identical bosons}) = 2! \times \text{Prob}(\text{two distinguishable particles}) = 2! \times 2P = 4P$$

as $2! = 1 \times 2 = 2$.

9.2.2 And Another Way...

There is another, rather illuminating, way of looking at this result. We start by calculating the probability of a particle W scattering into our detector, predicated on another n bosons having already done this.

This is not a completely new calculation. We already have everything we need to get the job done. After all, particle W is just an $(n+1)$th particle to plug into our earlier formula, i.e.:

$$\text{Prob}(n+1 \text{ identical bosons}) = (n+1)! \left| \langle 1|A\rangle\langle 2|B\rangle\langle 3|C\rangle \cdots \langle n|N\rangle\langle w|W\rangle \right|^2 \mathcal{S}^{n+1}$$

Separating all the terms specifically to do with W:

$$\text{Prob}(n+1 \text{ identical bosons}) = \left\{ (n+1) \left|\langle w|W\rangle\right|^2 \mathcal{S} \right\} \times n! \left| \langle 1|A\rangle\langle 2|B\rangle\langle 3|C\rangle \cdots \langle n|N\rangle \right|^2 \mathcal{S}^n$$

$$= \left\{ (n+1) \left|\langle w|W\rangle\right|^2 \mathcal{S} \right\} \times \text{Prob}(n \text{ identical bosons})$$

Now, $\left|\langle w|W\rangle\right|^2 \mathcal{S}$ is simply the probability of a single particle being scattered into the detector, with no prior scatterings, as the label W is a 'dummy' which could stand for any particle. Hence, we can simplify this expression further:

$$\text{Prob}(n+1 \text{ identical bosons}) = (n+1) \times \text{Prob}(1 \text{ boson}) \times \text{Prob}(n \text{ identical bosons})$$

This is a remarkable outcome.

To see how extraordinary it is, contrast this result with that for tossing a fair coin. The chance of the coin coming up heads is 50:50 (or the probability is 0.5) each time the coin is tossed. Although it might be astonishing to have a run of 50 or more heads in a row, *that doesn't alter the chance of the next one coming down heads* (no matter what your instinct might tell you). The probability doesn't change every time you toss the coin. The overall probability of experiencing a sequence of 51 heads in a row is:

$$\text{Prob}(51 \text{ heads in a row}) = 0.5 \times 0.5 \times \cdots 0.5 = (0.5)^{51} = 0.5 \times (0.5)^{50}$$

$$= 0.5 \times \text{Prob}(50 \text{ heads in a row})$$

Rewriting this to make the contrast with the identical boson result more apparent:

$$\text{Prob}(51 \text{ heads in a row}) = \text{Prob}(\text{heads}) \times \text{Prob}(50 \text{ heads in a row})$$

If you think of scattering and detecting a particle as analogous to getting heads when you toss a coin, then you would expect the probability of detecting $(n+1)$ scattered particles to be:

$$\text{Prob}(n+1 \text{ particles}) = \text{Prob}(1 \text{ particle}) \times \text{Prob}(n \text{ particles})$$

which does not work if the particles are identical bosons. If n identical bosons have already been scattered and detected, *the probability of the next one doing the same is increased by a factor of* $(n+1)$:

$$\text{Prob}(n+1 \text{ identical bosons}) = \{(n+1)\times \text{Prob}(1 \text{ boson})\} \times \text{Prob}(n \text{ identical bosons})$$

It's as if the number of heads already tossed *increases* the chance of the next toss being a head:

$$\text{Prob}(1 \text{ more boson joining } n \text{ already present}) = (n+1)\times \text{Prob}(1 \text{ boson on its own})$$

If this result is not sufficiently counter-intuitive as it stands, then consider the following thought: *how does the next particle to scatter know how many others have already been scattered?*

When you cut away all the experimental detail, the fundamental scenario starts with a collection of identical bosons spread across various states and ends up with all of them in the same (new) state. Now, as they're not simply identical but *quantum identical,* we absolutely can't tell the difference between them. *We can't even track their individual 'routes' to the final state.*

Each time we add another boson to those already in the final state, we have to accept that it's going to get confused with the others, even if they have gone 'before'. The more identical bosons that there are making the same trip to the final state, *the greater the number that the next boson can get confused with.* Hence the probability of it doing the same thing is increased.

Although we have derived this result in the specific context of identical bosons scattering into a detector, it is an example of a more widely based rule:

9.3 TRANSITIONS AWAY FROM STATES

Atoms emit photons when one of their electrons changes its energy state. Stating the process like this, sometimes conjures a specific picture: that the atom is a 'bag' containing protons, neutrons, electrons, and photons, and hence when the bag 'bursts' a photon flies out. In truth, the photon is created only at the moment of transition and does not in any sense exist before the electron's energy change.

As photons belong to the boson family, we expect them to be bound by the rules derived in the previous sections. Specifically, if we have a collection of atoms emitting photons into the same state, then the greater the number of photons already in that state, the greater the probability that the next photon will get emitted into the *same* state (Figure 9.2).

In general, the amplitude governing the transition from an initial state to a final state is $\langle \text{final state} | \text{initial state} \rangle$. So, if we start with no photons having been emitted, the initial state is $|0\rangle$ and the amplitude for the first photon to be emitted is $\langle 1|0\rangle$. Without the identical nature of the photons getting in the way, we would expect that the amplitude for a photon to be emitted would be the same no matter how many were knocking about in the system already. In other words, $\langle n+1|n\rangle = \langle 4|3\rangle = \langle 1000|999\rangle$ etc. However, the photons *are* identical bosons, which has consequences.

In the previous section, we worked our way to the rule:

$$\text{Prob}(1 \text{ more boson joining } n \text{ already present}) = (n+1)\times \text{Prob}(1 \text{ boson on its own})$$

IMPORTANT RULE 10: SOCIAL BOSONS

The probability of a boson entering a state already occupied by n bosons of the same type is proportional to $(n+1)$

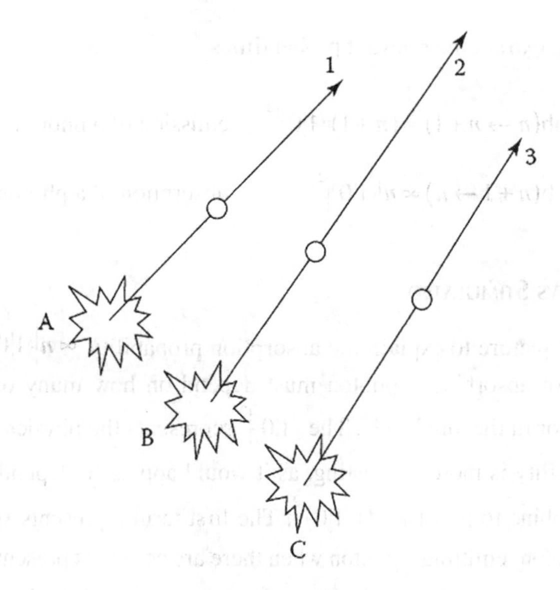

FIGURE 9.2 Atoms emitting photons: the probability of a photon being emitted with a given state depends on the number of photons that are already present in that state.

Applying it to this case gives:

$$\text{Prob}(1 \text{ more photon joining } n \text{ already present}) = (n+1) \times |\langle 1|0\rangle|^2$$

Hence, the amplitude for an atom emitting a photon into a state that already contains n photons is:

$$\text{Amplitude}(1 \text{ more photon joining } n \text{ already present}) = \langle n+1|n\rangle = \sqrt{(n+1)} \times \langle 1|0\rangle$$

The situation is, however, slightly more complicated than this simple analysis would indicate. After all, atoms can absorb photons as well as emit them. The amplitude for an atom absorbing a single photon from a collection of $(n+1)$ photons all in the same state would be $\langle n|n+1\rangle$. We can relate this absorption amplitude to the emission one, as Quite Important Rule 6 connects the two: $\langle \text{final state}|\text{initial state}\rangle = \langle \text{initial state}|\text{final state}\rangle^*$, so that:

$$\langle n|n+1\rangle = \langle n+1|n\rangle^* = \left[\sqrt{(n+1)} \times \langle 1|0\rangle\right]^*$$

Having established this relationship, we can re-jig it into a more conventional form by starting with n photons in the initial state (rather than $n+1$). In which case, the final state after a photon has been absorbed is $|n-1\rangle$. Hence:

$$\langle n-1|n\rangle = \sqrt{n} \times \langle 1|0\rangle^*$$

This switch enables us to write the amplitudes for a photon to be emitted or absorbed by an atom, when there are other photons present in the same state, in a comparable form:

$$\langle n+1|n\rangle = \sqrt{(n+1)} \times \langle 1|0\rangle \qquad \text{emission of a photon}$$

$$\langle n-1|n\rangle = \sqrt{n} \times \langle 1|0\rangle^* \qquad \text{absorption of a photon}$$

From here it is simple to extract the relevant probabilities:

$$\text{Prob}(n \rightarrow n+1) \propto (n+1)\left|\langle 1|0\rangle\right|^2 \quad \text{emission of a photon}$$

$$\text{Prob}(n+1 \rightarrow n) \propto n\left|\langle 1|0\rangle\right|^2 \quad \text{absorption of a photon}$$

9.3.1 Spontaneous vs Stimulated

Constructing a physical picture to explain the absorption probability, $\propto n\left|\langle 1|0\rangle\right|^2$, is not so difficult. After all, the chances of absorbing a photon must depend on how many of them are around to absorb, hence the n factor in the amplitude. The $\left|\langle 1|0\rangle\right|^2$ represents the physics of absorption itself.

The emission probability is more interesting, as it would appear to depend on two terms, $\left|\langle 1|0\rangle\right|^2$ and $n\left|\langle 1|0\rangle\right|^2$, which combine to give $(n+1)\left|\langle 1|0\rangle\right|^2$. The first term represents *spontaneous emission*, which is the same as an atom emitting a photon when there are no others present. The second, $n\left|\langle 1|0\rangle\right|^2$ term corresponds to an atom emitting a photon in the presence of other photons in the same state, a process called *stimulated emission*, which was first suggested by Einstein. This is a direct result of photons being classed as identical bosons, with all the implications that we have been exploring.

9.3.2 Lasers

In modern-day technology, the laser has become commonplace, with applications in Blue-Ray players, bar code readers, eye surgery and a wide range of other situations. In its contemporary applications, the laser takes on a range of different, often solid-state, forms. However, it is interesting to see how the original laser design functioned, and how this crucially depends on the physics of identical bosons.

As the acronym laser (light amplification by stimulated emission of radiation) hints, the physics underlying a laser's operation is the stimulated emission process referred to in the previous section. The first lasers used a cylindrical crystal of ruby surrounded by a high-intensity quartz flash tube (rather like a camera flashgun). Ruby is crystalline aluminium oxide with a 0.05% impurity of chromium, which gives the characteristic red colour. It is the chromium atoms that are central to the laser's function (Figure 9.3).

To trigger the laser, an electrical pulse fires the flash tube, illuminating the ruby with white light. Many of the blue and green wavelength photons in the light are absorbed by the chromium atoms, exciting them into higher-energy states, a process called *optical pumping*. The atoms in these excited states then rapidly lose some of this energy by transferring it into vibrational energy in the crystal (which warms up a little in the process), leaving them in another state with less energy (see Figure 9.4).

When the atoms drop from this level back into their lower-energy states, they emit photons of red light. The first few photons are emitted via the spontaneous process, but they then start to stimulate further emissions via the social boson effect. As the number of photons present in the ruby increases, the probability of further stimulated emissions also rises, and a cascade starts.

To give the photons the greatest chance of stimulating further emissions, the ruby is capped at both ends by mirrors, set at 90° to the long axis of the crystal, which reflect the photons. As the photons pass back and forth along the length of the ruby, they stimulate further emissions into the same state, which includes their energy, direction, polarization, and phase.

The mirror at one end is not completely silvered, so it allows some photons to escape and form the emitted beam. The light emerging from the laser is then composed of almost identical photons, giving the light its unique and useful properties.

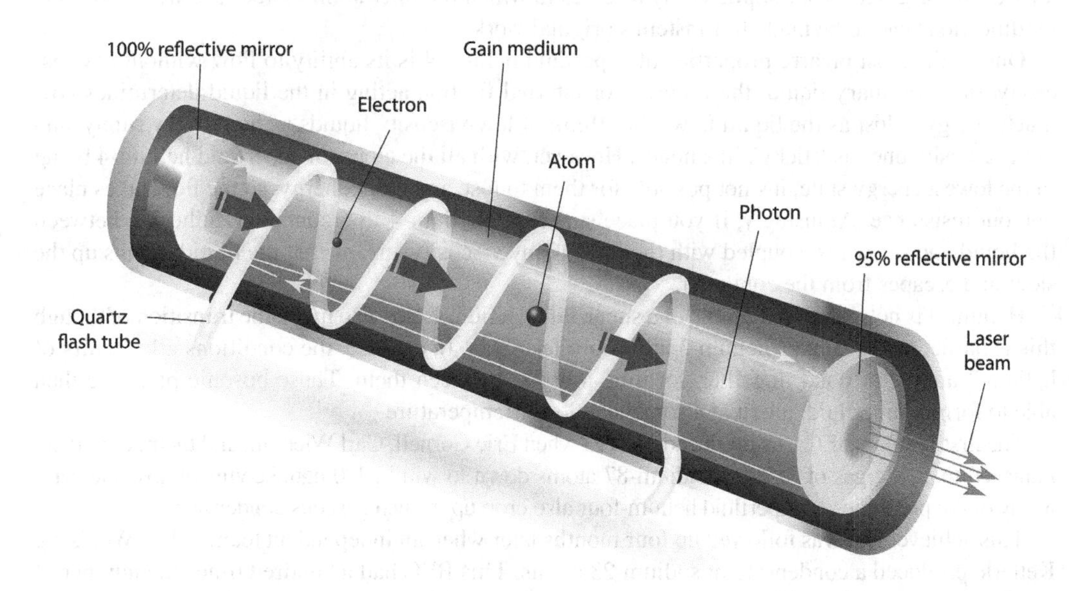

FIGURE 9.3 The basic construction of a ruby laser—cut-away view.

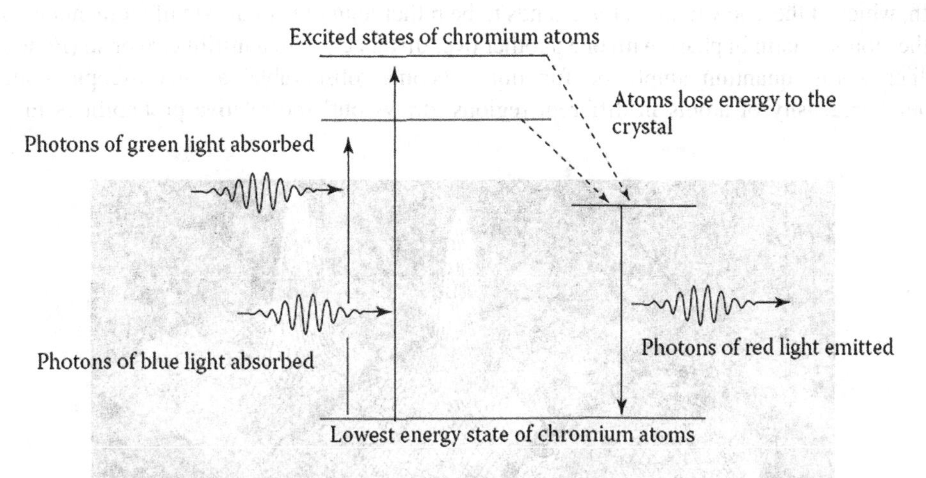

FIGURE 9.4 The process of photon emission in a ruby laser. Photons from the triggering flash are absorbed by atoms, lifting them into an excited state. The atoms then lose some energy, dropping them into a lower state from which they emit photons of red light.

9.4 BOSE–EINSTEIN CONDENSATES

In 1925 Einstein published an article,[4] building on the work of Satyendra Nath Bose, in which he predicted the existence of a new form of matter at low temperature. Under the right circumstances, a gas of bosonic atoms can be cooled to a critical temperature at which all the atoms collapse (condense) into their lowest possible energy state. With such a large number of atoms sharing the same state, quantum effects become visible on a macroscopic scale.

The first Bose–Einstein condensate (BEC) was discovered in 1938 by Pyotr Kapitsa[5] and separately the team of John Allen, and Don Misener.[6] As they cooled helium-4 down to temperatures below –271°, the helium made a sharp transition, becoming a new form of fluid, a *superfluid*, with many strange properties. Although it was soon realized that superfluid helium-4 was an example of a BEC, the original theory applied only to gases in which the interatomic forces are weak, so some modifications had to be made to Einstein's original work.

One of the most bizarre properties of superfluid helium-4 is its ability to flow without any viscosity. In an ordinary liquid, the *viscosity* or internal friction acting in the liquid determines how much energy is lost as the liquid flows. We think of low-viscosity liquids as being very runny, and high-viscosity ones as 'sticky', like honey. However, with all the atoms of superfluid helium-4 being in the lowest energy state, it's not possible for them to lose any more energy, so the flow takes place without resistance. Amusingly, if you place superfluid helium-4 in a container, adhesion between the liquid and the walls, coupled with the lack of any viscosity, means that the liquid climbs up the sides and escapes from the container.

Helium-4 is not the only example of a superfluid. Helium-3 can also make the transition, although this is harder to explain, as helium-3 atoms are fermions. In essence, the conditions allow pairs of helium-3 atoms to 'bond' together, forming a boson between them. These bosonic pairs are then able to form a superfluid, albeit at a somewhat lower temperature.

The first 'proper' BEC was produced in 1995 when Eric Cornell, Carl Wieman, and their co-workers managed to cool a gas of ~2000 rubidium-87 atoms down to within 170 nanokelvins of absolute zero. Many of the properties of superfluid helium-four also crop up in such gaseous condensates.

This achievement was followed up four months later when an independent team led by Wolfgang Ketterle produced a condensate of sodium-23 atoms. This BEC had a hundred times the number of atoms as the Cornell and Wieman version, so Ketterle was able to observe some important new physics. As all the atoms in a BEC share the same quantum state, they have the same de Broglie wavelength, which at these low temperatures tends to be rather long (due to the small momenta involved) and the atoms remain in phase with one another over distances up to a millimetre or so (Figure 9.5). In other words, quantum amplitude functions become 'observable' at macroscopic scales, as the number density of atoms in different regions 'draws out' the relative probabilities involved.

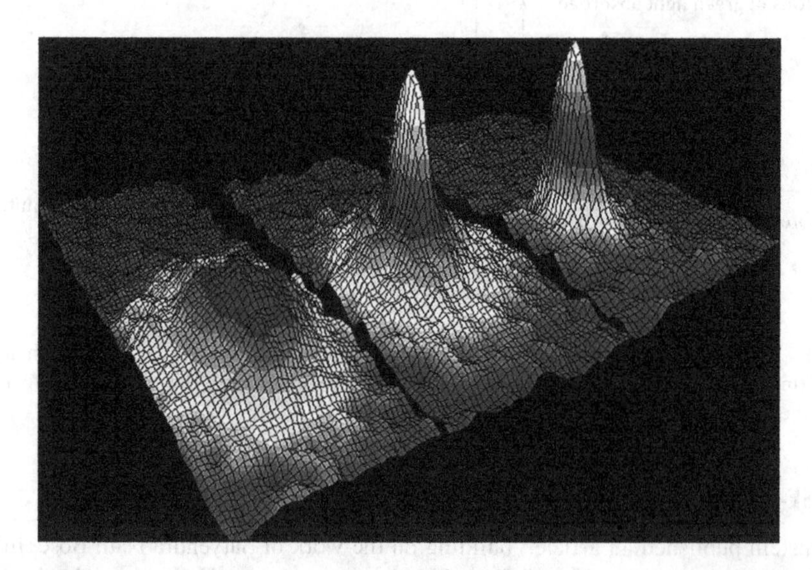

FIGURE 9.5 A 3-dimensional representation of the velocity distribution in a gas of rubidium atoms. The left-hand image shows a broad, flat distribution. The sharp central peak develops in the middle image as the gas is cooled to the point at which a Bose–Einstein condensate appears. This indicates that the broad majority of atoms are now in the same quantum state and hence have the same velocity. The right-hand image shows the distribution after some further atoms have been evaporated off, leaving almost pure condensate. (Image credit: Mike Matthews/NIST/JILA/CU-Boulder.)

These remarkable features allowed Ketterle to observe interference effects between two different condensates. Cornell, Wieman, and Ketterle shared the Nobel Prize in physics in 2001.

Since these initial discoveries, a thriving branch of research has grown up around working with BECs and trying to make them at ever lower temperatures. Many of the features demonstrated by these exotic forms of matter are yet to be fully explained, so theoreticians are being kept interested as well. The lure of observing quantum effects at macroscopic scales is a compelling motivation for this research.

9.4.1 Einstein's Argument

To understand some aspects of Einstein's original argument for the existence of BECs, consider the following simplified approach.

Imagine a collection of four *distinguishable* atoms, each of which has two possible states available, $|0\rangle, |1\rangle$. The overall state of the four atoms, assuming that they are independent of one another, can then come in several different configurations as in Table 9.1:

TABLE 9.1
Configurations of Four Distinguishable Atoms with Two States Available to Them

$|1\rangle|1\rangle|1\rangle|1\rangle$	$|0\rangle|1\rangle|1\rangle|1\rangle$	$|0\rangle|0\rangle|1\rangle|1\rangle$	$|1\rangle|0\rangle|0\rangle|0\rangle$	$|0\rangle|0\rangle|0\rangle|0\rangle$
	$|1\rangle|0\rangle|1\rangle|1\rangle$	$|1\rangle|1\rangle|0\rangle|0\rangle$	$|0\rangle|1\rangle|0\rangle|0\rangle$	
	$|1\rangle|1\rangle|0\rangle|1\rangle$	$|1\rangle|0\rangle|1\rangle|0\rangle$	$|0\rangle|0\rangle|1\rangle|1\rangle$	
	$|1\rangle|1\rangle|1\rangle|0\rangle$	$|0\rangle|1\rangle|1\rangle|0\rangle$	$|1\rangle|1\rangle|1\rangle|0\rangle$	
		$|0\rangle|1\rangle|0\rangle|1\rangle$		
		$|1\rangle|0\rangle|0\rangle|1\rangle$		
1 configuration	4 configurations	6 configurations	4 configurations	1 configuration

If the two states, $|0\rangle$ and $|1\rangle$, are of the same energy, then they should be equally likely to crop up. If we take an ensemble collection of multiple sets of four atoms, then each atomic state should be seen equally often across the ensemble. This is provided that an equilibrium can be established due to atoms bouncing off each other and triggering a change in state. However, were we to observe any one set (gas) of atoms, we would most likely find a situation in which the same number of atoms are in states $|0\rangle$ and $|1\rangle$, simply because there are more configurations of that form (six) than any other. The statistics of the situation tells us that we are, by and large, going to find the two states equally distributed between the atoms.

If we move to a more realistic gas with N atoms, then there are 2^N possible configurations (as each atom has two possible states independent of what the other atoms are doing). The list of configurations with equal numbers of atoms in $|0\rangle$ and $|1\rangle$ will now dominate over all other configurations, so that is how we will find the gas.

The situation is rather different if we try the same thing for a gas of identical bosons. In the case of just four atoms, the configurations are (Table 9.2):

TABLE 9.2
Configurations of Four Indistinguishable Bosons with Two States Available to Them

$|4 \text{ in state } 1\rangle$	$|1 \text{ in state } 0, 3 \text{ in state } 1\rangle$	$|2 \text{ in state } 0, 2 \text{ in state } 1\rangle$	$|3 \text{ in state } 0, 1 \text{ in state } 1\rangle$	$|4 \text{ in state } 0\rangle$
1	1	1	1	1

Once again, each atomic state is equally likely but as there are only five possible configurations, we can't say that we're going to find the gas with equal numbers of atoms in each state. We are just as likely to find them all in the same state.

If we now move to a gas of N identical bosons, things start to get very interesting. For a start there are radically fewer configurations available to the boson gas compared with our distinguishable N atom gas. In fact, just $(N+1)$ of them, which is a lot fewer than 2^N. To see this, think of a configuration with all the atoms in state $|0\rangle$, then change one of the atoms to $|1\rangle$. That's one configuration you have added. Now change another atom to state $|1\rangle$, and you have added another configuration. You can do this N times in total, until all the atoms are in state $|1\rangle$. Of course, you have to remember to add the original configuration $\left(\text{all } |0\rangle\right)$, so that makes $(N+1)$ in total. Even with a very large number of atoms in our boson gas, we are still just as likely to find all of them in the same state as we are to find any other distribution of states.

Now let's change the situation slightly and assume that $|1\rangle$ has a slightly higher energy than $|0\rangle$. This makes the probability of an individual atom being in state $|1\rangle$ slightly *less* than that of finding it in state $|0\rangle$. In the case of our normal gas, this affects things somewhat. A configuration with most of the atoms in state $|1\rangle$ becomes less likely, and that with most of the atoms in state $|0\rangle$ becomes comparatively more likely. However, we will still observe the gas to have roughly equal numbers of atoms in each state, as that is the situation with the largest number of configurations available to it. The number of configurations helps to balance the lower probability of a configuration.

The same balancing act can't rescue a boson gas. In this case, there is only one configuration for each count of states. *Now the most likely outcome, at a low enough temperature, is that we will find all but a few of the atoms in the lowest energy state.*

This, in rough outline, is how a BEC comes about when a gas of bosons is cooled so that there are only a few energy states available.

NOTES

1 Much of this chapter is heavily based on an argument presented by Richard Feynman in his Lectures on Physics, Vol. 3. I devoured The Feynman Lectures on Physics as a teenager, and his unique view of quantum mechanics has been a great influence on my own thinking.

2 In brief, there are an infinite number of exact mathematical points in any finite area, no matter how small. Hence the probability of arriving at one of these points ~ $1/\infty$.

3 Assuming that moving A_1 around does not change the amplitudes.

4 See Section 16.4.4 for a further discussion.

5 Kapitza, P. (1938). Viscosity of liquid helium below the λ-point. *Nature*. 141 (3558): 74.

6 Allen, J. F.; Misener, A. D. (1938). Flow of liquid helium II. *Nature*. 142 (3597): 643.

10 Spin

10.1 FERMIONS, BOSONS, AND STERN–GERLACH MAGNETS

We have some loose ends lying around.

For one thing, I have yet to explain why an electron passing through an SG magnet can take only one of two paths. Up to now, I have simply assumed that the electron has some physical property that determines which path it will take through the SG magnetic field. I have not named nor explained this property. Of course, as this is a quantum mechanical property, we have seen how the behaviour of the electron depends on the context of the experiment as well.

In Chapter 8 we looked at multiparticle states and discovered that states composed of identical particles could be either *symmetrical* under the switching of any two particles, in which case the particles are called *bosons*, or *antisymmetrical* making them *fermions*. However, we have not said what it is that makes a boson behave differently from a fermion.

In this chapter, I am going to start to pull these threads together by introducing a new physical property called *spin*. The process will take us into the next chapter as well, where we are going to look at fermion states in general and how they react to particle exchanges.

The distinguishing factor between a fermion and a boson is the respective spin of the particles. Being a charged particle, the electron's spin gives it a tiny magnetic field, which interacts with the SG field, causing the electrons to be deflected. However, this classical picture has significant limitations, as we will come to appreciate. Fundamentally, spin is a quantum mechanical physical property with no exact classical equivalent. Having said that, we must start somewhere and as good a place as any is with *angular momentum*.

10.2 ANGULAR MOMENTUM

Momentum plays a very important role in classical mechanics:

- The total momentum in the world is *conserved*, meaning that the sum total for all objects is a fixed quantity.
- The change in momentum per second experienced by an object is a measure of the force applied to it.

Angular momentum plays a very similar role in the classical mechanics of rotating objects. It is also a conserved quantity, and its rate of change per second is a measure of the turning force (sometimes called *torque*) applied.

Calculating angular momentum is very straightforward, at least in the case of a point mass moving in a circular path:

Angular momentum[1] $L = r \times mv$ where r is the radius of the path and mv the magnitude of the point masses' linear momentum.

If we have a more complicated situation, such as a spinning top, we imagine that the top is constructed of a collection of point masses and add up the angular momentum of each one to get the angular momentum of the whole object.

Linear momentum is a vector quantity, as it inherits the direction of the velocity involved. Similarly, angular momentum has a vector nature, although given that the linear direction of the

DOI: 10.1201/9781003225997-12

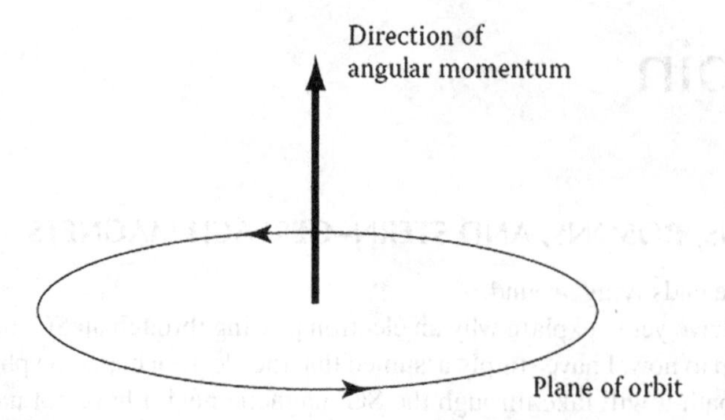

FIGURE 10.1 For a point mass moving in a circular path, the angular momentum is given a direction (vertical arrow) at right angles to the plane of the orbit. The head of the arrow always points away from the plane and toward your eye if you are looking down from above and the orbit is anticlockwise from that perspective.

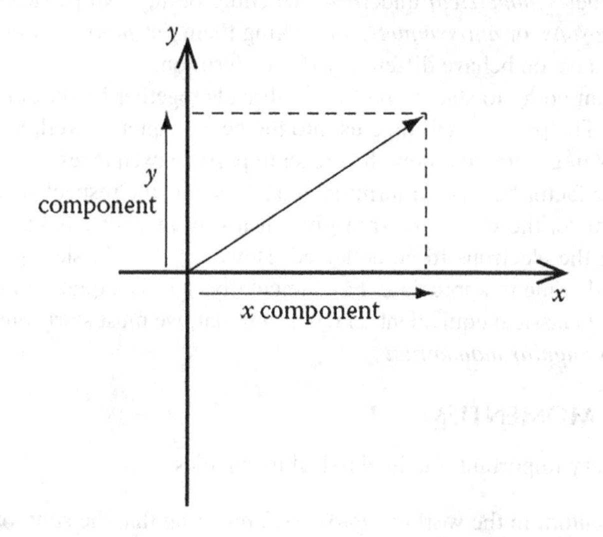

FIGURE 10.2 A angular momentum (large arrow) can be split into x and y components.

object is constantly changing, its direction is not intuitively obvious. The direction of angular momentum is at right angles to the plane of the orbit (see Figure 10.1)[2].

If the angular momentum has direction, then it must have components as with any other directed quantity. This idea is shown in Figure 10.2, in which an angular momentum is placed inside a 2-D reference system. In this situation, you can consider the angular momentum to have x and y components as illustrated.

Most likely, we would be applying angular momentum in a 3-D context, so it would have x, y, and z components:

$$L^2 = L_x^2 + L_y^2 + L_z^2$$

with L being the total angular momentum (symbolized by the length of the thick arrow in Figure 10.2) and $\left(L_x, L_y, L_z\right)$ the components in the three reference directions.

Figure 10.3. illustrates a gyroscope, which can spin about its own axis, giving it an angular momentum, L_1, with direction at right angles to the flywheel part of the gyroscope. In addition, the gyroscope can be set to precess about the point of contact with the stand. This is a second angular momentum, L_2. The total angular momentum in the system is the sum of these two, with L_1 and L_2 being components of the total. In this particular instance, it would make more sense to split L_1 and L_2 into horizontal and vertical components and to add them up to get the horizontal and vertical components of L.

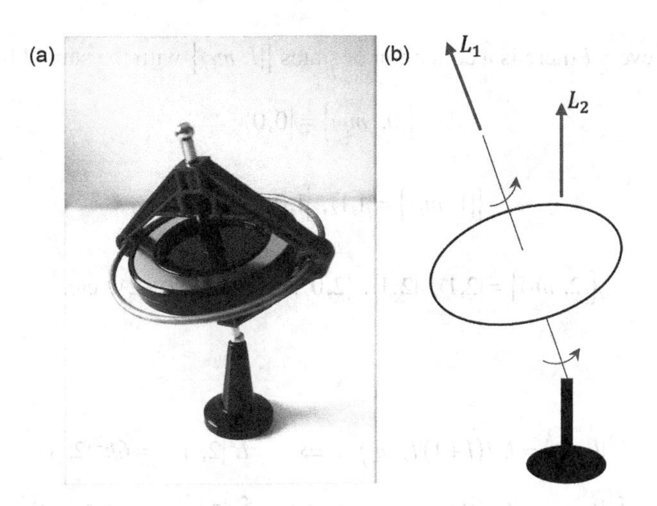

FIGURE 10.3 A gyroscope can spin on its own axis as well as rotate (precess) about the point of contact with its support. The total angular momentum in the system is then the sum of the two angular momenta as illustrated on the left-hand side.

10.2.1 Angular Momentum in Quantum Theory

To represent angular momentum in quantum mechanics, we have to replace L with a suitable operator $\hat{L}$. A sensible starting guess for the operator would be:

> Angular momentum operator
>
> $$\hat{L} = \hat{r} \times \hat{p}$$
>
> where $\hat{r}$ is a version of the position operator and $\hat{p}$ the ordinary linear momentum operator[3].

We can't be any more specific about $\hat{L}$ at this stage as we lack mathematical forms for $\hat{p}$ or $\hat{r}$. Indeed, the version of the operator quoted is not always the easiest one to deal with, especially when we have to worry about components.

We need three operators for x, y, and z components as well as an operator for the total angular momentum:

$$\widehat{L^2} = \widehat{L_x^2} + \widehat{L_y^2} + \widehat{L_z^2}$$

As with all operators, eigenstates of $\widehat{L^2}$ have a definite value of total angular momentum. The same is true for eigenstates of the component operators $\left(\hat{L}_x, \hat{L}_y, \hat{L}_z\right)$. However, an eigenstate of $\widehat{L^2}$ is NOT also an eigenstate of the three components. You can only have an eigenstate of $\widehat{L^2}$ and one of the three components at any one time. This surprising feature is the key piece needed to explain what is happening in the SG experiment.

10.2.2 Eigenstates of Angular Momentum

You can construct mutual eigenstates of $\widehat{L^2}$ and any of one of the three component operators, but by convention, we usually work with $\widehat{L^2}$ and $\hat{L}_z$. An eigenstate of both of these operators is characterized by two integer *quantum numbers* l and m_l where:

$$l = 0, 1, 2, 3, \ldots$$

$$m_l = 0, \pm 1, \pm 2, \cdots \pm l$$

Consequently, for every l there is a collection of states $\{|l,\ m_l\rangle\}$ with the same l but different m_l:

$$\{|0,\ m_l\rangle\} = |0,0\rangle$$

$$\{|1,\ m_l\rangle\} = |1,1\rangle,\ |1,0\rangle,\ |1,-1\rangle$$

$$\{|2,\ m_l\rangle\} = |2,1\rangle,\ |2,1\rangle,\ |2,0\rangle,\ |2,-1\rangle,\ |2,-2\rangle,\ \text{etc.}$$

and eigenvalues:

$$\hat{L}^2|l,\ m_l\rangle = \hbar^2 l(l+1)|l,\ m_l\rangle \quad \Rightarrow \quad \hat{L}^2|2,\ 1\rangle = 6\hbar^2|2,\ 1\rangle$$

$$\hat{L}_z|l,\ m_l\rangle = \hbar m_l|l,\ m_l\rangle \quad \Rightarrow \quad \hat{L}_z|2,-\ 1\rangle = -\hbar|2,\ -1\rangle$$

The properties of angular momentum eigenstates follow directly from the full mathematical form of the operators concerned, including the fact that angular momentum is *quantized*, i.e., it can appear only in values taken from a restricted set of whole number multiples (see Figure 10.4).

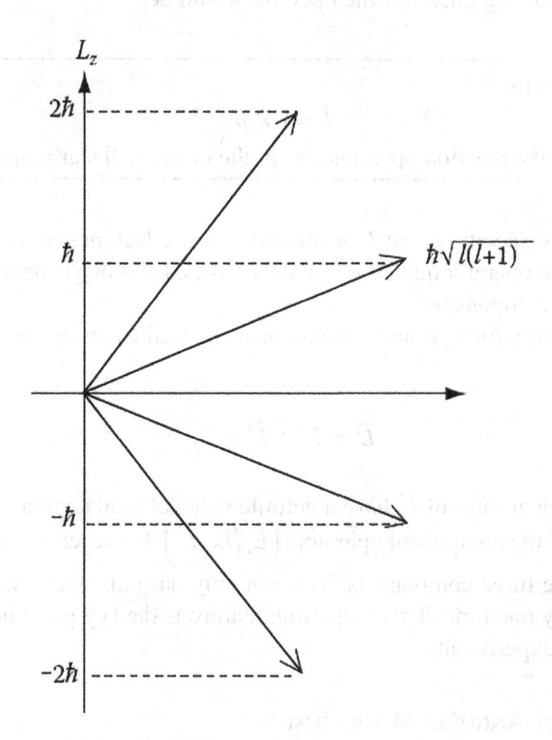

FIGURE 10.4 The behaviour of angular momentum eigenstates $\{|l, m_l\rangle\}$. For each l the total angular momentum takes on the value $L = \hbar\sqrt{l(l+1)}$. Each arrow here is of length $\hbar\sqrt{l(l+1)}$. Within the collection of states with the same L are states with different $L_z = m_l\hbar$. These are the different z components, shown by the length along the z axis. As the state $|l,\ m_l\rangle$ is an eigenstate of $\hat{L}^2$ and $\hat{L}_z$ only, the values of $\hat{L}_x, \hat{L}_y$ are indefinite. Hence this representation is sometimes shown as a series of cones around the z axis, to illustrate the lack of 'horizontal' components.

The first person to guess that angular momentum might be quantized in this way was Niels Bohr, who used it in his celebrated model of the atom; a piece of physics that is one of the historical foundation stones of quantum theory.

10.2.3 Magnetic Moments

As I mentioned at the start of Chapter 2, the original SG experiment was performed with a beam of silver atoms rather than with electrons. Stern and Gerlach expected the beam to be deflected by a nonuniform magnetic field, as they knew that each silver atom has its own magnetic field.

Any moving object with an electrical charge will produce a magnetic field. A crude (pre-quantum) picture of the hydrogen atom has an electron travelling in a circular orbit around the nucleus, as in Figure 10.5. This situation bears some comparison with a single loop of wire carrying a current. As the electron is a moving charge, it will produce a magnetic field circulating through the plane of the orbit.

The direction of this field is determined by the sense in which the current (electron) travels. In this case, the electron is moving anticlockwise when viewed from above, so the magnetic field will be pointing downward (Figure 10.6).

More complicated atoms often don't have an overall magnetic field, as the fields due to the various electrons manage to cancel out each other. Silver was selected for the SG experiment as it was easier to work with than hydrogen and had an 'unpaired' electron in its outermost orbit, and so a magnetic field to work with. The size of the field resulting from this orbital motion is related to the angular momentum of the orbiting electron via the *magnetic moment.*

Classically, the magnetic moment, μ, is defined in terms of the turning force experienced by a magnet in an external field[4]:

$$\text{turning force} = \text{magnetic moment} \times \text{external field strength}$$

$$\tau = \mu \times B$$

A simple bar magnet has a magnetic moment $\mu = pd$ where p is the strength of its magnetic poles and d the distance between them.

For a current loop, this becomes $\mu = IA$ with I being the magnitude of the current and A the area of the loop. Assuming the loop to be circular, this is readily adapted to the notion of an orbiting charge, q, with mass m as:

$$I = qv/2\pi r \qquad A = \pi r^2$$

giving:

$$\mu = \frac{qv\pi r^2}{2\pi r} = \frac{qL}{2m}$$

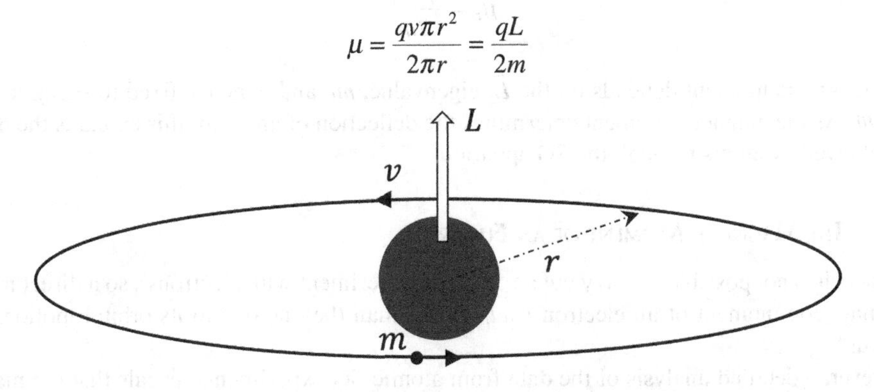

FIGURE 10.5 An electron moving in a fixed path around a nucleus is essentially an electrical current and should therefore give the atom a magnetic field. Conventionally, the current direction is taken as the flow of positive charge, so in this case, the current is in the opposite sense to the electron's motion. The angular momentum of the electron is $L = mvr$.

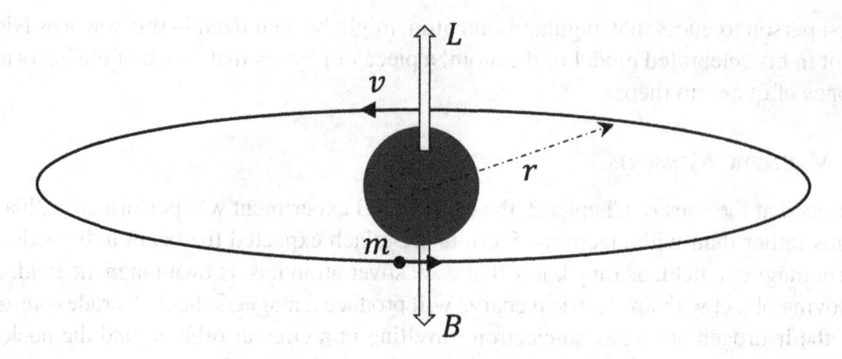

FIGURE 10.6 The sense of the orbital magnet is classically determined by the direction of the electron's movement. In this case, the magnetic field is oriented as if there was a north pole above the plane of the orbit.

where L is the angular momentum of the orbiting object (mvr).

The magnetic field strength, B, at the centre of a current loop is then:

$$B = \frac{\mu_0 I}{2\pi r} = \frac{\mu_0 qv}{4\pi^2 r^2} = \frac{\mu_0 \mu}{2\pi^2 r^3}$$

We can incorporate the results from Section 10.2.2:

$$L = \hbar\sqrt{l(l+1)}$$

$$L_z = \hbar m_l$$

to suggest that the magnetic moment due to the electron's orbital path is:

$$\mu = \frac{q\hbar\sqrt{l(l+1)}}{2m}$$

with a z component:

$$\mu_z = \frac{q\hbar m_l}{2m}$$

So, the magnetic moment depends on the $\hat{L}_z$ eigenvalue, m_l, and is hence fixed to being multiples of $q\hbar/2m$. As the magnetic moment determines the deflection of an atom, this explains the discrete paths followed by atoms through the SG apparatus.

10.2.4 The Magnetic Moment of an Electron

In practice, it is not possible to carry out an SG type experiment with electrons[5], so a direct measure of the magnetic moment of an electron *itself*, rather than the one due to its orbital motion, is not open to us[6].

However, a detailed analysis of the data from atomic SG experiments reveals that the magnetic field of the atom is a combination of the field due to the circulating current of an unpaired electron in orbit around the nucleus and a further field attributed to the electron's own *intrinsic angular momentum*. This is especially easy to deal with in silver atoms as the unpaired outer electron has

$l = 0$, so that only the intrinsic field is visible[7]. Hence another reason why silver is an excellent choice for these experiments.

Crudely (and ultimately highly inaccurately) it would seem that the electron spins on its own axis, like a top[8], and that this motion produces an intrinsic magnetic moment.

The results of atomic SG experiments have allowed the z-component of electron's intrinsic magnetic moment to be established as:

$$\mu_e = \frac{e\hbar}{2m_e}$$

If we assume that the result:

$$\mu_z = \frac{q\hbar m_l}{2m}$$

which we derived for orbital angular momentum, applies to all forms of angular momentum, including the putative top-like 'spin' of the electron, then to reproduce the measured value we would have:

$$\mu_z = \frac{e\hbar m_s}{2m_e}$$

introducing m_s as the z -component of the spin, leading to $m_s = 1$. However, this can't be right. According to our rules, if $m_s = 1$ then there should be $m_s = 0, -1$ components as well and hence *three* SG beams for electrons, rather than two.

The mystery is solved by deploying Dirac's relativistic equation for electrons (see, in brief, Section 21.1), which corrects the *predicted* magnetic moment to:

$$\mu_{zD} = \frac{e\hbar m_s}{m_e}$$

Giving, for consistency with the measured result, $m_s = \pm\frac{1}{2}$, $S = \frac{1}{2}$ and only two beams.

However, the purpose of this argument was not to get an exact value of angular momentum for the electron as much as to illustrate the following thought sequence:

- Atoms get deflected by SG magnets as they have a magnetic field.
- This field partly comes from the orbiting electrons.
- A detailed analysis of the results of SG experiments reveals a further field attributed to the electrons *themselves*, rather than any atomic motion they are involved in.
- The source of a magnetic field is a charge in motion[9]
- Electrons must have their own angular momentum.

10.2.5 Intrinsic Angular Momentum

This is where we start to get into trouble.

It's very easy to try and compare an electron to a classical object that has intrinsic angular momentum, for example, a spinning top. If we think of an electron as a tiny ball of charge spinning about some axis, we can believe that it would have a magnetic field. After all, the rotation of the charge will generate a magnetic effect.

Unfortunately, it can't be that simple. For one thing, the radius of the electron, coupled with the rate of its rotation, would determine the intrinsic angular momentum and hence the magnetic field.

The trouble is, given any reasonable size estimate for an electron, its edges would have to be moving faster than the speed of light to generate the measured magnetic field, which is in clear contradiction to Einstein's theory of special relativity.

Another reason why the simple picture won't work goes back to the SG results. For a classically rotating ball, there would appear to be no reason why it should not spin about any axis we fancy. As the magnetic field is aligned to the spin axis, we ought to be able to prepare electrons with a magnetic field in any given direction. At the very least, if we do nothing to constrain the rotation axis, a beam of electrons should contain an assortment of directions at random.

Not according to the SG experiment.

After all, if a general SG state $|\varphi\rangle$ collapses into either $|U\rangle$ or $|D\rangle$, the electron's magnetic field can be pointing only UP or DOWN and not at any intermediate angle. Hence the electron's spin axis can *only* be pointing UP or DOWN, which is not very classical behaviour.

Finally, the classical picture can't explain why all electrons have *exactly* the same spin. We know that this must be the case, as they are all deflected by the same amount in the SG field (it might be upward or downward, but it's the same distance[10]).

We can alter the angular momentum of a classically spinning sphere in a number of ways. All we have to do is apply a suitable force to speed it up or slow it down. However, *nothing affects an electron's spin*.

Given these difficulties, it might be better to give up the idea that an electron's magnetic field was related to any sort of angular momentum and try and find some different physics to explain what is going on. However, there are independent pieces of evidence that point to electrons having angular momentum of their own, no matter how difficult it might be to account for that.

When an electron orbits an atomic nucleus, the magnetic field it sets up interacts with the electron's own magnetic field, a process known as *spin–orbit coupling*. This interaction influences the energy of the atom, with measurable consequences, especially to the spectrum of light emitted by the atom. Also, a thorough analysis of the possible orbital motions shows that you can't conserve angular momentum for the atom as a whole *just by using the orbital angular momentum*, the spin has to be taken into account as well.

Finally, the operators that we introduced in Chapter 5 to describe the SG measurement process, $\hat{S}_x$, $\hat{S}_y$ and $\hat{S}_z$ turn out to have exactly the same mathematical properties as $\left(\hat{L}_x, \hat{L}_y, \hat{L}_z\right)$, strongly suggesting that they must represent some form of angular momentum as well. This last point is pursued further in the next section where we will start to explore these mathematical properties.

So, we are left with a very odd situation. On the one hand, electrons appear to behave as if they have an intrinsic angular momentum leading to a magnetic moment. On the other hand, the attempt to work out a theory in which an electron is spinning like a top leads to absurd conclusions.

Physicists have long resigned themselves to the notion that spin is a quantum mechanical effect *that has no direct correspondence to anything that we see in the large-scale world.*

The British astrophysicist Sir Arthur Stanley Eddington believed that the term 'spin' was so misleading we ought to replace it. Calling on his knowledge of Lewis Carroll's famous nonsense verse *Jabberwocky,* Eddington plucked the word *gyre* from the line "did gyre and gimble in the wabe" and suggested using that as an alternative, as it would be just about as exact and much less misleading than spin. Unfortunately, Eddington's suggestion did not catch on and we got stuck with spin, for better and for worse.

10.3 SPIN OPERATORS

In Chapter 5 I wrote down an operator corresponding to the (UP, DOWN) SG measurements and gave it eigenvalues of ±1:

$$\hat{S}_z|U\rangle = +1|U\rangle \qquad \hat{S}_z|D\rangle = -1|D\rangle$$

Actually, $\hat{S}_z$ is part of a set along with $\hat{S}_x$, $\hat{S}_y$ and $\hat{S}^2 = \hat{S}_x^2 + \hat{S}_y^2 + \hat{S}_z^2$ with the same mathematical properties as the angular momentum operators from earlier. This means that we ought to be able to construct sets of eigenstates similar to those listed in Section 10.2.2. In other words, there is a set $\{|s, m_s\rangle\}$ specified by two quantum numbers:

$$\hat{S}^2|s,\ m_s\rangle = \hbar^2 s(s+1)|s,\ m_s\rangle$$

$$\hat{S}_z|s,\ m_s\rangle = \hbar m_s|s,\ m_s\rangle$$

All we need is a clue from experimental results to help figure out what values s and m_s take.

Earlier on I quoted the measured value of the electron's magnetic moment, $\mu_e = e\hbar/2m_e$ and compared it with the theoretical calculation (as tweaked by Dirac), $\mu_{zD} = \frac{e\hbar m_s}{m_e}$, from which we deduced that $m_s = +\frac{1}{2}$.

So, the *total* spin angular momentum, of the electron must be $s = \frac{1}{2}$. The SG experiment tells us that there are two *components* $m_s = +\frac{1}{2}$ and $m_s = -\frac{1}{2}$ corresponding to $|U\rangle$ and $|D\rangle$ respectively. So, we can finally make the connection:

$$|U\rangle = \left|\frac{1}{2}, \frac{1}{2}\right\rangle$$

$$|D\rangle = \left|\frac{1}{2}, -\frac{1}{2}\right\rangle$$

Consequently:

$$\hat{S}_z|U\rangle = \hat{S}_z\left|\frac{1}{2},\ \frac{1}{2}\right\rangle = \frac{1}{2}\hbar\left|\frac{1}{2}, \frac{1}{2}\right\rangle = \frac{1}{2}\hbar|U\rangle$$

$$\hat{S}_z|D\rangle = \hat{S}_z\left|\frac{1}{2},\ -\frac{1}{2}\right\rangle = -\frac{1}{2}\hbar\left|\frac{1}{2}, -\ \frac{1}{2}\right\rangle = -\frac{1}{2}\hbar|D\rangle$$

My earlier statement, that the eigenvalues of $\hat{S}_z$ are ± 1, should now be interpreted as being in units of $\hbar/2$.

10.3.1 Spin Matrices

The relative simplicity of the spin 1/2 quantum system gives us the opportunity to dip a gentle toe into a specific pond: the *matrix representation of quantum theory*. Mathematically, we can represent the pair of spin 1/2 states, $|U\rangle$ and $|D\rangle$, as *column vectors*:

$$|U\rangle = \begin{pmatrix} 1 \\ 0 \end{pmatrix} \qquad |D\rangle = \begin{pmatrix} 0 \\ 1 \end{pmatrix}$$

Strictly, the numbers in the column vectors represent the amplitudes to be in the states, rather than the states themselves, but it is common to use rather more loose terminology in this context.

Any other state $|\psi\rangle = a|U\rangle + b|D\rangle$ is then:

$$|\psi\rangle = a\begin{pmatrix} 1 \\ 0 \end{pmatrix} + b\begin{pmatrix} 0 \\ 1 \end{pmatrix} = \begin{pmatrix} a \\ b \end{pmatrix}$$

Back in Section 5.4.1, I suggested that turning a ket into a bra involved slightly more than just taking the complex conjugate. To work correctly in a matrix representation, bras must be represented by a *row vector*:

$$\langle U| = \begin{pmatrix} 1 & 0 \end{pmatrix} \qquad \langle D| = \begin{pmatrix} 0 & 1 \end{pmatrix} \qquad \langle \psi| = \begin{pmatrix} a^* & b^* \end{pmatrix}$$

The process of tipping over a column vector is called taking the *transpose*, hence a bra is the *conjugate transpose* of its ket, $\langle \psi| = \left(|\psi\rangle^*\right)^T$. This ensures that we can deploy the rules of *matrix multiplication* when calculating:

$$\langle \psi|\psi\rangle = \begin{pmatrix} a^* & b^* \end{pmatrix}\begin{pmatrix} a \\ b \end{pmatrix} = a^*a + b^*b = |a|^2 + |b|^2$$

In this representation, the various spin operators can be denoted by 2×2 *matrices* of the following form:

$$\hat{S}_x = \frac{1}{2}\hbar\sigma_1 \qquad \hat{S}_y = \frac{1}{2}\hbar\sigma_2 \qquad \hat{S}_z = \frac{1}{2}\hbar\sigma_3$$

with the set $\{\sigma_i\}$ being the *Pauli Matrices*:

$$\sigma_1 = \begin{pmatrix} 0 & 1 \\ 1 & 0 \end{pmatrix} \qquad \sigma_2 = \begin{pmatrix} 0 & -i \\ i & 0 \end{pmatrix} \qquad \sigma_3 = \begin{pmatrix} 1 & 0 \\ 0 & -1 \end{pmatrix}$$

A little exploration using matrix multiplication confirms the obvious relationships, such as:

$$\hat{S}_z|U\rangle = \frac{1}{2}\hbar\begin{pmatrix} 1 & 0 \\ 0 & -1 \end{pmatrix}\begin{pmatrix} 1 \\ 0 \end{pmatrix} = \frac{1}{2}\hbar\begin{pmatrix} 1\times 1 + 0\times 0 \\ 0\times 1 + (-1)\times 0 \end{pmatrix} = \frac{1}{2}\hbar\begin{pmatrix} 1 \\ 0 \end{pmatrix} = \frac{1}{2}\hbar|U\rangle$$

$$\hat{S}_z|D\rangle = \frac{1}{2}\hbar\begin{pmatrix} 1 & 0 \\ 0 & -1 \end{pmatrix}\begin{pmatrix} 0 \\ 1 \end{pmatrix} = \frac{1}{2}\hbar\begin{pmatrix} 1\times 0 + 0\times 1 \\ 0\times 0 + (-1)\times 1 \end{pmatrix} = \frac{1}{2}\hbar\begin{pmatrix} 0 \\ -1 \end{pmatrix} = -\frac{1}{2}\hbar|D\rangle$$

We can also confirm that these matrices, which correspond to observables, are Hermitian (Section 5.4.1). Taking the transpose of a matrix switches rows with columns:

$$\begin{pmatrix} a & b \\ c & d \end{pmatrix}^T = \begin{pmatrix} a & c \\ b & d \end{pmatrix}$$

so, with $\hat{S}_y$, for example:

$$\hat{S}_y^* = \frac{1}{2}\hbar\begin{pmatrix} 0 & i \\ -i & 0 \end{pmatrix}$$

Hence:

$$\left(\hat{S}_y^*\right)^T = \frac{1}{2}\hbar\begin{pmatrix} 0 & i \\ -i & 0 \end{pmatrix}^T = \frac{1}{2}\hbar\begin{pmatrix} 0 & -i \\ i & 0 \end{pmatrix} = \hat{S}_y$$

In general, if $\hat{O} = \begin{pmatrix} A & B \\ C & D \end{pmatrix}$, then $\hat{O}^\dagger = \begin{pmatrix} A^* & C^* \\ B^* & D^* \end{pmatrix}$, hence for $\hat{O}$ to be Hermitian, A and D must be real-valued, and if $B = a + ib$, $C = a - ib$.

The Pauli matrices are unitary (Section 5.5) as well as being Hermitian[11]. Using σ_2 as an example:

$$\sigma_2 = \begin{pmatrix} 0 & -i \\ i & 0 \end{pmatrix} \qquad \sigma_2^\dagger = \begin{pmatrix} 0 & -i \\ i & 0 \end{pmatrix}$$

$$\sigma_2\sigma_2^\dagger = \begin{pmatrix} 0 & -i \\ i & 0 \end{pmatrix}\begin{pmatrix} 0 & -i \\ i & 0 \end{pmatrix} = \begin{pmatrix} 1 & 0 \\ 0 & 1 \end{pmatrix}$$

We can incorporate $|L\rangle$ and $|R\rangle$ into this scheme by writing:

$$|R\rangle = \frac{1}{\sqrt{2}}\begin{pmatrix} 1 \\ 1 \end{pmatrix} \qquad |L\rangle = \frac{1}{\sqrt{2}}\begin{pmatrix} 1 \\ -1 \end{pmatrix}$$

The $1/\sqrt{2}$ factor is the normalisation constant, as:

$$\langle R| = \frac{1}{\sqrt{2}}\begin{pmatrix} 1 & 1 \end{pmatrix} \qquad \langle L| = \frac{1}{\sqrt{2}}\begin{pmatrix} 1 & -1 \end{pmatrix}$$

giving:

$$\langle R|R\rangle = \frac{1}{\sqrt{2}}\begin{pmatrix} 1 & 1 \end{pmatrix}\frac{1}{\sqrt{2}}\begin{pmatrix} 1 \\ 1 \end{pmatrix} = \frac{1}{2}(1\times 1 + 1\times 1) = \frac{1}{2}(2) = 1$$

the same being true for $\langle L|L\rangle$. For $|U\rangle$ and $|D\rangle$ we have:

$$\langle U| = \begin{pmatrix} 1 & 0 \end{pmatrix} \qquad \langle D| = \begin{pmatrix} 0 & 1 \end{pmatrix}$$

so that:

$$\langle U|U\rangle = \begin{pmatrix} 1 & 0 \end{pmatrix}\begin{pmatrix} 1 \\ 0 \end{pmatrix} = 1\times 1 + 0\times 0 = 1$$

etc.

Having $|L\rangle$ and $|R\rangle$ allows us to tackle the following calculations:

$$\hat{S}_x|R\rangle = \frac{1}{2}\hbar\sigma_1|R\rangle = \frac{1}{2}\hbar\begin{pmatrix} 0 & 1 \\ 1 & 0 \end{pmatrix}\frac{1}{\sqrt{2}}\begin{pmatrix} 1 \\ 1 \end{pmatrix} = \frac{1}{2}\hbar\frac{1}{\sqrt{2}}\begin{pmatrix} 0\times 1 + 1\times 1 \\ 1\times 1 + 0\times 1 \end{pmatrix} = \frac{1}{2}\hbar\frac{1}{\sqrt{2}}\begin{pmatrix} 1 \\ 1 \end{pmatrix} = \frac{1}{2}\hbar|R\rangle$$

$$\hat{S}_x|L\rangle = \frac{1}{2}\hbar\sigma_1|L\rangle = \frac{1}{2}\hbar\begin{pmatrix} 0 & 1 \\ 1 & 0 \end{pmatrix}\frac{1}{\sqrt{2}}\begin{pmatrix} 1 \\ -1 \end{pmatrix} = \frac{1}{2}\hbar\frac{1}{\sqrt{2}}\begin{pmatrix} 0\times1+1\times-1 \\ 1\times1+0\times-1 \end{pmatrix} = -\frac{1}{2}\hbar|L\rangle$$

showing that $|R\rangle$ and $|L\rangle$ are eigenstates of $\hat{S}_x$. Note that neither $|U\rangle$ nor $|D\rangle$ are eigenstates of $\hat{S}_x$ as, for example:

$$\hat{S}_x|U\rangle = \frac{1}{2}\hbar\sigma_1|U\rangle = \frac{1}{2}\hbar\begin{pmatrix} 0 & 1 \\ 1 & 0 \end{pmatrix}\begin{pmatrix} 1 \\ 0 \end{pmatrix} = \frac{1}{2}\hbar\begin{pmatrix} 0 \\ 1 \end{pmatrix} = \frac{1}{2}\hbar|D\rangle$$

Equally, $|R\rangle$ and $|L\rangle$ are not eigenstates of $\hat{S}_z$. However, while we have shown that these eigenstates of $\hat{S}_x$ are not eigenstates of $\hat{S}_z$, and vice versa, there may still be another state lurking somewhere, $|\phi\rangle$ which is an eigenstate of both. In which case:

$$\hat{S}_z|\phi\rangle = \epsilon_z\,|\phi\rangle \qquad \hat{S}_x|\phi\rangle = \epsilon_x\,|\phi\rangle$$

and so:

$$\hat{S}_z\left(\hat{S}_x|\phi\rangle\right) = \hat{S}_z\left(\epsilon_x\,|\phi\rangle\right) = \epsilon_z\epsilon_x|\phi\rangle$$

$$\hat{S}_x\left(\hat{S}_z|\phi\rangle\right) = \hat{S}_x\left(\epsilon_z\,|\phi\rangle\right) = \epsilon_x\epsilon_z|\phi\rangle$$

suggesting that $\hat{S}_z\hat{S}_x = \hat{S}_x\hat{S}_z$. Effectively, this is the condition for our state being a mutual eigenstate of both spin directions. A little matrix manipulation shows that the operators do *not* obey this rule:

$$\hat{S}_z\hat{S}_x = \frac{1}{4}\hbar^2\begin{pmatrix} 1 & 0 \\ 0 & -1 \end{pmatrix}\begin{pmatrix} 0 & 1 \\ 1 & 0 \end{pmatrix} = \frac{1}{4}\hbar^2\begin{pmatrix} 0 & 1 \\ -1 & 0 \end{pmatrix}$$

$$\hat{S}_x\hat{S}_z = \frac{1}{4}\hbar^2\begin{pmatrix} 0 & 1 \\ 1 & 0 \end{pmatrix}\begin{pmatrix} 1 & 0 \\ 0 & -1 \end{pmatrix} = \frac{1}{4}\hbar^2\begin{pmatrix} 0 & -1 \\ 1 & 0 \end{pmatrix}$$

Actually $\hat{S}_z\hat{S}_x = -\hat{S}_x\hat{S}_z$. In fact, for any $i \neq j$, $\hat{S}_i\hat{S}_j = -\hat{S}_j\hat{S}_i$. Consequently, $|\phi\rangle$ does not exist. This feature of the operators has deep physical significance, as we will see in Chapter 13.

With each of the Pauli matrices being unitary and Hermitian:

$$\sigma_1^2 = \sigma_2^2 = \sigma_3^2 = \begin{pmatrix} 1 & 0 \\ 0 & 1 \end{pmatrix} = I$$

and as $\hat{S}^2 = \hat{S}_x^2 + \hat{S}_y^2 + \hat{S}_z^2$, it follows:

$$\hat{S}^2 = \frac{1}{4}\hbar^2\sigma_1^2 + \frac{1}{4}\hbar^2\sigma_2^2 + \frac{1}{4}\hbar^2\sigma_3^2 = \frac{3}{4}\hbar^2\hat{I}$$

which is correct, given our earlier identity of: $\hat{S}^2|s,\ m_s\rangle = \hbar^2 s(s+1)|s,\ m_s\rangle = \hbar^2\frac{1}{2}\left(\frac{1}{2}+1\right)\left|\frac{1}{2}, m_s\right\rangle = \frac{3}{4}\hbar^2\left|\frac{1}{2}, m_s\right\rangle$ for $s = \frac{1}{2}$. We also see from this that as $\hat{S}^2 = \frac{3}{4}\hbar^2\hat{I}$ our eigenstates of $\hat{S}_z$ will also be eigenstates of $\hat{S}^2$, as is the case for the eigenstates of $\hat{S}_x$ as well.

Summarizing, our spin operators have the properties:

> Spin operator properties
>
> $$\text{For any } i \neq j,\ \hat{S}_i\hat{S}_j = -\hat{S}_j\hat{S}_i$$
>
> $$\text{For any } i,\ S_i^2 = \hat{I}$$
> $$\hat{S}^2 = \hat{S}_x^2 + \hat{S}_y^2 + \hat{S}_z^2$$
>
> These are characteristic features of *any* angular momentum operators.

10.3.2 Fermions and Bosons

The picture that I have built up describes electrons as particles with a *system property* called spin, which has the value $\hbar\sqrt{s(s+1)}$ with $s = \frac{1}{2}$, and a *state property,* the component of spin, which has eigenvalues $m_s\hbar = \pm\frac{1}{2}\hbar$ along any direction that we point an SG magnet, as illustrated in Figure 10.7.

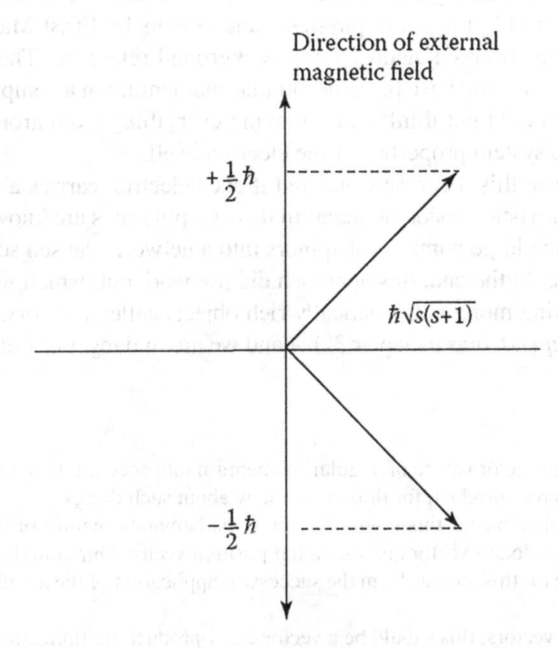

FIGURE 10.7 In the presence of a magnetic field, a preferred z -axis is defined, along the field direction. A spin 1/2 particle would then have a measured z -component of spin of either $+\frac{1}{2}\hbar$ or $-\frac{1}{2}\hbar$. This happens irrespective of the direction in which the magnetic field is pointing. The arrows represent the direction of the particle's spin axis. In truth, these arrows could be pointing in any direction around a cone of the correct angle, which is consistent with the indefinite nature of the x and y spin components. The magnitude of the spin is given by the formula $\hbar\sqrt{s(s+1)}$ where $s = \frac{1}{2}$ in this case.

Electrons are known as spin 1/2 particles because of their s value. All fermions have spins that are odd multiples of 1/2, that is, 3/2, 5/2 etc. In Chapter 11, we will see how this spin property relates to the antisymmetry of fermion multiparticle states.

Bosons have whole number spins: 1, 2, 3, etc. This is also intimately linked to the symmetry of their multiparticle states.

10.4 QUANTUM SCALE, SPIN, SPINORS AND TWISTORS

The *reduced Planck's constant*, $\hbar = h/2\pi$, is often taken to be a measure of the 'scale' of a quantum effect, as it determines the size of the energy in a photon and the 'wavelength' of a particle. In our imaginations, we can change the value of $\hbar$ and see what happens, which is often an interesting exercise. If we systematically *reduce* the value of $\hbar$ then quantum features frequently blend seamlessly into classical physics. This is *not* the case for spin. If we let $\hbar \rightarrow 0$, then the spin of an electron vanishes, which is a further indication that *spin belongs solely to the quantum world and has no extension into classical physics.*

As it is a purely quantum property, there is some frustration involved in writing about spin: I am not able to give a clear physical picture of what spin is. It is all very well calling it 'intrinsic angular momentum,' but as this angular momentum doesn't seem to be linked to any motion in a *classical sense*, that doesn't help very much. The next chapter will take care of the other aspects of spin and the way in which it is linked to the symmetry properties of fermion and boson states, but if anything, that adds to the mystery.

What is really needed is a way of building up from the fundamental quantum property of spin and the quantization of angular momentum to the classical world and its rotating objects. Some promising work has been done in this regard, notably by Roger Penrose[12]. The motivation for Penrose's approach goes back to an old problem in physics made famous by Ernst Mach,[13] who pointed out how rotation can only be observed against some background reference. That being the case, it is hard to see how an electron could have intrinsic angular momentum in a completely empty universe. On the other hand, you would not think that removing everything from around an object, i.e., the electron could affect the system properties of the electron itself.

In an attempt to resolve this, Penrose wondered if each electron carries a small piece of 'proto-space' with it. In the relativistic version of quantum theory, spin states are known as *spinors*. Penrose discovered that combining large numbers of spinors into a network, caused some of the basic properties of space to emerge. In the end, this approach did not work out, which was part of the motivation for Penrose's inventing more mathematically rich objects called *twistors*. Now it turns out that twistors are related to *superstrings* (Chapter 32)... and we are in danger of falling off the deep end.

NOTES

1 If we need to take the vector nature of angular momentum into account, then this simple multiplication becomes the vector cross product, for those who know about such things.
2 Ultimately, the justification for this comes from the mathematical nature of the vector cross-product used to combine the velocity vector and the radial position vector. Our confidence in the cross product as the correct choice for this, comes from the successful application of the angular momentum vector to experimental data.
3 As both p and r are vectors, this should be a vector cross-product multiplication.
4 This should fully be a vector equation involving the vector cross product, but the details are not required for this argument.
5 See endnote 2 of Chapter 2 for an explanation of why we can't do these experiments in practice.
6 At the time of writing, strenuous experimental efforts are being directed to measuring the magnetic moment of a particle known as the *muon*, which is a heavy sister particle to the electron.
7 *How* an 'orbiting' electron can have $l = 0$ is another matter. See, for some elucidation, Sections 15.2 - 15.2.2.

8 To belabour the point, nothing at all like a top...
9 Or a time-changing electrical field, but that instance is not relevant here.
10 To within a variation easily accounted for by normal experimental fluctuations.
11 As the spin matrices are simply multiples of the Pauli matrices, it follows that the latter are Hermitian.
12 Sir Roger Penrose, FRS (1931–), Emeritus Rouse Ball, professor of mathematics at Oxford University and Nobel Prize winner in 2020.
13 Ernst Mach (1838–1916) was a Bohemian-Austrian physicist and philosopher. His writings greatly influenced Einstein. In modern physics the Mach number, as in speeds relative to the speed of sound, is named after him.

11 Fermion States

11.1 STATES, NORMALIZATION, AND PHASE

When a state is constructed by expanding over a basis formed from the eigenstates of an observable operator, it is important that the expansion is *complete*, i.e., that all possible eigenstates of the observable are included. This is for two connected reasons. Firstly, we want all possible outcomes of an experiment to be represented and secondly, if we do not include all the eigenstates in the basis, we will not be able to ensure that the expanded state is properly normalized. For any continuous basis, y, this means:

$$\int \langle \psi(y) | \psi(y) \rangle dy = 1$$

where the range of integration is taken over all possible values of the state variable, y. If this is not the case, then not only will our calculation of the relative probability of certain results be off, but also the expectation value:

$$\langle \hat{O} \rangle = \int \langle \psi(y) | \hat{O} | \psi(y) \rangle dy$$

will be wrong as well.

To ensure the correct normalization of a state, the expansion is generally set up with an arbitrary constant, A, known as the *normalization constant*. The value of A is then fixed by carrying out the integration and adjusting A in order that it comes correctly to 1:

$$|\psi(y)\rangle = A \int \psi(y) | y \rangle dy$$

$$|A|^2 \int \psi^*(y) \psi(y) \langle y | y \rangle dy = 1$$

As a consequence of this normalization process, any multiple of a state, $B|\psi\rangle$, is physically the same state as $|\psi\rangle$, as the multiple is 'factored out' during normalization. In other words, if we take the state $B|\psi\rangle$ rather than $|\psi\rangle$, we are simply adjusting the ultimate value of the normalization constant inside $|\psi\rangle$.

As the normalization constant is generally a complex number, and hence of the form $A = Re^{i\vartheta}$, the overall phase of a state (wave function) is non-physical. After all, when we construct:

$$|A|^2 \int \psi^*(y) \psi(y) \langle y | y \rangle dy = 1$$

$$|A|^2 = A^* A = \left(Re^{-i\vartheta}\right)\left(Re^{i\vartheta}\right) = R^2$$

so that the phase, ϑ, disappears from the calculation.

While the overall phase has no physical implications for a state, the *relative phase* between one state and another, especially inside an expansion, is crucially important as it leads to interference.

DOI: 10.1201/9781003225997-13

The action of an operator, $\hat{O}$, on a state can disturb the relative phase within the state's expansion. For example, with a simple state $|\phi\rangle = A\big(a|i\rangle + b|j\rangle\big)$, with normalization constant A, and:

$$\hat{O}|i\rangle = c|j\rangle \qquad \hat{O}|j\rangle = d|i\rangle$$

we get[1]:

$$\hat{O}|\phi\rangle = \hat{O}A\big(a|i\rangle + b|j\rangle\big) = A\big(ac|i\rangle + bd|j\rangle\big)$$

If $c = R_1 e^{i\vartheta_1}$ and $d = R_2 e^{i\vartheta_2}$ then:

$$\hat{O}|\phi\rangle = A\big(aR_1 e^{i\vartheta_1}|i\rangle + bR_2 e^{i\vartheta_2}|j\rangle\big) = AR_1 e^{i\vartheta_1}\left(a|i\rangle + b\frac{R_2}{R_1}e^{i(\vartheta_2-\vartheta_1)}|j\rangle\right)$$

The term $AR_1 e^{i\vartheta_1}$ is 'removed' on normalization, and the term $\frac{R_2}{R_1}e^{i(\vartheta_2-\vartheta_1)}$ introduces a new phase difference between $|i\rangle$ and $|j\rangle$. The upshot is that we can always write:

$$\hat{O}|\phi\rangle = A'\big(a|i\rangle + kb|j\rangle\big)$$

(where $k = \frac{R_2}{R_1}e^{i(\vartheta_2-\vartheta_1)}$)

a trick that will be exploited later in this chapter.

11.2 EXCHANGE AND ROTATION

In Chapter 8, I suggested that identical fermion multiparticle states are antisymmetric under an exchange of any two of the particles. Exploring this idea fully would involve us in a detailed study of quantum field theory, which is further than we can go without a much higher level of mathematical knowledge. Nevertheless, some of the ideas are important from a conceptual and philosophical standpoint, so for this chapter we are forced to engage in a more 'hand waving' form of argument. Look on it as a nice change of pace.

On further reflection, the antisymmetry of fermion states presents us with something of a puzzle. After all, if the particles are identical, we're not supposed to be able to tell which is which; yet exchanging any two does make a difference to the state. Clearly the state can, in some manner, tell what has happened, even if we can't. Put in a different way: if the particles are identical, then the physical situation after the exchange ought to be identical to that beforehand, yet the state is different. This would appear to be something of a paradox.

Exchanging two particles is a simple thing in *symbolic terms*. All we must do is rub out some letters and write in some different ones instead. In practice, however, the process involves rather more than simply snapping our fingers and watching particles teleport from place to place. We would have to pick one particle up and move it to a new place and move the other one back again.

Figure 11.1a shows two clothes pegs linked by a ribbon and laid out on my dining-room table. For the purpose of this argument, the pegs represent two identical particles in a multiparticle state. The ribbon plays the role of the rather more abstract geometrical 'connection' between the particles (an idea that we will build on as the argument progresses).

In the next step (Figure 1.1b), I have picked up the right-hand particle (peg) and moved it across the other (re-centring the photograph as well). In other words, I have swapped (exchanged) the position of the two particles, putting a loop into the ribbon in the process. Now if I take the bottom of

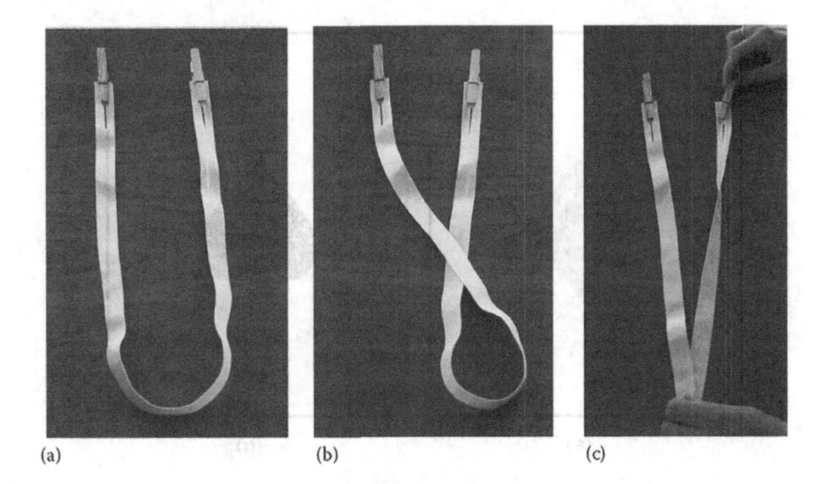

FIGURE 11.1 (a) Two particles (pegs) are connected by a 'ribbon.' (b) The particle on the right is moved across to the left side. (c) Smoothing down the ribbon so that it is like (a), shows that a twist has been introduced in the right side. The same effect could be produced by leaving the particles in place and simply rotating the right-hand side particle through 2π radians.

this loop and pull it tight, the arrangement becomes remarkably similar to that in Figure 1.1a; *I have just gained a twist in the ribbon under the (new) right-hand particle.*

Given that the pegs are supposed to represent identical particles, if we were to simply look at the pegs, *we couldn't tell the difference between Figures 11.1a and c.*

However, the connecting ribbon tells another story, as now it has a twist in it beneath the right-hand particle. This twist is the record of the switch having taken place.

Interestingly (and significantly) the same effect could have been achieved by leaving all the pegs in place, and simply twisting the ribbon under the right-hand one, *a twist that can be generated by rotating the right-hand peg through* 2π *radians on the spot.*

This is an extraordinary prediction: switching two identical particles has the same physical effect as leaving them where they are and rotating one of them on the spot.

There is no easy way of visualizing this without the aid of something like Figure 11.1. The best I can do is to remind you that any switch, in practice, would have to be done by moving and sliding particles about, something that would 'distort' their 'connections' to the surroundings and to one another.

Geometrically, we have shown that this sort of exchange distortion is the same as a single rotation. Furthermore, as we have already seen that exchanging fermions causes a multiparticle state $|\varphi\rangle$ to go to $-|\varphi\rangle$ (phase inversion), *we can conclude that the same thing must happen if we simply rotate a fermion through* 2π. Finally, as any change made as a result of the rotation must be localized to one of the particles, it follows that a 2π rotation of any *single-particle* fermion state should invert the phase of that state.

11.3 ROTATIONAL SYMMETRY OF STATES

Sometimes when you rotate an object through an angle less than 2π, things look just the same as they did before. This is called *rotational symmetry*. The basic idea is illustrated in Figure 11.2, which shows two different playing cards. The Ace of Diamonds has π rotational symmetry: if we rotate the card by π it will look exactly the same.[2] No smaller rotation than this will work. If we turn it through $\pi/2$ it will look completely different. Of course, bigger rotations can work. A 2π rotation is just as good. After all, that is two π rotations in sequence. Any whole number multiple of π will be fine.

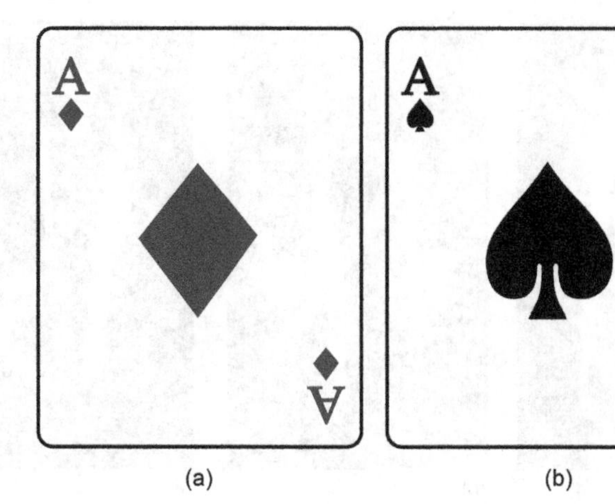

FIGURE 11.2 (a) This object has π rotation symmetry, that is, if you turn it through π about one corner, you can't tell the difference. (b) This object has 2π symmetry—it requires a full 2π rotation before it looks the same.

However, the Ace of Spades does not have π rotational symmetry, as the spade symbol looks different when upside down. This card has 2π rotational symmetry, but nothing smaller than that will work.

You might think that every object must have 2π rotational symmetry. Indeed, most do, but surprisingly it is possible to set up a situation where 4π *is the smallest rotation angle that will restore an object*. I will give a 3-D example, using a coffee mug, in Section 11.3.2.

The fermion situation is even more puzzling. After all, we are supposed to be dealing with a particle that can be treated as an infinitesimal point. Yet, according to the argument of the previous section, if we rotate a fermion through 2π its state will change. Physically something is different. To see how this can come about, we need to think more about the nature of rotation. However, given the counter-intuitive nature of the prediction, we need some experimental confirmation before going further.

11.3.1 Reversing the Polarity of the Neutron Flow[3]

In the Stern–Gerlach experiment, electrons are deflected because their tiny magnetic field interacts with the SG magnet's field. The poles of the SG magnet are specially shaped to generate the non-uniform field needed to deflect an electron's path, but in the presence of any magnetic field, moving electrons reorient their magnetic axis so that it aligns with the external field (Figure 11.3).

So, if we vary the field strength of a magnet as an electron passes by, we should be able to alter the alignment of the electron's field and *rotate the electron in the process.*

Figure 11.4 illustrates one way of doing this. The dotted arrow represents the orientation of an electron's magnetic field as we alter an externally applied combination of fields (black arrows). The electron's field will try to align itself with the two fields, but more towards the direction of the stronger field (the longer black arrow of the pair). By varying the strength of the two fields, we can force the electron to rotate about a vertical axis. In this case, the electron has rotated through π about a vertical axis (perpendicular to the page as we look at it) by the time the sequence has ended.

In truth, this sort of experiment is impractical with electrons, as their tiny mass makes them overly sensitive to small variations in a magnetic field. The magnetic field of the Earth can influence them enough to spoil the results. Fortunately, the principle can be effectively demonstrated using a somewhat heavier fermion: the neutron.

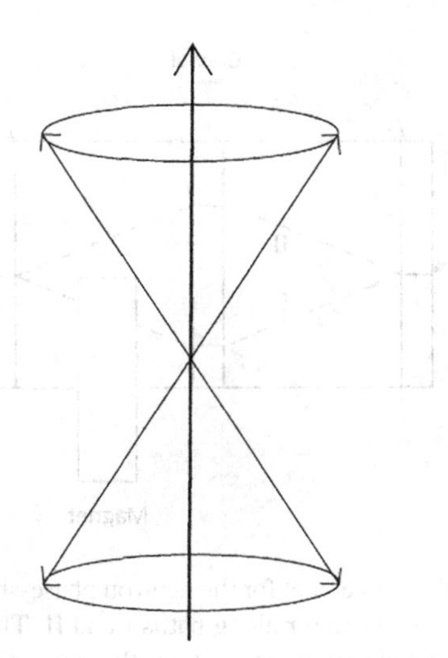

FIGURE 11.3 The vertical arrow represents an externally applied magnetic field. In the presence of such a field, the magnetic axis of a particle (which is the same as its spin axis) rotates about the magnetic field direction.

At first sight, neutrons would seem to be an odd choice. Given that they have no electrical charge, you would not expect neutrons to have a magnetic field to interact with an external field. However, neutrons are composed of *quarks* which do have electrical charges. As it happens, in the case of the neutron, the net charge of the quarks inside is zero, hence the neutron is electrically neutral. However, the residual effect of the quark charges provides a magnetic field, even though there is no overall electrical field.

The first experimental results confirming the rotational behaviour of fermion states were published in *Physics Letters* in October 1975.[4] The team involved used a specially developed crystal of silicon to split a beam of neutrons, emerging from a nuclear reactor, into two separate paths (Figure 11.5, paths I and II). Neutrons travelling along each path were then reflected by a second crystal so that their tracks merged again and interfered. Neutrons on path I passed through an electromagnet, which rotated their orientation, prior to paths I and II merging again. By appropriately adjusting the electromagnet, the team could control the amount of rotation applied to the path I neutrons, and hence observe the effect this rotation had on the interference. Finally, detectors measured the intensity of the emerging beams (H and O) as the rotation of the neutrons in path I was adjusted.

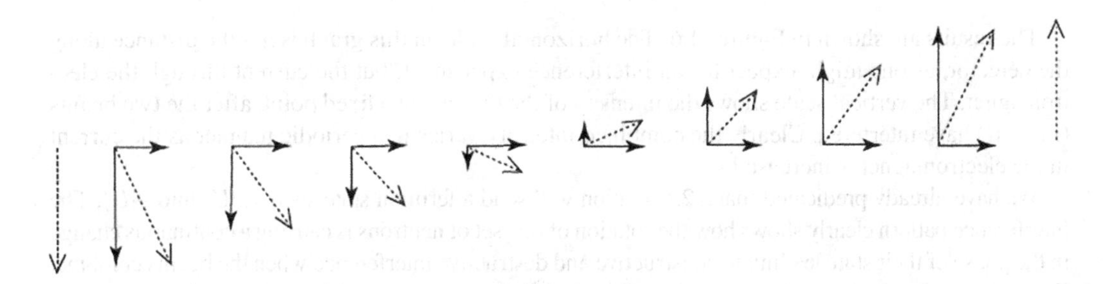

FIGURE 11.4 Rotating an electron in a magnetic field. The dotted arrow represents the magnetic field direction of an electron. The two thick lines are externally applied fields, which change strength along the sequence.

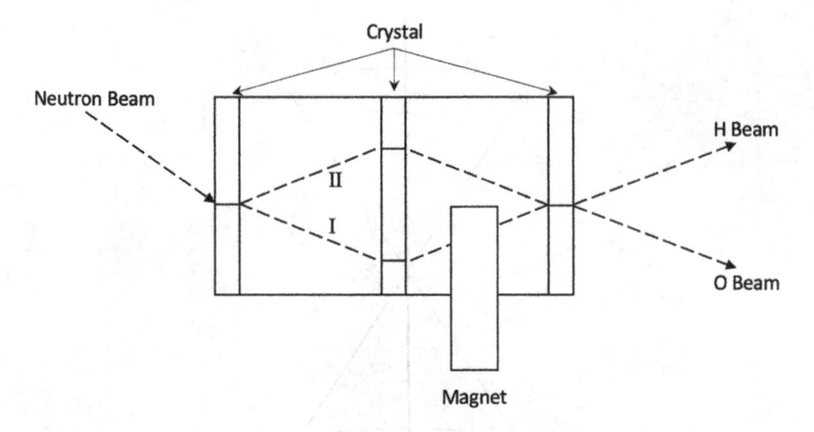

FIGURE 11.5 The experimental arrangement for the neutron phase-shift experiment, as viewed from above. A crystal splits a neutron beam from a reactor along paths I and II. The second chunk of crystal reflects the split beams towards one another. An electromagnet rotates the state in path I, leaving path II unaffected. The O and H beams are formed from the interference of the I and II beams in the final piece of crystal. Finally, the intensity of both beams is measured as the current varies. The results show that as the current changes, the intensity of both beams varies but out of phase with one another.

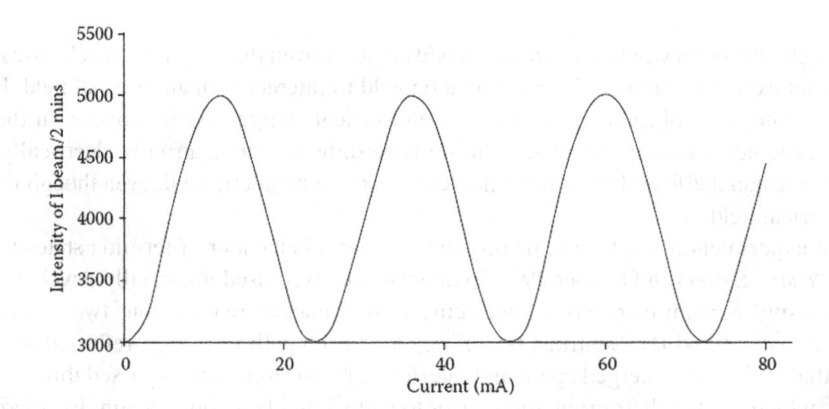

FIGURE 11.6 Typical results from the neutron-interference experiment. The vertical scale is the intensity of the combined I beam and the horizontal scale shows the current through the electromagnet, which is equivalent to a rotation angle for the neutron state. A 2π change in neutron state rotation brings a minimum intensity to a maximum and vice versa. The measured periodicity from the experiment was $(704 \pm 38)°$. More modern experiments have tightened this result more closely to $720° = 4\pi$ radians.

The results are shown in Figure 11.6. The horizontal scale on this graph is *not* the distance along the detector, as one might expect for an interference experiment, but the current through the electromagnet. The vertical scale shows the intensity of the O beam at a fixed point, after the two beams (I and II) have interfered. Clearly the combined intensity varies in a periodic manner as the current in the electromagnet is increased.

We have already predicated that a 2π rotation will send a fermion state such as $|U\rangle$ into $-|U\rangle$, The interference pattern clearly shows how the rotation of one set of neutrons is causing a continuous change in the phase of their state leading to constructive and destructive interference when the beams combine. When the current is changed sufficiently, a 2π rotation comes about, which sends the phase to -1 and destructive interference is produced. Increasing the current further, so that an additional 2π rotation takes place, sends the phase back again to $+1$ and the interference state goes back to what it was.

Given neutrons in state $|U\rangle$, and an operator $\hat{R}(\vartheta)$, responsible for rotating a state about some predefined axis, it would appear that $\hat{R}(\vartheta)|U\rangle = e^{i\vartheta/2}|U\rangle$, giving:

$$\hat{R}(2\pi)|U\rangle = e^{i\pi}|U\rangle = (\cos(\pi) + i\sin(\pi))|U\rangle = -|U\rangle$$

This is a truly remarkable result, which, once again, raises questions about the nature of quantum states. By and large, everyday physical objects look exactly the same if we rotate them through 2π. Yet the quantum state of a fermion is not *entirely* the same. Any probability that we calculate from the state is unchanged, no matter what the rotation angle, as:

$$|U'\rangle = e^{i\vartheta/2}|U\rangle \qquad \langle U'| = e^{-i\vartheta/2}\langle U|$$

giving:

$$\langle U'|U'\rangle = \left(e^{-i\vartheta/2}\langle U|\right)\left(e^{i\vartheta/2}|U\rangle\right) = \langle U|U\rangle$$

However, as we have seen from the results of the neutron experiment, the state is different in a physically meaningful way. The interference that it generates with another state is altered. Perhaps there *is* something 'real' about complex amplitudes.

11.3.2 Coffee Mugs and Quantum States

When we talk about rotating an object, such as a particle, or exchanging two particles (as we did in Chapter 8), we must have some fixed points of reference, or we can't tell what has happened. Normally we use whatever system of coordinates has been set up as a background to our whole theory. However, in more general terms, the mathematical consequences of what we are doing can't depend on the *specific* system of reference that we have chosen.

Consider one of my favourite objects, a coffee mug, floating in empty space. I am using the pattern of distant stars as a reference to orient myself, and hence the alignment of the mug with respect to me. In a moment of weakness, I leave the coffee mug for a period and return to find that a colleague claims to have rotated the mug through 2π about its vertical axis. If I return myself to the same spot and orientation, the 2π rotation of the mug appears to have no effect. It would seem that I have no way of verifying my colleague's claim.

A mathematician would suggest the following. Imagine that we have used a set of strings to anchor the coffee mug to a collection of fixed reference points (e.g., the stars). These strings represent the abstract mathematical links between an object and the fixed reference environment. As such, the strings can be stretched, bent, and twisted, provided their connection points stay fixed to the mug and the reference stars. Cutting a string would break the connection, so that is also not allowed.

If the mug is rotated through 2π about its vertical axis, a twist is placed in all the strings, which can't be removed no matter how we fiddle with things. The only way of untwisting them is to rotate the mug back again. The 'connections' between the mug and the reference coordinate system have been distorted by the rotation and retain a 'memory' of what has happened (see Figure 11.7). The presence, or absence, of twists in the strings would allow us to verify the single 2π rotation of the mug.

However, if the mug is rotated again by 2π in the same direction, we now have a double twist in the strings, which can be removed by taking the strings over the top of the mug. We are allowed this manoeuvre as the connection points stay fixed. Consequently, *double twisted bands are 'connectively' the same as untwisted bands.*

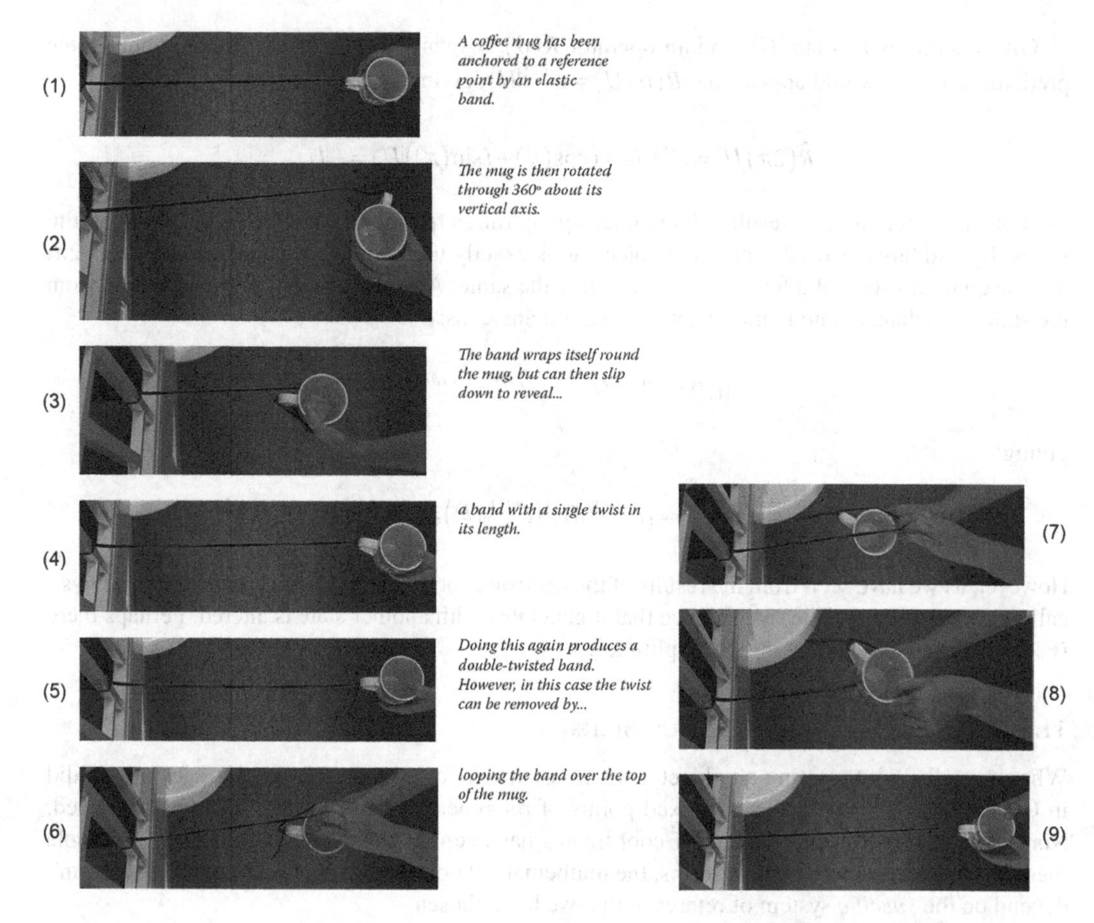

(1)
A coffee mug has been anchored to a reference point by an elastic band.
(2)
The mug is then rotated through 360° about its vertical axis.
(3)
The band wraps itself round the mug, but can then slip down to reveal...
(4)
a band with a single twist in its length.
(5)
Doing this again produces a double-twisted band. However, in this case the twist can be removed by...
(6)
looping the band over the top of the mug.
(7)
(8)
(9)

FIGURE 11.7 The connection between an object and its surroundings, as illustrated by a captive coffee mug.

Here then is a concrete example, as promised earlier, of a 3-D situation which does *not* have 2π rotation symmetry. The twist in the string prevents that. However, a second 2π rotation in the same sense changes things to a situation that can then be easily returned to the starting configuration.

Although we've been talking about coffee mugs, the same sort of argument can be applied to a particle, such as an electron or neutron. As we have seen, a 2π rotation of a fermion makes a physical difference to its state, and a second 2π put things back again.

So, although it looks as if an object must have 2π rotational symmetry, there may be some internal physics that is able to keep track of the situation.

11.3.3 Spin, Symmetry, and Exchanges

Looking back, we have now accumulated three pieces of information about fermions:

1. They have fractional spin: electrons for, example, are spin 1/2.
2. Their identical multiparticle states are antisymmetric under an exchange of any two particles.
3. The phase of a fermion state is inverted by a 2π rotation.

All these features are related to one another.

We have already demonstrated the connection between the phase inversion on rotation and the antisymmetry of the multiparticle state, but we have not related this to spin.

A full understanding of the link between spin and rotation would mean developing a significant amount of mathematics and a familiarity with the theory of relativity, which is sad as tying together the spin of a particle with the exchange properties of its state is one of the most profound and beautiful results in physics.

Fortunately, it is possible to gain a rough understanding of how this works by using a less rigorous but nevertheless subtly elegant argument. To develop this line of thought, we first need to return to the topic of *time*, in the context of quantum theory.

11.4 TIME

The nature of time is one of the great mysteries in physics. It has also confounded philosophical understanding for centuries. Fortunately, for this argument, we only need a loose understanding of time.[5]

One of the distinct characteristics of time is its apparent one-way flow. Time marches on relentlessly from the past towards the future[6]. This has occasionally led physicists to speculate about what might happen if you could reverse the flow of time. Mathematically this is very easy to do; you just replace t with $-t$ in equations that describe physical processes and see what happens. In quantum physics, we even have an operator to do the job.

It is instructive to explore what happens to the spin components of a fermion when we imagine reversing the flow of time. To do that, we must work with the dangerously classical idea that electrons have spin because they are tiny rotating spheres of charge.

I know that we have already shown how this picture can't account for all the quantum properties of spin, and indeed that spin is purely a quantum property without a classical analogue. Unfortunately, unless we try to delve deeply into quantum field theory, we won't get anywhere unless we use this semi-classical picture, with some care and several reservations.

To make the idea of reversing the flow of time clearer, let's propose that we can film the rotation of an electron at a fast enough rate that we can see the tiny sphere turning from one frame to the next. Then we try to visualize what it would look like if we played the film back in the opposite direction.

In Chapter 10, we specified the angular momentum as being in a direction at right angles to the plane of an orbit. For a rotating sphere, the angular momentum is defined by considering a point on the equator of the sphere and thinking of that as a particle orbiting the centre (see Figure 11.8). The arrow points upward if the spin is anticlockwise when viewed from above, and downward if the spin is clockwise. Indeed, this is where the terms 'spin UP' and 'spin DOWN' come from.

As Figure 11.8 illustrates, playing our film backwards will make the electron appear to be rotating in the opposite direction, so it will turn spin UP into spin DOWN. In terms of the basis used to describe spin states, we can write this as follows:

$$\hat{T}|U\rangle = |D\rangle$$

$$\hat{T}|D\rangle = |U\rangle$$

with $\hat{T}$ being the symbol used for the operator[7] that acts to "reverse the direction of time in the following state."

However, I've been a bit hasty with these two equations. I should allow the possibility that $\hat{T}$ will alter the *phase* of the state. Using the argument of Section 11.1, I am allowed to pick one of the

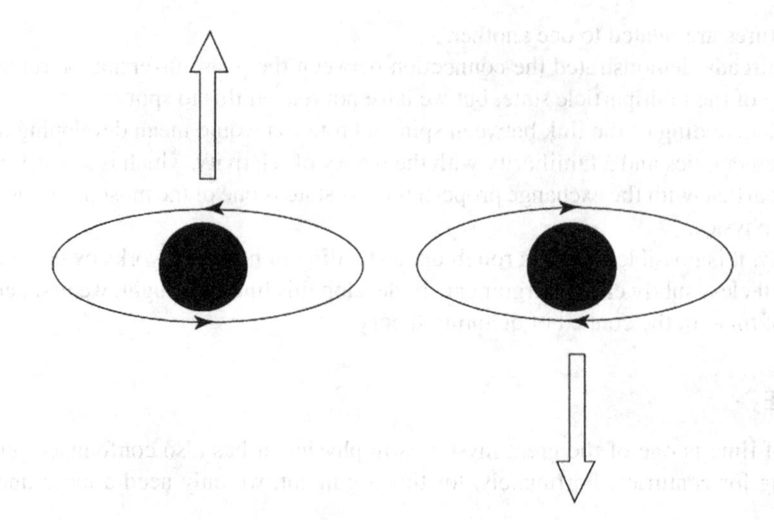

FIGURE 11.8 The electron on the left-hand side is 'spin UP' $\left(+\frac{1}{2}\hbar\right)$ and that on the right-hand side is 'spin DOWN' $\left(-\frac{1}{2}\hbar\right)$. Filming the electron on the left-hand side and then playing the tape backwards would make it look identical to the one on the right-hand side.

states to multiply by a phase factor, as other changes can be absorbed into phase differences and normalization constants. So:

$$\hat{T}|U\rangle = |D\rangle$$

$$\hat{T}|D\rangle = k|U\rangle$$

where k is a complex number of magnitude 1 and phase ϑ.

To obtain the value of k, we apply $\hat{T}$ to a different state such as $|L\rangle$ or $|R\rangle$. Writing $|L\rangle$ and $|R\rangle$ in terms of $|U\rangle$ and $|D\rangle$:

$$|L\rangle = \frac{1}{\sqrt{2}}\big(|U\rangle - |D\rangle\big) \qquad |R\rangle = \frac{1}{\sqrt{2}}\big(|U\rangle + |D\rangle\big)$$

and then applying our time-reversal operator gives:

$$\hat{T}|L\rangle = \hat{T}\left[\frac{1}{\sqrt{2}}\big(|U\rangle - |D\rangle\big)\right] = \frac{1}{\sqrt{2}}\big(\hat{T}|U\rangle - \hat{T}|D\rangle\big) = \frac{1}{\sqrt{2}}\big(|D\rangle - k|U\rangle\big)$$

Given what we said about time reversing the $|U\rangle$ and $|D\rangle$ states, you'd guess that $\hat{T}|L\rangle = |R\rangle$ (at least to within a constant). As $|R\rangle = \frac{1}{\sqrt{2}}\big(|U\rangle + |D\rangle\big)$ we have,

$$\hat{T}|L\rangle = |R\rangle = \frac{1}{\sqrt{2}}\big(|U\rangle + |D\rangle\big) = \frac{1}{\sqrt{2}}\big(|D\rangle - k|U\rangle\big)$$

from which it follows that $k = -1$.

Equally:

$$\hat{T}|R\rangle = \hat{T}\left[\frac{1}{\sqrt{2}}\left(|U\rangle + |D\rangle\right)\right]$$

$$= \frac{1}{\sqrt{2}}\left(\hat{T}|U\rangle + \hat{T}|D\rangle\right)$$

$$= \frac{1}{\sqrt{2}}\left(|D\rangle + k|U\rangle\right) = \frac{1}{\sqrt{2}}\left(|D\rangle - |U\rangle\right)$$

$$= -|L\rangle$$

As one final point, consider the following quick calculation:

$$\hat{T}\left[\hat{T}|U\rangle\right] = \hat{T}|D\rangle = -|U\rangle$$

$$\hat{T}\left[\hat{T}|D\rangle\right] = \hat{T}|U\rangle = -|D\rangle$$

The same applies to the $|R\rangle$ and $|L\rangle$ states.

$$\hat{T}\left[\hat{T}|L\rangle\right] = \hat{T}|R\rangle = -|L\rangle$$

$$\hat{T}\left[\hat{T}|R\rangle\right] = \hat{T}\left(-|L\rangle\right) = -\hat{T}\left(|L\rangle\right) = -|R\rangle$$

Hence, we have demonstrated that a double time-reversal flips the phase of a state.

You would expect that reversing the flow of time and then reversing it back again would put things back just the way they were. In this case, evidently not.

11.4.1 Spinning Things Round

In geometry, we study the properties of points, lines, shapes, and angles. The branch of mathematics called *topology* takes this a step further by looking at how regions of space are connected.

Topologically, any two shapes that can be deformed into another without cutting or tearing are regarded as being the same. To a topologist there would be no difference between a circle and a square as one can be changed into another with a bit of bashing but without cutting any of the lines. However, it would not be possible to change a sphere into a doughnut shape as getting the hole in the middle would require cutting the surface.[8]

Physicists have taken an interest in topology, as it provides the right sort of mathematics to describe how particles relate to one another within space–time.

If we mark the progress of a particle through space and time by a series of 'points' (or more properly *events)* and then connect them, the result is known as a 'world line.' The world line of an electron travelling in a straight line at constant speed would be something like the tape in Figure 11.9a, taking the progress of time to be vertically up the table and the x-direction in space to be horizontally across the table. If we apply two time reversals in sequence, the tape would change to look like that in Figure 11.9b. If you now take either end of the tape and pull it out gently, then at its full extension the tape has a 2π twist in it (Figure 11.9d).

This rather curious demonstration shows that a double-time reversal is topologically equivalent to the electron rotating through 2π as it proceeds on its way.

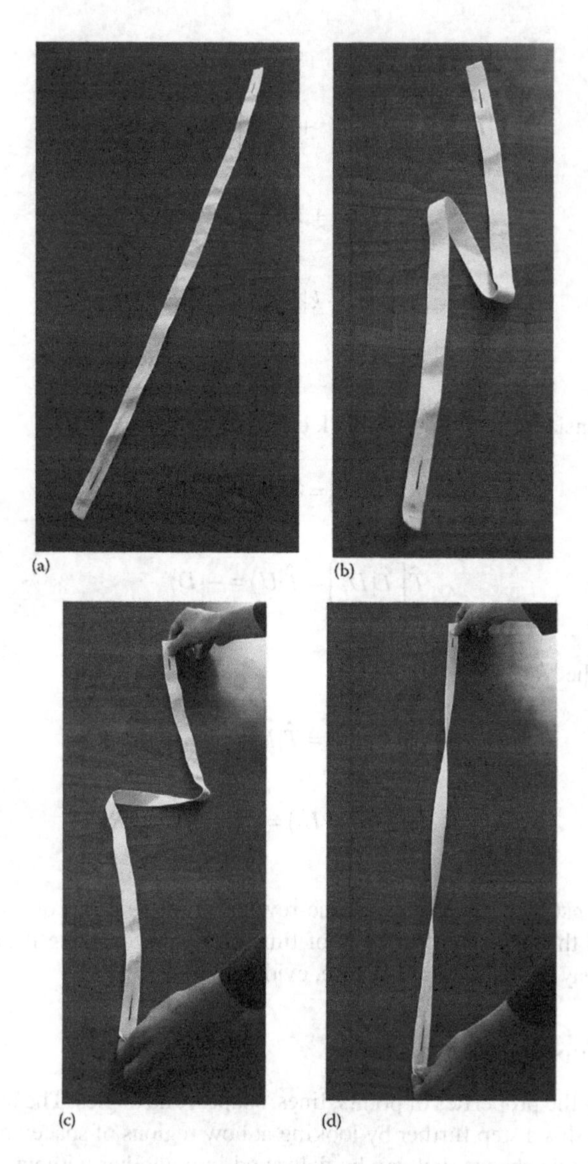

FIGURE 11.9 (a) Take a strip of tape marked at either end and lay it out flat on a surface. (b) Introduce a 'doubling back' twice so that the ends of the tape maintain their orientation. (c) Gradually pull the tape out until it is taut. (d) The tape now has a 2π twist in it.

Another way to think about this is to imagine that the electron has 'feelers' spreading out into space which are its connections to other parts of the universe (we are back to the discussion in Section 11.3.2 again). The continuity of the tape in Figure 11.9 illustrates this to some extent as this is modelling the electron's world line as a continuous line, not a collection of disconnected points. Double time reversing the electron's path will inevitably put 'twists' in these feelers in a manner similar to the electron spiralling through space.

So, the conclusion to this argument is that *a double-time reversal in the world line of an electron produces the same topological effect as rotating the electron through* 2π.

As we have just seen that a double-time reversal introduces a phase flip, *we are entitled to conclude that a* 2π *rotation will produce a phase flip.* In other words, as $\hat{R}(2\pi)|U\rangle = \hat{T}\hat{T}|U\rangle$ and as $\hat{T}\hat{T}|U\rangle = -|U\rangle$, then $\hat{R}(2\pi)|U\rangle = -|U\rangle$.

Of course, we have already shown experimentally that this phase shift on rotation takes place. The purpose of the argument is to establish the π phase shift by examining the effect of a double-time reversal on a spin state, then to 'prove' topologically that double-time reversal is the same as 2π rotation, hence putting some theoretical justification behind why the phase shift happens on rotation.

One rather curious aspect of the argument though is that a rotation can take place through any angle in a continuous range, with a corresponding phase shift. It is not possible however (at least as far as we know!) to reverse time in any partial sense, so the two situations are not completely equivalent.

11.4.2 Rotation for More Fun and Profit

In Section 11.3.1 I suggested that $\hat{R}(\vartheta)|U\rangle = e^{i\vartheta/2}|U\rangle$, which is correct for a rotation about the z axis. For the more general case, it turns out that:

ROTATION OPERATORS

$$\hat{R}_x(\vartheta) = \exp\left(\frac{i\vartheta S_x}{\hbar}\right) = \exp\left(\frac{i\vartheta\sigma_1}{2}\right) \qquad \hat{R}_y(\vartheta) = \exp\left(\frac{i\vartheta S_y}{\hbar}\right) = \exp\left(\frac{i\vartheta\sigma_2}{2}\right) \qquad \hat{R}_z(\vartheta) = \exp\left(\frac{i\vartheta S_z}{\hbar}\right) = \exp\left(\frac{i\vartheta\sigma_3}{2}\right)$$

with the S_x, S_y, S_z being our spin operators related to the Pauli matrices by $S_x = \frac{1}{2}\hbar\sigma_1$ etc.

It may not be entirely obvious how we interpret taking the exponential of a matrix. For that, we need to call on the expansion:

$$e^x = 1 + x + \frac{x^2}{2} + \frac{x^3}{6} + \cdots + \frac{x^n}{n!} + \cdots = \sum_n \frac{x^n}{n!}$$

From the properties of the Pauli matrices, $\sigma_i^2 = I$, hence:

$$\sigma_i^n = I \quad \text{for even values of } n$$

$$\sigma_i^n = \sigma_i \quad \text{for odd values of } n$$

This is very neat, for now, we can apply this to our exponential expansions, for example:

$$\hat{R}_z(\vartheta) = \exp\left(\frac{i\vartheta\sigma_3}{2}\right) = \sum_n \frac{\left(\frac{i\vartheta\sigma_3}{2}\right)^n}{n!}$$

$$= \left[\sum_n \frac{(-1)^n(\vartheta/2)^{2n}}{(2n)!}\right] I + \left[\sum_n \frac{i(-1)^n(\vartheta/2)^{2n+1}}{(2n+1)!}\right]\sigma_3$$

$$= \cos\left(\frac{\vartheta}{2}\right) I + i\sin\left(\frac{\vartheta}{2}\right)\sigma_3$$

if you know[9] your expansions for $\cos x$ and $\sin x$. Using the appropriate matrices:

$$\hat{R}_z(\vartheta)|U\rangle = \left[\cos\left(\frac{\vartheta}{2}\right)I + i\sin\left(\frac{\vartheta}{2}\right)\sigma_3\right]|U\rangle$$

$$= \left[\cos\left(\frac{\vartheta}{2}\right)\begin{pmatrix} 1 & 0 \\ 0 & 1 \end{pmatrix} + i\sin\left(\frac{\vartheta}{2}\right)\begin{pmatrix} 1 & 0 \\ 0 & -1 \end{pmatrix}\right]\begin{pmatrix} 1 \\ 0 \end{pmatrix}$$

$$= \cos\left(\frac{\vartheta}{2}\right)\begin{pmatrix} 1 \\ 0 \end{pmatrix} + i\sin\left(\frac{\vartheta}{2}\right)\begin{pmatrix} 1 \\ 0 \end{pmatrix}$$

$$= e^{i\vartheta/2}|U\rangle$$

as previously stated. For a rotation about the x-axis, a similar argument constructs:

$$\hat{R}_x(\vartheta)|U\rangle = \left[\cos\left(\frac{\vartheta}{2}\right)I + i\sin\left(\frac{\vartheta}{2}\right)\sigma_1\right]|U\rangle$$

$$= \left[\cos\left(\frac{\vartheta}{2}\right)\begin{pmatrix} 1 & 0 \\ 0 & 1 \end{pmatrix} + i\sin\left(\frac{\vartheta}{2}\right)\begin{pmatrix} 0 & 1 \\ 1 & 0 \end{pmatrix}\right]\begin{pmatrix} 1 \\ 0 \end{pmatrix}$$

$$= \cos\left(\frac{\vartheta}{2}\right)\begin{pmatrix} 1 \\ 0 \end{pmatrix} + i\sin\left(\frac{\vartheta}{2}\right)\begin{pmatrix} 0 \\ 1 \end{pmatrix}$$

$$= \cos\left(\frac{\vartheta}{2}\right)|U\rangle + i\sin\left(\frac{\vartheta}{2}\right)|D\rangle$$

The discussion in this section has, up to now, been based entirely on a spin $1/2$ context. That step was taken at the outset when we made the link between S_x, S_y, S_z and the Pauli matrices via $S_x = \frac{1}{2}\hbar\sigma_1$ etc. The introduction of the 2×2 matrices and the factor of $\frac{1}{2}\hbar$ ultimately lead to the $\cos(\varphi/2)$, $\sin(\varphi/2)$ and $e^{i\vartheta/2}$ terms that have abounded in our arguments. Consequently, *all of the states we have dealt with have 4π rotational symmetry.* All of them will phase flip (change sign) under a 2π rotation.

However, our rotation operators are correct for any spin system. All we need is the appropriate set of matrices to deal with the specific context. For example, $\hat{R}_z(\vartheta) = \exp\left(\frac{i\vartheta S_z}{\hbar}\right)$ is always true, but for a spin 1 system $S_z = \hbar\Sigma_3 = \hbar\begin{pmatrix} 1 & 0 & 0 \\ 0 & 0 & 0 \\ 0 & 0 & -1 \end{pmatrix}$, ultimately leading to 2π rotational symmetry[10]…

11.4.3 So Spin Is?

…in which case, *the rotational symmetry of a particle determines its spin.*

Just as we defined the quantum version of momentum in terms of how it advanced the phase of a state over distance, and the energy as the phase change of a state over time, *spin becomes the phase change due to rotation*. A fermion, i.e., a particle with spin equal to an odd integer multiple of $\frac{1}{2}\hbar$, has 4π rotational symmetry, whereas a boson, i.e., a particle with spin equal to an integer multiple of $\hbar$, has 2π rotational symmetry, as we will see.

The link between rotational symmetry, phase flip, time reversal and ultimately the exchange properties of states runs something like this:

- Switching particles in an identical multiparticle state has the same effect on the state as rotating one of them, in place, through 2π.
- Double time reversal applied to an example set of fermion states $(|U\rangle \text{and} |D\rangle)$ causes a phase flip in the states.
- A double time reversal is topologically the same as a 2π rotation.
- Hence, switching fermions in an identical multiparticle state must cause a phase flip
- Hence, fermion multiparticle states are antisymmetric.
- Hence, no two fermions of the same type can ever be in the same state.

This is the link between spin and fermion nature. At root, we could say that it is the behaviour of fermion states under time reversal that leads to the antisymmetric nature of their multiparticle states.

The weak point in the argument is the third bullet point. I have not as yet demonstrated that double time reversals don't do the same thing for bosons. That's the subject of the next section.

11.5 BOSON SPIN STATES

A spin 1 charged particle passing through an SG apparatus produces *three* emerging beams. As with the spin 1/2 case, one beam would be deflected upward (but with more of a deflection) and one beam downward. The big difference to the spin 1/2 example is the presence of the third beam that proceeds straight through without being deflected.

The existence of three beams follows directly from our discussion of angular momentum in the previous chapter. For a spin 1 particle, $s = 1$, thus $m_s = 0, \pm 1$ giving three emerging beams $\{|s, m_s\rangle\} = |1,1\rangle, |1,0\rangle, |1,-1\rangle$. I am going to abbreviate these three states as $|+1\rangle$, $|0\rangle$ and $|-1\rangle$, respectively, dropping the reference to the overall spin, s, in the state's ket.

The semiclassical interpretation of these states is similar to our picture of the spin 1/2 components (see Chapter 10) and is illustrated in Figure 11.10.

Each of the components $\{m_s\} = +1, 0, -1$ is with reference to a specific direction in space. If we wish to experiment using these different m_s states, we need to prepare our particles appropriately, most likely by passing a beam through an SG magnet which separates the m_s along different paths, but with reference to the direction of the SG field. If we use a vertical SG magnet, the three paths will be in a vertical plane, but in a horizontal plane if we use a horizontal SG field. We need an indication of the direction of the preparing field in our state symbol. Something such as $|V, +1\rangle$, $|V, 0\rangle$, $|V, -1\rangle$ and $|H, +1\rangle$, $|H, +1\rangle$, $|H, 0\rangle$, $|H, -1\rangle$. The state $|V, +1\rangle$ would then be the +1 component emerging from a vertically aligned SG magnet. Alternatively, $|H, -1\rangle$ is the −1 component from a horizontal SG magnet.

Now imagine that we have a particle, which we know to be in the state $|V, 0\rangle$, approaching a horizontally aligned SG magnet. To deal with such a situation we would have to write the $|V, 0\rangle$ state in terms of the three horizontal states that act as a basis in this case:

$$|V, 0\rangle = a|H, +1\rangle + b|H, 0\rangle + c|H, -1\rangle$$

where a, b, and c are the complex amplitudes determining the relative probabilities of the three possible outcomes.

I'm not all that interested in the specific values of a, b, and c as I want to try and figure out why it is that spin 1 particle (and by extension bosons, in general) don't flip their signs under exchange.

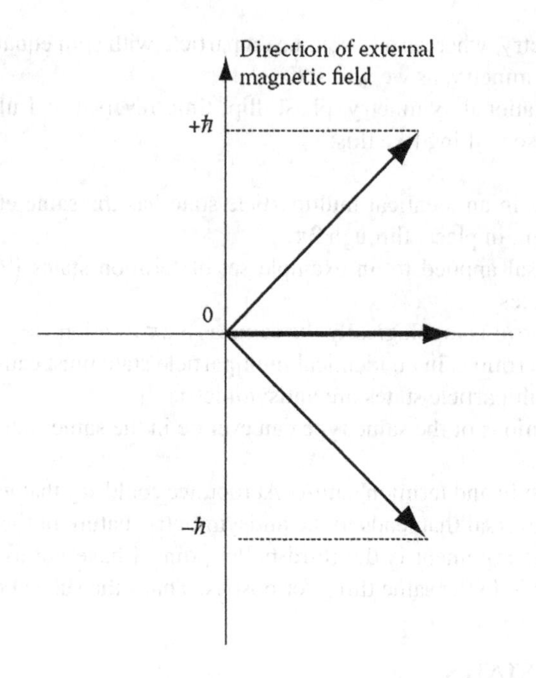

FIGURE 11.10 A spin 1 object can have three possible components with respect to an applied magnetic field.

I can do this by considering what happens to the states when we apply the time-reversal argument as before, and for that I do not need specific values for the amplitudes.

11.5.1 More on Time Reversal

Before we can pursue what happens to boson states when they are 'time reversed' we need to clarify the effect of applying $\hat{T}$ to a general state multiplied by an amplitude. As an illustration of this, consider a free-particle momentum eigenstate:

$$|\Psi\rangle = \int A\exp\left(\frac{i}{\hbar}(p_x x - Et)\right)|x\rangle dx$$

where p_x is the momentum of the particle and E the energy. The amplitude multiplying each basis state $|x\rangle$ is a complex number of the form $z = Ae^{i\varphi} = Ae^{\frac{i}{\hbar}(p_x x - Et)}$, which will be a useful thing to keep in mind for later.

If we apply a time reversal to the state $|\Psi\rangle$, we need to deal with the effect of $\hat{T}$ on the amplitudes and the basis states. We imagine that $|x\rangle$ does not change as the precise point where we might find the particle is not altered, if the dynamics is unchanged by a time reversal.[11] Equally, the value of E will be unaffected as it is a scalar property. However, $p_x \Rightarrow -p_x$ as our 'tape played backward' would show the particle travelling in the opposite direction. Of course, the other alteration is, $t \Rightarrow -t$. Putting these together, the upshot is:

$$\hat{T}|\Psi\rangle = \hat{T}\int A\exp\left(\frac{i}{\hbar}(p_x x - Et)\right)|x\rangle dx$$

$$= \int A\exp\left(\frac{i}{\hbar}(-p_x x + Et)\right)\hat{T}\,|x\rangle dx$$

$$= \int A\exp\left(-\frac{i}{\hbar}(p_x x - Et)\right)|x\rangle dx$$

For any complex number, $z = Ae^{i\vartheta}$, we have $z^* = Ae^{-i\vartheta}$, which tells us that:

$$\hat{T}|\Psi\rangle = \hat{T}\left(\int z(x, t)|x\rangle dx\right) = \int z^* |x\rangle dx$$

$$\hat{T}(A|\varphi\rangle) = A^*|\varphi\rangle$$

Although we have not formally proven this relationship, it is true for the general case.

11.5.2 Time-Reversed Boson States

Armed with a slightly deeper understanding of how $\hat{T}$ works on a state, we can now consider the simplest of the boson states—that of a spin 1 particle in a state with no vertical spin component.

It makes reasonable common sense to propose that

$$\hat{T}|V,0\rangle = k|V,0\rangle$$

because filming a particle spinning 'on its side' and playing the film backwards will still show a particle that is spinning 'on its side.'

To be sure, the direction of the spin will be reversed in this process, but in terms of the *component* of spin (see Figure 11.10) it will still be $|V,0\rangle$. The factor k has been introduced to remind us that although the state may be the same, the process of applying $\hat{T}$ can always multiply the state by an amplitude of magnitude 1 and any phase.

So, following this argument, we should have:

$$\hat{T}|V,0\rangle = k|V,0\rangle = k\left[a|H,+1\rangle + b|H,0\rangle + c|H,-1\rangle\right]$$

where I have expanded the $|V,0\rangle$ state in terms of the set of states appropriate for a horizontal spin measurement (a, b, and c being the appropriate amplitudes for this expansion).

It seems reasonable to further suggest that:

$$\hat{T}|H,+1\rangle = A|H,-1\rangle \quad \hat{T}|H,0\rangle = B|H,0\rangle \quad \hat{T}|H,-1\rangle = C|H,+1\rangle$$

following an argument similar to that applied when we considered time-reversed spin 1/2 states. Here A, B, and C are the phases introduced into the states, as k was earlier.

The next step needs to be carried out carefully:

$$\begin{aligned}\hat{T}|V,0\rangle &= \hat{T}\left[a|H,+1\rangle + b|H,0\rangle + c|H,-1\rangle\right] \\ &= a^*\hat{T}|H,+1\rangle + b^*\hat{T}|H,0\rangle + c^*\hat{T}|H,-1\rangle \\ &= a^*A|H,-1\rangle + b^*B|H,0\rangle + c^*C|H,+1\rangle\end{aligned}$$

Earlier we argued that $\hat{T}|V,0\rangle = k|V,0\rangle$ so:

$$\begin{aligned}k|V,0\rangle &= a^*A|H,-1\rangle + b^*B|H,0\rangle + c^*C|H,+1\rangle \\ &= k\left[a|H,+1\rangle + b|H,0\rangle + c|H,-1\rangle\right]\end{aligned}$$

If this is to work, $ka = c^*C$, $kb = b^*B$, and $kc = a^*A$, from which we can deduce that:

$$A = kc / a^* \quad B = kb / b^* \quad C = ka / c^*$$

Now we are nearly where we want to be, as the next step is to consider what happens if we double time reverse the various states. For example:

$$\hat{T}\hat{T}|H,+1\rangle = \hat{T}(A|H,-1\rangle) = A^*\hat{T}|H,-1\rangle = A^*C|H,+1\rangle$$

$$A^*C = \left(\frac{k^*c^*}{a}\right)\left(\frac{ka}{c^*}\right) = k^*k$$

so that $\hat{T}\hat{T}|H,+1\rangle = k^*k|H,+1\rangle$.

Working this through for the other states reveals that applying $\hat{T}\hat{T}$ to any of the spin states produces k^*k times the original state. Furthermore, as $k^*k = |k|^2 \geq 0$, there is no phase flip (sign change) on double time reversal. Furthermore, applying $\hat{T}\hat{T}$ to any expansion of states such as $|V,\ 0\rangle = a|H,\ +1\rangle + b|H,\ 0\rangle + c|H,\ -1\rangle$ multiplies each part of the expansion by k^*k, shifting the phase of each part by the same amount, which is not physically relevant.

Hence, as a double time reversal is the same as a 2π rotation and a 2π rotation is the same as a swap between two particles, we end up demonstrating that swapping two particles in a boson state does not produce a sign flip. Multiparticle boson states are symmetric.

The key difference between the boson and fermion states is the presence of the 0 components (e.g., $|V,\ 0\rangle$ and $|H,\ 0\rangle$). These states are, in principle, present in any expansion of another state, yet do not change when time reversal is applied. What this does is to 'lock in place' the phase relationships between the other states, which in turn fixes the exchange behaviour.

11.6 DEEP WATERS

The relationship between spin and rotation is one of the deepest results in quantum theory. Unfortunately, we do not have the time (or space) to explore it further. Many of the arguments that I have presented in this chapter lack rigour, but they can be replaced by more formal mathematical discussions in the context of quantum field theory. Nevertheless, we have glimpsed one of the most intriguing aspects of the quantum realm – the extent to which symmetry plays a part in determining both the properties of particles and the span of different species of particles that can exist. If we recall the, very rough, attribution of fermions to 'matter' and bosons to 'field', then we must accept that symmetry helps to delineate the basic structure of reality. The world starts to look more like the expression of abstract mathematical relationships.

NOTES

1 This may seem like a somewhat artificial example, but it is going to pay off later.
2 About a vertical axis through the face of the card that is.
3 To get the reference, study your classic Dr. Who…e.g., *The Sea Devils* (1972) and, https://physicstoday.scitation.org/do/10.1063/PT.5.010323/full/
4 H. Rauch, A. Zeilinger, G. Badurek, A. Wilfing, W. Bauspiess, U. Bonse, *Phy. Lett.*, 1975, October: 425–427.
5 Something roughly along the lines of "time is God's way of making sure that not everything happens at once…."

6 The philosopher Wittgenstein would probably point out that when we use language such as "time marches towards the future" we are implying that the future is already out there for us to approach. Language can bewitch and confuse, but it's all we have.
7 This is not a member of the class of operator that leads to a physical observable.
8 Hmmm, doughnuts….
9 If not, they are easily looked up on the internet: https://people.math.sc.edu/girardi/m142/handouts/10sTaylorPolySeries.pdf
10 Oh, very well… $\Sigma_1 = \frac{1}{\sqrt{2}}\begin{pmatrix} 0 & 1 & 0 \\ 1 & 0 & 1 \\ 0 & 1 & 0 \end{pmatrix}, \Sigma_2 = \frac{1}{\sqrt{2}i}\begin{pmatrix} 0 & 1 & 0 \\ -1 & 0 & 1 \\ 0 & -1 & 0 \end{pmatrix}$.
11 Again, aside from perhaps some multiplying factor that alters the overall phase, which is not physically relevant in an amplitude.

12 Continuous Bases

12.1 REPRESENTATIONS

To write down a complete expression for the state of a system in any circumstance, we need a list of all its physical variables.[1] For a free electron this would be energy, momentum, position, and spin, giving the state:

$$|\Phi\rangle = \left\{ A \sum_n \exp\left(\frac{i}{\hbar}(px_n - Et) \right) |x_n\rangle \right\} \otimes \left\{ a|U\rangle + b|D\rangle \right\}$$

To a mathematician, the '⊗' symbolizes a *tensor product*, but for us, it simply indicates that two independent aspects of the state must be combined to get the full picture.

In general terms one would write:

$$|\Phi\rangle = |\Psi\rangle \otimes |S\rangle$$

where $|S\rangle$ is the spin state of the system and $|\Psi\rangle$ the *spatial state,* the part that deals with energy, momentum, etc. Note that the spin information is not contained within the spatial state and must be dealt with separately.[2]

As we have previously seen, $|S\rangle$ can be expanded over (UP, DOWN), (LEFT, RIGHT) or any other selected basis. These would be different *representations* of the same spin state. Equally $|\Psi\rangle$ can, in general, be expanded over a position basis, momentum basis or energy basis, each of which would be a different representation of that state.

Constructing the different representation of a state is somewhat akin to resolving a vector[3] in traditional physics.

Any vector in a three-dimensional space can be represented by a triplet of *components* (a, b, c) implicitly referring to three reference directions along the axes of a 3-D coordinate system. Axis directions are denoted by a set of *unit-vectors,* $\{\mathbf{i}, \mathbf{j}, \mathbf{k}\}$, subject to the conditions $\mathbf{i}\cdot\mathbf{j} = \mathbf{i}\cdot\mathbf{k} = \mathbf{j}\cdot\mathbf{k} = 0$ and $\mathbf{i}\cdot\mathbf{i} = \mathbf{j}\cdot\mathbf{j} = \mathbf{k}\cdot\mathbf{k} = 1$ where ' ' is the vector dot product.[4] Any vector $\mathbf{V}$ can hence be written:

$$\mathbf{V} = a\mathbf{i} + b\mathbf{j} + c\mathbf{k}$$

It is both sensible and conventional to ensure that the three selected axes are orthogonal (i.e., they are all arranged at 90° to each other), but aside from that, the specific orientations chosen are at the whim and convenience of an individual physicist. For that reason, the properties of the vector $\mathbf{V}$ must be independent of the specific set of axes chosen.

If another physicist chose the vectors $\{\mathbf{l}, \mathbf{m}, \mathbf{n}\}$ as axes, then the same vector would be:

$$\mathbf{V} = A\mathbf{l} + B\mathbf{m} + C\mathbf{n}$$

subject to the constraint:

$$V^2 = A^2 + B + C^2 = a^2 + b^2 + c^2$$

It would not be difficult to figure out the set of rules that mapped $\{\mathbf{i}, \mathbf{j}, \mathbf{k}\} \rightarrow \{\mathbf{l}, \mathbf{m}, \mathbf{n}\}$ and in the same vein $(a, b, c) \rightarrow (A, B, C)$. Our vector would then have two (or more) representations,

DOI: 10.1201/9781003225997-14

(a, b, c), (A, B, C),etc., but *the physics of the situation would be entirely independent of the representation chosen.* Note also that the properties of the vector are contained with the values of the components, appropriate for the axes chosen.

It is frequently the case that one representation is more convenient for analysis purposes. Perhaps some symmetries of the situation, or some significant directions, strongly suggest themselves as a means of simplifying the calculations. For example, when dealing with the (standard) problem of a box sliding down a slope, the directions perpendicular and parallel to the slope are convenient for resolving the forces acting on the box.

In quantum theory, the same state $|\psi\rangle$ can be expanded over different bases $\{|q_i\rangle\},\{|Q_i\rangle\}$:

$$|\psi\rangle = \sum_i a_i |q_i\rangle = \sum_i A_i |Q_i\rangle$$

which form different representations of the state. The analogy with the mathematics of vectors is clear:

3-D VECTORS	QUANTUM STATES
Vector components (a, b, c)	Amplitudes $\{a_i\}$
Unit vectors $\{\mathbf{i}, \mathbf{j}, \mathbf{k}\}$	Basis $\{|q_i\rangle\}$

In some developments, $|\psi\rangle$ is called the *state vector* to make the point explicit. Just as the properties of an ordinary vector are contained in the components, so *the nature of the state is expressed via the amplitudes.*

Switching between bases is facilitated by expanding one basis over the states of the other:

$$|Q_i\rangle = \sum_n \langle q_n | Q_n \rangle |q_n\rangle$$

which is essentially the same process as moving from one axis set to another.

As we have seen, there are clear points of comparison between the vector and state situations, but there are also fundamental differences:

- The unit-vectors $\{\mathbf{i}, \mathbf{j}, \mathbf{k}\},\{\mathbf{l}, \mathbf{m}, \mathbf{n}\}$ etc. specify directions in physical 3-d space
- The bases $\{|q_i\rangle\},\{|Q_i\rangle\}$ etc. represent 'directions' in a multi-dimensional abstract space, known in the trade as *Hilbert space.* There may be some directionality associated with the basis states, for example, a $|U\rangle$ and $|D\rangle$ are aligned with the axis of an SG magnet, but the states themselves have no physical direction. They do not exist within a directly measurable space.
- In some instances, there are an infinite number of basis states in the set. For example, a position basis might be limited in the range of positions relevant (say from $x = 0$ to $x = 1$), but as position is a continuous variable, that is still an infinite set. This means that the Hilbert space has an infinite number of 'dimensions'. We will see how to cope with these situations in the next section.
- The choice of basis is influenced by any proposed measurements on the state. The choice of axes for a vector representation is not tied to any formal measurement.
- Operators play a crucial role in quantum theory, in that they 'symbolize' the process of measurement to extract information from a state. The mathematical form of an operator

depends on the basis being used. We will see specific examples of this in Section 12.3.3. Operators can be transformed from one representation to another, although the process is not something that we need to discuss.

The mathematical theory of representations was developed by Dirac (of bra-ket notation fame). Given the range of crucial contributions that he made to both quantum theory and quantum field theory, it is interesting that the theory of representations was the work that he was reportedly most proud of.

12.2 TWO ISSUES

We must now deal with two pressing and related issues regarding state expansions.

Many of the physical variables that we need to deal with are continuous in nature, and so require a basis set with an infinite number of members for their expression. Previously, when dealing with position, we have fudged things by dividing the region of interest into a large but finite number of chunks (the eventual cure for our problem will be a development of this idea), but momentum and energy, for example, would be equally problematic. The infinite basis set is one of the two issues that we need to contend with. The other is a little more subtle.

12.2.1 Probability Density

If we ask for the probability that a particle will be found at a precisely specified location, say $x = 0.7$ from the range $0 \le x \le 1$, then, perhaps surprisingly, the answer is *zero.* After all, we are asking for the probability of one instance out of an infinite number of possibilities. Although 1 / infinity is not an allowed calculation, we can certainly see that 1 / very large number $\rightarrow 0$ as the number gets larger and larger.

The thought that we can get around this because a particle must occupy a small range of positions, by dint of its size, is rendered moot as quantum theory and quantum field theory consider fundamental particles, like the electron, as inherently point-like.[5]

In essence, asking for the probability of finding a point-like particle at a precise location is *an ill-framed question*. This is not an issue confined to quantum theory. Standard probability theory must deal with the same situation, so a solution is already on the table.

Instead of trying to find the probability at a point, which is not well defined, we need to calculate the probability of finding our particle within a small range of positions, $x_1 \le x \le x_1 + \Delta x$, centred on the specified location.

Let's take a region of space e.g., L in Figure 12.1 and cut it up into many chunks of equal size Δx. One such chunk is the rectangular box in Figure 12.1, starting at $x = x_1$ and ending at

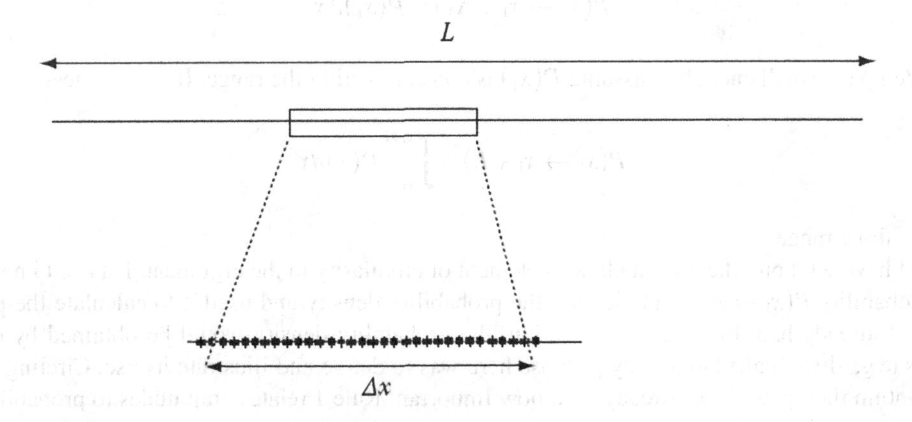

FIGURE 12.1 A range of possible positions L can be cut up into many smaller chunks, Δx, inside which the probability for finding a particle is about the same for every point.

$x = x_1 + \Delta x$. Supposed that we know that the probability of finding a particle within that region is $P(x_1 \rightarrow x_1 + \Delta x)$. What is to stop us from dividing that probability by Δx to find the probability of a particle being at any individual spot in that range?

Well, firstly, dividing a probability by a distance does not yield a probability.

Secondly, it is very likely that the probability is not evenly distributed across the whole region. If we cut Δx into even smaller slices, δx, we would discover that the probability of locating the particle was different, slice by slice, even if only by a small amount. However, if we make δx progressively smaller, by cutting Δx into more and more pieces, with each piece starting at x_i and ending at $x_i + \delta x$, and calculate $P(x_i \rightarrow x_i + \delta x) / \delta x$ then something interesting happens. As we gradually make δx smaller, so the probability $P(x_i \rightarrow x_i + \delta x)$ gets smaller as well. Mathematically, would write:

$$\text{as } \delta x \rightarrow 0, \text{ so } P(x_i \rightarrow x_i + \delta x) \rightarrow 0$$

which is consistent with our argument that the probability at a single point is zero. However, even if $\delta x \rightarrow 0$ and $P(x_i \rightarrow x_i + \delta x) \rightarrow 0$, that does *not* mean that $\dfrac{P(x_i \rightarrow x_i + \delta x)}{\delta x} \rightarrow 0$. In fact, within a broad class of 'well behaved' situations, $\dfrac{P(x_i \rightarrow x_i + \delta x)}{\delta x} \rightarrow$ finite value. This process of taking successive stabs at calculating a quantity $\dfrac{P(x_i \rightarrow x_i + \delta x)}{\delta x}$ as $\delta x \rightarrow 0$ is known as *taking the limit*. The procedure is like making a series of approximations, successively tightening the estimated value towards some ultimate figure. With this in mind, we define the *probability density* $\mathcal{P}(x)$ by:

$$\mathcal{P}(x) = \lim_{\delta x \rightarrow 0} \left(\frac{P(x \rightarrow x + \delta x)}{\delta x} \right)$$

which can, and does, have a finite, non-zero value at any specific isolated x. Unlike a probability, which has no associated unit, the probability density is measured in m^{-1}.

Having obtained the probability density, we are then free to calculate the probability of finding a particle in a region Δx about x_1:

$$P(x_1 \rightarrow x_1 + \Delta x) = \mathcal{P}(x_1)\Delta x$$

provided Δx is small enough to assume $\mathcal{P}(x_1)$ is constant within the range. If not, we need:

$$P(x_1 \rightarrow x_1 + L) = \int_{x_1}^{x_1 + L} \mathcal{P}(x)dx$$

for any finite range.

As I have set it out, there is an obvious element of circularity to the argument. I started knowing the probability $P(x_1 \rightarrow x_1 + \Delta x)$, defined the probability density and used it to calculate the probability I already had. In a realistic situation, the probability density would be obtained by other means (e.g., direct calculation), my purpose here was to define and illustrate its use. Circling back to quantum theory, we have already seen how Important Rule 1 relates amplitudes to probabilities, in the case of discrete bases:

Important Rule 1: The Born Rule

If

$$|\varphi\rangle = a_1|A\rangle + a_2|B\rangle + a_3|C\rangle + a_4|D\rangle + \cdots$$

then:

$$\text{Prob}\big(|\varphi\rangle \to |A\rangle\big) = a_1^* \times a_1 = |a_1|^2$$

$$\text{Prob}\big(|\varphi\rangle \to |B\rangle\big) = a_2^* \times a_2 = |a_2|^2$$

Now we need to extend its scope into the continuous case, e.g.:

$$P(x_1 \to x_1 + L) = \int_{x_1}^{x_1+L} |\langle x|\varphi\rangle|^2\, dx$$

which, given our previous statement:

$$P(x_1 \to x_1 + L) = \int_{x_1}^{x_1+L} \mathcal{P}(x)dx$$

indicates that the probability density in quantum theory is related to the amplitude:

$$\mathcal{P}(x) = |\langle x|\varphi\rangle|^2$$

This, in turn, gives us a useful clue for dealing with continuous bases.

12.2.2 Infinite State Expansions

The problem with a continuous basis, e.g., position, is that we must move from a sum over a very large, but finite, basis set $|\Phi\rangle = \sum_n a_n|x_n\rangle$ into a continuous situation, somewhat like $|\Phi\rangle = \sum_x a(x)|x\rangle$, which may not be well defined. After all, if $a(x) > 0$ for all x, then the sum could well be infinite, which would lead to a strange probability (to say the least).

When we set out to fudge the issue over a position basis, we followed a similar procedure to the calculation in the previous section. Taking a region of length L between x_1 and $x_1 + L$, we divide it into sections of width δx. If these sections are small enough, we can assume that a_i, the amplitude for finding a particle in that section, is constant between[6] $x_1 + i\delta x$ and $x_1 + (i+1)\delta x$. As each section is identified with a basis state $|x_1 + i\delta x \to x_1 + (i+1)\delta x\rangle$, a state expansion takes the form:

$$|\Phi\rangle = \sum_{i=1}^{i=L/\delta x} a_i |x_1 + i\delta x \to x_1 + (i+1)\delta x\rangle = \sum_{i=1}^{i=L/\delta x} a_i|x_i\rangle$$

abbreviating $|x_1 + i\delta x \to x_1 + (i+1)\delta x\rangle$ to $|x_i\rangle$.

We need to manage the transition between this summation and the continuous case as $\delta x \to 0$. As part of that, we would expect the amplitude a_i to become the amplitude function $a(x)$ with the resulting identity $|a(x)|^2 = \mathcal{P}(x)$. Given that the probability density is measured in m^{-1}, our amplitude function must have the units m$^{-1/2}$. To ensure that the unitless amplitude becomes the amplitude function measured in m$^{-1/2}$, we write:

$$|\Phi\rangle = \sum_{i=1}^{i=L/\delta x} a_i |x_i\rangle = \sum_{i=1}^{i=L/\delta x} \left(\frac{a_i}{\sqrt{\delta x}}\right)\left(\frac{|x_i\rangle}{\sqrt{\delta x}}\right)\delta x$$

with the presumption that:

$$\lim_{\delta x \to 0}\left(\frac{a_i}{\sqrt{\delta x}}\right) = a(x) \text{ and } \lim_{\delta x \to 0}\left(\frac{|x_i\rangle}{\sqrt{\delta x}}\right) = |x\rangle$$

as we saw with the probability density. In the process:

$$|\Phi\rangle = \lim_{\delta x \to 0}\left(\sum_{i=1}^{i=L/\delta x}\left(\frac{a_i}{\sqrt{\delta x}}\right)\left(\frac{|x_i\rangle}{\sqrt{\delta x}}\right)\delta x\right) = \int_{x_1}^{x_1+L} a(x)|x\rangle dx$$

using the definition of the Riemann integral, which may be familiar to you.

While this is not a rigorous argument, it does establish the correct result. In particular, the formal alignment between the amplitude function and the probability density:

$$P(x_1 \to x_1 + L) = \int_{x_1}^{x_1+L} \mathcal{P}(x)dx = \int_{x_1}^{x_1+L} |\langle x|\varphi\rangle|^2 dx$$

is an aspect of the Born Rule and so axiomatic in some interpretations of quantum theory.

12.2.3 The Identity Operator

For any properly normalized state $|\varphi\rangle, \langle\varphi|\varphi\rangle = 1$ (Section 4.2.1), so if we expand the state over a discrete basis $\{|i\rangle\}$:

$$|\varphi\rangle = \sum_i \langle i|\varphi\rangle|i\rangle \quad \text{and} \quad \langle\varphi| = \sum_j \langle\varphi|j\rangle\langle j|$$

we get:

$$\langle\varphi|\varphi\rangle = \left(\sum_j \langle\varphi|j\rangle\langle j|\right)\left(\sum_i \langle i|\varphi\rangle|i\rangle\right) = \sum_{i,j}\langle\varphi|j\rangle\langle i|\varphi\rangle\langle j|i\rangle$$

Remembering that $\langle j|i\rangle = \delta_{ij}$ (Section 4.2.1), this reduces to:

$$\langle\varphi|\varphi\rangle = \sum_i \langle\varphi|i\rangle\langle i|\varphi\rangle = \langle\varphi|\left(\sum_i |i\rangle\langle i|\right)|\varphi\rangle = 1$$

which suggests a useful identity:

$$\sum_i |i\rangle\langle i| = \hat{I}$$

with $\hat{I}$ being the *identity operator* which, as you would guess, is an operator that leaves anything that it acts on unchanged. It's more useful than you might think…

A similar argument can be constructed by expanding $|\varphi\rangle$ over a continuous basis, e.g., position[7]:

$$|\varphi\rangle = \int_{x_1}^{x_1+L} \langle x|\varphi|x\rangle dx \quad \text{and} \quad \langle\varphi| = \int_{y_1}^{y_1+L} \langle\varphi|y\rangle\langle y|dy$$

This leads to:

$$\langle\varphi|\varphi\rangle = \int_{x_1}^{x_1+L}\int_{y_1}^{y_1+L} \langle\varphi|y\rangle\langle x|\varphi\rangle\langle y|x\rangle dxdy$$

The equivalent of $\langle j\,|\,i\rangle = \delta_{ij}$ for the continuous situation is $\int_{y_1}^{y_1+L} \langle y|x\rangle dy = \delta(x-y)$ where $\delta(x-y)=1$ if $x=y$ and $\delta(x-y)=0$ if $x \neq y$. The term $\delta(x-y)$ is known as the *Dirac delta function*, and we will define it a little more cautiously in Section 12.3.3.

This reduces our equation to:

$$\langle\varphi|\varphi\rangle = \int_{x_1}^{x_1+L} \langle\varphi|x\rangle\langle x|\varphi\rangle dx = \langle\varphi|\left(\int_{x_1}^{x_1+L} |x\rangle\langle x|dx\right)|\varphi\rangle = 1$$

suggesting that:

$$\int_{x}^{x+L} |x\rangle\langle x|dx = \hat{I}$$

None of the steps in this argument have specifically depended on us using the position basis, so *the conclusion should apply to all continuous basis sets*. For example, if we chose a momentum basis in the x-direction:

$$\int_{-\mathbb{P}}^{\mathbb{P}} |p_x\rangle\langle p_x|dp_x = \hat{I}$$

The integration limits have been selected to respect the vector nature of momentum, so that p can take any value between $-\mathbb{P}$ and $+\mathbb{P}$ in a full basis set.

Armed with this, we can simply write down expressions such as:

$$|\varphi\rangle = \int_{-\mathbb{P}}^{\mathbb{P}} |p_x\rangle\langle p_x|\varphi\rangle dp_x \quad \text{and} \quad \langle\varphi| = \int_{-\mathbb{P}}^{\mathbb{P}} \langle\varphi|\,p_x\rangle\langle p_x|dp_x\rangle$$

and

$$P(-\mathbb{P} \le p \le \mathbb{P}) = \int_{-\mathbb{P}}^{\mathbb{P}} \left|\langle p_x\,|\,\varphi\rangle\right|^2 dp_x$$

12.2.4 A Short Aside: Projection Operators

Time for a short aside to develop some ideas which will be very useful to us later…

In any of these expansions:

$$\sum_i |i\rangle\langle i| = \hat{I}$$

$$\int_{x}^{x+L} |x\rangle\langle x| dx = \hat{I}$$

$$\int_{-\mathbb{P}}^{\mathbb{P}} |p_x\rangle\langle p_x| dp_x = \hat{I}$$

the individual terms, $|i\rangle\langle i|, |x\rangle\langle x|, |p_x\rangle\langle p_x|$ are known as *projection operators*:

$$\hat{P}_i = |i\rangle\langle i|$$

as they effectively 'resolve' or 'project' any state onto the basis:

$$\hat{P}_i|\psi\rangle = |i\rangle\langle i\,|\,\psi\rangle = \langle i\,|\,\psi\rangle|i\rangle = a_i\,|\,\psi\rangle$$

where a_i comes from:

$$|\psi\rangle = \sum_i a_i|\psi\rangle = \sum_i \langle i|\psi\rangle\,|\,i\rangle$$

One amusing property is that $\hat{P}_i^2 = \hat{P}_i$, as:

$$\hat{P}_i^2 = \left(|i\rangle\langle i|\right)\left(|i\rangle\langle i|\right) = |i\rangle\langle i|i\rangle\langle i| = |i\rangle\langle i|$$

Also:

$$\hat{P}_i^* = \left(|i\rangle\langle i|\right)^* = \hat{P}_i$$

If $|\varphi\rangle$ is an eigenvector of $\hat{P}_i$, then $\hat{P}_i|\varphi\rangle = \lambda\,|\,\varphi\rangle$. Also $\hat{P}_i^2|\varphi\rangle = \hat{P}_i\hat{P}_i|\varphi\rangle = \lambda^2|\varphi\rangle$. However, as $\hat{P}_i^2 = \hat{P}_i$ then $\hat{P}_i^2|\varphi\rangle = \lambda^2|\varphi\rangle = \hat{P}_i|\varphi\rangle = \lambda|\varphi\rangle$. So, $\lambda = 1, 0$.

It is easy to see that the eigenstates of the projection operators are the relevant individual basis states:

$$\hat{P}_i|i\rangle = |i\rangle\langle i\,|\,i\rangle = 1|i\rangle$$

$$\hat{P}_i|k\rangle = |i\rangle\langle i\,|\,k\rangle = 0|i\rangle$$

and the eigenvalues are clearly +1, 0. Note that as the only eigenvalues are real, this makes $\hat{P}_i$ Hermitian (Section 5.4.1) and hence potentially linked to an observable. The Hermitian nature of $\hat{P}_i$ can also be confirmed by a simple calculation:

$$\left(\hat{P}_i\right)^\dagger = \left(|i\rangle\langle i|\right)^\dagger = \left((|\,i\rangle)^*(\langle i\,|)^*\right)^T = \left((|\,i\rangle)^*\right)^T\left((\langle i\,|)^*\right)^T = |i\rangle\langle i| = \hat{P}_i$$

Projection operators belonging to any basis set necessarily commute with each other. It is easy to show this. From the basis set $\{|\,i\rangle\}$ we pick out the projection operators $\hat{P}_i = |i\rangle\langle i|$ and $\hat{P}_j = |j\rangle\langle j|$. Consequently:

$$\hat{P}_i\hat{P}_j - \hat{P}_j\hat{P}_i = |i\rangle\langle i|j\rangle\langle j| - |j\rangle\langle j|i\rangle\langle i|$$

If $i \neq j$ then $\langle i|j\rangle = \langle j|i\rangle = 0$ and $\hat{P}_i\hat{P}_j - \hat{P}_j\hat{P}_i = 0$.
If $i = j$, then $\langle i\,|\,i\rangle = \langle j\,|\,j\rangle = 1$ and $\hat{P}_i\hat{P}_j - \hat{P}_j\hat{P}_i = 0$.

We can summarize this, along with the property $\hat{P}_i^2 = \hat{P}_i$ by writing $\hat{P}_i\hat{P}_j = \delta_{ij}\hat{P}_j$.

Finally, *any operator $\hat{O}$ associated with a physical variable can be written in terms of projection operators formed from the eigenstates of $\hat{O}$*. Let the basis set $\{|\,\varphi_i\rangle\}$ be the eigenstates of $\hat{O}$. The eigenvalues are o_i according to $\hat{O}|\varphi_i\rangle = o_i\,|\,\varphi_i\rangle$. We expand an arbitrary state:

$$|\Psi\rangle = \sum_k a_k\,|\,\varphi_k\rangle$$

and act on it with $\hat{O}$:

$$\hat{O}|\Psi\rangle = \hat{O}\sum_k a_k|\varphi_k\rangle = \sum_k a_k\hat{O}|\varphi_k\rangle = \sum_k a_k o_k|\varphi_k\rangle$$

Now we write:

$$\hat{O} = \sum_k o_k|\varphi_k\rangle\langle\varphi_k| = \sum_k o_k\hat{P}_k$$

and act on $|\Psi\rangle$ producing:

$$\hat{O}|\Psi\rangle = \left(\sum_k o_k|\varphi_k\rangle\langle\varphi_k|\right)|\Psi\rangle = \sum_k o_k|\varphi_k\rangle\langle\varphi_k|\Psi\rangle = \sum_k o_k a_k\,|\,\varphi_k\rangle$$

as before, establishing the result. We can confirm this by calculating the expectation value, $\hat{O}$, as:

$$\langle\hat{O}\rangle = \langle\psi|\hat{O}|\psi\rangle = \langle\psi\,|\sum_k o_k\hat{P}_k|\psi\rangle = \sum_k o_k a_k\langle\psi\,|\,\varphi_k\rangle$$

$$= \sum_k o_k a_k \sum_l a_l^*\langle\varphi_l\,|\,\varphi_k\rangle = \sum_k o_k|a_k|^2$$

This opens a different perspective on the measurement process, as it can now be viewed as a collation of specific yes/no questions. Each projection operator, $\hat{P}_k$, with eigenvalues 1 (for yes) and 0 (for no), poses the question 'are you in state $|\,k\rangle$?' or equivalently 'do you have value o_k'.

As the projection operators always commute, they represent a set of yes/no 'measurements' that can be conducted, at the same time without being subject to the strictures of the uncertainty principle.

We will return to this approach in Chapters 26 and 30.

12.3 STATE FUNCTIONS AND WAVE FUNCTIONS

In the earliest forms of quantum theory, before Dirac introduced the bra-ket notation, our amplitude function $a(x) = \langle x\,|\,\varphi\rangle$ would have been written as $\varphi(x)$ and went by the name *state function*. In many situations, both position and time will be of relevance, in which case the *wave function* $\Psi(x,t) = \langle x,\,t\,|\,\varphi\rangle$ is used instead.

In the early quantum theoretical work on the wave nature of electrons, physicists were searching for a physical wave that could either guide (De Broglie) or be (Schrödinger) the 'true' nature of an electron. By a physical wave, I mean something existing in real space, spread out or condensed in some region, with a measurable amplitude (rather like the height of a water wave). Although they had no idea what this physical wave might be, they called it the wave function, gave it the symbol Ψ, and searched for an equation that would describe its behaviour.

Schrödinger was the first person to write down a suitable equation for $\Psi(x,t)$ (January 1926). His great triumph was to use the Schrödinger Equation to derive $\Psi(x,t)$ for the electron orbiting a hydrogen nucleus and then show how many of the known properties of hydrogen atoms (such as the spectrum of light they emit) could be explained from this wave function.

Meanwhile, working in parallel, Dirac was developing his theory of states and amplitudes based on some of Heisenberg's early work. In turn, Heisenberg, together with Born and Jordan, had been expanding his early ideas using matrices.

With all this work going on in parallel and providing similar results, it was inevitable that people would search for links between the different approaches. Schrödinger contributed an outline of how his own 'wave mechanics' and the more abstract Heisenberg/Jordan/Born 'matrix mechanics' could relate to one another (May 1926). Despite this breakthrough, Schrödinger still felt that the wave function was a physical wave of some form.

The decisive step was taken by Max Born who demonstrated how quantum theory could be placed on an entirely consistent footing by interpreting $|\Psi(x,t)|^2 \Delta x$ as the probability of finding a particle in the region of x at time t (October 1926).

In many ways, Ψ is no more significant than any other amplitude that we've talked about. It happens to be the subject of the Schrödinger equation and so tends to be the amplitude function we reach for first in any given situation. Once armed with $\Psi(x,t)$, we can use Dirac's rules to transfer to any other related basis.

In some books on quantum theory, states are not referenced at all, in a formal way. The theory can be developed entirely by working with the wave function, which is not tied to any state expansion as we have done. However, as our goal is to introduce quantum field theory, a thorough understanding of states, bases and state expansions is a preferable starting point.

12.4 OBSERVABLES

So far, we have mentioned some operators that relate directly to physical variables (such as $\hat{S}_x$, $\hat{S}_y$, $\hat{S}_z$, $\hat{p}_x$ for momentum, and $\hat{E}$ for energy) and one that rolls a quantum state into the future, $\hat{U}$. However, we haven't discussed why physical variables must be treated in this manner in quantum physics. To explore this, we're going to start with a consideration of momentum.

12.4.1 The Problem of Momentum

In classical physics the momentum of an object in any given direction has a strict definition, which is evidently and clearly related to observable properties:

$$p_x = m\left[\frac{x_2 - x_1}{t_2 - t_1}\right]$$

where p_x is the momentum in the x-direction, m the mass of the object, x_1 its position at time t_1, and x_2 its position at t_2. From this definition, it's easy to imagine a simple set of measurements used to obtain momentum. One suitable procedure would involve taking two photographs at times t_1 and t_2 to observe the position of the object against some prearranged length scale (ruler). Translating this into an experimental arrangement suitable for the quantum realm, however, is not straightforward.

For starters, we would need a collection of identical particles in the same state $|\varphi\rangle$, which is presumably not an eigenstate of position. Next, we would have to carry out the first position measurement on each particle at the same time, t_1. This produces a set of measured positions $\{x_i\}$ with each possible value appearing a certain number of times, depending on the probability of the position being measured. The final step would then be to measure the position again, at time t_2. However, and here is the catch, after the first position measurement we no longer have a collection of particles in the same state. The position measurement has collapsed the same state $|\varphi\rangle$ for each particle into a collection of states $\{|x_i\rangle\}$ with a subcollection of particles in each possible state.

One solution would be to take only the particles with a specific position, say x_1, after the first measurement and do a repeated position measurement on them. However, as we have seen previously, a position eigenstate cannot be a momentum eigenstate. We can, if we wish, take a second position measurement on this subcollection of particles and then employ $p_x = m\left[\dfrac{x_2 - x_1}{t_2 - t_1}\right]$ to calculate something…but it is not clear what the result represents.

Our second position measurement, carried out only on the particles with state $|x_1\rangle$ after the first, will yield another range of possible positions $\{x_2\}$, so we will get a wide variation of p values at the end of this calculation. And, lest we forget, we have used only a fraction of the particles that we started with. What can we say about the momentum of the others?

The root of our problem is, once again, the inappropriateness of trying to accuse a quantum particle of having a definite path.

Yet for all this, there must be some way of dealing with momentum, for no other reason than the imperative that quantum theory yields classical physics in a suitably limited domain.

The problem doesn't lie with the *idea* of momentum, but with the *classical definition* enthroned in $p_x = m\left[\dfrac{x_2 - x_1}{t_2 - t_1}\right]$. The classical definition cannot be quantized. We need another approach. In essence we need a quantum reinterpretation of what we mean by momentum.

12.4.2 Momentum in Quantum Theory

Looking back through our development of quantum theory so far, we can spot an intriguing possibility that might yield a suitable quantum approach to momentum. Back in Section 6.2.3, we introduced the De Broglie relationship, $p_x = h/\lambda$ which relates momentum to the wavelength of the wave aspect of a quantum particle. From Section 7.2.3 we also have the free particle state:

$$|\psi(x_n,t)\rangle = \sum_n a(x_n, t)|x_n\rangle = A\sum_n e^{i\left(\frac{2\pi}{\lambda}x_n - 2\pi ft\right)}|x_n\rangle$$

which generalizes into a continuous basis:

$$|\psi(x,t)\rangle = A\int e^{i\left(\frac{2\pi}{\lambda}x - 2\pi ft\right)}|x\rangle dx$$

It follows that the amplitude to find the particle at any chosen x_i is:

$$\langle x_i \,|\, \psi(x,t)\rangle = A\int e^{i\left(\frac{2\pi}{\lambda}x - 2\pi ft\right)}\langle x_i|x\rangle dx = Ae^{i\left(\frac{2\pi}{\lambda}x_i - 2\pi ft\right)}$$

Applying the de Broglie relationships yields:

$$\Psi(x_i, t) = Ae^{\frac{i}{\hbar}(p_x x_i - Et)}$$

as the wavefunction of the particle.

The wavelength, if you remember, is defined as the shortest distance between any two points along x that have a phase difference of 2π (Figure 12.2). Now we can see that this phase difference

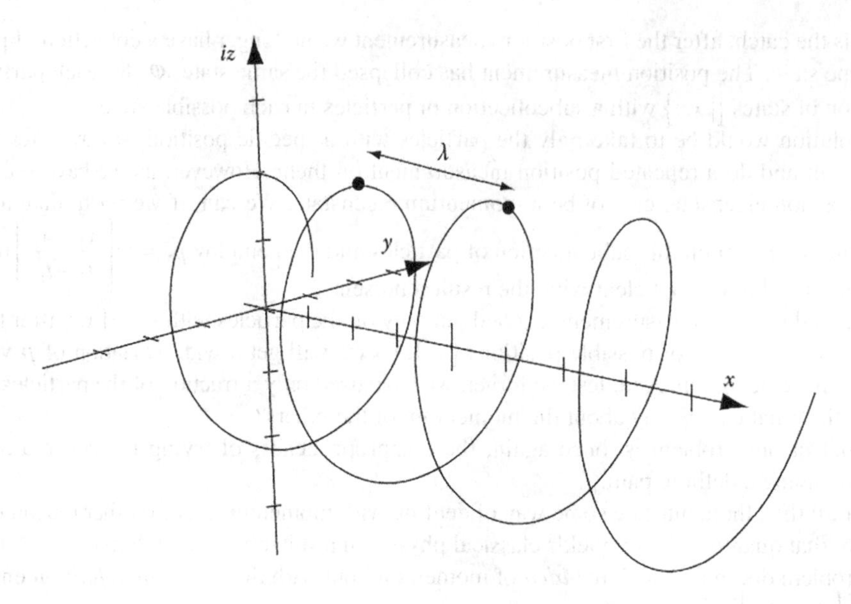

FIGURE 12.2 The momentum eigenstate wave function at a fixed time.

is generated by the momentum of the particle. As the phase of our wave function is given by $\phi = \frac{1}{\hbar}(p_x x - Et)$, if we compare the points x and $x+1$, we will see a phase difference of $p_x/\hbar$ between them, provided we look at the two points at the same time. It would be appropriate to say that the 'rate' at which the phase changes with distance is given by $p_x/\hbar$.

In everyday speech, 'rate' is used in the context of something changing with time. Mathematicians, on the other hand, never leave a good idea alone if they can generalize it. Mathematically, a rate of change refers to how much one quantity changes as you vary another. In this case we're looking at the change in phase as we pick different points on the x-axis. Remember, though, that this comparison between different points is only going to work if we look at them *at the same time*, otherwise the Et part of the phase is going to get in the way. This fixed time condition helps to point out that our use of 'rate' is not referencing time.

In the branch of mathematics known as *calculus*, the rate of change of one quantity, P, with another, Q, is calculating by *differentiating*, i.e., rate of change of P with $Q = dP/dQ$.

In this case, we are looking for the rate of change of phase, ϕ, with distance, x:

$$\frac{d\phi}{dx} = \frac{d}{dx}\left(\frac{1}{\hbar}(p_x x - Et)\right) = \frac{p_x}{\hbar}$$

Taking this one step further, if we differentiate our wave function:

$$\frac{d\Psi(x,t)}{dx} = A\frac{d}{dx}\left(e^{\frac{i}{\hbar}(p_x x_i - Et)}\right) = \frac{i}{\hbar}p_x A e^{\frac{i}{\hbar}(p_x x_i - Et)} = \frac{i}{\hbar}p_x \Psi(x,t)$$

using the characteristic property of the exponential, $\frac{d}{dx}\left(Ae^{bx}\right) = Abe^{bx}$.

The result $\frac{d\Psi(x,t)}{dx} = \frac{i}{\hbar}p_x\Psi(x,t)$ is reminiscent of an operator acting on an eigenstate, $\hat{O}|\psi\rangle = o\,|\psi\rangle$, suggesting that, if we do a little rearrangement, we will create an operator to represent momentum.

In other words:

$$\hat{p}_x = -i\hbar \frac{d}{dx}$$

A similar argument yields the energy operator:

$$\hat{E} = i\hbar \frac{d}{dt}$$

Summarizing, we have the following quantum mechanical interpretations (Table 12.1):

TABLE 12.1
Quantum Mechanical Interpretations of Classical Concepts

Classical Concept	Quantum Interpretation	Operator
Momentum	Rate of change of phase with distance	$\hat{p}_x = -i\hbar \frac{d}{dx}$
Energy	Rate of change of phase with time	$\hat{E} = i\hbar \frac{d}{dt}$

Both operators are Hermitian, but as they are not in a matrix form, the proof requires an expanded definition.

HERMITIAN OPERATOR 2

The definition of a Hermitian operator which is not expressed in matrix form is:

$\hat{O}$ is Hermitian, if:

$$\int_{-\infty}^{\infty} \varphi^* \left(\hat{O}\varphi\right) dx = \int_{-\infty}^{\infty} \varphi \left(\hat{O}\varphi\right)^* dx$$

Applying this to either the momentum or energy operator requires us to integrate by parts using the formula:

$$\int u\, dv = uv - \int v\, du$$

so that, for $\hat{p}_x$:

$$I = \int_{-\infty}^{\infty} \varphi^* \left(\hat{O}\varphi\right) dx = \int_{-\infty}^{\infty} \varphi^* \left(-i\hbar \frac{d\varphi}{dx}\right) dx$$

$$= -i\hbar \int_{-\infty}^{\infty} \varphi^* \left(\frac{d\varphi}{dx}\right) dx$$

Taking $u = \varphi^*$, $dv = \frac{d\varphi}{dx} dx$ gives $du = \frac{d\varphi^*}{dx} dx$, $v = \varphi$. Hence:

$$I = -i\hbar\left[\varphi^*\varphi\right]_{-\infty}^{\infty} - (-i\hbar)\int_{-\infty}^{\infty}\varphi\left(\frac{d\varphi^*}{dx}\right)dx$$

$$= \int_{-\infty}^{\infty}\varphi\left(i\hbar\frac{d\varphi^*}{dx}\right)dx = \int_{-\infty}^{\infty}\varphi\left(-i\hbar\frac{d\varphi}{dx}\right)^* dx$$

as required. The first term is taken to vanish as $\varphi \to 0$ as $x \to \infty$ for a localized particle.

12.4.3 Operators and Representations

The job of an operator such as $\hat{p}_x$ is to 'pull out' the eigenvalue from one of its eigenstates:

$$\hat{p}_x|p_1\rangle = p_1|p_1\rangle$$

but this is rather an abstract task. After all, states don't have specific mathematical forms and the operator $\hat{p}_x$ is not given a structure in this equation. However, we can perform a simple trick as follows:

$$\langle x|\left(\hat{p}_x|p_1\rangle\right) = \langle x|\left(p_1|p_1\rangle\right) = p_1\langle x|p_1\rangle|$$

Or, in terms of state functions $\varphi_{p_1}(x)$:

$$\hat{p}_x\varphi_{p_1}(x) = p_1\varphi_{p_1}(x)$$

which shows the momentum operator *acting on a momentum eigenstate that has been written in a position representation*. Just to be clear, $\varphi_{p_1}(x)$ is the state function corresponding to the position state expansion of a momentum eigenstate:

$$|p_1\rangle = \int \varphi_{p_1}(x)|x\rangle\, dx$$

As the $\hat{p}_x$ operator is, in this case, acting on the position representation of a state function, it had better have a position representation as well. Of course, we extracted this position representation in the previous section: $\hat{p}_x = -i\hbar\frac{d}{dx}$.

The point that I'm trying to make is all to do with representations. Any state of a system can be written in various representations, meaning that the state can be expanded over any suitable basis. If it is a position basis, then we have a position representation of the state. Operators that represent physical variables must be able to act and get the same results no matter what representation the state is in. Hence the same operator must have different representations as well.

So far, I have shown you just the position representation of $\hat{p}_x$. The other commonly used representation involves expanding over momentum eigenstates:

$$|\chi\rangle = \int A(p_x)\,|\,p_x\rangle dp_x$$

So that $|\chi\rangle$ could be a position eigenstate, or simply an arbitrary state. Incidentally, Dirac introduced a representation of the position eigenstate using the *Dirac delta function*,[8] $\delta(x-\alpha)$:

$$\delta(x-\alpha) = \frac{1}{2\pi\hbar}\int_{-\infty}^{\infty} e^{\frac{i}{\hbar}p(x-\alpha)}\, dp$$

$$|x=\alpha\rangle = \frac{1}{2\pi\hbar}\int_{-\infty}^{\infty} e^{\frac{i}{\hbar}p(x-\alpha)}\,|\,p\rangle dp$$

With any expression expanded over momentum eigenstates, the position representation of the momentum operator would not be appropriate. Converting operators from one representation to another requires a set of mathematical techniques that lie beyond our scope (e.g., Fourier transformations). Hence, I can only quote the different versions for you (Table 12.2)

TABLE 12.2
Different Representations of Common Operators

Operator	Position Representation	Momentum Representation
$\hat{x}$	$x\hat{I}$	$-i\hbar\dfrac{d}{dp_x}$
$\hat{p}_x$	$-i\hbar\dfrac{d}{dx}$	$p_x\hat{I}$

In this table, I have indicated that the position operator, $\hat{x}$, takes the form $\hat{x}=x\hat{I}$, a step that is going to be more stringently justified in the next section.

12.4.4 Expectation Values Again

Important Rule 7 sets out the role of operators, $\hat{O}$, in representing physical variables and how an operator can be used to calculate the expectation value, $\langle\hat{O}\rangle=\langle\psi|\hat{O}|\psi\rangle$, of a series of measurements made on a collection of systems in the same state $|\psi\rangle$.

In standard statistical theory, the average value of a variable, v, can be calculated from the relevant probability density:

$$\langle v\rangle = \int v\mathcal{P}(v)dv$$

which is a continuous generalization of the argument that we used in Section 5.3. Imagine we had a probability density (not necessarily quantum derived) describing the likelihood of finding a particle at a position in space. The average position would then be:

$$\langle x\rangle = \int x\mathcal{P}(x)dx$$

Translating this into the quantum equivalent, given our identification of $\mathcal{P}(x)=|\langle x\,|\,\psi\rangle|^2=\psi^*(x)\psi(x)$, gives us:

$$\langle x\rangle = \int x|\langle x\,|\,\psi\rangle|^2\,dx = \int x\psi^*(x)\psi(x)dx$$

Comparing this with our standard formulation of an expectation value:

$$\langle x\rangle = \langle\psi|\hat{x}|\psi\rangle = \int\langle\psi\,|\,x\rangle\hat{x}\langle x\,|\,\psi\rangle dx = \int\psi^*(x)\hat{x}\psi(x)dx$$

suggests that we should make the identification $\hat{x} = x\hat{I}$ as we indicated in the previous section.

Applying the momentum operator instead, we can illustrate how an expectation value would be calculated for momentum. Using the position representation, we start by expanding our state:

$$|\psi\rangle = \int \langle x|\psi\rangle\, |\, x\rangle dx = \int \psi(x)|x\rangle dx$$

and

$$\langle\psi| = \int \langle\psi|y\rangle\langle y|dy = \int \psi^*(y)\langle y\,|\,dy$$

where I have used y to indicate a position in the second expansion so that in the final calculation the role of the two separate position variables is clear.[9]

Next, we insert the expansions into the formula for the expectation value:

$$\langle \hat{p}_x\rangle = \langle\psi|\hat{p}_x|\psi\rangle = \iint \psi^*(y)\langle y\,|\left(-i\hbar\frac{d}{dx}\right)\psi(x)\,|\,x\rangle dxdy$$

As the operator acts on the amplitude/state function, $\psi(x)$, rather than the basis state $|\,\psi\rangle$, we can push the basis state $\langle y|$ through the operator to give:

$$\langle \hat{p}_x\rangle = \iint \psi^*(y)\left(-i\hbar\frac{d}{dx}\right)\psi(x)\langle y\,|\,x\rangle dxdy$$

$$= -i\hbar\int \psi^*(y)\frac{d\psi(x)}{dx}dx$$

as $\langle y\,|\,x\rangle = 0$ unless $x = y$.

To proceed any further, we would need a mathematical expression for the specific $\psi(x)$. It is worth practising by trying the calculation for the momentum eigenstate $\psi(x) = A\exp\left\{\frac{i}{\hbar}(px - Et)\right\}$.

12.4.5 Operators and Variables

By now you may be feeling a certain state of confusion. After all, we have liberally used x and p to represent the state variables' position and momentum *inside* our state functions, and now we are using operators $\hat{x}$ and $\hat{p}_x$ to *act on* the state functions. Why do the same ideas appear to have two separate mathematical expressions in the theory?

These are very deep waters, and I'm not sure that orthodox quantum physics, in the end, has a satisfactory answer to give. Modern day quantum theories of gravity undermine our intuitive view of space so that we start to puzzle over what we mean by x anyway. That might, in the end, be what rescues us from this quandary. However, the question deserves a more immediate answer, so here we go.

When we use terms such as x, p_x, E, and t inside wave functions, we take them as *parameters* determining the state function/wave function. After all, that is what we mean by a state variable. The corresponding operators $\hat{x}$, $\hat{p}_x$ and $\hat{E}$ are mathematical tools that allow us to extract information of a given sort out of a state. They determine the *measurable observables* that can be extracted. In almost all cases, the parameter is observable, but the link is not guaranteed.[10]

That's a very brief and only partially satisfactory answer to the question, but I will have another stab when we get into Section 14.3.1

NOTES

1 Or at least all the ones that we know of...
2 Although, clearly, one can influence the other—as in the direction through which an electron passes out of an SG magnet.
3 The term 'vector' can have a variety of different meanings in different contexts. In this case, I am referring to a physical variable with both magnitude and direction.
4 The dot product of two vectors $\mathbf{a} = (a_1, a_2, a_3)$ and $\mathbf{b} = (b_1, b_2, b_3)$ is $\mathbf{a} \cdot \mathbf{b} = a_1 b_1 + a_2 b_2 + a_3 b_3$. The formalism extends to vectors with more than three components.
5 which gives quantum field theorists some very specific, and non-finite, headaches...
6 I am using i here as an index, not $i = \sqrt{-1}$.
7 The choice to use y in the bra expansion is simply to make the subsequent steps clearer, it is not denoting a different axis or an entirely different physical variable.
8 Although it is called the delta *function*, this object is not that well defined mathematically as a function. It is better thought of as a limit of distributions.
9 It's because of the structure of the sum. If I used x in both cases, that would indicate that it is the same value of x term for term in both sums. Yet when I multiply two sums together, I do not get term for term matching like that. Later in the calculation it will reduce to term matching, but that is because of the physics involved - specifically that $\langle y \mid x \rangle = 0$ unless $x = y$.
10 $\hat{t}$ is a source of some worry that I don't want to dip my, or even your, toe into just now.

13 Uncertainty

13.1 EXPECTATION IS NOT ENOUGH

If we take a collection of quantum objects, which are all in the same state $|\varphi\rangle$, and measure the value of the same physical property, O, for each of them, then we will get a collection of values $\{o_i\}$. The average value of these measurements will be equal to the expectation value of the operator, $\hat{O}$, associated with this property.

The expectation value, however, does not tell the whole story. We also need to know how widely spaced the range of values are. In one instance, every member of the collection $\{o_i\}$ could be very close to the average value. In another, some of the $\{o_i\}$ might be considerably different from the average, not because of some experimental mistake, but simply due to the nature of the quantum state.

Whenever physicists think about a perfect set of experimental values, they dream of something called the *normal distribution curve*, as illustrated in Figure 13.1.

Imagine that Figure 13.1 is a graph showing the results of identical measurements made on a collection of objects. They do not have to be quantum systems as the normal distribution applies to experimental results found in classical physics as well, although for different physical reasons.

The x-axis spans the values of the results obtained and the y-axis the number of occasions on which a specific value turned up (e.g., the result 5 was obtained 40 times). Also, as seems evident, the average value over all readings was 5. However, there were quite a few occasions on which a value significantly different from 5 was obtained.

Now consider Figure 13.2.

In this case, we have the same peak value, 5, and average value, 5, as before, but the spread of values is much narrower. On this curve, a much higher proportion of the results are closer to the average. If we simply focused on the average value (i.e., the expectation), then we would have missed an important characteristic of the distribution. We need a measure of the *spread* of values as well. Statistical theory provides such as measure.

Given a set of values, $\{o_i\}$, with an average $\langle o\rangle$ we can calculate the difference between each value and the average, $o_i - \langle o\rangle$. Adding this up across the whole range of values should give a total of zero:

$$\sum_{i=1}^{i=n} o_i - \langle o\rangle = 0$$

if the distribution is symmetrical about the average. As this is not a helpful measure of the width of the distribution, we alter the calculation to:

$$\sum_{i=1}^{i=n} \frac{\left(o_i - \langle o\rangle\right)^2}{n}$$

which is the average of the square of the difference between each value and the overall average. This ensures that each term is positive and hence successive terms accumulate rather than tending to cancel out.

Finally, to compensate for taking squares, we take the square root:

$$\Delta o = \sqrt{\sum_{i=1}^{i=n} \frac{\left(o_i - \langle o\rangle\right)^2}{n}}$$

DOI: 10.1201/9781003225997-15

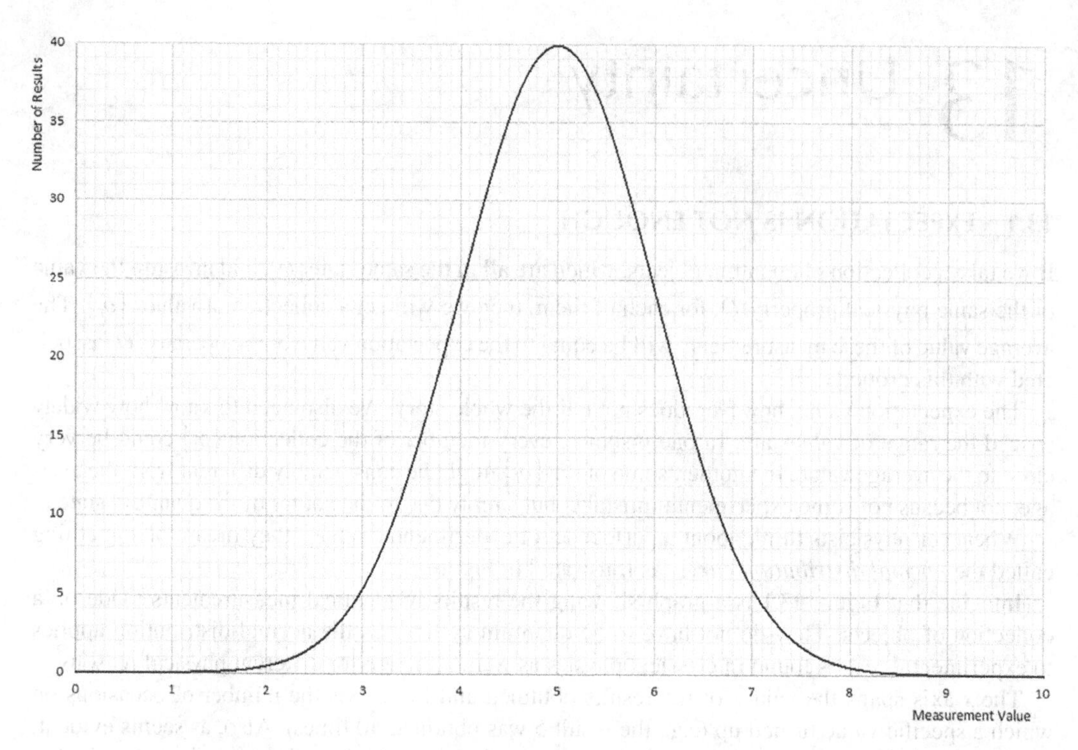

FIGURE 13.1 The normal distribution curve.

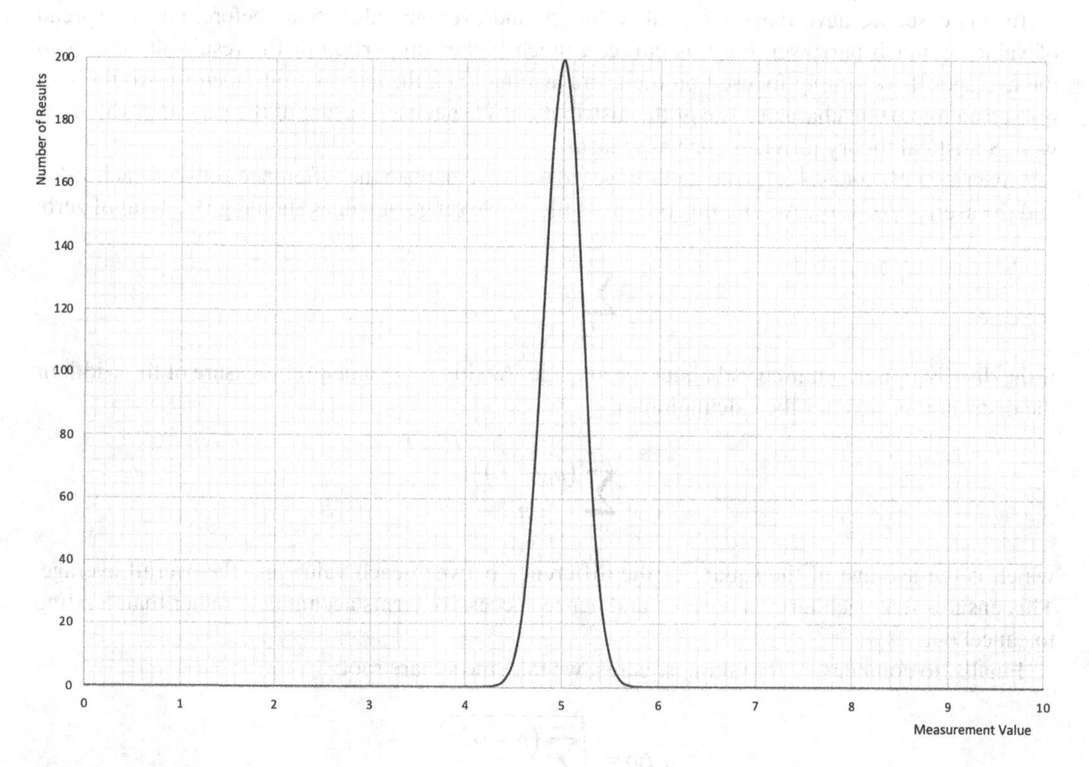

FIGURE 13.2 The same normal distribution curve, but with a much narrower spread. Note how the vertical scale has increased to accommodate the same number of results in a narrower spread.

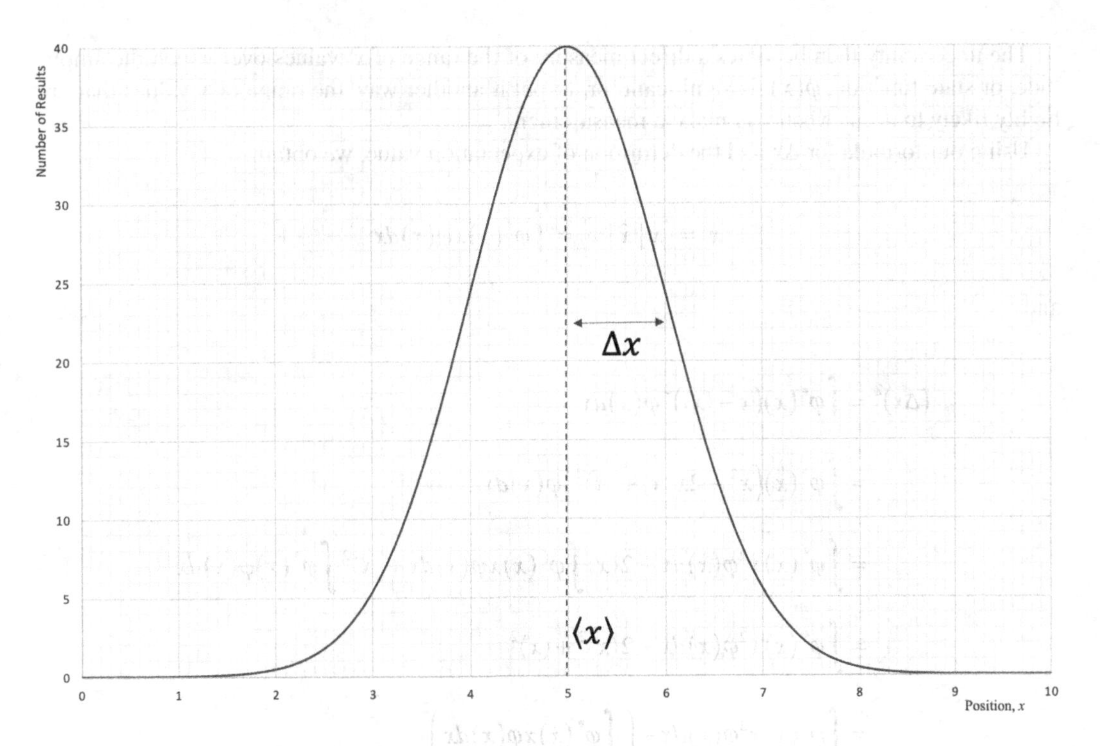

FIGURE 13.3 The uncertainty in a set of experimental results using a collection of systems which are not in a position eigenstate.

Mathematicians call this the *standard deviation*, but to be consistent with quantum theory I will use the term *uncertainty* (Figure 13.3).

13.1.1 Developing Uncertainty

In classical physics, the width, or uncertainty, in a set of experimental results is generally attributable to experimental variations, due to small differences in the objects being tested, or variations in the apparatus used, instrumental factors, environmental factors etc.

Experiments in quantum physics also suffer from all these sorts of variations. Consequently, even if we have a collection of systems, all carefully prepared in the same eigenstate of the physical variable in question, we will not always get exactly the same measured value. The spread of this distribution is down to *experimental* uncertainty.

If the prepared systems are not in an eigenstate, then the measurements will have a quantum variation, even if the systems are quantum identical.

To illustrate this, we can start with a state $|\varphi\rangle$ given by:

$$|\varphi\rangle = \int \langle x \,|\, \varphi\rangle \,|\, x\rangle dx = \int \varphi(x) \,|\, x\rangle dx$$

In which case, making position measurements on a set of particles in this state results in each possible x value, within a very small range δx, occurring according to the probability $\varphi^*(x)\varphi(x)\delta x = |\varphi(x)|^2 \delta x$.

Charting the data produces a curve, such as Figure 13.3, with a corresponding width or uncertainty Δx. In other words, the measurement distribution will draw out $|\varphi(x)|^2 \delta x$ for us.

The uncertainty then becomes a direct measure of the range of x values over which the amplitude, or state function, $\varphi(x)$, is significant, or, to put it another way, the range of x values that are highly likely to occur when you make a measurement.

Using our formula for Δx and the definition of expectation value, we obtain:

$$\langle x\rangle = \langle x\,|\,\hat{x}\,|\,x\rangle = \int \varphi^*(x)x\varphi(x)dx$$

and

$$\begin{aligned}(\Delta x)^2 &= \int \varphi^*(x)(x-\langle x\rangle)^2\varphi(x)dx \\ &= \int \varphi^*(x)\left(x^2 - 2x\langle x\rangle + \langle x\rangle^2\right)\varphi(x)dx \\ &= \int \varphi^*(x)x^2\varphi(x)dx - 2\langle x\rangle\int \varphi^*(x)x\varphi(x)dx + \langle x\rangle^2\int \varphi^*(x)\varphi(x)dx \\ &= \int \varphi^*(x)x^2\varphi(x)dx - 2\langle x\rangle^2 + \langle x\rangle^2 \\ &= \int \varphi^*(x)x^2\varphi(x)dx - \left(\int \varphi^*(x)x\varphi(x)dx\right)^2 \\ &= \langle x^2\rangle - \langle x\rangle^2\end{aligned}$$

(I have calculated $(\Delta x)^2$ here rather than Δx as it is visually more elegant).

Of course, the same state could equally be expanded over a momentum basis $\{|\,p_i\rangle\}$ and the experimental results of momentum measurements also charted to reveal the uncertainty in this variable Δp_i. The formulas would easily generalize.

In the case of a momentum eigenstate, we can predict with absolute certainty what the result of a measurement of momentum will be. A chart of the results will then show no width, aside from the fluctuations that you can't avoid in experiments, giving $\Delta p_i = 0$. After all:

$$\langle p\rangle = \langle p_I|\hat{p}|\,p_I\rangle = p_I\langle p_I\,|\,p_I\rangle = p_I$$

and then:

$$\Delta p = \langle p^2\rangle - \langle p\rangle^2 = p_I^2 - (p_I)^2 = 0$$

(Note that I have used I to indicate a specific value p_I out of the range of possible momenta $\{p_i\}$.)

However, there is a catch. As we have already said, a momentum eigenstate can't also be a position eigenstate and vice versa. When you expand a momentum eigenstate over a position basis you have to take, in principle, an infinite number of possible positions:

$$|p_I\rangle = \int_{-\infty}^{\infty} \langle x|p_I\rangle\,|\,x\rangle dx$$

The calculation of $\langle x \rangle$ for such a state would proceed as follows:

$$\langle x \rangle = \langle p_I | \hat{x} | p_I \rangle = \langle p_I | x | p_I \rangle$$

$$= \left\{ \int_{-\infty}^{\infty} \langle p_I \mid y \rangle \langle y \mid dy \right\} x \left\{ \int_{-\infty}^{\infty} \langle x \mid p_I \rangle \mid x \rangle dx \right\}$$

$$= \int \int_{-\infty}^{\infty} \langle p_I \mid y \rangle \langle x \mid p_I \rangle \langle y \mid x \rangle x \, dx \, dy$$

$$= \int \int_{-\infty}^{\infty} \langle p_I \mid y \rangle \langle y \mid x \rangle \langle x \mid p_I \rangle x \, dx \, dy$$

Then, as $\int_{-\infty}^{\infty} |y\rangle\langle y| dy = \hat{I}$

$$\langle x \rangle = \int_{-\infty}^{\infty} \langle p_I \mid x \rangle \langle x \mid p_I \rangle x dx = \int_{-\infty}^{\infty} \varphi^*(x) x \varphi(x) dx$$

using $\langle y|x \rangle = 0$ unless $x = y$. The momentum eigenstate is our free particle state:

$$\varphi(x) = A\exp\left[\frac{i}{\hbar}(px - Et)\right] = Ae^{\frac{i}{\hbar}(px-Et)}$$

so:

$$\langle x \rangle = \int_{-\infty}^{\infty} \varphi^*(x) x \varphi(x) dx$$

$$= A^2 \int_{-\infty}^{\infty} e^{-\frac{i}{\hbar}(px-Et)} x e^{\frac{i}{\hbar}(px-Et)} dx$$

$$= A^2 \int_{-\infty}^{\infty} x dx = 0$$

Where as:

$$\langle x^2 \rangle = \int_{-\infty}^{\infty} \varphi^*(x) x^2 \varphi(x) dx$$

$$= A^2 \int_{-\infty}^{\infty} e^{-\frac{i}{\hbar}(px-Et)} x^2 e^{\frac{i}{\hbar}(px-Et)} dx$$

$$= A^2 \int_{-\infty}^{\infty} x^2 dx = \text{ not well defined}$$

Hence $(\Delta x)^2 = \langle x^2 \rangle - \langle x \rangle^2$ diverges.

13.2 HEISENBERG'S PRINCIPLE

This is all very good, but what about those states (surely in the majority) that are neither position nor momentum eigenstates? Well, given such a state $|\psi\rangle$ we can expand over a position basis, yielding an appropriate expectation value and uncertainty:

$$|\psi\rangle = \int \varphi(x)\,|\,x\rangle dx \Rightarrow \langle x\rangle, \Delta x$$

or over a momentum basis, with its appropriate measures:

$$|\psi\rangle = \int \psi(p)\,|\,p\rangle dp \Rightarrow \langle p\rangle, \Delta p$$

As it's the same state in both representations, there is a connection between Δx and Δp_x. This was first demonstrated by Heisenberg and is now named after him:

IMPORTANT RULE 8: HEISENBERG'S UNCERTAINTY PRINCIPLE

$$\Delta x \times \Delta p \geq \frac{\hbar}{2}$$

The key part of this expression is the "≥" sign, which is telling us that the product $\Delta x \times \Delta p$ can't be less than $\hbar/2$. In some situations, the measurement results might be such that $\Delta x \times \Delta p$ is much greater than $\hbar/2$, which is fine; but you can never find a situation where $\Delta x \times \Delta p$ is less than $\hbar/2$. Quantum theory places a limit on how precise our two measurements can be. Note that this principle relates to the *quantum uncertainty*, we must blend in normal experimental uncertainty as well.

Putting it in a different way: Δx measures the range of x values over which $\varphi(x)$ is significantly large and Δp, the corresponding range of $\psi(p)$, if you expanded the same state over a momentum basis. Hence, the uncertainty principle is relating the scope of $\varphi(x)$ to that of $\psi(p)$. If you choose to make a position measurement you will most likely get a value within Δx of the expectation value; if you make a momentum measurement *instead* (not *after*) then you will most likely get a value within Δp of $\langle p\rangle$.

That last statement is important. If you make a position measurement you will collapse the state into something close to a position eigenstate (in practice you will get a wave packet, which are constructions that we come across in Chapter 15), which needs a different momentum expansion to the original state. The uncertainty principle relates Δx and Δp *for the same state.*

Some immediate thoughts spring to mind, not the least of which is: how does the uncertainty principle deal with an object that is not moving? Is this not sitting perfectly still with a precise momentum of zero, at a precise spot? Well, putting to one side the notion that nothing can be perfectly still (we are orbiting the sun after all), the immediate point is that the uncertainty principle *prevents such a state from happening in the microworld.*

13.2.1 So What?

People have not been afraid of drawing philosophical conclusions from the uncertainty principle, almost from the moment the result was first published. Heisenberg himself launched one of the earliest sallies, his target being the principle of *causality.*

Causality, the notion that everything that happens has been caused by something happening previously, has been central to physics for hundreds of years. The success of classical physics, based on the notion of strict causality, leads one naturally to believe that a sufficiently powerful computer, fed with detailed information about the positions, momenta, and forces at work on every particle in the universe, would be able to predict, as accurately as you want, the future of the universe. Such thinking easily leads one to doubt such things as free will, the human mind, morality, and pretty much the whole ball game in the end.[1]

Experiments in the quantum world started to undermine causality as they revealed apparently unpredictable outcomes (see Section 1.4.2). Then came the uncertainty principle, which chipped away at a fundamental aspect of physics: our ability to measure with unlimited precision. The uncertainty principle prevents us from knowing the combined position and momentum of a particle to any accuracy we deem sufficient, rendering prediction impossible. To a physicist at the time, this would have been outrageous.[2] Nature seemed to be built on a set of principles that ran contrary to the approach that had served science well for generations. It was almost like a betrayal.

Actually, the modern theory of chaos (Section 3.2) deals a more significant blow to our predicting computer. It would be possible to regard the quantum challenge as being mounted in the world of the small, of no relevance to the big world in which we live. However, the study of chaotic systems has shown us that small effects do not necessarily lead to small consequences and that no measurement (quantum or otherwise) can, even in principle, yield sufficiently accurate results to do the job.

Nevertheless, there is still an important philosophical point to tackle, namely the ontological status of the uncertainty principle. All our discussion has been couched in terms of measurement. We could take the view that strict causality was preserved, despite our inability to measure precisely enough to know what Nature was working up to. Provided Nature knows the positions and momenta of the particles, to a level of precision that we can't match, all is well.

However, there is an alternative view. The uncertainty principle places a limit on our measurement capabilities *because there is nothing there to measure any more precisely.*

13.2.2 I'm Not Sure What You Mean by Uncertainty...

Many paragraphs of text have been written about the meaning and significance of the uncertainty principle. As you might expect, it depends on what side of the realist/instrumentalist divide you stand and on exactly what shade of interpretation you favour. However, the most important philosophical aspect of the uncertainty principle is whether you take it to be a limitation on what we can *know about the world* or an inherent *limitation in the world.* In philosophical speak, is it an epistemological statement, or is it ontological?

Whatever Heisenberg may have thought at the time, one of his first illustrations of the uncertainty principle had an epistemological tinge. This is the famed 'gamma ray microscope'.

Heisenberg envisaged a microscope constructed to use gamma rays rather than light waves. Such an instrument could be used to determine the position of something as small as an electron by detecting any gamma rays that bounced off it. Gamma rays lead to greater precision than normal light waves, due to their much smaller wavelength.

As is well known in optics, there is a limitation on the detail that can be seen by a microscope. This limitation depends on the wavelength of the waves being used. To view a level of detail sufficient to locate something as small as an electron, quite high-energy gamma rays (very short wavelength) are needed. However, this is not the end of the story, as such a gamma ray bouncing off an electron would be bound to transfer a significant amount of energy. As a result, the electron would recoil in an unspecified direction, hence affecting its momentum (it's basically the Compton effect from Section 1.4).

There would be no way of pinning down the energy transferred and the subsequent disturbance in the electron's momentum. If you wanted to do that, you would have to measure the gamma ray before it hit, and that process would be bound to disturb the gamma ray, so that what you measured would not be what hit the electron.

Heisenberg showed that a detailed analysis of the gamma ray microscope's measurement capabilities leads directly to the uncertainty principle.

There's been a good deal of controversy about the gamma ray microscope and many refinements to Heisenberg's original calculations. However, these are technical issues that shouldn't distract us. The key point is this: the gamma ray microscope argument assumes that the electron *has* a position and a momentum; the problem is that the interaction with the measuring device means that we cannot accurately *determine* what these quantities are.

Every time a measuring device makes a measurement it must interact with the thing that it's measuring. In classical physics the scale of these interactions is either too small to worry us or can be calculated and accounted for in a sensible way. That can't be said in quantum physics. The scale of the interaction is too great and causes profound changes to the properties of the object being measured. Also, the interaction is inherently unpredictable. Any attempt to stick another measuring device in to judge the scale of the interaction won't work as the second instrument will itself have to interact in a manner that can't be detected. An infinite regression follows, which, like most infinite regressions, gets us nowhere.[3]

This discussion cuts, once again, to the nature of a quantum state. If an electron happens to be in an eigenstate of position (or momentum), we can say with certainty what the result of measuring that position (momentum) will be.[4] If the electron is not in an eigenstate, we can determine the various values that *might* be revealed by a measurement, and their relative probabilities, but what does that tell us? What is actual position (or momentum) of an electron in this situation?

There is a possible way forward that preserves causality. Perhaps some state variables are not currently accessible to us. There could be 'hidden variables'- variables we do not know yet how to measure - which take different values for different particles in the collection. When a measurement is made, these variables determine the outcome, but because we don't know what they are, or their values, it all looks random to us. The electrons have perfectly well-determined positions and momenta, but we can't know in advance what they will be as we can't access the physics that determines the values.

Such a 'hidden variable' approach would view the uncertainty principle as expressing the *epistemological limits* of current science. However, this approach comes at something of a price as the restriction of ensuring that a hidden variable theory matches the predictions of orthodox quantum theory dictates some unpalatable consequences for the hidden variables (this point is developed further in Chapter 23).

Heisenberg took the view that the uncertainty principle was *ontological*, at least in his later writings. To him the uncertainty principle expressed the limits to which classical *concepts* of momentum, or position, can be applied in a specific situation. This harks back to our earlier discussion (Section 12.3.1) about transferring the classical definition of momentum into quantum physics. If we take this view, then we have to say that an object that is not in an eigenstate of position simply *does not have a position*. The concept has no meaning in such a state. Heisenberg would have said that the position of the particle was *latent* or a potential property. When a measurement takes place, this latency becomes actualized and (some of) our classical ways of thinking can now be applied.

While this is very much a realistic view it does acknowledge that our prejudices about what should be classed as *real* may have to be enlarged. To Heisenberg latent properties should be awarded some degree of reality.

The gamma-ray microscope was used by Heisenberg as a way of introducing the uncertainty principle. However, he held back in the original paper (partly under pressure from Bohr) from stating that it was an ontologically valid description of what would happen. His opinions on the matter became more evident later.

13.3 YET MORE UNCERTAINTY

I have tried to show how the uncertainty principle arises as a natural consequence of our ability to expand a state over different bases. Although I have illustrated this with position and momentum, there are clearly lots of other physical variables with basis sets that we can expand a state over.

Unless position and momentum have some special significance, there must be versions of the uncertainty principle that apply to other physical variables as well.

In some circumstances, a state can be the eigenstate of more than one physical variable. The free particle expansion over $|x_n\rangle$, for example, is an eigenstate of energy *and* momentum (we first noted this back in Chapter 7). However, it is never possible to be in an eigenstate of position and momentum at the same time. Position and momentum are said to be *conjugate variables.* The various spin and angular momentum components are also conjugate variables, which is why you can't construct an eigenstate of both the total angular momentum and more than one of its components.

Let's consider two physical variables represented by operators $\hat{O}_1$ and $\hat{O}_2$, which are *not* conjugates.[5] Suppose that the state $|\varphi\rangle$ is an eigenstate of both. Acting on $|\varphi\rangle$ with both operators, one by one, gives us:

$$\hat{O}_1|\varphi\rangle = o_1|\varphi\rangle$$

$$\hat{O}_2 o_1|\varphi\rangle = o_2 o_1|\varphi\rangle$$

Or, if we operate with $\hat{O}_2$ first:

$$\hat{O}_2|\varphi\rangle = o_2|\varphi\rangle$$

$$\hat{O}_1 o_2|\varphi\rangle = o_1 o_2|\varphi\rangle$$

which shows that $\hat{O}_2\hat{O}_1|\varphi\rangle = \hat{O}_1\hat{O}_2|\varphi\rangle$, or:

$$\left[\hat{O}_2\hat{O}_1 - \hat{O}_1\hat{O}_2\right]|\varphi\rangle = 0$$

The square bracket in this relationship is important and frequently arises in quantum theory, so it is given its own name:

THE COMMUTATOR:

$$\left[\hat{O}_2,\hat{O}_1\right] = \left[\hat{O}_2\hat{O}_1 - \hat{O}_1\hat{O}_2\right]$$

If two operators *commute,* then $\left[\hat{O}_2,\hat{O}_1\right] = 0$, and they *can* have simultaneous eigenstates.[6] If they don't commute, simultaneous eigenstates are not possible, and an uncertainty relationship follows.

13.3.1 The Generalized Uncertainty Principle

Given any two non-commuting operators, $\hat{O}_1$ and $\hat{O}_2$ the following uncertainty principle applies:

IMPORTANT RULE 9: GENERALIZED UNCERTAINTY PRINCIPLE

$$\Delta O_1 \times \Delta O_2 \geq \left|\frac{i}{2}\left\langle\left[\hat{O}_2,\hat{O}_1\right]\right\rangle\right|$$

Unfortunately, we do not have the depth of experience within quantum theory to prove this rule,[7] so I am going to ask you to take it on trust. However, we can do a quick calculation to show how this generalized rule can be applied.

Setting $\hat{O}_1 = \hat{p} = -i\hbar\dfrac{\partial}{\partial x}$ and $\hat{O}_2 = \hat{x} = x$ we have:

$$\left\langle \left[\hat{O}_2,\hat{O}_1\right]\right\rangle = i\hbar\int \psi^*\left(-x\frac{\partial}{\partial x}+\frac{\partial}{\partial x}x\right)\psi\, dx$$

$$= i\hbar\left\{-\int \psi^* x\frac{\partial \psi}{\partial x}dx + \int \psi^*\frac{\partial}{\partial x}(x\psi)dx\right\}$$

$$= i\hbar\left\{-\int \psi^* x\frac{\partial \psi}{\partial x}dx + \int \psi^* x\frac{\partial \psi}{\partial x}dx + \int \psi^*\frac{\partial x}{\partial x}\psi\, dx\right\}$$

$$= i\hbar\int \psi^*\psi\, dx = i\hbar$$

Which leads us to:

$$\Delta O_1 \times \Delta O_2 \geq \left|\frac{i}{2}\left\langle\left[\hat{O}_2,\hat{O}_1\right]\right\rangle\right| \quad \Rightarrow \quad \Delta p \Delta x \geq \left|\frac{i}{2}(i\hbar)\right| = \frac{\hbar}{2}$$

Confirming the Heisenberg Uncertainty relationship as we have already seen it.

NOTES

1 The alternative view, which I subscribe to, is that the (defensible) intuitive idea that we have free will indicates that classical physics cannot be a completely encompassing picture of the world.
2 If only psychology had been more developed at the time, I am quite sure that a study of the founding fathers would have yielded some fascinating results about how people dealt with challenges to their basic conceptions of life.
3 Or possibly everywhere?
4 Neglecting the inherent experimental uncertainty.
5 This argument mimics the one that we had in Section 10.3.1 with regard to spin operators.
6 *Can,* not *must.* However, this is a distinction that will not worry us here.
7 If you are interested, then a proof can be found in many excellent books on quantum theory.

14 The Equations of Quantum Theory

14.1 THE SCHRÖDINGER EQUATIONS

In classical physics, if you want to calculate what's going to happen in a given situation, you basically have two options open to you. You can carefully specify the positions and momenta of all the particles involved, list the forces acting, and then plug the lot into Newton's laws of motion and calculate away. Alternatively, you can inspect all the kinetic energies, potential energies, and any other form of energy involved and work through a calculation based on energy conservation.

As it happens, the second approach, using energy, is the more amenable to a straightforward quantum development.

Every form of energy that we come across in mechanics (particles in motion) can be related back to position or momentum. For example, kinetic energy is $\frac{1}{2}mv_x^2$ (assuming that the object is travelling in the x-direction), or, as momentum is $p_x = mv_x$ we can write $KE = p_x^2/2m$. Potential energies come in various forms. Gravitational potential energy is mass × strength of gravity × height above the ground, mgh, and h can be expressed in terms of position. The elastic energy stored in a stretched spring is $\frac{1}{2}kx^2$, where k is the strength of the spring and x the amount of stretch (position again). All of which illustrates the basic notion that mechanical energies depend on position or momentum. This is very important, as we already have quantum mechanical operators for position and momentum, so we should be able to build operators for any energy we want based on them.

In classical physics, the sum of all energies involved in a given situation is known as the *Hamiltonian*:

$$H = \frac{p_x^2}{2m} + V(x)$$

where the expression $V(x)$ denotes the sum of the potential energies possessed by the system (restricting things to a 1-dimensional treatment for ease). Using a simple translation, with the quantum operators that we already have, should result in an appropriate Hamiltonian operator (in the position representation):

$$\hat{H} = \frac{\hat{p}_x^2}{2m} + \hat{V}(x) = \frac{1}{2m}\left(-i\hbar\frac{d}{dx}\right)\left(-i\hbar\frac{d}{dx}\right) + \hat{V}(x)$$

$$= -\frac{\hbar^2}{2m}\frac{d^2}{dx^2} + \hat{V}(x)$$

with the operator $\hat{V}(x)$ working like the position operator: i.e., $\hat{V}(x) = V(x)\hat{I}$

Clearly, if we act on a state that is an eigenstate of the Hamiltonian operator, we ought to get one of the eigenvalues, E_n:

$$\hat{H}|E_n\rangle = E_n|E_n\rangle$$

DOI: 10.1201/9781003225997-16

if our formulation is correct.

Using the argument from the start of Section 12.3, we can convert this state expression into one involving state functions instead:

$$\langle x|\hat{H}|E_n\rangle = \hat{H}\langle x|E_n\rangle = E_n\langle x|E_n\rangle$$

with $\varphi_n(x) = \langle x|E_n\rangle$.

This is the famed *Schrödinger equation*, which written out in full looks like this:

TIME-INDEPENDENT SCHRÖDINGER EQUATION

$$-\frac{\hbar^2}{2m}\frac{d^2\varphi_n(x)}{dx^2} + V(x)\varphi_n(x) = E_n\varphi_n(x)$$

Solutions to this equation give us the state function for a given situation, specified by the form of $V(x)$ appropriate to the physics of the system.

The time-independent Schrödinger equation is very useful in specific situations when we know that the system must be in an energy eigenstate. However, there needs to be another equation to deal with more general cases, for example, when the potential energy depends on time as well as position, $V(x,t)$.

It will have occurred to you that we now have *two* energy operators $\hat{H}$ and $\hat{E} = i\hbar\frac{d}{dt}$ from Section 12.3.2. The difference between them is explored in more detail in the next section, but if our theory is to make sense, then we must get the same result when we apply either operator to the same state, i.e.:

$$\hat{H}|\Psi(x,t)\rangle = \hat{E}|\Psi(x,t)\rangle$$

which, when written out in the position representation, take the form:

TIME-DEPENDENT SCHRÖDINGER EQUATION

$$-\frac{\hbar^2}{2m}\frac{d^2\Psi(x,t)}{dx^2} + V(x,t)\Psi(x,t) = i\hbar\frac{d\Psi(x,t)}{dt}$$

Solving this equation for a given $V(x,t)$ will produce the wave function for the system. For example, if we plug in the appropriate $V(x,t)$ for an electron in an atom, we can solve the time-dependent Schrödinger equation to get the wave function of an electron in orbit around that atom, something of vital importance in atomic physics and chemistry.[1] We will follow up on this in Chapter 15.

The discovery of the Schrödinger equation(s) was a key moment in the history of quantum physics. Previously, quantum calculations required the use of mathematical techniques that were unfamiliar to many physicists at the time (matrix manipulations). The Schrödinger equation had a recognisable mathematical form, that of a wave equation, and conventional techniques could be applied to its solution.

There is one crucial point though. None of the steps we used to produce the time-dependent or time-independent versions of the Schrödinger equation, were forced on us. There is no proof we can construct from first principles to derive them. We can argue for their plausibility, but having done that, we rely on using them to make predictions which are then checked against reality. Their successful application is their only compelling justification. Of course, the same is true for all the laws of nature.

14.1.1 $\hat{E}$ AND $\hat{H}$

I want to make a couple of remarks about the different roles of $\hat{E}$ and $\hat{H}$ in quantum theory.

The *energy operator* is $\hat{E}$, which always represents energy measurement. The Hamiltonian $\hat{H}$ is a *model* of the system, constructed by considering a particular $\hat{V}(x,t)$. This model may be an approximate one, as the details are too difficult to include. However, if the model is reasonably correct and we set $\hat{H}\Psi(x,t) = \hat{E}\Psi(x,t)$, which is effectively the time-dependent Schrödinger equation, then we can solve this to give the wave function $\Psi(x,t)$ for that model of that system. Using the wave function, we can make various predictions about what the system is likely to do and pass them on to the experimental experts. If their results agree with our predictions, then our model is accurate.

Put it another way: $\hat{E}$ is the total energy of the system, and $\hat{H}$ is the list of what we think that total energy is made up of. If we have got this list correct, then $\hat{H}\Psi = \hat{E}\Psi$.

14.1.2 STATIONARY STATES

Having a pair of Schrödinger equations seems to be an embarrassment of riches: when do we use one and when the other?

The key point in answering this question is to see that the time-dependent equation will convert into the time-independent version when:

$$i\hbar \frac{d\Psi(x,t)}{dt} = E_n \Psi(x,t)$$

that is, $\Psi(x,t)$ is the position representation of an eigenstate of $\hat{E}$. Physically the implication is that a set of repeated energy measurements on the same system yield the same value each time. The energy is not changing with time. The system is in a *stationary state*. If a system is going to change its energy, then it can only do this by leaking some energy into (or from) its surroundings, so the only chance of producing and maintaining a stationary state is to isolate the system from its environment. If we set about using the time-independent equation, then we are searching for any stationary states that might exist.

For the mathematically more inclined, any state function $\varphi(x)$ that is a solution of the time-independent equation must also be a solution of the time-dependent version. If we write, $\Psi(x, t) = \varphi(x)\phi(t)$ then:

$$-\phi(t)\frac{\hbar^2}{2m}\frac{d^2\varphi(x)}{dx^2} + V(x,t)\varphi(x)\phi(t) = E\varphi(x)\phi(t)$$

So that:

$$i\hbar\varphi(x)\frac{d\phi(t)}{dt} = E\phi(t)\varphi(x)$$

Rearranging and integrating[2]:

$$\int \frac{d\phi(t)/dt}{\phi(t)}dt = -\frac{i}{\hbar}\int E\,dt$$

$$\log(\phi(t)) = -\frac{iEt}{\hbar} + C$$

$$\phi(t) = Ae^{-\frac{i}{\hbar}Et}$$

Consequentially, all stationary states have the form $\Psi(x,t) = A\varphi(x)e^{-\frac{i}{\hbar}Et}$.

14.2 EHRENFEST'S THEOREM

In Section 12.4.5 we started a discussion about the different roles of system variables, such as x, p_x, E, etc. and the operators $\hat{x}, \hat{p}_x, \hat{E}$. To develop the point further, it is interesting to consider an important theorem first derived by Ehrenfest[3] in 1927.

We start, once again, with the expectation value of an operator, $\langle \hat{O} \rangle = \langle \psi | \hat{O} | \psi \rangle$, and consider how this quantity changes with time:

$$\frac{d\langle \hat{O} \rangle}{dt} = \frac{d}{dt}\left(\langle \psi | \hat{O} | \psi \rangle\right)$$

Applying the basic rules of differential calculus, we get:

$$\frac{d\langle \hat{O} \rangle}{dt} = \left(\frac{d\langle \psi |}{dt}\right)\hat{O}|\psi\rangle + \left\langle \psi \left| \left(\frac{d\hat{O}}{dt}\right) \right| \psi \right\rangle + \langle \psi | \hat{O}\left(\frac{d|\psi\rangle}{dt}\right)$$

On another tack, a slight re-arrangement of the Schrödinger equation produces:

$$\frac{d|\psi\rangle}{dt} = -\frac{i}{\hbar}\hat{H}|\psi\rangle$$

and also:

$$\frac{d\langle \psi |}{dt} = \frac{i}{\hbar}\langle | \hat{H}^{\dagger} = \frac{i}{\hbar}\langle \psi | \hat{H}$$

as, like any observable, $\hat{H}$ is Hermitian (Section 5.4.1).

After inserting these expressions into our calculation, we obtain:

$$\frac{d\langle \hat{O} \rangle}{dt} = \frac{i}{\hbar}\langle \psi | \hat{H}\hat{O} | \psi \rangle + \left\langle \psi \left| \left(\frac{d\hat{O}}{dt}\right) \right| \psi \right\rangle - \frac{i}{\hbar}\langle \psi | \hat{O}\hat{H} | \psi \rangle$$

$$= \frac{i}{\hbar}\langle \psi | \hat{H}\hat{O} - \hat{O}\hat{H} | \psi \rangle + \left\langle \psi \left| \left(\frac{d\hat{O}}{dt}\right) \right| \psi \right\rangle$$

$$= \frac{i}{\hbar}\left\langle \psi \left| \left[\hat{H}, \hat{O}\right] \right| \psi \right\rangle + \left\langle \psi \left| \left(\frac{d\hat{O}}{dt}\right) \right| \psi \right\rangle$$

using the commutator, $\left[\hat{H},\hat{O}\right]=\hat{H}\hat{O}-\hat{O}\hat{H}$. In a more compact format, we have:

HEISENBERG'S EQUATION OF MOTION

$$\frac{d\langle\hat{O}\rangle}{dt}=\frac{i}{\hbar}\left\langle\left[\hat{H},\hat{O}\right]\right\rangle+\left\langle\frac{d\hat{O}}{dt}\right\rangle$$

Note that the second term on the right-hand side will be zero if the operator $\hat{O}$ does not have an explicit time dependence – something we deploy in each of the following calculations:

$$\hat{O}\neq\hat{O}(x,t)\Rightarrow\frac{d\hat{O}}{dt}=0$$

To reach Ehrenfest's Theorem, we apply Heisenberg's equation of motion to the specific examples of $\hat{x}$ and $\hat{p}_x$. As a preliminary step, we calculate $\left[\hat{H},\hat{x}\right]\Psi$ and $\left[\hat{H},\hat{p}_x\right]\Psi$:

$$\left[\hat{H},\hat{x}\right]\Psi=\left(-\frac{\hbar^2}{2m}\frac{d^2}{dx^2}+V(x)\right)x\Psi-x\left(-\frac{\hbar^2}{2m}\frac{d^2}{dx^2}+V(x)\right)\Psi$$

$$=-\frac{\hbar^2}{2m}\frac{d}{dx}\left(\frac{d(x\Psi)}{dx}\right)+V(x)x\Psi+x\frac{\hbar^2}{2m}\frac{d^2\Psi}{dx^2}-xV(x)\Psi$$

$$=-\frac{\hbar^2}{2m}\frac{d}{dx}\left(\frac{dx}{dx}\Psi+\frac{xd\Psi}{dx}\right)+x\frac{\hbar^2}{2m}\frac{d^2\Psi}{dx^2}$$

$$=-\frac{\hbar^2}{2m}\frac{d\Psi}{dx}-\frac{\hbar^2}{2m}\frac{dx}{dx}\frac{d\Psi}{dx}-x\frac{\hbar^2}{2m}\frac{d^2\Psi}{dx^2}+x\frac{\hbar^2}{2m}\frac{d^2\Psi}{dx^2}$$

$$=-\frac{\hbar^2}{m}\frac{d\Psi}{dx}$$

Hence:

$$\left[\hat{H},\hat{x}\right]=-\frac{\hbar^2}{m}\frac{d}{dx}=\frac{\hbar}{m}\left(-\hbar\frac{d}{dx}\right)=\frac{\hbar}{m}\frac{\hat{p}_x}{i}=-\frac{i\hbar}{m}\hat{p}_x$$

The second calculation involves the following steps:

$$\left[\hat{H},\hat{p}_x\right]\Psi=\left(-\frac{\hbar^2}{2m}\frac{d^2}{dx^2}+V(x)\right)\hat{p}_x\Psi-\hat{p}_x\left(-\frac{\hbar^2}{2m}\frac{d^2}{dx^2}+V(x)\right)\Psi$$

$$=\left(-\frac{\hbar^2}{2m}\frac{d^2}{dx^2}+V(x)\right)\left(-i\hbar\frac{d\Psi}{dx}\right)-\left(-i\hbar\frac{d}{dx}\right)\left(-\frac{\hbar^2}{2m}\frac{d^2\Psi}{dx^2}+V(x)\Psi\right)$$

$$=\frac{i\hbar^3}{2m}\frac{d^3\Psi}{dx^3}-i\hbar V(x)\frac{d\Psi}{dx}-\frac{i\hbar^3}{2m}\frac{d^3\Psi}{dx^3}+i\hbar\frac{dV(x)}{dx}\Psi+i\hbar V(x)\frac{d\Psi}{dx}$$

$$=i\hbar\frac{dV(x)}{dx}\Psi$$

$$\therefore \left[\hat{H}, \hat{p}_x\right] = i\hbar \frac{dV(x)}{dx}$$

Now we can use these results to obtain the two parts of the theorem that we are looking for. Firstly:

$$\frac{d\langle\hat{x}\rangle}{dt} = \frac{i}{\hbar}\left\langle\left[\hat{H}, \hat{x}\right]\right\rangle + \left\langle\frac{d\hat{x}}{dt}\right\rangle$$

giving:

$$\frac{d\langle\hat{x}\rangle}{dt} = \frac{i}{\hbar}\left\langle-\frac{i\hbar}{m}\hat{p}_x\right\rangle = \frac{\langle\hat{p}_x\rangle}{m} \quad \Rightarrow \quad \langle\hat{p}_x\rangle = m\frac{d\langle\hat{x}\rangle}{dt}$$

Secondly:

$$\frac{d\langle\hat{p}_x\rangle}{dt} = \frac{i}{\hbar}\left\langle\left[\hat{H}, \hat{p}_x\right]\right\rangle + \left\langle\frac{d\hat{p}_x}{dt}\right\rangle \quad \Rightarrow \quad \frac{d\langle\hat{p}_x\rangle}{dt} = -\left\langle\frac{dV(x)}{dx}\right\rangle$$

It is easy to lose track of the beauty and significance of this result while wading through the mathematical detail, so I should elevate the theorem appropriately:

EHRENFEST'S THEOREM

$$\langle\hat{p}_x\rangle = m\frac{d\langle\hat{x}\rangle}{dt} \qquad \frac{d\langle\hat{p}_x\rangle}{dt} = -\left\langle\frac{dV(x)}{dx}\right\rangle$$

14.2.1 The Classical Limit

The first Ehrenfest relationship should be compared with the classical formulation $p_x = m\frac{dx}{dt}$. The second with $\frac{dp_x}{dt} = -\frac{dV(x)}{dx} = F$ i.e., Newton's second law of motion[4]. Here we see an indication of how the ephemeral quantum world blends into the more tangible construction that we know as 'ordinary' reality. While it is obviously a relief to see classical laws emerge from quantum calculations, we should not get carried away for two reasons:

1. As the quantum versions involve expectation values, we can't use them to describe the motion of an individual particle. We are still constrained by the repeated indication that individual particles do not have well-defined trajectories, with all that implies.
2. We have, in a sense, folded the result out of the assumptions, as we built the Hamiltonian operator out of the classical version. It would be surprising if a few lines of calculation later, we did not get something familiar out of the other end.

At the very least, this argument should confirm that our quantum interpretations, e.g., that momentum is the rate of change of phase with distance, are at least self-consistent. While this is certainly a somewhat mysterious way of thinking about momentum, as experience and education have programmed us to think about it in the more concrete terms of mass × velocity, in some manner, when considering expectation values at least, the classical version emerges.

None of this provides a definitive answer to the open questions surrounding the realist/instrumentalist debate over the interpretation of quantum theory. Undoubtedly, some physicists see the

Ehrenfest results with their reliance on expectation values, as confirmation that the state refers to a collection of objects, rather than having some ontological reality for each individual. However, this does not rule out a Heisenberg-like interpretation based on latent properties. The expectation values would then be showing how the latencies typically manifest themselves across a collection of objects.

14.2.2 Constants of Motion

Moving on from Ehrenfest's theorem, let us revisit Heisenberg's equation of motion and explore one of its significant consequences:

$$\frac{d\langle\hat{O}\rangle}{dt} = \frac{i}{\hbar}\left\langle\left[\hat{H}, \hat{O}\right]\right\rangle + \left\langle\frac{d\hat{O}}{dt}\right\rangle$$

The left-hand side of this relationship is the time rate of change of the expectation value of the operator. Let's think carefully about what that means. A set of measurements of $\hat{O}$ on a collection of objects in the same state, at the same time, will produce a range of values with average $\langle\hat{O}\rangle$ (provided the state is not an eigenstate, of course). Given this natural fluctuation, which is the inevitable quantum consequence of not having an eigenstate, it is not easy to tell if the state is changing with time. We can't even test the same objects sometime after the first measurement, as they will no longer be in the same state.

A suitable procedure would be to have a large collection of objects in the same state, measure a subset at a certain time, and then another subset at a later time, and another even later etc. If the expectation value drifts over time, then we can see that the state is evolving.

The right-hand side of the equation is the driver of this change. If the operator is not explicitly dependent on time, $\hat{O} \neq \hat{O}(x, t)$, then $\frac{d\hat{O}}{dt} = 0$ and hence $\left\langle\frac{d\hat{O}}{dt}\right\rangle = 0$ as well. That just leaves $\frac{i}{\hbar}\left\langle\left[\hat{H}, \hat{O}\right]\right\rangle$. If $\hat{O}$ commutes with $\hat{H}$, this term will be zero and hence the rate of change of expectation will also be zero – the 'value' of that variable will not change with time.

Take an obvious example, energy, represented for a system by $\hat{H}$. Clearly $\left[\hat{H}, \hat{H}\right] = 0$, so $\frac{d\hat{H}}{dt} = 0$ and the expectation value of the total energy of the system does not change with time (is a *constant of the motion*). The system would then be in a stationary state, à la Section 14.2.1.

However, should the potential energy of the system, V, have some explicit time dependence, $V = V(x, t)$, then $\left\langle\frac{d\hat{V}}{dt}\right\rangle \neq 0$ and $\frac{d\langle\hat{H}\rangle}{dt} \neq 0$ as well. In which case, the system is clearly leaking energy to some external sink, presumably in the surroundings. Expanding the definition of the system to include this sink would result in $V \neq V(x, t)$, and energy being constant in the encompassing system.

14.3 THE ENERGY-TIME INEQUALITY

In the previous section, I was slightly hasty about the procedure for demonstrating that a system had evolved over time. Even if we did follow the outlined steps, testing subsamples out of a collection of systems, we would not see sets of values that gradually increased over time. There would always be a distribution generated, partly by quantum variations and partly from experimental uncertainty as well. Consider two sets of measurements from two appropriate sub-samples, tested at different times. If the experiments were to generate a set of results like those in Figure 14.1, we would not

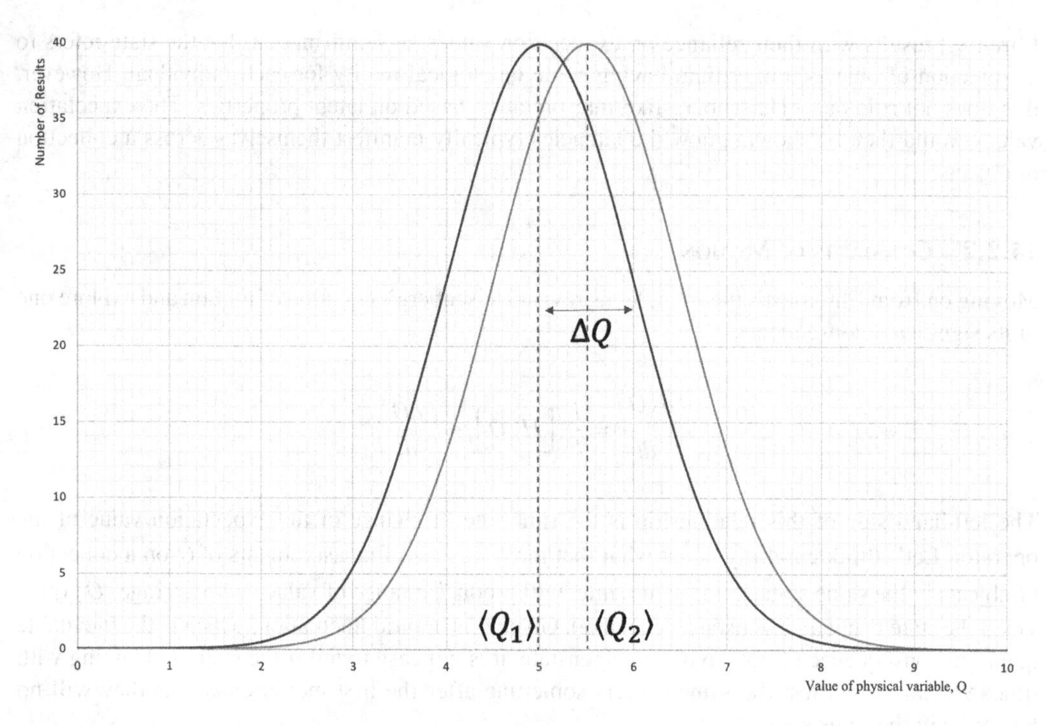

FIGURE 14.1 Two normal distributions, generated by successive measurements. In this case, the two distributions are not sufficiently separated for us to be certain that the system had evolved. A result from the second set could easily be smaller than several results from the first set.

be justified in concluding that $\langle Q_1 \rangle$ was different to $\langle Q_2 \rangle$, given the width (uncertainty), ΔQ, in the distributions.

Presumably, if we left the system for longer between measurements, the system would evolve further so that the distributions ended up more clearly separated. In Figure 14.2, the two expectation values are separated by the uncertainty in the distribution, so that $\langle Q_2 \rangle - \langle Q_1 \rangle \geq \Delta Q$.

The question now turns to a related issue: how long would we have to wait between measurements so that we could be certain that the resulting distributions were sufficiently distinct? If we set this time interval equal to δt, then we can obtain its value from the combination of ΔQ and $d\langle Q \rangle / dt$, evaluated at the same moment[5]:

$$\delta t = \frac{\Delta Q}{d\langle Q \rangle / dt}$$

In turn, the rate of change of the expectation value with time is already determined by Heisenberg's equation of motion:

$$\frac{d\langle \hat{O} \rangle}{dt} = \frac{i}{\hbar}\left\langle \left[\hat{H}, \hat{O}\right] \right\rangle + \left\langle \frac{d\hat{O}}{dt} \right\rangle$$

Assuming the operator is not explicitly time-dependent, we get:

$$\delta t = \frac{-i\hbar \Delta Q}{\left\langle \left[\hat{H}, \hat{O}\right] \right\rangle}$$

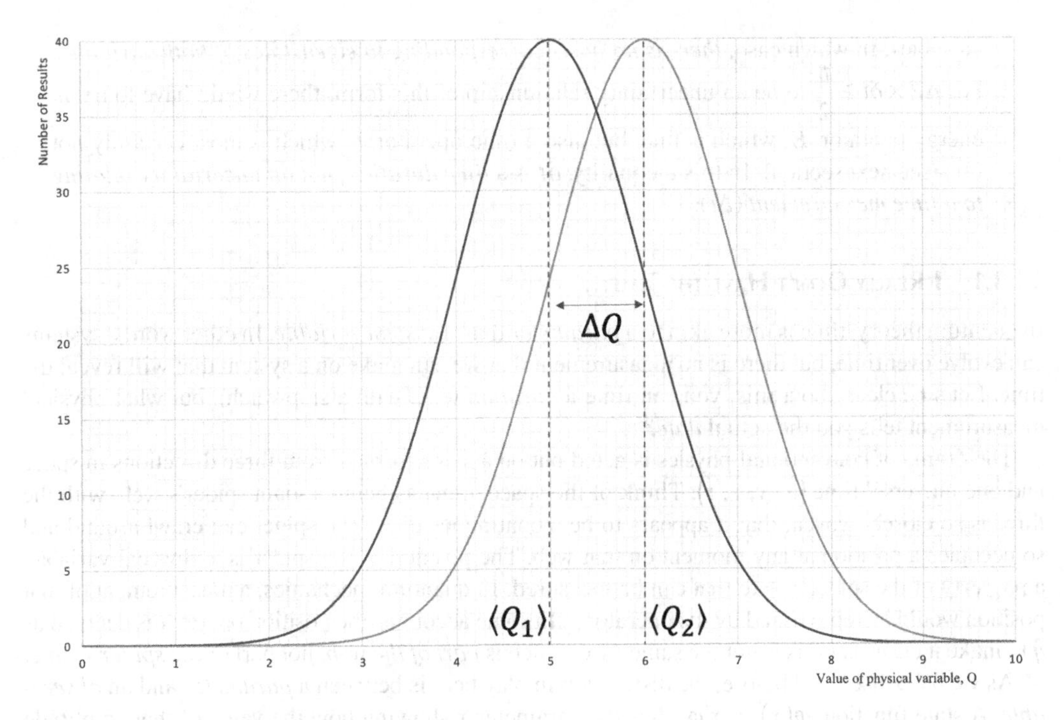

FIGURE 14.2 In this case, the two measurement distributions are statistically distinguishable.

Switching tack, we now reconsider the generalized uncertainty principle:

$$\Delta O_1 \times \Delta O_2 \geq \left| \frac{i}{2} \left\langle \left[\hat{O}_2, \hat{O}_1 \right] \right\rangle \right|$$

Setting $\hat{O}_1 = \hat{E}$ and $\hat{O}_2 = \hat{Q}$, which is by no means an obvious next move, gives:

$$\Delta E \times \Delta Q \geq \left| \frac{i}{2} \left\langle \left[\hat{H}, \hat{Q} \right] \right\rangle \right|$$

Now we can plug in our equation for δt:

$$\Delta E \times \Delta Q \geq \left| \frac{i}{2} \frac{(-i\hbar \Delta Q)}{\delta t} \right| \quad \Rightarrow \quad \Delta E \times \delta t \geq \frac{\hbar}{2}$$

This is an astonishing and subtle result, for a few connected reasons:

- It is remarkable that the time we need to wait, δt, to be sure that two sets of measurements made on *any system variable* will reveal that the system has evolved, is related to the uncertainty that would be obtained in energy measurements, ΔE.
- Consequently, if the system is in an energy eigenstate, $\Delta E = 0$ then $\delta t \rightarrow \infty$, i.e., we would have to wait for an indefinite time before we could be sure that the system had evolved. In other words, the system is not evolving in this case.
- While the inequality $\Delta E \times \delta t \geq \frac{\hbar}{2}$ resembles the uncertainty relationships discussed in Chapter 13, *it is not an uncertainty relationship.* As we clearly demonstrated in Chapter 13, uncertainty relationships come about when you are dealing with conjugate

variables. In which case, *there is no basis corresponding to eigenstates of both variables.* For $\Delta E \times \delta t \geq \frac{\hbar}{2}$ to be an uncertainty relationship of this form, there would have to be an energy operator, $\hat{E}$, which is fine, but also a time operator, $\hat{t}$, which is most certainly not fine (see next section). In this inequality, δt is a *time duration, not an uncertainty relating to a time measurement* (Δt).

14.3.1 I Really Don't Have the Time...

In quantum theory, time is more akin to a *parameter* than a *system variable.* In other words, systems can evolve over time, but there is no measurement that we can make on a system that will reveal the time. Let's be clear about this: you can time a *duration* (e.g., with a stopwatch), but what physical measurement tells you the actual *date?*

The drama of conventional physics is acted out on a stage formed from three directions in space and one universal time (x, y, z, t). Think of the space arena as being a giant spider's web, with the threads so closely woven, that it appears to be a continuous disk. The spider can crawl around and so occupies a position at any moment on that web. The position of the spider is a physical variable, a property of the spider's state that can be measured. In quantum mechanics, a measurement of that position would be represented by the operator $\hat{x}$. In some accounts, the position operator is denoted as $\hat{q}$ to make it clear that this is not the same as x, which is *part of the web,* not part of the *spider's state.*

As we have discussed before, the distinction in play here is between a *parameter* and an *observable.* A state function, $\varphi(x) = \langle x|\varphi\rangle$, has the parameter x showing how the value of that amplitude varies along the x-axis. However, the position of the particle is extracted from the state via the position operator, $\hat{x}$, and especially the expectation value $\langle\hat{x}\rangle = \langle\varphi|\hat{x}|\varphi\rangle$. The operator represents the measurable observable property of the state. In classical physics, we tended to blithely assume that the two, parameter and observable, were clearly the same thing, but this is not necessarily the case in quantum theory. At the very least, it is unclear whether a particle has a location until the latent property is manifest, as a state collapses (within the width of a wave packet, as we shall see later). However, that does not stop there from being values of the x parameter where the particle might be about to appear.

Time is evidently rather different from this. If we can picture it at all, we think of it as a line stretching out to infinity in both directions.[6] Picturesquely, the spider would be sliding along this line as time passes. We can't conceive of an experiment that would enable us to 'measure' the spider's position on the timeline. Time itself is not a property of the spider's state, hence there is no time operator.

But what of a system that is changing with time, a falling ball, for example? Here the momentum of the ball increases as time passes, suggesting that it is a 'time variable'. However, in this case, the physical variable that is changing does not change at a uniform rate (the momentum increases at an increasing rate), and in any case, when it hits the ground, time as measured by the momentum would stop, although t continues.

Of course, there are some systems in which a physical variable corresponds in a more satisfactory way with the passage of time. We tend to call them clocks. The second hand of an analogue clock has an angular position ϑ, which changes with time. Measuring this angle gives us a measure of time for that system. In any real clock, there will also be energy, which will tend to depend on ϑ, so that we can define an energy operator $\hat{E}$ and a *time operator* $\hat{\vartheta}$, which will not commute and so lead to an uncertain relationship. However, ϑ is only a physical variable of that clock, and so can't be applied as a 'time operator' to any other system. In any case, a quantum measurement of the system if it is not in an eigenstate of time (whatever that would mean!) must reveal a scattering of values about some expectation, not a monotonically increasing result. The ticking of a clock corresponds to a passage of time, δt, not a range of time measurements spread over a normal distribution with uncertainty Δt.

If you need one further point to convince you, then consider this – does time have a probability density the way that position does?

In quantum field theory, the situation is somewhat worse, as it is doubtful if a position operator can be satisfactorily defined for all cases, so the situation with time is not unique.

14.3.2 Energy/Time Uncertainty

Many horrors have been attempted in the name of the energy–time uncertainty relationship; one of the worst being the notion that a system can 'borrow' some energy (from where?) ΔE and provided that the energy is 'paid back' (!) before time Δt has elapsed, then conservation of energy is not violated. This travesty is often used to justify how the exchange of particles can give rise to an interaction force—a process that we will discuss more correctly (albeit briefly) in Chapter 32.

14.4 TIME EVOLUTION

Back in Chapter 5, I introduced the time evolution operator $\hat{U}(t)$ which moved a state from T to $T+t$. I indicated that the action of the operator was continuous, so that for a time interval, δt, $\hat{U}(\delta t)|\Psi(T)\rangle = |\Psi(T+\delta t)\rangle \approx |\Psi(T)\rangle$. If we wanted to be a little more precise about this, then:

$$\hat{U}(\delta t)|\Psi(T)\rangle = \hat{U}(\delta t)\sum_i a_i(T)|i\rangle = \sum_i a_i(T+\delta t)|i\rangle$$

Using a fundamental result from differential calculus, $a_i(T+\delta t) \approx a_i(T) + \frac{da_i}{dt}\delta t$:

$$\begin{aligned}|\Psi(T+\delta t)\rangle &\approx \sum_i a_i(T+\delta t)|i\rangle = \sum_i \left(a_i(T) + \frac{da_i}{dt}\delta t\right)|i\rangle \\ &= \sum_i a_i(T)|i\rangle + \sum_i \frac{da_i}{dt}\delta t|i\rangle \\ &= |\Psi(T)\rangle + \delta t\frac{d}{dt}|\Psi(T)\rangle\end{aligned}$$

where we understand that the differential applied to the state is acting on the amplitudes within any decomposition of the state. Working in terms of the wave function, $\Psi(x, T) = \langle x, T|\Psi\rangle$, suggests that:

$$\Psi(x, T+\delta t) \approx \Psi(x, T) + \delta t\frac{d\Psi(x, T)}{dt}$$

The time-dependent Schrödinger equation can be re-arranged to give:

$$\frac{d\Psi(x,t)}{dt} = \frac{1}{i\hbar}\hat{H}\Psi(x,t) = -\frac{i}{\hbar}\hat{H}\Psi(x,t)$$

which, inserted into our time evolution produces:

$$\Psi(x, T+\delta t) \approx \Psi(x, T) - \delta t\frac{i}{\hbar}\hat{H}\Psi(x,T) = \left(1 - \delta t\frac{i}{\hbar}\hat{H}\right)\Psi(x,T)$$

Should we want to make another small step forward in time:

$$\Psi(x,\ T+2\delta t) \approx \Psi(x,\ T+\delta t) - \delta t \frac{i}{\hbar} \hat{H}\Psi(x, T+\delta t)$$

$$\approx \left(1 - \delta t \frac{i}{\hbar} \hat{H}\right) \Psi(x, T+\delta t)$$

$$\approx \left(1 - \delta t \frac{i}{\hbar} \hat{H}\right)^2 \Psi(x, T)$$

In fact:

$$\Psi(x,\ T+n\delta t) \approx \left(1 - \delta t \frac{i}{\hbar} \hat{H}\right)^n \Psi(x, T)$$

There is another helpful result from pure mathematics that we can now call upon:

$$e^x = \exp(x) = \lim_{n\to\infty} (1 + x/n)^n$$

To use this, we set $\delta t = t/n$, giving:

$$\Psi(x,\ T+t) \approx \left(1 - \left(t \frac{i}{\hbar} \hat{H}\right)/n\right)^n \Psi(x, T)$$

Now we let $n \to \infty$:

$$\Psi(x,\ T+t) = \lim_{n\to\infty:} \left(1 - \left(t \frac{i}{\hbar} \hat{H}\right)/n\right)^n \Psi(x, T)$$

$$= \exp\left(-it\hat{H}/\hbar\right) \Psi(x, T)$$

The prospect of taking the exponential of an operator is a curious one, but it can be interpreted as applying the sequence $e^x = \sum_{k=0}^{\infty} \frac{x^k}{k!}$, so that:

$$e^{-it\hat{H}/\hbar} = \sum_{k=0}^{\infty} \frac{\left(-it\hat{H}/\hbar\right)^k}{k!}$$

One proviso on this development is that the Hamiltonian is not an explicit function of time, $H \neq H(t)$. So, I have made good on my earlier promise, in Chapter 5, to show how the time evolution operator is related to the energy in the system if energy is not itself explicitly dependent on time:

$$\hat{U}(t) = \exp\left(-it\hat{H}/\hbar\right)$$

14.5 CONCLUSIONS

Although we have the Schrödinger equation in both forms, it's quite another matter to attempt to solve them for a given situation. Doing that would take us into realms of mathematics that lie outside the parameters of this book. Textbooks on quantum mechanics normally get to the Schrödinger equations quite quickly and spend many of their pages outlining how to solve them and what those solutions can tell us. I have aimed at developing a thorough understanding of states, and now we have seen, in principle, how wave functions can be calculated. In the next chapter, we will tackle some important, if elementary, examples.

NOTES

1. Most often the equation for electrons in atoms is too complicated to give an exact solution. All we can do in such circumstances is use approximations or computer models to see roughly how things must be.
2. Using the standard result $\int \frac{f'}{f} dx = \log f + c$.
3. Paul Ehrenfest (1880–1933) Austrian and Dutch theoretical physicist. The proof is contained in the paper Remark about the approximate validity of classical mechanics within quantum mechanics. *Zeitschrift für Physik* 45, 455, 1927.
4. In classical field theory, the spatial gradient of the potential is associated with the applied force, e.g., in gravitation.
5. In order to make a point, which will be clearer in a short while, I am breaking the convention of using Δ to indicate a change or duration and δ as an infinitesimal change or infinitesimally short duration.
6. Although given the Big Bang theory, we now believe that time itself started with the Big Bang some 13.7 billion years ago.

15 Constrained Particles

15.1 A PARTICLE IN A BOX

One of the classic first problems in quantum theory textbooks is that of the particle trapped in a box. Despite the apparent artificiality of the situation, it has quite a few interesting features that we can quickly apply to some more physically plausible situations.

The particle in a box is literally an object, such as an electron, trapped inside a 'box.' In a realistic physical example, this 'box' would be a volume of space bound by some constraining force acting on the particle. For example, an atom is a 'box' as far as the orbiting electrons are concerned, with the electrostatic force between the positive protons in the nucleus and the negative electrons providing the constraint.

However, to find our feet, we are going to work with the idea of a simple rectangular box made from walls that the electron can't penetrate. To set this up mathematically, the walls are represented by a region where the potential energy is infinite. To penetrate such a region, the electron would have to have infinite kinetic energy, which is impossible. Of course, such a setup, with infinite potential energy, can't be constructed in practice. We'll come to a more realistic situation in Section 15.2 when we take a look at the hydrogen atom.

While the walls represent an impenetrable barrier, within the box itself, the electron is free and uninfluenced by any forces. Figure 15.1 shows a simple one-dimensional (1-D) box bounded by walls at $x = 0$ and $x = L$.

The first move is to assume that the wave function for any particle in the box will have separate spatial and temporal dependencies, i.e., that it is of the form:

$$\Psi(x, t) = \varphi(x)\phi(t)$$

The specific advantages of looking to see if there are any appropriate solutions of this form will become apparent later.

We are not always at liberty to make this assumption. In some cases, such a factorization of the wave function is not possible. For example, if the energy has some spatial and temporal dependence, as per a potential energy of some kind which is varying over time. In which case, we must set up the time-dependent Schrödinger equation and use mathematical cunning to solve it. Here, we are simply trying to cut corners by making educated guesses.

Given the free nature of the particle within the volume of the box, a sensible starting point is to work with the free particle state as discussed in Chapter 7 and generalized to a continuous basis in Chapter 12. There we said that:

$$|\psi(x,t)\rangle = A\int e^{i\left(\frac{2\pi}{\lambda}x - 2\pi ft\right)}|x\rangle dx$$

which, using the De Broglie relationships can be converted to:

$$|\psi(x,t)\rangle = A\int e^{\frac{i}{\hbar}(px - Et)}|x\rangle dx$$

so that the wave function is:

$$\psi(x, t) = \langle x, t|\psi(x,t)\rangle = Ae^{\frac{i}{\hbar}(px - Et)} = Ae^{\frac{i}{\hbar}px}e^{-\frac{i}{\hbar}Et}$$

DOI: 10.1201/9781003225997-17

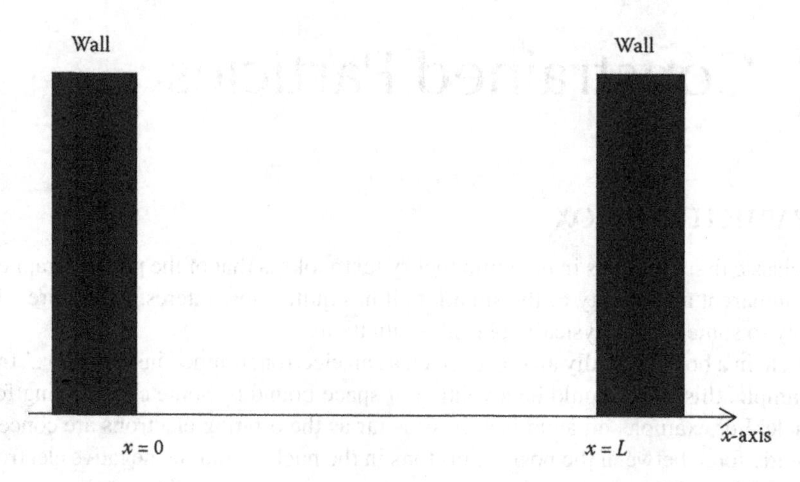

FIGURE 15.1 A one-dimensional box with impenetrable walls.

clearly factorising, as we hoped[1].

However, there is one further point to consider. After all, momentum is a vector, so even if two particles have the same *magnitude* of momentum, the *direction* can take either sign. This means that there are two possible free particle functions:

$$\psi_+(x, t) = A_+ e^{\frac{i}{\hbar}px} e^{-\frac{i}{\hbar}Et} \quad \text{positive momentum}$$

$$\psi_-(x, t) = A_- e^{-\frac{i}{\hbar}px} e^{-\frac{i}{\hbar}Et} \quad \text{negative momentum}$$

Within our box, the particle could be travelling in either direction, so we ought to consider a trial wave function which is the sum of the two momentum functions:

$$\Psi(x, t) = \psi_+ + \psi_- = \left\{ A_+ e^{\frac{i}{\hbar}px} + A_- e^{-\frac{i}{\hbar}px} \right\} e^{-\frac{i}{\hbar}Et}$$

This combined wave function is still factorizable, as the only energy is kinetic, which, classically, takes the form $E = p^2/2m$, i.e., the same for both signs of momentum. The constants A_+ and A_- will, in general, be complex numbers that will have to be calculated somehow. As it turns out, we can do this by considering what happens at the walls.

15.1.1 Another Brick in the Wall...

Putting temporal matters to one side for the moment, so that we can concentrate on the spatial dependence (state function), we can use the elegant relationship $e^{i\theta} = \cos(\theta) + i\sin(\theta)$ to write:

$$\varphi(x) = A_+\left(\cos\left(\frac{px}{\hbar}\right) + i\sin\left(\frac{px}{\hbar}\right)\right) + A_-\left(\cos\left(-\frac{px}{\hbar}\right) + i\sin\left(-\frac{px}{\hbar}\right)\right)$$

which, I acknowledge, does not look like much of an improvement. However, we press on by gathering terms together, in the process craftily using:

$$\cos(-\theta) = \cos(\theta) \qquad \sin(-\theta) = -\sin(-\theta)$$

to give:

$$\varphi(x) = \{A_+ + A_-\}\cos\left(\frac{px}{\hbar}\right) + i\{A_+ - A_-\}\sin\left(\frac{px}{\hbar}\right)$$

The walls of our box are located at $x = 0$ and $x = L$ (Figure 15.1) and as we have set them up to be impenetrable if follows that $\psi(0, t) = 0$ and $\psi(L, t) = 0$. As this must be true for all moments in time, it becomes an issue for the spatial dependence only, hence:

$$\varphi(0) = \varphi(L) = 0$$

Taking the case $x = 0$ first:

$$\varphi(0) = \{A_+ + A_-\}\cos(0) + i\{A_+ - A_-\}\sin(0) = A_+ + A_-$$

as $\cos(0) = 1$; $\sin(0) = 0$. We can guarantee that the state function vanishes by setting:

$$A_- = -A_+$$

Applying this back in the state function:

$$\varphi(x) = \{A_+ - A_+\}\cos\left(\frac{px}{\hbar}\right) + i\{A_+ + A_+\}\sin\left(\frac{px}{\hbar}\right)$$

$$= A\sin\left(\frac{px}{\hbar}\right)$$

with $A = 2iA_+$.

At the other end of the box, $x = L$, the state function must also vanish:

$$\varphi(L) = A\sin\left(\frac{pL}{\hbar}\right) = 0$$

but now we have run out of constants to 'tweak' to make this happen. If we were to set $A = 0$, then the state function would vanish across the whole box, which somewhat defeats our object. Instead, we consider values of $\frac{pL}{\hbar}$ such that $\sin\left(\frac{pL}{\hbar}\right) = 0$, which are $\frac{pL}{\hbar} = n\pi$ for integer values of n. Although $n = 0$ is a *mathematical* possibility, we can discount it on *physical* grounds as discussed shortly in Section 15.1.4.

Hence, we have a constraint on our momentum, which is consequently *quantized*, i.e., restricted to a collection $\{p_n\}$ of discrete values indexed by the *quantum number*, n:

$$\frac{p_n L}{\hbar} = n\pi$$

or:

$$p_n = \frac{n\pi\hbar}{L}$$

This, in turn, means that we have more than one possible state function:

$$\varphi_n(x) = A_n \sin\left(\frac{n\pi x}{L}\right)$$

with associated wave functions:

$$\Psi_n(x, t) = A_n \sin\left(\frac{n\pi x}{L}\right) e^{-\frac{i}{\hbar}E_n t} \qquad 0 \le x \le L \qquad n \ge 1$$

Note that the wave function is zero everywhere outside the box. If the box had 'softer' walls, i.e., a very 'high' potential barrier at each wall, rather than an infinite one in this case, then there would be a non-zero wave function outside the box.

15.1.2 Normalization

At first glance, it is surprising that we ran out of constants so quickly when we were constraining our state function to the boundary conditions $\varphi(0) = 0$; $\varphi(L) = 0$. After all, two constraints and two constants (A_+, A_-) sound perfect. In truth though, there are *three* constraints, as we must *normalize* each wave function so that:

$$\int_0^L |\psi_n(x, t)|^2 \, dx = 1$$

which is going to fix the values of A_n for us. This is one place where the factorizability of the wave function pays dividends, as $|\psi_n(x, t)|^2 = |\phi(t)|^2 |\varphi_n(x)|^2$ gives us:

$$\int_0^L |\psi_n(x, t)|^2 \, dx = \int_0^L |\phi(t)|^2 |\varphi_n(x)|^2 \, dx$$

$$= |\phi(t)|^2 \int_0^L |\varphi_n(x)|^2 \, dx = 1$$

Furthermore with $|\phi(t)|^2 = \left(e^{-\frac{i}{\hbar}E_n t}\right)\left(e^{+\frac{i}{\hbar}E_n t}\right) = 1$, the work reduces to:

$$\int_0^L |\varphi_n(x)|^2 \, dx = A_n^2 \int_0^L \sin^2\left(\frac{n\pi x}{L}\right) dx = 1$$

This integral has a standard solution, so the calculation proceeds:

$$A_n^2 \int_0^L \sin^2\left(\frac{n\pi x}{L}\right) dx = \frac{A_n^2 L}{n\pi}\left[\frac{n\pi x}{2L} - \frac{1}{4}\sin\left(\frac{2n\pi x}{L}\right)\right]_0^L$$

$$= \frac{A_n^2 L}{n\pi}\left[\left(\frac{n\pi L}{2L} - \frac{1}{4}\sin\left(\frac{2n\pi L}{L}\right)\right) - 0 - \frac{1}{4}\sin(0)\right]$$

$$= \frac{A_n^2 L}{2n\pi}\frac{n\pi L}{L} = \frac{A_n^2 L}{2} = 1$$

resulting in:

$$A_n = \sqrt{\frac{2}{L}}$$

Our complete, normalized wave function is then[2]:

$$\Psi_n(x, t) = \sqrt{\frac{2}{L}} \sin\left(\frac{n\pi x}{L}\right) e^{-\frac{i}{\hbar}E_n t}$$

Now we can do some physics.

15.1.3 Energy within the Box

As the particle is free within the box, the only form of energy is kinetic. Assuming that the particle is not travelling at relativistic speeds, we can use $\text{KE} = p_n^2/2m$ to give:

$$E_n = \text{KE} = \frac{p_n^2}{2m} = \frac{n^2\pi^2\hbar^2}{2mL^2} = \frac{n^2\pi^2\hbar^2}{2mL^2}$$

Having discounted $n = 0$, and congruent with the confined momentum being non-zero, we see that there is a *minimum energy for the confined electron*:

$$E_1 = \frac{\pi^2\hbar^2}{2mL^2} = \frac{6.02 \times 10^{-38}}{L^2} \text{ J}$$

which is the *ground state* for an electron within the box.

If we take an ad-hoc criterion for the electron being relativistic, i.e., that its kinetic energy is ~ 85% of its intrinsic (rest mass) energy, $E = m_e c^2$, then we can see that:

$$\frac{6.02 \times 10^{-38}}{L^2} \leq 0.85 \times 9.11 \times 10^{-31} \text{kg} \times \left(3.00 \times 10^8 \text{ms}^{-1}\right)^2$$

Hence:

$$L \geq \sqrt{\frac{6.02 \times 10^{-38}}{6.97 \times 10^{-14}}} = 9.29 \times 10^{-13} \text{m}$$

For comparison purposes, the diameter of a typical atom ~ 10^{-10} m, so we would not expect, on this criterion, that relativity should play a major part in the physics of atomic electrons.

Returning to the particle in a box, given the ground state energy value, E_1, we can re-cast our formula for any other energy level as:

$$E_n = n^2 E_1$$

which allows us to explore what will happen at very large n, say $n \geq 10^6$.

The value of the $n = 10^6$ energy level will be $E_{10^6} = \left(10^6\right)^2 E_1 = 10^{12} E_1$.

The next energy level up has $n = 10^6 + 1$, so its energy is $E_{10^6+1} = \left(10^6+1\right)^2 E_1 = \left(10^{12} + 2\times 10^6 + 1\right) E_1$.

Taking the difference:

$$E_{10^6+1} - E_{10^6} = \left(2 \times 10^6 + 1\right) E_1 \sim 2 \times 10^6 E_1$$

which is only a 2×10^{-4} % increase on E_{10^6}.

As the energy levels climb to very large n values, *the difference between one level and the next becomes insignificant compared with the energy already in the level.* For all practical purposes, the electron no longer appears to be restricted to specific values of energy. The levels are so closely spaced it seems as if it can take any energy it wants. This is exactly what you might expect from classical physics, and it shows how quantum effects can become unobservable under the right circumstances. This is an example of the *correspondence principle*: the predictions of quantum theory must mirror those of classical physics in the limit of large quantum numbers.

Given that we have established a single specific energy $E_n = \frac{n^2\pi^2\hbar^2}{2mL^2}$ for each wave function $\Psi_n(x, t)$, we would suspect that the $\Psi_n(x, t)$ are energy eigenstates.

This is readily confirmed by applying the energy operator (Chapter 12), a job made easier by the factorizability of the wavefunction:

$$\hat{E}\Psi_n(x, t) = \hat{E}\{\varphi_n(x)\phi_n(t)\} = \varphi_n(x)\hat{E}\phi_n(t)$$

$$= \sqrt{\frac{2}{L}}\sin\left(\frac{n\pi x}{L}\right)i\hbar\frac{d}{dt}\left(e^{-\frac{i}{\hbar}E_n t}\right)$$

$$= \sqrt{\frac{2}{L}}\sin\left(\frac{n\pi x}{L}\right)i\hbar\left(\frac{-i}{\hbar}\right)E_n e^{-\frac{i}{\hbar}E_n t}$$

$$= E_n\Psi_n(x, t)$$

Just to round things off in terminological matters, the wave function, in terms of energy eigenstates, is $\Psi_n(x, t) = \langle x, t|E_n\rangle$ and the state function $\varphi_n(x) = \langle x|E_n\rangle$.

15.1.4 Momentum in the Box

One of the key steps in obtaining our wave function was the quantization of the momentum:

$$p_n = \frac{n\pi\hbar}{L}$$

which was forced on us in order to satisfy the boundary conditions of the system. If we take $n = 0$, the immediate consequence is that $p_0 = 0$. Note however, that this is *not representative of a particle stationary within the box*, as if $n = 0$ *the entire wave function vanishes* – there is no particle in the box. Spinning it round, we come to a strange quantum conclusion: *it is not possible to confine a particle, such as an electron, to a localized region of space and have it stationary as well*[3].

One could view this as an application of the uncertainty principle, $\Delta x\Delta p \geq \hbar/2$. As the box has width L, we know that for any object within the box the maximum $\Delta x = L$, which means that the minimum $\Delta p = \hbar/2L$. The momentum distribution for the electron within the box should be centred on zero (i.e., the average momentum from a series of measurements would be zero, as the particle would be equally likely to be travelling to the left as to the right), but it has a non-zero width, $\Delta p = \hbar/2L$, so there must be at least some individual momentum measurements that are not zero.

In the free particle case, energy eigenstates are also momentum eigenstates, but not so here:

$$\hat{p}\Psi_n(x, t) = \hat{p}\{\varphi_n(x)\phi_n(t)\} = \hat{p}\varphi_n(x)\phi_n(t)$$

and

$$\hat{p}\varphi_n(x) = -i\hbar\frac{d}{dx}\left(\sqrt{\frac{2}{L}}\sin\left(\frac{n\pi x}{L}\right)\right)$$

$$= -i\hbar\sqrt{\frac{2}{L}}\left(\frac{n\pi}{L}\right)\cos\left(\frac{n\pi x}{L}\right)$$

$$= -\frac{i\hbar n\pi\sqrt{2}}{L^{3/2}}\cos\left(\frac{n\pi x}{L}\right)$$

$$\neq p_n\varphi_n(x)$$

This should not come as a surprise, as we explicitly built the wave function by summing over *two* basis states which were momentum eigenstates, but in opposite senses. A measurement of momentum on φ_n would cause collapse into either ψ_+ or ψ_- with an appropriate momentum value $\pm p_n$. This would happen with equal probability, as we determined $A_- = -A_+$.

15.1.5 Spatial Distribution

The probability that the electron will be found within a region of space $x_1 \to x_1 + \Delta x$ within the box is:

$$P(x_1 \to x_1 + \Delta x) = \int_{x_1}^{x_1+\Delta x} |\psi_n(x, t)|^2 \, dx$$

and as $|\psi_n(x, t)|^2 = |\varphi_n(x)|^2 |\phi(t)|^2 = |\varphi_n(x)|^2$, we have:

$$P(x_1 \to x_1 + \Delta x) = \frac{2}{L}\int_{x_1}^{x_1+\Delta x} \sin^2\left(\frac{n\pi x}{L}\right) dx$$

making the probability density (see Chapter 12):

$$\mathcal{P}_n(x) = \frac{2}{L}\sin^2\left(\frac{n\pi x}{L}\right)$$

Note that both the probability and the probability density are independent of time for our energy eigenstates. Having a wave function that could be factorized has resulted in the time parameter dropping out of the final probability. States of this form are said to be *stationary* (see also Section 18.3.2).

Picking out $n = 4$, for illustrative purposes, and charting the probability density gets us to Figure 15.2.

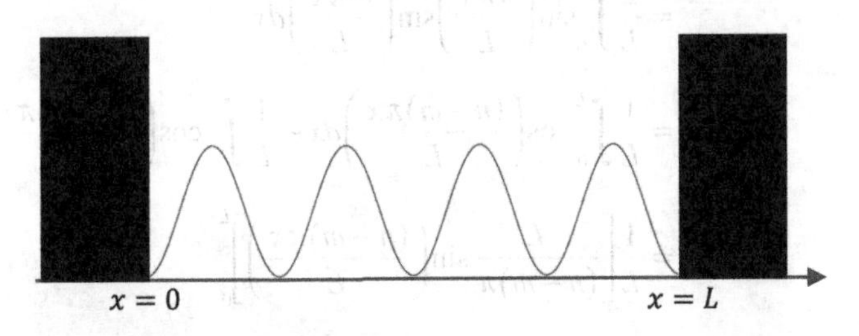

FIGURE 15.2 The probability density within a box for $n = 4$. Note that the vertical height here is relative probability. The height of the walls is non-representational.

As we would expect, we see zero probability that the electron will be found in close proximity to the walls, but more surprisingly there are three other locations, called *nodes*, across the span of the box where the probability density is zero. Classically, there is no reason for the probability to be anything other than uniformly distributed within the box. The existence of the nodes is a direct result of interference between the two momentum states that are superposed in the state function. An indefinitely repeated series of position measurements, using a collection of identical particle/box combinations, would never find the particles at the nodal locations (subject to experimental uncertainty, of course).

Given the existence of these nodes, which are present in all the state functions with $n > 1$, it is hard to see how the electron can move from one side of the box through the other while never passing through the nodal points. As with many of these sorts of issues, it's a mistake in the assumptions that is causing the problem.

Firstly, measuring an electron to be 'here' and then measuring to be 'there' does not give you the right to conclude anything about its 'whereness' in between times, *if indeed such a concept is applicable.* Once again, we are running aground when trying to think in terms of trajectories for a quantum object.

Secondly, the state function is an eigenstate of *energy*; it doesn't represent the position of the electron from moment to moment. After all, we have just seen that it's a stationary state, *so it can't represent anything that is changing with time.* If we pin down the electron's position by observing inside the box, this will cause the state to collapse into a localized region. From that point onward, *the state will have changed*, and the wave function will no longer be the same. To construct a representative wave function for this new state, we must take a new combination of energy eigenstates and any further measurements will be on that state, not the original one.

15.1.6 Wave Packets

The energy eigenstates that we've been working with don't confine the electron within a small region of the box. Yet it must be possible to construct a state that localizes things more than this. After all, we haven't placed any restrictions on the size of the box: it could be the size of a room, or even a galaxy. If we make a measurement of position inside the box, we collapse the state to one that localizes the electron to within bounds set by the apparatus.

Our set of energy eigenstates should form a complete basis, from which we can construct any appropriate state for the particle in a box system.

An extended argument would be needed to demonstrate that our energy eigenstates have all the necessary characteristics for a complete basis, but we can indicate how that would be done. For example, the states in any basis $\{|E_n\rangle\}$ must be orthonormal, $\langle E_n|E_m\rangle\rangle = \delta_{nm}$, or in state function terms[4]:

$$\int_0^L \langle E_n|x\rangle\langle x|E_m\rangle dx = \int_0^L \left(\sqrt{\frac{2}{L}}\sin\left(\frac{n\pi x}{L}\right)\right)\left(\sqrt{\frac{2}{L}}\sin\left(\frac{m\pi x}{L}\right)\right)dx$$

$$= \delta_{nm}$$

$$= \frac{2}{L}\int_0^L \sin\left(\frac{n\pi x}{L}\right)\sin\left(\frac{m\pi x}{L}\right)dx$$

$$= \frac{1}{L}\int_0^L \cos\left(\frac{(n-m)\pi x}{L}\right)dx - \frac{1}{L}\int_0^L \cos\left(\frac{(n+m)\pi x}{L}\right)dx$$

$$= \frac{1}{L}\left[\frac{L}{(n-m)\pi}\sin\left(\frac{(n-m)\pi x}{L}\right)\right]_0^L$$

$$-\frac{1}{L}\left[\frac{L}{(n+m)\pi}\sin\left(\frac{(n+m)\pi x}{L}\right)\right]_0^L$$

For either of these integrals, when $n \neq m$, $x = 0 \rightarrow \sin(0) = 0$ and $x = L \rightarrow \sin(N\pi) = 0$.

However, when $n = m$, the first integral, I_1, becomes:

$$\frac{1}{L}\int_0^L \cos(0)dx = \frac{1}{L}\int_0^L dx = 1$$

demonstrating the orthonormality.

Having established that our energy eigenstates form a valid basis, we are at liberty to combine them in subtle and creative ways:

$$|\Phi\rangle = \sum_n c_n |E_n\rangle$$

so that the corresponding wave function would be:

$$\langle x,\, t|\Phi\rangle = \sum_n c_n \langle x,\, t|E_n\rangle = \sum_n c_n \sin\left(\frac{n\pi x}{L}\right) e^{-\frac{i}{\hbar}E_n t}$$

Such a combination is not factorizable as before, nor is it a stationary state as each component has a phase that rotates with time at rate determined by E_n.

The values of c_n determine the 'shape' of the resulting wave function. They would be calculated via solving the Schrödinger equation, or from adapting them to circumstances. The only fundamental constraint is that $\sum |c_n|^2 = 1$, which is normalization.

If we sum over a range of relatively high n values and adopt the right values of c_n, then we can build a wave function corresponding to a wave packet, as mentioned in Chapter 7. For completeness, something like:

$$\Phi(x,\, t) = \sum_{n=n_0-\Delta n}^{n_0+\Delta n} K\cos^2\left[\frac{(n-n_0)\pi}{2(\Delta n+1)}\right] e^{-i(n-n_0)\pi/2} \sin\left(\frac{n\pi x}{L}\right) e^{-\frac{i}{\hbar}E_n t}$$

(where $n_0 - \Delta n \leq n \leq n_0 + \Delta n$, $\Delta n \ll n_0$ and K is a normalization constant).
gives a smooth, peaked probability distribution within the box (Figure 15.3).

As time advances, so the shifting interference between components of the wave function causes the peak to move to the right. When the packet impacts the right-hand wall, it reflects so that now there is interference between the incoming aspects of the packet and those that have been reflected by the wall (Figure 15.4):

With further reflections from each wall, and the general tendency for a wave packet to spread out, no matter how localized it was initially, the distribution steadily changes to something like that in Figure 15.5. However, the wave packet does not continue to spread across the box as time passes. Remarkably, it will eventually reform in precisely its initial configuration.

At $t = 0$, the wave function $\Phi(x,\ 0)$ takes the form:

$$\Phi(x,\, 0) = \sum_{n=n_0-\Delta n}^{n_0+\Delta n} K\cos^2\left[\frac{(n-n_0)\pi}{2(\Delta n+1)}\right] e^{-i(n-n_0)\pi/2} \sin\left(\frac{n\pi x}{L}\right)$$

as all the energy-dependent phase terms vanish. It is then of interest to see if there is some later time, T_r, when the initial function is 'revived', i.e., $\Phi(x, T_r) = \Phi(x, 0)$. This will happen whenever all the phase factors $e^{-\frac{i}{\hbar}E_n t}$ are equal[5]. The condition for this is:

$$\frac{E_n T_r}{\hbar} = 2\pi N_n$$

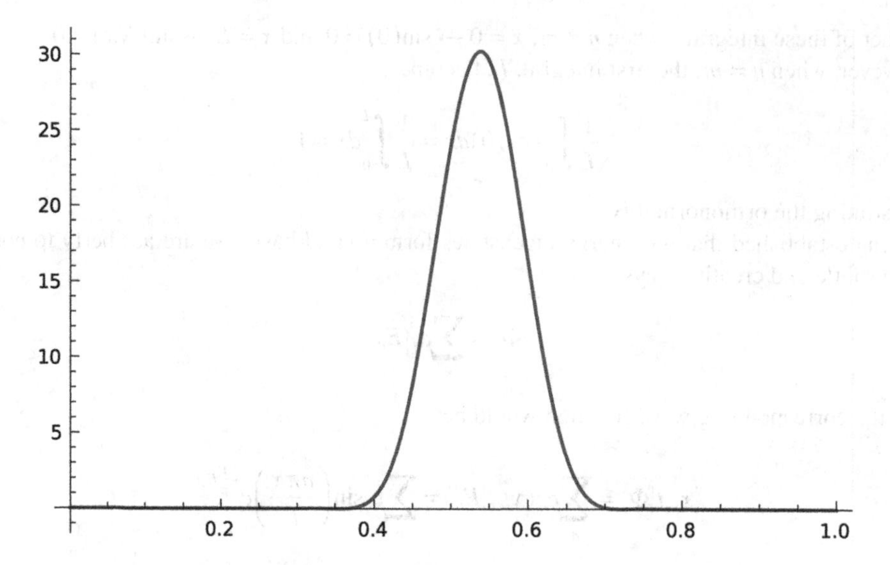

FIGURE 15.3 A wave packet within a box generated via a sum over particle in a box states. In this rendering of the probability distribution, $t = 4$ s. Here, and in all subsequent figures, $n_0 = 100$ and $\Delta n = 10$. The walls of the box are not displayed but lie at $x = 0$ and $x = 1$. (Image credit for images 22.3–22.6, images produced by Mathematica code kindly provided by Michael Seifert, Associate Professor of Physics, Connecticut College.)

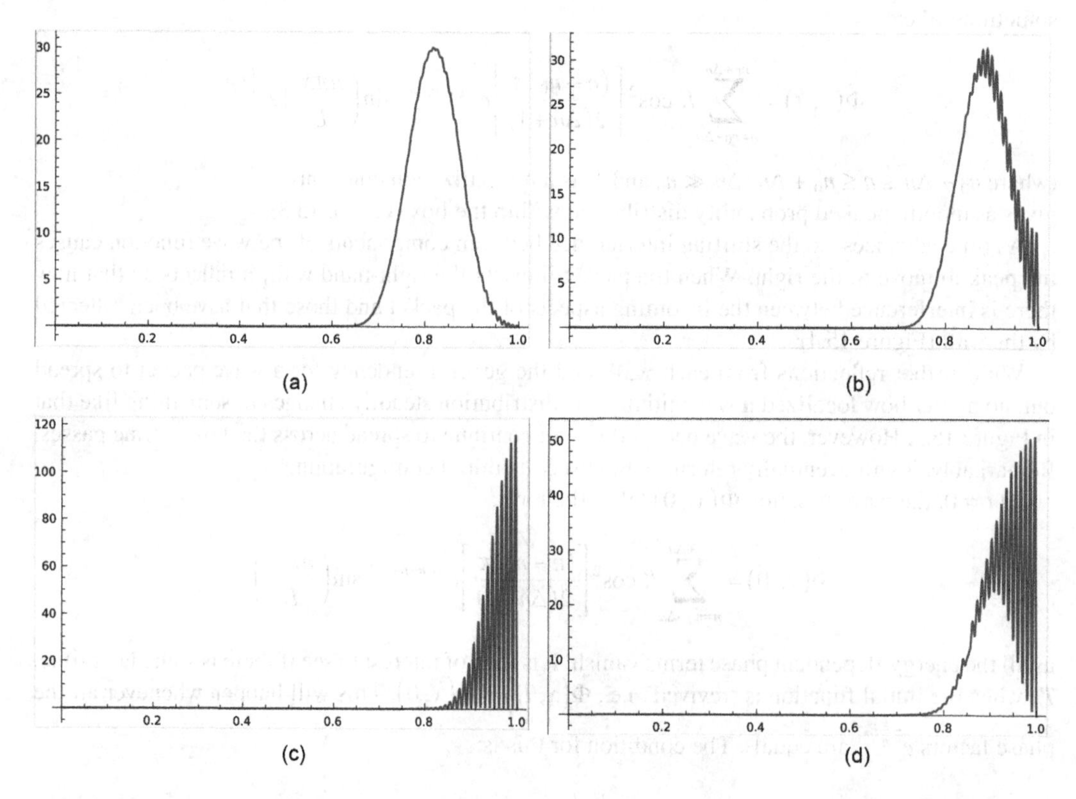

FIGURE 15.4 The progression of a wave packet reflecting off the wall of a rigid box. Top left, $t = 25$ s, top right $t = 30$ s, bottom left $t = 40$ s, bottom right $t = 45$ s.

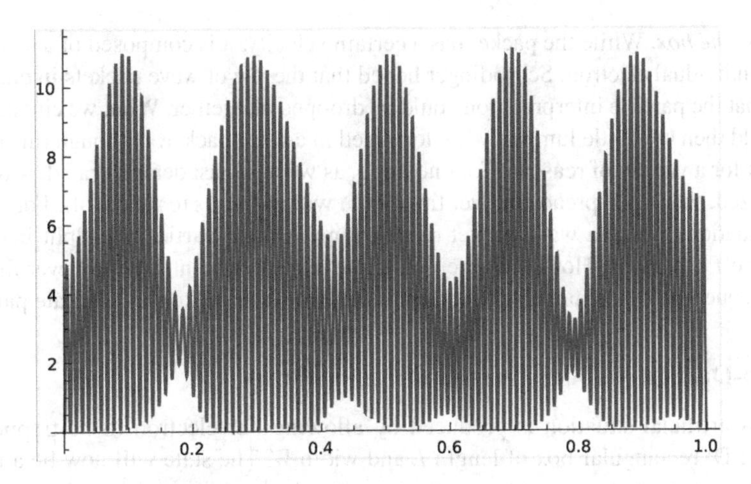

FIGURE 15.5 The wave packet probability distribution after many reflections from each wall: $t = 20,500$ s.

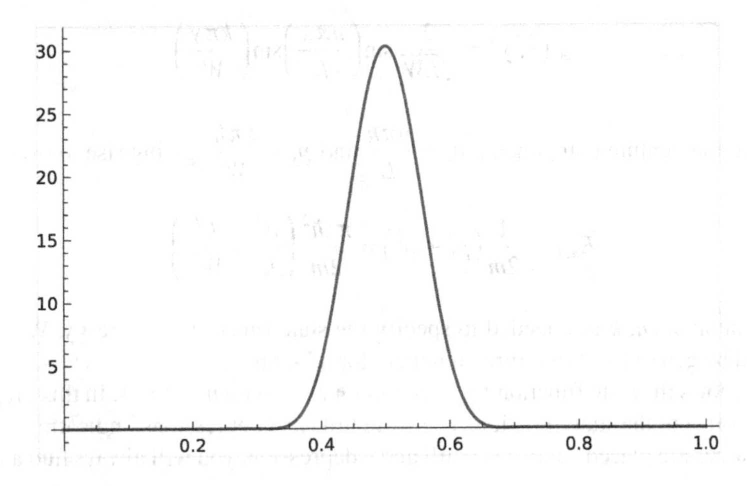

FIGURE 15.6 The wave packet shown in Figures 15.3–15.5 at time $t = 31,416$ s, which is the refresh time calculated for the parameters used in the simulation.

where N_n is an integer, that can be different for each n. If we substitute the equation for the energy levels, we obtain:

$$T_r = 2\pi\hbar N_n\left(\frac{2mL^2}{n^2\pi^2\hbar^2}\right) = \left(\frac{4mL^2}{\pi\hbar}\right)\frac{N_n}{n^2}$$

For T_r to become independent of n (hence that there is a *single* refresh time), $N_n = (\text{integer})n^2$. The smallest consistent value has the integer equal to one, hence that $N_n = n^2$ giving $T_r = \dfrac{4mL^2}{\pi\hbar}$. This refresh is shown in Figure 15.6.

It's important to realize that the evolution that we see in these successive diagrams is the $\hat{U}$ evolution that we first mentioned back in Chapter 5. While the wave packet has been constructed to resemble a particle moving through space, the distributions we see are the *unobserved evolution of a state over time.* If we make a measurement, to see what's going on, we will disturb this $\hat{U}$ process and collapse the state. So, *it is not valid to interpret the packet as mimicking the movement of the electron across the box*. While the packet has a certain velocity, it is composed of a sum over energy states for the individual electron. Schrödinger hoped that the use of wave packets in quantum theory might mean that the particle interpretation could be dropped altogether. What we classically knew as a particle would then be a little lump of wave localized in a wave packet. Although this is a nice idea, it fails to work for a variety of reasons. For one thing, as we have just demonstrated, wave packets do not stay localized, they will spread out over time, even without walls to reflect off. For another, there are certain situations where a wave packet representing a single particle can split into two pieces, which wander off separately. However, when we make a measurement, we only ever find one whole particle. Consequently, we are not able to interpret the two wave packets as separate particles.

15.1.7 Two-Dimensional and Three-Dimensional Boxes

A slightly less artificial situation is produced by allowing our electron to be trapped in a two-dimensional (2-D) rectangular box of length L and width W. The state will now be a function of x and y and we would look for functions of the form $\langle x, y|\varphi_n\rangle = \langle x|\varphi_n\rangle\langle y|\varphi_n\rangle$. The most immediate suggestion is to simply multiply the state functions for 1-D boxes along the perpendicular axes:

$$\varphi_{nm}(x, y) = \frac{2}{\sqrt{LW}}\sin\left(\frac{n\pi x}{L}\right)\sin\left(\frac{k\pi y}{W}\right)$$

using quantized momentum components $p_x = \dfrac{n\pi\hbar}{L}$ and $p_y = \dfrac{k\pi\hbar}{W}$, giving rise to overall energy:

$$E_{n,k} = \frac{1}{2m}\left(p_x^2 + p_y^2\right) = \frac{\pi^2\hbar^2}{2m}\left(\frac{n^2}{L^2} + \frac{k^2}{W^2}\right)$$

Two quantum numbers n, k are needed to specify the state function and energy. We would expect that this should be generalized into three numbers for a 3-D box.

Figure 15.7 shows the state function for a rectangular box with $n = k = 4$. In this graph, the vertical scale is the value of the state function. There are hills and depressions, indicating where positive and negative values are placed. Between a hill and a depression, you will always find a node, as with the 1-D case.

Having more than one quantum number leads to an interesting possibility: more than one combination of n, k leading to the same energy. This is easily illustrated in the case of a square box with $W = L$, for then:

$$E_{nk} = \frac{\pi^2\hbar^2}{2mL^2}\left(n^2 + k^2\right)$$

and so, for example, $n, k = 2, 1$ or $n, k = 1, 2$ will lead to different state functions with the same energy. If you have two or more states with the same energy, they are called *degenerate states*.

The extension from a 2-D box to a three-dimensional (3-D) box is straightforward, aside from the small fact that the state functions can't be drawn out in full. Normally when we draw a graph of the function, we need one dimension to indicate the size of the function at a particular point. For a 3-D box, this would require access to a fourth dimension, and we have not cracked that problem yet.

15.2 THE HYDROGEN ATOM

The jump in complexity from a particle in a box to a calculation of the state function for an electron in even the simplest atom, hydrogen, is considerable. Having said that, some of the general principles established for the box do transfer to the atomic situation.

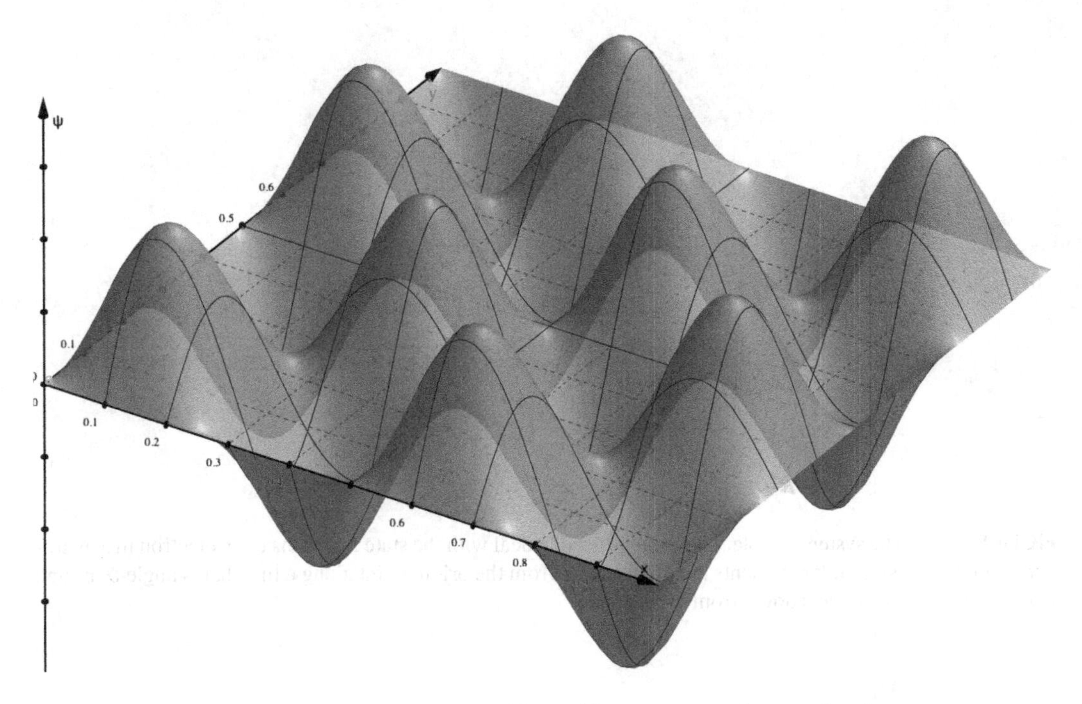

FIGURE 15.7 The state function for a 2-D square box with quantum numbers $n = k = 4$.

1. The state functions of the hydrogen atom are energy eigenstates, although in some cases they can be *degenerate*, that is, more than one state can have the same value of energy.
2. The ground state of the hydrogen atom is not a zero-energy state. There is a minimum energy.
3. There are nodes in the spatial distribution of the state function.

When we talked about 2-D boxes in the previous section, we assumed that our state function could be broken down into the product of two functions, one for each dimension in space: $\langle x, y|\varphi_n\rangle = \langle x|\varphi_n\rangle\langle y|\varphi_n\rangle$. The obvious extension to 3-D is $\langle x, y, z|\varphi_n\rangle = \langle x|\varphi_n\rangle\langle y|\varphi_n\rangle\langle z|\varphi_n\rangle$, an arrangement that works well if the living space for the particle lies inside a rectangular block. This is not true for an electron in an atom. In that situation it makes more sense to factor the state function differently:

$$\langle r, \vartheta, \phi|\varphi\rangle = \langle r|\varphi\rangle\langle \vartheta|\varphi\rangle\langle \phi|\varphi\rangle$$

using the system of coordinates shown in Figure 15.8 (*spherical polar coordinates*).

When the Schrödinger equation is set up in this context, it looks somewhat more complicated, but much of the detail is down to the new coordinate system:

$$\frac{-\hbar^2}{2m}\frac{1}{r^2\sin\vartheta}\left[\sin\vartheta\frac{\partial}{\partial r}\left(r^2\frac{\partial\varphi}{\partial r}\right)+\frac{\partial}{\partial\vartheta}\left(\sin\vartheta\frac{\partial\varphi}{\partial\vartheta}\right)+\frac{1}{\sin\vartheta}\frac{\partial^2\varphi}{\partial\phi^2}\right]+V(r)\varphi = E\varphi$$

Given that V is the electrostatic potential energy between the proton in the nucleus and the orbiting electron:

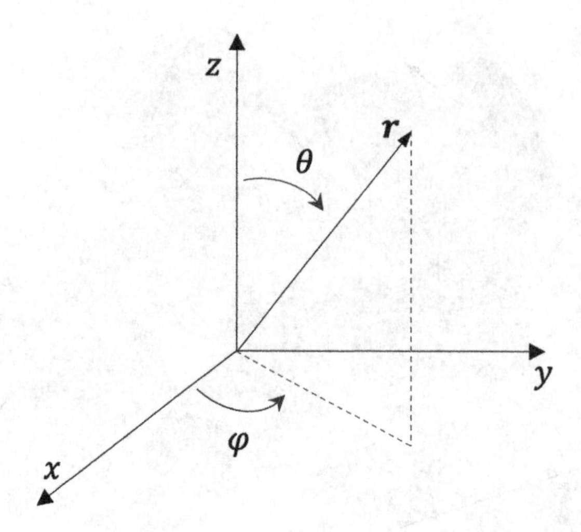

FIGURE 15.8 The system of polar coordinates used to deal with the state functions of an electron in a hydrogen atom. In this system, r represents the distance out from the origin point along a line that is angle ϑ around from the x-axis and angle ϕ down from the z-axis.

$$V(r) = \frac{-e^2}{4\pi\varepsilon_0 r}$$

and hence a function of r only, we can factor the state function along the lines we hoped, writing:

$$\varphi(r,\vartheta,\phi) = R(r)Q(\phi)P(\vartheta)$$

Inserting this into the equation, making sure that the differentials are only acting on the correct functions and bringing all terms to one side results in:

$$\frac{-\hbar^2}{2m}\frac{1}{r^2\sin\vartheta}\left[QP\sin\vartheta\frac{\partial}{\partial r}\left(r^2\frac{\partial R}{\partial r}\right)+RQ\frac{\partial}{\partial\vartheta}\left(\sin\vartheta\frac{\partial P}{\partial\vartheta}\right)+\frac{PR}{\sin\vartheta}\frac{\partial^2 Q}{\partial\phi^2}\right]+\left(V(r)-E\right)PQR=0$$

Multiplying out, and dividing through by PQR :

$$\frac{1}{R}\sin\vartheta\frac{\partial}{\partial r}\left(r^2\frac{\partial R}{\partial r}\right)+\frac{1}{P}\frac{\partial}{\partial\vartheta}\left(\sin\vartheta\frac{\partial P}{\partial\vartheta}\right)+\frac{1}{Q\sin\vartheta}\frac{\partial^2 Q}{\partial\phi^2}-\frac{2mr^2\sin\vartheta}{\hbar^2}\left(V(r)-E\right)=0$$

Now, if we multiply by $\sin\vartheta$ and gather terms:

$$\frac{1}{R}\sin^2\vartheta\frac{\partial}{\partial r}\left(r^2\frac{\partial R}{\partial r}\right)+\frac{\sin\vartheta}{P}\frac{\partial}{\partial\vartheta}\left(\sin\vartheta\frac{\partial P}{\partial\vartheta}\right)-\frac{2mr^2\sin^2\vartheta}{\hbar^2}\left(V(r)-E\right)=-\frac{1}{Q}\frac{\partial^2 Q}{\partial\phi^2}$$

On the right-hand side, we have functions of ϕ only and on the left, functions of r,ϑ. As (r,ϑ,ϕ) are completely independent variables, a split like this can only work for all possible values of the variables if the two sides are both equal to the same constant. That is:

$$-\frac{1}{Q}\frac{\partial^2 Q}{\partial \phi^2} = m_l^2 \tag{15.1}$$

$$\frac{1}{R}\sin^2\vartheta\frac{\partial}{\partial r}\left(r^2\frac{\partial R}{\partial r}\right)+\frac{\sin\vartheta}{P}\frac{\partial}{\partial\vartheta}\left(\sin\vartheta\frac{\partial P}{\partial\vartheta}\right)-\frac{2mr^2\sin^2\vartheta}{\hbar^2}(V(r)-E)=m_l^2$$

Representing the constant as m_l^2 is a matter of convention and convenience, which reveals itself later in the calculation.

Taking the second equation, dividing by $\sin^2\vartheta$ and re-arranging to taste gives:

$$\frac{1}{R}\frac{\partial}{\partial r}\left(r^2\frac{\partial R}{\partial r}\right)-\frac{2mr^2}{\hbar^2}(V(r)-E)=\frac{m_l^2}{\sin^2\vartheta}-\frac{1}{P\sin\vartheta}\frac{\partial}{\partial\vartheta}\left(\sin\vartheta\frac{\partial P}{\partial\vartheta}\right)$$

This is pulling the same trick, with functions of r on the left and functions of ϑ on the right. The same argument allows us to set each side equal to a constant, with the choice this time being $l(l+1)$.

$$\frac{m_l^2}{\sin^2\vartheta}-\frac{1}{P\sin\vartheta}\frac{\partial}{\partial\vartheta}\left(\sin\vartheta\frac{\partial P}{\partial\vartheta}\right)=l(l+1) \tag{15.2}$$

$$\frac{1}{R}\frac{\partial}{\partial r}\left(r^2\frac{\partial R}{\partial r}\right)-\frac{2mr^2}{\hbar^2}(V(r)-E)=l(l+1) \tag{15.3}$$

We are now the proud owners of three separate equations, each of which can be solved to give the dependence of the state function on the coordinates. The first equation, Eq. 15.1, is easily dealt with:

$$-\frac{1}{Q}\frac{\partial^2 Q}{\partial \phi^2} = m_l^2$$

$$Q(\phi) = A_1 e^{im_l\phi} + A_2 e^{-im_l\phi}$$

with A_1, A_2 being integration constants to be determined. When $\phi = 0$, $Q = A_1 + A_2$, so we are at liberty to set $A_2 = 0$ as one constant is enough for normalization purposes.

Given the nature of the coordinate system, ϕ and $\phi + 2\pi$ locate the same plane in space, hence:

$$Q(\phi) = Q(\phi + 2\pi) \rightarrow Ae^{im_l\phi} = Ae^{im_l(\phi+2\pi)} = Ae^{im_l\phi}e^{im_l 2\pi}$$

and so $e^{im_l 2\pi} = 1$, which constrains m_l to be $\{0, \pm1, \pm2, \pm3\ldots\}$.

It is much trickier to find solutions to the other equations.

Equation 15.2 turns out to have solutions only if l is an integer and $l \geq |m_l|$, which reads back into a constraint on m_l such that $m_l = \{0, \pm1, \pm2, \pm3\ldots\pm l\}$. The solutions to Eq. 15.3, for the radial part of the state function, are also complicated but restricted so that either $E > 0$, in which case the electron is not bound to the nucleus or one of a set of values $E_n = -\frac{me^4}{32\pi^2\varepsilon_0^2\hbar^2}\left(\frac{1}{n^2}\right)$. These energy level values match those calculated by Bohr using a much less developed and historically earlier quantum approach (Section 18.3.2)

15.2.1 Quantum Numbers for Hydrogen

As we have seen, there are three quantum numbers n, l, and m_l, which pick out a particular state function from the family. The *principal quantum number, n*, specifies the energy of the state function. We will see in Chapter 18 that Bohr introduced the principal quantum number as part of his early work on quantization. The role of the *orbital quantum number, l*, is not so immediately obvious but can be clarified by the following argument.

The total energy, E, of the electron can be decomposed into three parts: the electrostatic potential energy, the radial kinetic energy (of motion along the radius arm of the co-ordinates) and the orbital kinetic energy (for motion at 90° to the radial arm):

$$E = \frac{-e^2}{4\pi\varepsilon_0 r} + T_r + T_O$$

Although Eq. 15.3 looks to involve r only, some angular dependence sneaks in via the orbital kinetic energy. Putting the total energy into Eq. 15.3 gives:

$$\frac{1}{R}\frac{\partial}{\partial r}\left(r^2\frac{\partial R}{\partial r}\right) + \frac{2mr^2}{\hbar^2}(T_r + T_O) = l(l+1)$$

so now, if we set:

$$\frac{2mr^2 T_O}{\hbar^2} = l(l+1)$$

the terms cancel and the equation reduces to:

$$\frac{1}{R}\frac{\partial}{\partial r}\left(r^2\frac{\partial R}{\partial r}\right) + \frac{2mr^2}{\hbar^2}T_r = 0$$

which involves kinetic energy in the radial direction only. Furthermore, as $T_O = \frac{1}{2}mv_r^2$ and the angular momentum $L = mrv_r$, $T_O = L^2/2mr^2$. This now means that:

$$\frac{2mr^2 T_O}{\hbar^2} = \frac{\cancel{2mr^2}L^2}{\hbar^2\,\cancel{2mr^2}} = l(l+1)$$

or $L^2 = \hbar^2 l(l+1) \rightarrow L = \hbar\sqrt{l(l+1)}$, which, as we have seen before in Section 10.2.2, is angular momentum quantization. We also recall from that section that while l specifies the total angular momentum, m_l determines the component along the z direction. Given the spherically symmetrical nature of the hydrogen atom, it is hard to understand why the arbitrary z direction, or any direction for that matter, should be picked out in any physically meaningful way. This becomes easier to see when considering that an electron in motion about a nucleus still represents a current and hence has an associated magnetic field. In the presence of an external field, the electron's field will interact and hence the direction of the external field becomes crucial – this is the imposed z direction.

By convention, we specify angular momentum states using a letter, according to the scheme:

$l = 0$	1	2	3	4	5	6
s	p	d	f	g	h	

The first few letters are derived from the terminology developed to refer to spectral lines of hydrogen: s, *sharp*; p, *principal*; d, *diffuse*; f, *fine*. Thereafter it becomes alphabetical. Coupled with using the principal quantum number, there is a shorthand for specifying a state function, as shown in Table 15.1.

In normal circumstances, and neglecting some corrections brought about by applying the Dirac equation rather than Schrödinger's equation to the hydrogen atom, all state functions with the same value of n are degenerate, i.e., they have the same energy. In a magnetic field, this degeneracy is split, with the energy now being also dependent on the value of m_l. This results in the spectral lines of hydrogen being split in the presence of a magnetic field, an effect that was first observed[6] in 1896.

15.2.2 Visualising Hydrogen State Functions

It is clearly of some interest to visualize the probability of finding an electron at certain points in the vicinity of the nucleus. The probability density $|\varphi(r,\vartheta,\phi)|^2 = |R(r)|^2|Q(\phi)|^2|P(\vartheta)|^2$ would be tricky to display on one graph, as it depends on three variables and would need a fourth axis to display the value. It helps that $Q(\phi) = Ae^{im_l\phi}$ as then $|Q(\phi)|^2 = A^2e^{im_l\phi}e^{-im_l\phi} = A^2$ so that when we normalize the function:

$$\int_0^{2\pi} |Q(\phi)|^2 \, d\phi = 2\pi A^2 = 1$$

we see that the probability of finding the electron at any angle ϕ is the same, and hence the probability distribution for any state function must be symmetrical about the z axis.

Consequently, we can concentrate on the r and ϑ dependencies. Rather than working with the probability density, we can work with the probability in a small volume element. As we are dealing with polar coordinates, the volume element is not $dV = dxdydz$, but $dV = r^2 dr \sin\vartheta d\vartheta d\phi$. To calculate, the probability of finding an electron between r and $r + \Delta r$, no matter what the angle we need to add up across all angles, which is effectively integrating, so:

$$\text{Prob} = r^2|R(r)|^2 \Delta r \int_0^{\pi} |Q(\phi)|^2 \sin\vartheta d\vartheta \int_0^{2\pi} |P(\vartheta)|^2 \, d\phi$$

TABLE 15.1
The Naming Convention for the First Few State Functions of Hydrogen

	s	p			d				
	$l = 0$	$l = 1$			$l = 2$				
	m_l	m_l			m_l				
	0	−1	0	1	−2	−1	0	1	2
$n = 1$	1s								
$n = 2$	2s	$2p_{-1}$	$2p_0$	$2p_1$					
$n = 3$	3s	$3p_{-1}$	$3p_0$	$3p_1$	$3d_{-2}$	$3d_{-1}$	$3d_0$	$3d_1$	$3d_2$

Assuming that $P(\vartheta)$ is also normalized, this reduces to Prob $= r^2|R(r)|^2\,\Delta r$, which is something that we can plot as a function of r.

In the case of the $1s_0$ state function, $R = \frac{2}{a_0^{3/2}} e^{-r/a_0}$ where a_0 is the *Bohr radius* (see Section 18.3.2), making our probability $\frac{4}{a_0^3} r^2 e^{-2r/a_0}\,\Delta r$. This is shown in Figure 15.9 along with the probability distributions for the $2s$ and $2p$ state functions.

In the case of the $1s$ state function, the peak probability occurs at $r = a_0$, and the average radial distance for an electron in this state is $1.5a_0$.

As with other state functions in the hydrogen set, the $2s$ state has a node at $r = 2a_0$. This is like the nodes found for the particle in a box, except in this case we have a curved nodal plane, like the surface of a sphere, surrounding the nucleus at this radius. There is a definite probability that the electron can be found nearer to the nucleus ($r < 2a_0$) and further ($r > 2a_0$), but not at this specific distance. Newcomers to the subject will be tempted to ask how the electron can get from one region to the other without crossing the intermediate space. Once again, any attempt to think in terms of electron paths is confounded. Indeed, if we recall that these state functions are stationary states, the probability distributions are independent of time, so they cannot represent electron 'movement'.

The radial probability distribution is, however, only part of the picture as the state functions have angular dependencies as well. In the case of the s states, with $l = 0$, $m_l = 0$ this is simply a spherical distribution about the nucleus. Things get a little more interesting with the probability density for $2p_0$, which is shown in Figure 15.10.

Such a diagram needs careful interpretation. Most importantly, the length of a radial arm from the origin of coordinates to the surface of the probability density shows the relative value of that density, *not a physical distance in space*. There is no indication of the distance from the nucleus on such graphs. All we are seeing is a comparison between the chances of finding the electron in any

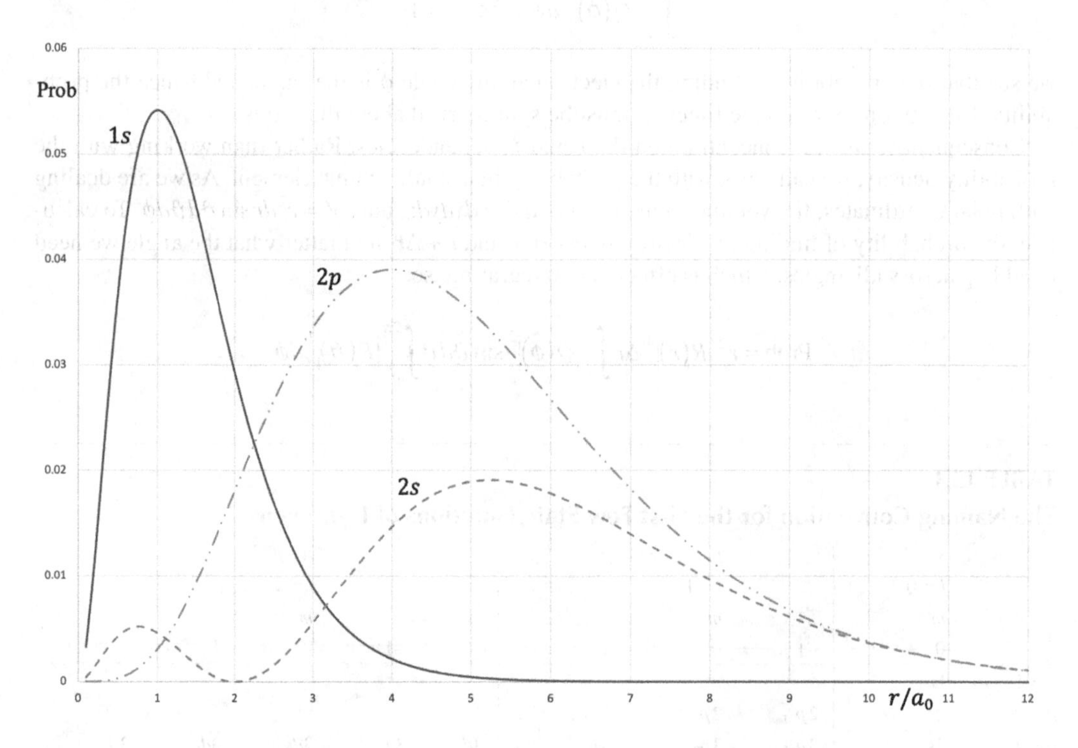

FIGURE 15.9 The probability that an electron will be found in a thin shell of radius r from the hydrogen nucleus for the first three state functions. Note the node present at $r = 2a_0$ for the $2s$ state function.

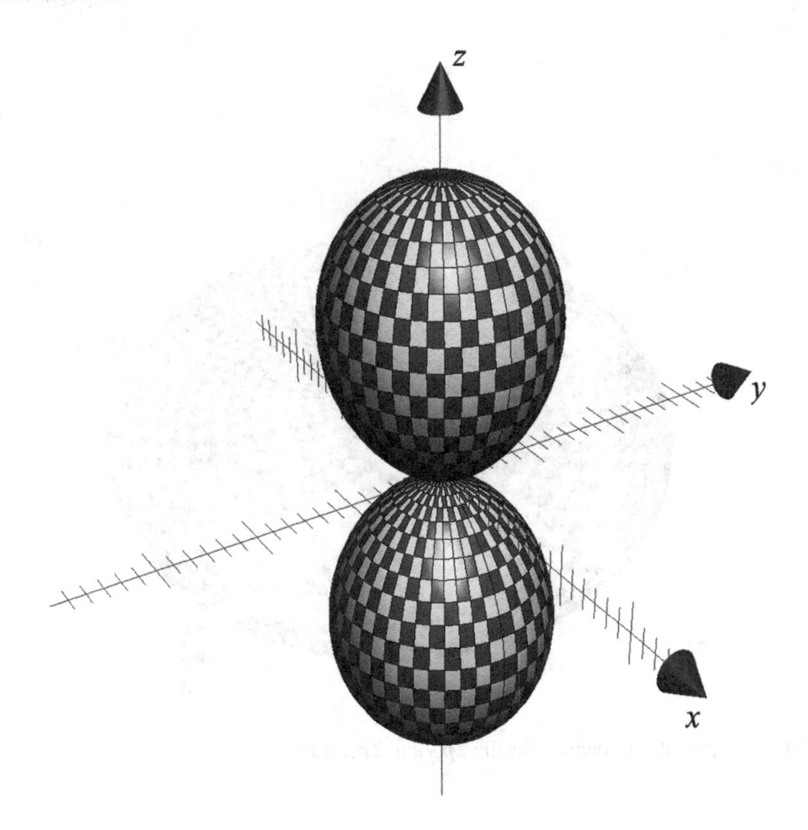

FIGURE 15.10 The probability density for a state function $2p_0$. The checked pattern is simply to highlight the curvature of the surfaces.

angular direction looking out from the nucleus. Note that in this case, there is a nodal plane lying through origin with $\vartheta = \pi/2; 0 \le \phi \le 2\pi$.

With the $2p_{-1}$ and $2p_{+1}$ states, the angular distribution of the probability density is identical and shown in Figure 15.11. Here there is a nodal line along the z -axis.

Note that both Figures 15.9 and 15.11 are showing the symmetry about the z axis that was mentioned earlier.

These renderings of the angular distributions of the probability densities, while being pertinent to our needs, are not the way the hydrogen electrons are often displayed in chemistry texts. The convention is to plot the *state function* rather than the *probability density*. This is easily done for the $2p_0$ state (which is referred to as $2p_z$) as it is a real-valued function. However, $2p_{-1}$ and $2p_{+1}$ are both complex-valued functions (in the sense that they contain products of $\sqrt{-1}$), making them hard to display on an ordinary graph. To overcome this, new state functions are constructed:

$$2p_x = \frac{1}{\sqrt{2}}\left(2p_1 + 2p_{-1}\right)$$

$$2p_y = \frac{1}{\sqrt{2}i}\left(2p_1 - 2p_{-1}\right)$$

both of which are real and hence can be plotted. As correctly constructed linear combinations of degenerate basis states also form bases, this is not such a drastic step as may first appear. In any case, these state functions are plotted in Figure 15.12. Notice that the combinations for $2p_x$ and $2p_y$ have broken the symmetry about the z -axis.

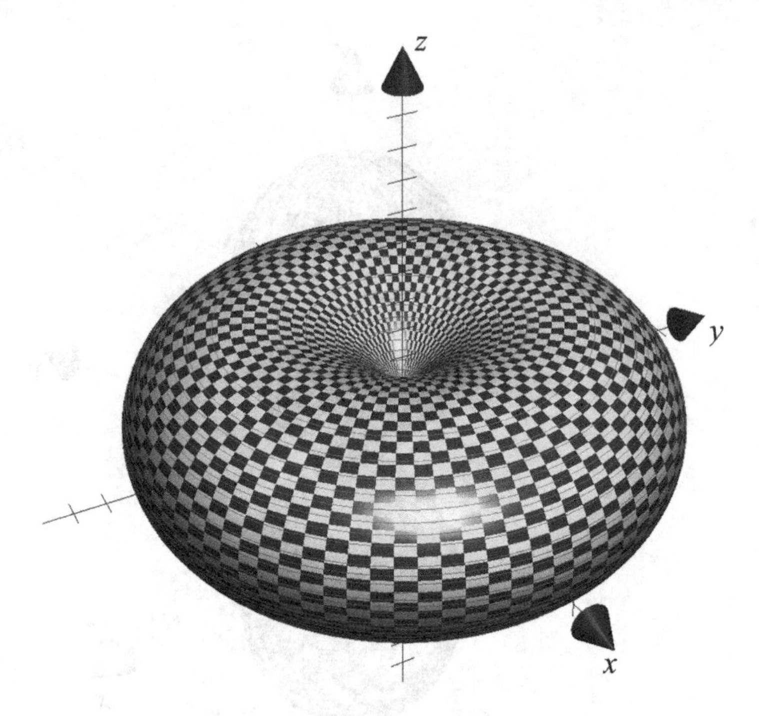

FIGURE 15.11 The probability density for the $2p_{-1}$ and $2p_{+1}$ states.

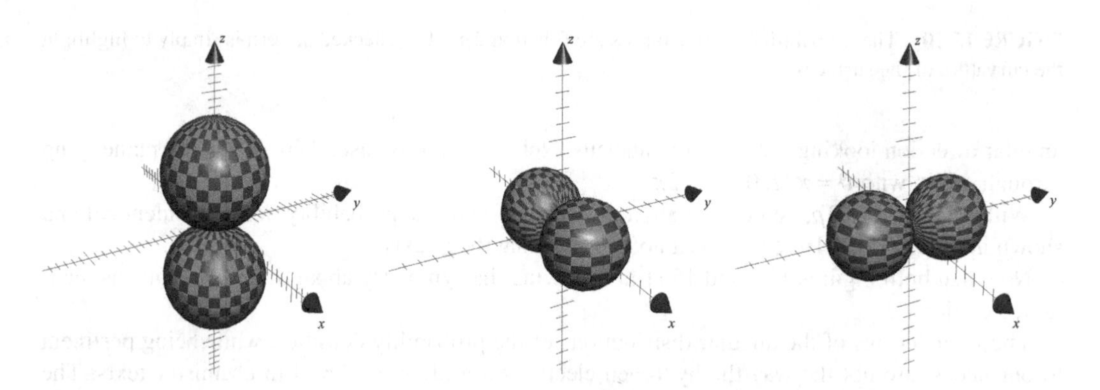

FIGURE 15.12 The $2p_z$ (a), $2p_x$ (b) and $2p_y$ (c) state functions.

In chemistry, the hydrogen state functions are referred to as *orbitals* and the conventional plots are helpful when it comes to visualising and calculating how certain types of chemical bonds are formed.

15.3 A BOX CONTAINING MORE THAN ONE ELECTRON

The idea of a particle rattling around in an otherwise empty box is very similar to the model we have of a classical gas. Of course, the obvious difference is that a gas contains a far more than just one particle. Nevertheless, the quantum mechanics of a particle in a box is a good starting point for developing a theory to cover gases under conditions where we can expect their quantum behaviour to show itself. So, it's worth considering what happens if we put a second particle in our box, and to build things up from there.

Firstly, we need to assume that the interaction between the particles (e.g., electrostatic repulsion if they are both electrons) is negligible, so we can assume that the states we derived before are still applicable. If they are fermions, then they can both occupy the lowest energy eigenstate only if they have opposite spins. Taking these to be $|U\rangle$ and $|D\rangle$ for a vertically aligned SG apparatus, the overall state of the two electrons would be:

$$|\Psi(x_1, x_2, t)\rangle = |\varphi_1(x_1)\rangle|\varphi_1(x_2)\rangle\left\{\frac{1}{\sqrt{2}}\left[|U_1\rangle|D_2\rangle - |D_1\rangle|U_2\rangle\right]\right\}e^{-iE_1t/\hbar}$$

which is asymmetrical under the exchange of particles 1 and 2, as required (see Chapter 8).

Adding another electron into the mix would force one of them to be in the next highest energy state. We would not be able to say which, as being quantum identical their identities would be indistinguishable.

Generalising from this to a situation with a moderately sized box, say about the size of a shoebox, containing many electrons, we can still start by assuming that the individual particle states be reasonable approximations. Such a collection of effectively non-interacting fermions is known as a *Fermi gas*.

If we decant N electrons into the box, they will occupy all the energy levels up to and including $n = N/2$, which is known as the *Fermi level*. At least that would be true if we could do this at absolute zero. With higher temperatures, some states below the Fermi level will be vacant as thermal excitation raises electron energies above the Fermi level (see later in Figure 15.13).

The free electrons in a metal provide an interesting example of such a system. At school, we learn that a metal conducts electricity because some atomic electrons can break away and wander through the structure. Within the body of the metal, such electrons are attracted equally on all sides. The net result is that they can move freely through the structure without being able to escape from it. The impact on their dynamics is such that they can be modelled as free electrons, with a slightly higher mass than normal.

If they were behaving classically, the electrons would tend to crowd down into the lowest possible energy state and due to thermal exchange between the metal and its surroundings, end up with an average energy very similar to that of a gas at room temperature. However, their fermion nature and the consequent requirement of an antisymmetric state means that is not possible. The electrons must pile up in spin doubles until they reach the vicinity of the Fermi level. In our shoebox example, we could continue to add in electrons, in which case the Fermi level was dependent on the number of electrons. In a metal, however, the number of free electrons is determined by the atomic properties, density of the material etc. so the Fermi level is a property of the metal. A detailed calculation shows the Fermi level to be:

$$E_F = \frac{\hbar^2}{2m_e}\left(3\pi^2\mathbb{N}\right)^{2/3}$$

where $\mathbb{N}$ is the number density of free electrons, which for a typical conductor lies between 10^{28} and 10^{29} electrons per m^3. This equates to an energy $\sim 9.6 \times 10^{-19}$ J, or in the more convenient unit for energies of this scale, the electron-volt[7], ~ 6 eV. As a result, the most energetic free electrons in the metal have energies far in excess of the particles in a classical gas at the same temperature.

In copper, for example, the energy of the Fermi level is roughly 6.3 eV, which compared with the typical kinetic energy of air molecules at room temperature, 0.025 eV, is ~ 250 times greater.

15.3.1 Temperature and the Fermi Gas

In school physics, we are taught that temperature is a measure of the average kinetic energy of particles in a material. Although this is true for a gas of non-interacting particles, it does cause problems when you consider something like a Fermi gas.

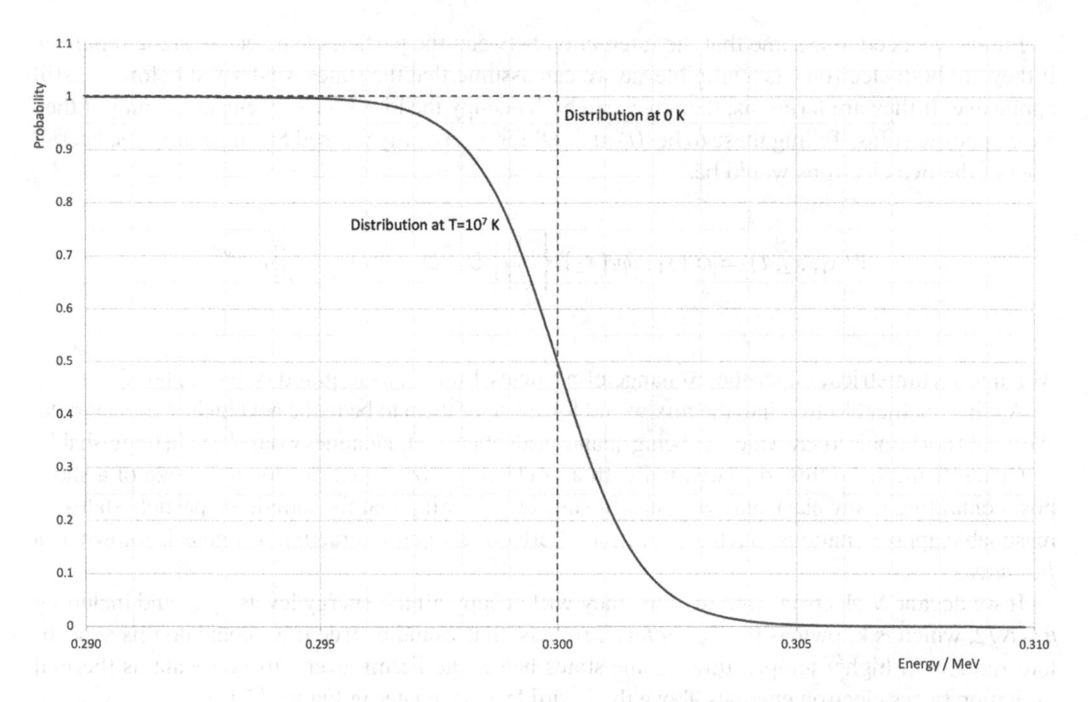

FIGURE 15.13 The occupation of energy levels in a Fermi gas in a white dwarf. The vertical axis represents the probability of a state being occupied and the horizontal axis in electron energy in MeV. The Fermi level here is 0.3 MeV.

Even if we could cool a Fermi gas to absolute zero, the particles would retain considerable kinetic energies, as the state antisymmetry requirement prevents them from all falling into the ground state. Consequently, their average energy is certainly not zero, which certainly sounds as if the gases absolute temperature is not zero.

A detailed answer to this puzzle would take us deep into the statistical theory of heat. However, to hint at the answer, consider that one of the essential features of thermal energy is its ability to flow from a hot object to a colder one. Temperature differences cause energy flows. In the case of a Fermi gas at absolute zero, all the energy levels are full. If a particle in the gas were to collide with another particle from its surroundings (e.g., in a container wall), it wouldn't be able to lose any energy as there is no vacant lower energy level for it to drop into. All the particles are locked in their energy levels. The energy in the Fermi gas is unavailable, and so it is not counted in the calculation of temperature.

At higher temperatures (Figure 15.13) some particles are excited into levels above the Fermi level, leaving gaps behind. As a result, energy exchanges with the surroundings can take place, the amounts of energy involved being *typical of the gaps* <u>between</u> *levels, not the total amount of energy* <u>in</u> *the level.* It is the scale of these exchange energies that are related to the temperature of the gas.

15.3.2 White Dwarf Stars

An ordinary, mid-lifetime, star like our sun, exists in a delicate balance between expansion and contraction. The huge mass of the star's own material exerts an enormous gravitational pull tending to collapse the star into itself. However, the star's material is a super-hot gas, maintained with energy from the nuclear reactions taking place at its core. The enormous kinetic energies involved result in an outward pressure as the gas tries to expand. While the star is stable, the thermal expansion balances the inexorable pull of gravity.

However, the situation can't last. Not only is the star producing a pressure to resist gravity, but it's also radiating energy away into space: this is the light that warms our planet. This energy is being continually replaced by the energy released from the nuclear core, but stars must eventually run out of fuel. When this happens, several different outcomes are possible. Many stars will gradually leak material into space (forming a *planetary nebula*), leaving behind a burnt-out ember called a *white dwarf*. Others will swell to enormous size (a *red giant*) and then either leak away into space or die in a gigantic explosion (a *supernova*) to leave behind a white dwarf. Still more massive stars will also supernova, but leave behind more exotic objects as their remnants, either *neutron stars* or *black holes*[8].

Even in the dead white dwarf embers of stars, the temperatures are still sufficiently high to strip all the electrons away from atoms in the gas. The material of the star is then composed of positively charged nuclei (mostly hydrogen and helium, but with some carbon and other elements) and electrons. However, the pressure that the gas of nuclei can exert is not sufficient to hold the star against its gravitational collapse. On the face of it, the stability of a white dwarf is something of a mystery. The explanation lies in the quantum physics of Fermi gases.

Within the white dwarf, gravity prevents nuclei from escaping and the nuclei's electrostatic attraction binds the electrons. As a result, electrons are held in a contained volume, so, they're a Fermi gas with a Fermi level that is very much higher than the thermal energies of the nuclei. The star is not at absolute zero, so some energy levels near the Fermi level are vacant, as the electrons have been excited into levels higher than the Fermi level (Figure 15.13). The probability of finding a fermion with energy ϵ in a Fermi gas with a Fermi level μ is dependent on the probability density[9]:

$$\mathcal{P}_F(\epsilon) = \frac{1}{e^{(\epsilon-\mu)/kT} + 1}$$

For completeness, the distribution for Bosons looks somewhat different, as we might expect:

$$\mathcal{P}_B(\epsilon) = \frac{1}{e^{\epsilon/kT} - 1}$$

The typical Fermi level for a white dwarf star is ~ 0.3 MeV, and the temperature ~ 10^7 K which are the figures used to calculate the distribution shown in Figure 15.13.

For an energy of ~ 0.298 MeV, the energy level occupation is 90%, so it is only the levels within ~ 0.6% of the Fermi level that are being affected by the temperature.

Under this sort of situation, where the overwhelming majority of the energy levels are unaffected by temperature, the Fermi gas is said to be 'cold' (!) and the gas is *degenerate*.

So, in a white dwarf star, we have a gas composed of atomic nuclei, behaving like a normal gas at very high temperatures, and a degenerate Fermi gas of electrons. Although these electrons are much less massive than the nuclei, their Fermi energy is high, so the pressure they muster *is much greater than the pressure exerted by the gas of atomic nuclei.*

It's this electron pressure that is resisting gravity in a white dwarf, something that it will be able to do indefinitely. With its nuclear reactions shut down, the star continues to radiate energy away (it is hot after all), but the energy is not being replaced. The normal gas will consequently cool, exert less pressure, and on its own would be unable to resist gravity. However, even if the Fermi gas was reduced somehow to absolute zero, it would still be capable of providing most of its pressure.

White dwarfs are remarkable objects. For a particle in a box, the energy levels depend on $1/L^2$:

$$E_n = \frac{n^2\pi^2\hbar^2}{2mL^2} \sim \frac{1}{L^2}$$

Similarly, as a star shrinks, the Fermi gas pressure goes up. Given that a white dwarf still musters a considerable mass, between 0.15 and 1.2 solar masses, the star must shrink to quite a small size so that the Fermi energy is high enough for the pressure to resist gravity.

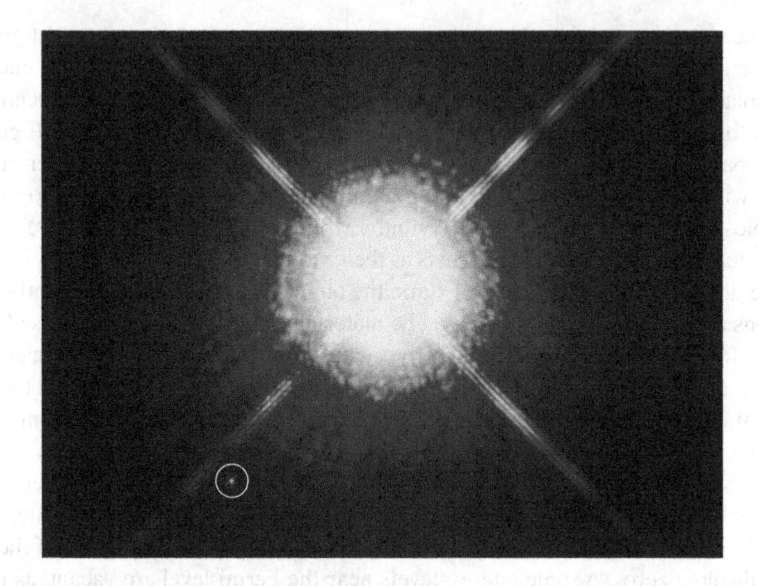

FIGURE 15.14 The white dwarf star, Sirius B (ringed), a smaller companion of the very bright star Sirius A. (Image credit: NASA, ESA, H. Bond (STScI), and M. Barstow (University of Leicester), http://www.spacetelescope.org/images/heic0516a, CC4: https://creativecommons.org/licenses/by/4.0/.)

The most well-known white dwarf star is Sirius B, the small companion star to Sirius A (Figure 15.14), which is a very bright star visible below the constellation of Orion in the Northern Hemisphere. Despite being roughly the same mass as our sun, this star is only about the size of the Earth.

The existence of white dwarf stars is a dramatic confirmation that quantum theory, despite its reputation as being a theory which only has relevance to the microworld, can apply in some extreme and important branches of physics.

Subrahmanyan Chandrasekhar[10] was the first person to work out a fully correct theory of white dwarf stars, building on the work of Arthur S. Eddington and Ralph H. Fowler, and incorporating relativity into his calculations.

Chandrasekhar demonstrated that the Fermi gas pressure would be sufficient to restrain gravitational collapse for any white dwarf star less than 1.4 solar masses. If the mass were greater than this limit, the pressure of the Fermi gas would not be able to support the dying star against gravity and the remnant collapses straight through the white dwarf phase to a much smaller size. In the process, the electrons react with the protons in the stellar material converting them into neutrons. After this phase, the dominant form of matter in the whole star is neutrons, which are also fermions. As a result, a Fermi gas of neutrons is formed, but with lower energies than before:

$$E_n = \frac{n^2\pi^2\hbar^2}{2m_N L^2} \sim \frac{1}{m_N}$$

and for a neutron $m_N \sim 200m_e$

The pressure of the neutron Fermi gas is sufficient to support the star against gravity, but only at a much smaller size. A typical neutron star of this type is 10–20 km across, although it can up to around twice the mass of the Sun.

However, this is not the end of the story. If the stellar remnant is more than three times the mass of the sun, then not even the neutron Fermi pressure can support it and the star continues to collapse through the neutron star phase. Nothing can resist the pull of gravity in such a star; it will continue to collapse until all the matter is converted into pure gravity, and a black hole is formed.

NOTES

1 Note that I have left out any explicit reference to the momentum, p, being in the x direction, taking it as evident, given the context.

2 Technically, we have only determined A_n to within an arbitrary phase, so $A_n = \sqrt{\frac{2}{L}}e^{i\phi}$. However, as we have already shown that a shift in absolute phase has no physical meaning, it is acceptable to set $\phi = 0$.

3 For those whose relativity sense is tingling, we are meaning stationary relative to the box.

4 I use the trig identity $2\sin A \sin B = \cos(A - B) - \cos(A + B)$ to evaluate the integrals.

5 The phase factors do not all have to be one, as happens when $t = 0$. They can all be $e^{-i\varphi}$ as ϕ and $e^{-i\varphi}\phi$ are the same wave function.

6 First observed by Pieter Zeeman, 1865–1943, who received the Nobel Prize in physics for this discovery in 1902.

7 Defined as 1 eV = energy gained when an electron is accelerated by a potential difference of 1 V = 1.6×10^{-19} J.

8 A neutron star is a stellar remnant composed mostly of neutrons. As neutrons are also fermions, their degeneracy pressure can support such objects against gravitational collapse, up to ~ $2\times$mass of the Sun. Above this mass, nothing can prevent the gravitational collapse and, in principle, the matter contracts into a point known as the *singularity* of zero size and infinite density. A black hole is a singularity screened from view by an *event horizon* – a region of space where the escape velocity becomes equal to the speed of light.

9 Our previous encounters with a probability density, $\mathcal{P}(x)$, involved a distribution over spatial co-ordinate x. Here the distribution is over closely packed energy levels.

10 Subrahmanyan Chandrasekhar 1910–1995, Nobel Prize, 1983.

Part 2

16 Genealogy

16.1 THE SCIENTIFIC COMMUNITY

Philosophers and historians of science have spent a long time pondering why science is successful. They talk about scientific methods, the principles of falsification, the nature of paradigms, and other rather technical matters. To my mind, there is one key aspect that's sometimes overlooked: science is a communal activity. If a scientist gets an idea or does a piece of work, they publish it. Once it is in the public arena it can be subjected to criticism. Someone else can try and verify the work, or pick up the idea and run with it, or shoot it down in flames. In principle it doesn't matter who you are: even the best can have a bad idea. Sooner or later the court of scientific peer review will sort that out.

The collective energy, intellect, and experience of the community of scientists act to refine ideas, develop points of view, and gradually hone theories and experiments to a successful level.

During a period from ca. 1900 to 1927, a collection of highly talented minds worked toward a communal understanding of the new quantum theory and the integrated experimental results. There were many wrong turns and not everyone who made decisive contributions fully agreed with the end result. Figure 16.1 attempts to summarize the chain of influences in terms of ideas leading to the end products, at least as far as 1927. Not everyone is represented here (Wolfgang Pauli, for example, made several important advances) as I have focussed on people whose ideas will be helpful to us in understanding the nature of quantum theory. Inevitably, not everyone will agree with my choice. Figure 16.1 is only an outline at best, and the number of arrows connecting a person to the rest of the group is not an entirely fair representation of how influential they were, as it takes no account of the importance of the ideas involved. However, as we work our way through this outline history, you might find it useful to refer to this figure from time to time to check your bearings.

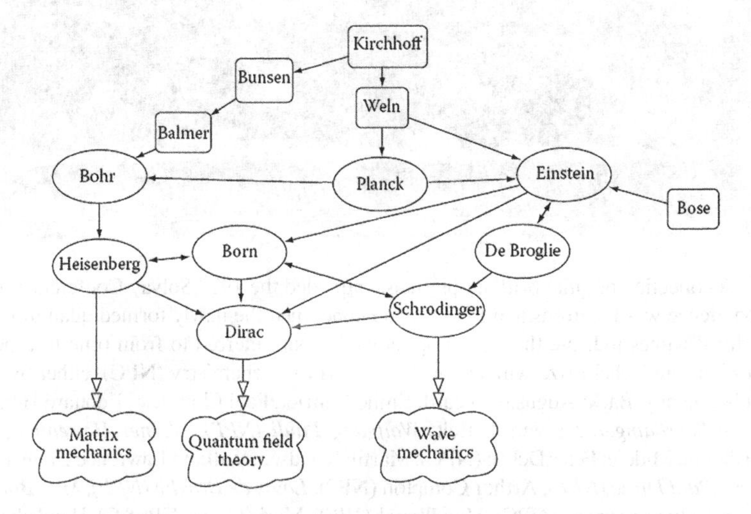

FIGURE 16.1 Some lines of influence in the history of quantum mechanics up to ca. 1927. (Based on Pais, A., *Subtle is the Lord*, Oxford, New York, 2005, p. 362.)

DOI: 10.1201/9781003225997-19

16.2 "IT WAS THE BEST OF TIMES, IT WAS THE WORST OF TIMES"[1]

> Whenever one solved one of the little problems, one could write a paper about it. It was very easy in those days for any second-rate physicist to do first-rate work. There has not been such a glorious time since then. It is very difficult now for a first-rate physicist to do second-rate work.
>
> *P A M Dirac*[2]

The founding fathers of quantum theory (some of whom are pictured in Figure 16.2) were by no means second rate; they were some of the greatest minds ever to consider issues in physics, and they found the going hard. Numerous memoirs of this time have been written by those close to the centre of developments. Frequently they refer to the profound psychological impact of grappling with the birth of quantum theory.

FIGURE 16.2 A collection of quite brilliant physicists attended the 1927 Solvay Conference on physics. The theme of the conference was Electrons and Photons and concerned the newly formed quantum theory. Ringed faces and underlined names indicate the 'founding fathers' group referred to from time to time in this book. Names in *itallics* indicate Nobel Prize winners in physics (NPP) or chemistry (NPC), either by the time of the conference or subsequently. Back: Auguste Piccard, Émile Henriot, Paul Ehrenfest, Édouard Herzen, Théophile de Donder, *Erwin Schrödinger*, JE Verschaffelt, *Wolfgang Pauli (NPP)*, *Werner Heisenberg (NPP)*, Ralph Fowler, Léon Brillouin. Middle: Peter Debye (NPC), Martin Knudsen, William Lawrence Bragg (NPP), Hendrik Anthony Kramers, *Paul Dirac (NPP)*, Arthur Compton (NPP), *Louis de Broglie (NPP)*, *Max Born (NPP)*, *Niels Bohr (NPP)*. Front: Irving Langmuir (NPC), *Max Planck (NPP)*, Marie Curie (NPP&C), Hendrik Lorentz (NPP), *Albert Einstein (NPP)*, Paul Langevin, Charles-Eugène Guye, CTR Wilson (NPP), Owen Richardson (NPP). (Image credit: Photograph by Benjamin Couprie. Courtesy of the Solvay Institutes, Brussels.)

> All my attempts, however, to adapt the theoretical foundation of physics to this knowledge failed completely. It was as if the ground had been pulled out from under one, with no firm foundation to be seen anywhere, upon which one could have built.
>
> *A Einstein*[3]

> I remember discussions with Bohr which went through many hours till very late at night and ended almost in despair; and when at the end of the discussion I went alone for a walk in the neighbouring park I repeated to myself again and again the question: Can nature possibly be so absurd as it seemed to us in these atomic experiments?
>
> *W Heisenberg*[4]

These were men who had gone through years of training and work in a complicated and technical subject area. They were used to solving problems by applying well-tried and successful methods. They had a view of the world that seemed secure and well-founded on experimental experience. Now, this was all crumbling in front of them. The period must have been disorienting, challenging, stimulating, and frustrating. Perhaps it was even good fun.

The process, however, involved a degree of 'soul searching' and a rethink about the nature of science and what it can tell us about reality. More so than probably any generation before them, the founding fathers were forced by experimental results, rather than simple interest, to debate fundamental questions about the nature of reality. Once things settled down, the debate went rather quiet. The philosophical issues surrounding quantum theory had not reached a conclusion that satisfied everyone, but the ongoing success of the theory was the focus of attention.

In more modern times, the debate has, to some degree, opened up again. Progress in technology has reached a point where we can test some of the more challenging aspects of quantum reality in the laboratory. The dividing line between the quantum world and the familiar macroscopic world is being pushed to greater and greater size scales, convincing some that the division may not in truth exist.

The current best hope in physics, string theory, is forcing us to reappraise the fundamentals of space and time. There are even those who are beginning to question our commitment to string theory, which, like the first creaky beginnings of quantum theory, could be a signpost on the way to a more radical view. It might just be that those incredibly fertile decades at the start of the twentieth century are about to be reprised as we move towards the middle of the 21st.

NOTES

1 The complete quotation is reasonably apposite: "It was the best of times, it was the worst of times, it was the age of wisdom, it was the age of foolishness, it was the epoch of belief, it was the epoch of incredulity, it was the season of Light, it was the season of Darkness, it was the spring of hope, it was the winter of despair, we had everything before us, we had nothing before us, we were all going direct to heaven, we were all going direct the other way," Charles Dickens, *A Tale of Two Cities.*

2 P. A. M. Dirac, *Directions in Physics*, John Wiley & Sons, 1978, p. 7.

3 A. Einstein, Autobiographical Notes, in *Albert Einstein: Philosopher-Scientist*, Ed P. A. Schilpp, La Salle, 1969.

4 W. Heisenberg, *Physics and Philosophy*, Penguin, 1989, p. 30.

17 Planck and Einstein

New scientific truth does not triumph by convincing its opponents and making them see the light, but rather because its opponents eventually die, and a new generation grows up that is familiar with it.

M Planck[1]

At a time like the present, when experience forces us to seek a newer and more solid foundation, the physicist cannot simply surrender to the philosopher the critical contemplation of theoretical foundations; for he himself knows best and feels more surely where the shoe pinches. In looking for a new foundation, he must try to make clear in his own mind just how far the concepts which he uses are justified, and are necessities.

A Einstein[2]

17.1 WHERE TO START?

Deciding where it all started can be a major difficulty when trying to trace the history of ideas, even within a reasonably confined area such as physics. Often you can find the first hints of an idea in various places; for instance, in lines of approach that seemed hopeless at one time, but which paid off later. Progress is rarely the nice clean progression summarized in textbooks. It is generally a lot messier than that.

However, a solid argument can be made for the key originator in the sequence of theoretical developments that eventually led to quantum theory being Max Karl Ernst Ludwig Planck.

17.2 PLANCK'S LIFE

Max Planck was born into intellectual life. His paternal grandfather and great grandfather were professors of theology in Göttingen, and his father, Johann Julius Wilhelm Planck, was a law professor, latterly in Munich where the family moved in 1867. Planck was born in 1858 to Johann and his second wife, Emma Patzig, the sixth child in the household (two of the other children being from Johann's first marriage).

Planck was a highly gifted musician, but he chose to study physics at the University of Munich. One of his professors there, Philipp von Jolly, advised him against following a career in physics because[3] "in this field, almost everything is already discovered, and all that remains is to fill a few holes." Ironically, some 20-odd years later, Jolly's student prised open one of these 'holes' until it was wide enough to swallow all of classical physics.

Having graduated in 1878, Planck was appointed as an associate professor of physics at the University of Kiel in 1885. He had spent the intervening years, while he was waiting for an academic appointment to appear, in a variety of ways. He briefly taught at his old school, worked for post-PhD qualifications and carried out unpaid research.

17.3 PLANCK ENTERS RESEARCH

In 1894, Planck's theoretical interests drew him into the branch of physics dealing with the interactions between matter and radiation, having been prompted by a commission from electricity companies who were trying to develop more efficient light bulbs.

DOI: 10.1201/9781003225997-20

Every object emits radiation[4] within a range of wavelengths and intensities determined by its temperature. For example, your body emits 'thermal radiation' as infrared waves, with a wavelength between 1 mm and 750×10^{-9} m.

At higher temperatures, objects start to glow as they are heated. At first, they are a dull red colour, but as they get hotter the colour becomes whiter until, at the highest temperatures, they are blue-white in radiance (provided they don't catch fire first, or melt). This change of colour comes about because the rate of energy emission (intensity) at each radiated wavelength changes as the temperature rises. At low temperatures, most of the energy is emitted as infrared waves, whereas at higher temperatures the intensity of the infrared drops, but that of the visible light goes up.

In most cases, the relationship between an object and the radiation it produces at a given temperature is rather complicated. However, things are particularly simple if the source is capable of absorbing electromagnetic radiation as efficiently as it can produce it, at all wavelengths. Such an object is called a *perfect thermal source,* or sometimes, rather misleadingly, a *black body.*[5] Under these circumstances, the spectrum of radiation produced doesn't depend on the material in the object, but only on its temperature. All black bodies at the same temperature will radiate in the same way.

In 1859, Gustav Kirchhoff challenged physicists to determine how a black body's radiation changed with temperature. In broad terms, a black body's *spectrum*[6] (Figure 17.1) has a characteristic hump where the intensity is greatest for a particular wavelength, determined only by the temperature of the black body. As it's impossible to find or build a *perfect* black body,[7] experimenters tended to work with the radiation emitted from a small hole in a closed oven, which does a reasonably good job as a black body over a restricted range of wavelengths. Hence, they were attempting to piece together the whole spectrum from segments of experimental evidence alongside theoretical developments.

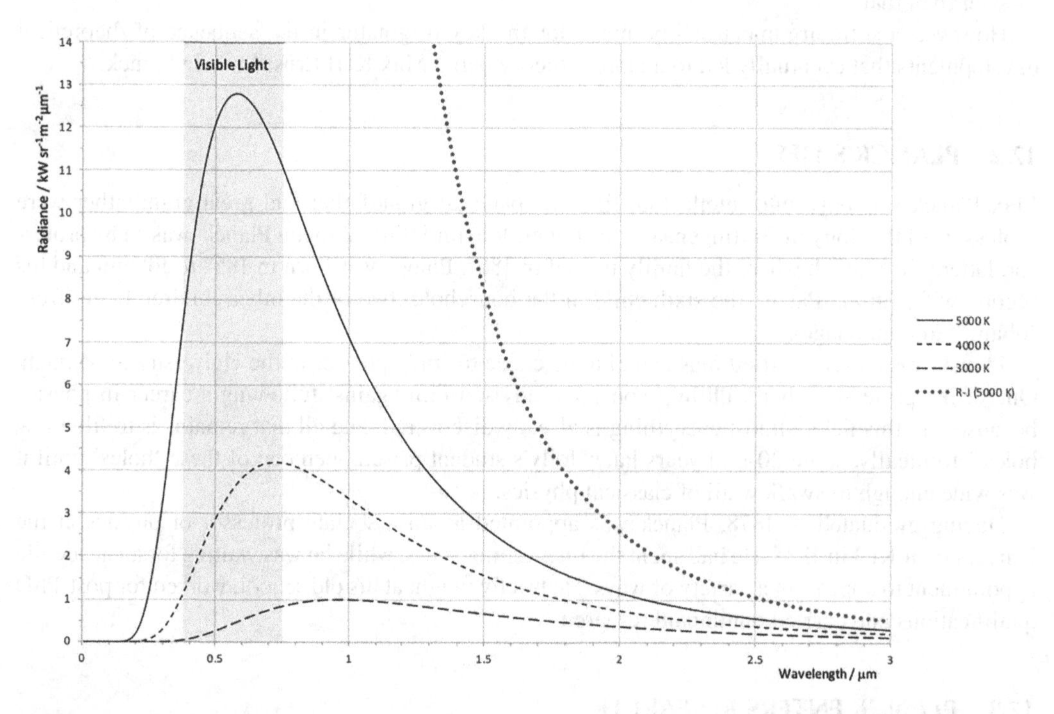

FIGURE 17.1 A black body's radiation spectrum at various temperatures. The horizontal axis shows the wavelength of the radiation produced in micrometres. The vertical axis shows the intensity produced at various wavelengths in arbitrary units. The three 'humped' curves are the measured products from a black body at different temperatures on the absolute (Kelvin) scale. The predicted (classical) spectrum based on standard theory (the Raleigh–Jeans law) is also shown. There is some agreement at long wavelengths but marked disagreement at shorter wavelengths.

When Planck entered the theoretical scene, the state of black body physics was a rather confused tangle of informed speculation and guesswork. It was clear that the spectrum depended only on temperature and wavelength,[8] but the exact form of the equation linking them was unknown. In 1896, Wilhelm Wein produced a formula that seemed to fit the measured spectrum perfectly, at least as it was known at the time. Planck's early work centred on developing a secure theoretical foundation for Wein's formula, a task he completed in 1899. The situation was then all but settled until 1900, when a new set of experimental results extended the available data into longer wavelengths (15–1000 μm), where Wein's formula failed to provide a correct description. As this undermined Planck's theoretical derivation, he went back in search of a mistake.

Independently, Lord Raleigh applied several straightforward and well-established ideas from classical physics to the problem. He modelled a black body as a collection of 'radiators' or 'oscillators' of different frequencies, which together produced the measured radiation[9]. Raleigh's calculations enabled him to show how many of these oscillators existed for each wavelength range and how the total energy of the black body ought to be divided among them. His result, published in 1900, totally disagreed with the early black body measurements (see Figure 17.1), a fact that Lord Raleigh was well aware of, prompting him to introduce a 'correcting factor'.

James Jeans then enters the story, as he found an error in Lord Raleigh's original article and published an amended version of the equation, which is now known as the Raleigh–Jeans law, in a letter to the journal *Nature* in 1905.

Interestingly, the 1900 experimental results that failed to match Wein's equation, and so disappointed Planck, agreed with the Raleigh–Jeans law at long wavelengths (without the correcting factor), but by this time Planck had produced a universally correct formula anyway.

The fact that the classically derived Raleigh–Jeans law was correctly worked out and presented *after* Planck had published the first 'quantum mechanical' result is a curious historical twist that is blurred by some accounts. In 1900, Planck only had Wein's law, which seemed to work at *short* wavelengths and experimental data for *long* wavelengths that displayed a much simpler dependence on the temperature. It was only later that the Raleigh-Jeans law could be seen to agree with the data at *long* wavelengths but diverged strongly from the experimental curves at *shorter* wavelengths (as shown in Figure 17.1). Any suggestion that Planck was specifically aiming to overcome the short-wavelength divergence in the Raleigh–Jeans law, rather dramatically known as the *ultraviolet catastrophe*, is not born out by historical analysis.

17.3.1 Planck's Formula for Black Body Spectra

Planck took a very practical approach and blended Wein's law at short wavelengths with the experimental results at long wavelengths to make a formula that worked for the whole spectrum. Apparently[10], Planck first wrote down this formula on the evening of Sunday, October 7, 1900–the same day he heard of the new experimental data. He went public with his result on October 19, 1900, at a meeting of the German Physical Society in Berlin. Fortunately, Heinrich Rubens, who was working on the black body spectrum[11], was present at the talk. Overnight, he compared Planck's formula with the detailed experimental results and confirmed that there was a beautiful fit. He called on Planck the next morning to convince him that he was on to something.

Planck was now in possession of a formula that perfectly matched the experimental spectrum, which sounds ideal. However, the result had been obtained by 'fitting' to the data, rather than derived from fundamental theory. Planck then devoted himself to finding a mathematical proof of his result, which he later referred to as "the most strenuous work of my life"[12].

Two months later, in December 1900, Plank had found a way to establish his formula, but only by "an act of desperation". He was forced to assume that the oscillator energies within the black body were restricted to whole number multiples of the oscillator's frequency. In other words, $E_f = nhf$ with f being the oscillator's frequency, and h a fundamental constant we now refer to as *Planck's constant.*

To see how new and bizarre this idea was, imagine that the oscillators present in a black body could be likened to a child's swing. Every swing has a natural *rate* of swinging (*frequency*), depending on the length of its rope. Now obviously with a playground swing, the *amount* of swing (*amplitude*) can vary depending on how hard it is pushed. If the swing is then left alone, it slowly reduces in amplitude and comes to a stop. Planck's early work on Wein's law assumed a similar effect for the oscillators, as their energy was emitted as radiation (the energy being replaced by whatever heat source was maintaining the black body's temperature). Now he was being forced to assume that the oscillators *could not have any amplitude they wanted.* It was as if, in the case of the child's swing, the amplitude could only be 1, 2, 3, 4 m, etc., with no fractional numbers in between.

Planck introduced the term *quanta* for these fixed values of energy. The word comes from the Latin *quantus,* for *how much.* Note, however, that Planck was not saying anything about how the oscillators emitted or absorbed energy, just the fixed values of energy within the oscillators themselves.

We are not sure to what extent Planck himself appreciated how radical the implications of this step would be and its ultimate historical significance. He was certainly uncomfortable with having to introduce the notion (his "act of desperation") and devoted some considerable further effort to trying to remove it or explain it on other grounds[13].

The early reaction to Planck's formula was enthusiastic, not least because more and more experimental data confirmed its accuracy across wider and wider wavelength ranges. However, there was much less reaction to the *manner* in which he had derived the formula and the 'quantum hypothesis' on which it was based. It was five years before another paper was published picking up on the quantum hypothesis and moving things decisively forward. In order to discuss that further, we need to introduce the next of our founding fathers.

17.4 EINSTEIN

In the same year that Planck produced his article on the spectrum of black body radiation, Albert Einstein was graduating from the Swiss Federal Polytechnic in Zurich.

Einstein had been born in 1879 to Herman Einstein and Pauline Koch. In 1880, the family moved from Ulm to Munich, where his father and uncle founded a company to manufacture direct current (DC) electrical equipment.

Einstein excelled in maths and physics from a young age. For example, aged 12, he taught himself algebra and Euclidean geometry over the summer and discovered his own proof of Pythagoras' theorem. At 13 he read the famously difficult philosophical work *Critique of Pure Reason*, by Immanuel Kant, who became his favourite philosopher.

In 1894, the family were forced to sell their factory having lost a bid to supply lighting to the city of Munich. While his parents moved to Italy in search of business, Einstein (by then 15 years old) stayed in Munich to finish his schooling. Having in 1895 failed the general part of the entrance examination for the Swiss Federal Polytechnic school in Zurich (he achieved high grades in maths and physics), Einstein took the advice of the school's principal and joined a school in Switzerland to complete his secondary education before re-applying. To avoid military service, and with his father's approval, he renounced his German citizenship in 1896. In the September of that year, he passed the Swiss secondary school 'Matura' with a range of decent grades and enrolled in the mathematics and physics teaching diploma course at the polytechnic.

However, his path after university was not conventional. He spent 1901 searching for a teaching job. In 1902 he finally, and famously, found a post at the Federal Office for Intellectual Property[14] in Bern where he was to examine patent applications that involved electromagnetic devices. Although some historians regard this arrangement as a waste of his talents, there is a contrary view that Einstein's subsequent interests and style of doing physics were influenced by the work he did in this office. Whatever be the historical truth, he was still working at the same institution when, in 1905, he published a series of brilliant papers:

- A pioneering article that paralleled Planck's work on black body radiation[15]. This was the only piece of work that Einstein himself regarded as truly revolutionary and is the next step paper referred to at the end of the previous section.
- A PhD thesis[16] in which he showed how to calculate the size of molecules and work out the number of molecules in a given mass of material, based on their motion.
- An article on *Brownian motion.*[17] In 1827, Robert Brown observed how tiny pollen grains suspended in water move about at random when viewed under a microscope. Einstein produced a full statistical analysis of how this motion was caused by the pollen grains being repeatedly struck by other particles (water molecules) that were too small to be seen. This paper can be seen as a crucial step in the cumulative case for the existence of atoms and molecules.
- His famous first work on relativity[18] which laid the theoretical groundwork for a new way of doing physics, subsuming Newton's laws of motion.
- An article on the equivalence of mass and energy[19] in which Einstein deduced the most famous expression in all of science: $E = mc^2$.

Separately, each of these five pieces of work that Einstein produced in this *annus mirabilis* are now regarded as extraordinary achievements. However, at the time they were not especially noticed. The relativity articles, for example, were not widely understood, due to their unusual philosophical approach rather than any mathematical complexity. Interestingly it is the first article, on matters relating to black body spectra, not his work on relativity, that is cited in Einstein's 1921 Nobel Prize nomination.

17.4.1 Quantization of Light

> According to the assumption to be contemplated here, when a light ray is spreading from a point, the energy is not distributed continuously over ever-increasing spaces, but consists of a finite number of energy quanta that are localized in points in space, move without dividing, and can be absorbed or generated only as a whole.[20]

In a breathtakingly simple argument, Einstein's 1905 paper connects the black body spectrum to a collection of point-like objects (energy quanta) exchanging energy with the 'oscillators' in the black body.

His first step was to derive a simple relationship for the probability that a collection of particles rattling around in a large volume would suddenly, by chance, find themselves all clustered in a smaller region within that volume[21]. Having done that, he then analysed Wein's formula, which he pointed out was approximately the same as Planck's version in the region of low temperatures and short wavelengths. Einstein focussed on a single wavelength in Wein's formula and showed that the probability of that radiation field's volume reducing was the same probability derived in the first step of the argument. In other words, the chance that *radiation* of a given wavelength (behaving according to Wein's law) would spontaneously jump to a smaller volume is the same as that for a collection of *particles* rattling around in a box. This also established a relationship between the wavelength of radiation and the energy of the quanta, as the probabilities only came out the same if their energy was $hf = hc/\lambda$.

Generalising the result to all wavelengths, Einstein connected Wein's law to a large-number collection of 'light quanta' with energies, hf.

Einstein was certainly aware of how radical this proposal was. After all, light had been established as a wave phenomenon since Young's interference experiment back in the early 1800s. As a result, he held back from declaring that light is a collection of particles and repeatedly emphasized that his argument applied only to low temperatures and short wavelengths, but he did go on to suggest how the idea could be checked.

17.4.2 The Photoelectric Effect

[The] bold, not to say reckless, hypothesis of an electro-magnetic light corpuscle of energy hv, which... flies in the face of thoroughly established facts of interference.[22]

In 1902, Philipp Eduard Anton von Lénárd[23] was studying the process by which electrons could be emitted from certain metals exposed to ultraviolet light (the *photoelectric effect*). Lénárd's delicate work produced some puzzling results. For a given brightness of UV, the electrons were emitted over a range of energies up to some maximum value. Making the UV brighter, did not alter the maximum energy of the electrons, but it did increase the *number* of electrons emitted. Equally baffling was the link between the energy of the electrons and the wavelength of the UV: the shorter the wavelength the greater the maximum energy of the electrons. Neither of these results could be explained by a wave theory, essentially as the energy of a wave depends on its intensity (which in the case of light we perceive as brightness) and not on its wavelength.

Einstein was aware of Lénárd's results and suggested an explanation based on his hypothesis of light quanta. He proposed that the brighter the UV the greater the *number* of quanta present. Given that the UV would be monochromatic (single wavelength) *all these quanta would be the same energy*. Brighter UV would then mean more quanta hitting the metal, producing more electrons of the same maximum energy. Altering the *wavelength* would change the energy of the quanta and so the maximum energy of the electrons coming off the metal.

Einstein went further and suggested that for the electrons:

$$KE_{max} = hf - \varnothing$$

where $\varnothing$, known as the *work function*, was the minimum amount of energy needed to break an electron free from a given type of metal. This relationship formed a measurable test of the light quanta hypothesis.

The general reaction to Einstein's article was not very positive. However, Robert Andrews Millikan was so convinced that it had to be wrong that he undertook a 10-year experimental program on the photoelectric effect. He found Einstein to be correct in every detail. As a consolation though, he was awarded the 1923 Nobel Prize for this work.

17.4.3 Enter the Photon

I therefore take the liberty of proposing for this hypothetical new atom, which is not light but plays an essential part in every process of radiation, the name photon.

Gilbert Newton Lewis[24]

Einstein was not finished with light quanta. His original 1905 article had made a passing reference to Planck's material oscillators but had worked with the radiation instead. In 1906, he returned to the topic and showed how Planck's formula was connected to his light quanta. The key was to extend Planck's idea. The oscillators in the black body are not just simply restricted to energy *values* that are multiples of hf, they can only gain and lose energy by emitting and absorbing in lumps of hf, i.e., by taking in a light quantum, or by emitting one. This was a further step that Planck was not willing to take. He was still struggling to see how his original idea could be made to fit with classical physics.

Einstein published two more articles on radiation and light quanta in 1916. It is quite clear by the tenor of these articles and his private writings that Einstein was starting to think of light quanta as real point-like particles, although their connection to interference experiments was still baffling him.

In 1917, Einstein extended the relationship between his quanta of radiation and the oscillators in a black body.[25] He assumed that the probability of an oscillator *absorbing* energy hf from the radiation was proportional to the number of quanta of that energy. He further assumed that the probability of an oscillator *emitting* a quantum hf was proportional to the number of oscillators (which makes sense) *and the number of quanta already existing in the radiation with that energy* (the *stimulated emission* term from Section 9.3.1). By balancing the rates of emission and absorption, as would happen in a real system at a constant temperature, Einstein was able to show that Planck's assumption that the oscillators had energy equal to multiples of hf directly followed, along with his formula for the black body spectrum. This was another brilliant and far-reaching result.

Another point brought out in the 1917 article (and earlier in 1916) was that light quanta interacting with matter were capable of transferring momentum, $p = h / \lambda$, as well as energy. This served to give light quanta even more of the attributes normally attached to particles. The wider community, however, was still understandably reluctant to accept the notion that light might after all be composed of particles. Undoubtedly one of the major turning points was Compton's 1923 series of experiments mentioned in Chapter 1 (Section 1.4), which showed how X-rays seemed to collide with electrons in the same manner as particles carrying energy and momentum. This cemented the two formula for light quanta:

$E = hf$ Photoelectric effect
$p = h/\lambda$ Compton Scattering

In 1929 G.N. Lewis coined the term *photon* and although his conception of light quanta was not quite on the right lines, the name stuck.

17.4.4 Bosons

In retrospect, Einstein's 1905 article introducing light quanta was a brilliantly fortuitous piece of improvisation. As I noted earlier, Einstein had to work with Wein's formula rather than the full Planck relationship and so restricted himself to short wavelengths and low temperatures. He later managed to get to the same formula as Planck by introducing stimulated emission in his 1917 article. Still, a full understanding did not arrive until an odd piece of luck happened in India.

Satyendra Nath Bose was presenting a lecture on the photoelectric effect at the University of Dakha. During the lecture, he intended to show how current theory yielded the wrong results for the black body spectrum when he unexpectedly derived the correct result! Tracing back, Bose discovered a numerical error in his work but a sufficiently deep mistake to set him thinking. What appeared to be a simple error in counting photons with identical energy had produced the correct result. Perhaps then, Bose thought, this was telling him something significant about the nature of photons.

The subject of identical particles was discussed in Chapter 8, and photons were identified as one of a class of quantum particles called *bosons* in Bose's honour. Using this new way of counting photons, Bose could derive Planck's formula but was unable to get his article published. Finally, he sent the work to Einstein, who grasped its significance and translated it into German. Einstein then wrote a companion article of his own in which he predicted the state of matter known as the Bose–Einstein condensate and sent both of them to the same journal. We now know that one of the properties of Bose particles is that they encourage stimulated emission processes similar to the one Einstein had guessed at.

17.5 FINAL THOUGHTS

Although Planck set the whole thing off, he took a long time to fully accept the implications of energy quantization and Einstein's light quanta. It is unclear what he thought at the time of his original article, but it seems likely that his 'desperate hypothesis' was introduced as a formal mathematical trick to get the result he desired. He was, however, a great supporter and admirer of Einstein's work in general, and in 1913 he was responsible for nominating Einstein as a member of the Prussian Academy of Science. Planck's opinions about light quanta shone through in his recommendation letter as he petitioned the members on Einstein's behalf: "That sometimes, as for instance in his hypothesis on light quanta, he may have gone overboard in his speculations should not be held against him."

Undoubtedly Einstein saw deeper than Planck and accepted that a whole new system of physics was in development. In Einstein's autobiographical notes (1951) he recalled his early impressions on reading Planck's article: "it was as if the ground had been pulled out from one, with no firm foundation to be seen anywhere, upon which one could have built."

The history of the photon is an interesting example of how science sometimes works to establish the 'reality' of a concept. The interplay between theory and experiment, starting with black body radiation and then moving through the photoelectric effect and Compton scattering, gradually built up a cumulative case for taking photons seriously. Of course, this still leaves open the question of the 'instrumental usefulness' versus the 'realistic existence' of the photon. Those of a realist bent would argue that the usefulness of the photon idea and the manner in which it tied together and explained previously obscure pieces of physics is a powerful indication of the underlying reality of the photon.

Einstein was undoubtedly a realist. He believed in an independent reality that can be accessed through mathematical physics. His big gripe with how quantum theory developed was its inability to produce precise predictions about experiments without resorting to probability. Although he eventually came to accept the success of quantum theory, he never gave up the hope that it would ultimately be replaced by a more fundamental theory that would allow the calculation of deterministic results. As part of this thinking, Einstein took the view that quantum states represented collections of objects rather than individuals.

Many authors consider Planck and Einstein to be among the last of the classical physicists rather than the first of the quantum generation, although their ideas were crucial in getting the enterprise off the ground.

NOTES

1 Wissenschaftliche Selbstbiographie. Mit einem Bildnis und der von Max von Laue gehaltenen Traueransprache. 35 pp. (Leipzig, 1948). Scientific Autobiography and Other Papers, trans. F. Gaynor (New York, 1949), pp. 33–34.
2 Physics and reality, J. Franklin Inst. 1936, 221(3).
3 From: Lightman, Alan P. (2005). The discoveries: great breakthroughs in twentieth-century science, including the original papers. Toronto: Alfred A. Knopf Canada. p. 8. ISBN 0-676-97789-8. Apparently Planck replied that he did not wish to discover new things, only to understand the known fundamentals.
4 Radiation is a general term for emission from a source. In popular parlance, *radiation* is taken to refer specifically to *ionising radiation* produced by radioactivity. However, in physics the term is much more general and includes things like thermal radiation (IR waves), visible light etc.
5 Confusingly because most black bodies are anything but black. Indeed, if they are hot enough, they will look white.
6 To a physicist the colours seen in a rainbow and when light passes through a prism or cut glass are each examples of a spectrum—the term generally being taken to refer to any range of wavelengths (or more generally any range of anything as in energy spectrum, etc.).
7 Er, that's not quite true. It seems that the early universe was rather a good black body and we can still measure the radiation from that early thermal source. It's called the cosmic microwaves background and forms one of the major pieces of evidence for the Big Bang creation of the universe. The thing about the early universe is that you can study it, but not really experiment on it.
8 That had been established by Gustav Kirchhoff in 1859.
9 It is worth pointing out that at this time, the idea that matter was composed of atoms had still not reached widespread acceptance in the physics community. Indeed, Planck was also highly skeptical of the atomic hypothesis in the early part of his career.
10 Abraham Pais, Introducing atoms and their nuclei, *Twentieth Century Physics*, IOPP, 1995, Vol. 1, p. 69.

11 Along with Otto Richard Lummer and Ernst Pringsheim from Physikalisch-Technische Reichsanstalt (PTR) in Berlin-Charlottenburg.
12 Max Planck, Nobel Lecture, June 2, 1920.
13 Another key step in Planck's derivation was the use of statistics to count the number of indistinguishable energy quanta in the system. This is an interesting partial move toward the eventual identification of the photon as a boson.
14 The Patent Office.
15 On a heuristic viewpoint concerning the production and transformation of light, *Annalen der Physik* 17: 132–148.
16 A new determination of molecular dimensions. PhD thesis completed April 30, submitted July 20.
17 On the motion—required by the molecular kinetic theory of heat—of small particles suspended in a stationary liquid, *Annalen der Physik* 17: 549–560.
18 On the electrodynamics of moving bodies, *Annalen der Physik* 17: 891–921.
19 Does the inertia of a body depend upon its energy content? *Annalen der Physik* 18: 639–641.
20 From Einstein's first 1905 paper referenced earlier.
21 Note for the technically more expert—the argument Einstein follows is actually related to the *entropy* of particles in a box and the entropy of a radiation field, which he derives from Wein's formula.
22 Robert Millikan, *Phys. Rev.* 1916, 7: 355–358, v has been used here to represent frequency rather than f.
23 Nobel Prize winner in 1905.
24 Gilbert N. Lewis, Letter to the editor of Nature, October 1929. Lewis was nominated 35 times for the Nobel Prize, but was never awarded.
25 On the quantum mechanics of radiation, *Physikalische Zeitschrift* 1917, 18: 121–128.

18 Bohr

There is no quantum world. There is only an abstract physical description. It is wrong to think that the task of physics is to find out how nature is. Physics concerns what we can say about nature …. What is it that we humans depend on? We depend on our words. Our task is to communicate experience and ideas to others. We are suspended in language. We must strive continually to extend the scope of our description, but in such a way that our messages do not lose their objective or ambiguous character.

Niels Bohr[1]

18.1 THE GODFATHER

Niels Bohr was crucial to the development of quantum theory in two distinct ways. His own theoretical work, initially in atomic theory and latterly in nuclear physics and the interpretation of quantum theory, moved the subject forward into new areas. Also, as the director of the Institute for Theoretical Physics (now named after him) in Copenhagen, Bohr acted as a spiritual centre and father confessor for a generation of talented physicists struggling to understand the mysterious world opening up before them.

18.2 EARLY LIFE

Niels Henrik David Bohr was born in Copenhagen in 1885. His parents' families were highly successful and influential in Danish society. Christian Bohr, his father, was a professor of physiology at the University of Copenhagen and Ellen Adler Bohr, his mother, came from a family prominent in banking and well known in Danish political life. Niels' older sister, Jenny, became a teacher while his younger brother, Harald Bohr, became a noted mathematician and was a member of Denmark's Olympic soccer team in 1908. Niels was also a passionate soccer player and acted as goalkeeper in a number of matches for Akademisk Boldklub.

Bohr left school in 1903 to attend Copenhagen University. He gained his master's degree in physics in 1909 and his PhD in 1911.

In the autumn of 1911 he travelled to the UK, specifically to Cambridge, where his original intention was to work at the Cavendish Laboratory under Sir J.J. Thompson.[2] Unfortunately, the two did not hit it off entirely, reputedly due to Bohr's trenchant criticism of Thompson's work.[3] However, while at Cambridge Bohr met Ernest Rutherford, famous for his key discoveries on the nature of the atomic nucleus, and in 1912 with Thompson's blessing, he moved to work at Rutherford's laboratory in Manchester. Rutherford subsequently had a profound influence on Bohr and his thinking.

In 1912, Bohr married Margrethe Nørlund having evidently found a perfect soul mate. Their marriage was long and happy, although two of their six children died young. The remaining children went on to have good careers. One, Aage Niels Bohr, also became a very successful physicist winning the Nobel Prize in 1975.

Between 1913 and 1914, Bohr held a lectureship in physics at Copenhagen University and in 1914–1916 a similar appointment at the Victoria University in Manchester. In 1916, he was appointed as professor of theoretical physics at Copenhagen University and from 1920 until his death in 1962 he ran the Institute for Theoretical Physics, established for him at Copenhagen. Aage Bohr succeeded Niels as the director of the institute.

DOI: 10.1201/9781003225997-21

18.3 ATOMIC THEORY

It's difficult for us to relate to the state of knowledge at the time when Bohr entered physics, as much that we now take for granted at school level about atoms and molecules was unknown when Bohr took his degree. Remember that in 1905, when Einstein was working through his *annus mirabilis*, the very existence of atoms was far from settled in the minds of many physicists.

The meeting between Bohr and Rutherford was timely. Only the previous May, Rutherford had published his model of the atom describing it as a very small central nucleus, accounting for most of the atom's mass, surrounded by a swarm of electrons. At the time, the best working model of the atom was that proposed by J.J. Thompson who envisaged the atom as a volume of positive material with electrons embedded within it. Now Rutherford was suggesting, on experimental grounds, that atoms were mostly empty space. However, Rutherford was well aware that his model had severe theoretical problems. According to this picture of the atom, the electrons were circulating the positively charged nucleus rather like planets orbiting the sun. Unfortunately, this can't quite work. Unlike planets, electrons have a negative electrical charge[4] so their motion around the nucleus should result in their continually radiating electromagnetic waves. In turn, this process would drain the electrons of energy and cause them to crash into the nucleus within a tiny fraction of a second after the atom formed.

In July 1912, Bohr returned to Denmark and in April 1913 he completed an article on the quantum theory of the hydrogen atom that changed everything. However, to understand the roots of this work, we need to take a short historical diversion.

18.3.1 Atomic Spectra

In the 1850s, Robert Bunsen invented the *Bunsen burner* familiar to all of us from school science labs. In a profitable collaboration with Kirchhoff (whose name has already come up in the context of black body radiation), Bunsen started to study the light emitted from burning elements. The development of the Bunsen burner gave them a huge advantage over previous workers, as they were able to heat their samples using a Bunsen flame that produced very little light of its own.

When a sample of an element such as sodium is placed in a Bunsen flame, the vapor produced starts to glow, specifically in the case of sodium with a yellow light. Passing this light through a prism doesn't produce a band of colours like a rainbow, you just get a thin line of yellow light.[5] Glowing vapours from other elements produce multiple colours, but never in a continuous rainbow band.[6] Kirchhoff and Bunsen started the process of classifying these *line spectra* and were even able to use this technique to find two new elements.[7] This work became the foundation of the science of *spectroscopy*.

Spectroscopy soon grew into a detailed series of measurements of the 'fingerprint' patterns of light produced by a wide variety of elements. In 1869 helium was discovered from a line in sunlight that couldn't be connected with any known element. Astronomers were also able to demonstrate that some of the glowing patches (or nebulae) seen in telescopes were actually huge clouds of luminous gas in our galaxy.[8]

As this work developed through the 1860s, the search began for a mathematical formula to connect the various measured frequencies of atomic light. Key to Bohr's later work was a beautiful and brilliant piece of improvization from a teacher in a girl's school in Basel. In 1885 and at the age of 60, Johan Balmer succeeded in producing the following simple mathematical formula for the wavelengths, λ_M, of the known lines in the spectrum of atomic hydrogen:

$$\lambda_m = B\left(\frac{M^2}{M^2 - 2^2}\right)$$

where M is a whole number from the sequence $(3, 4, 5 \ldots)$ and B a constant introduced by Balmer.

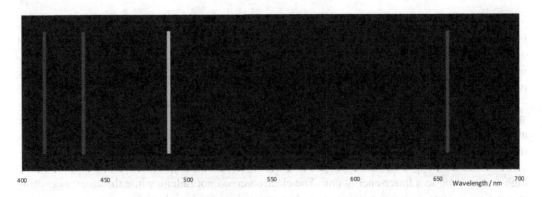

FIGURE 18.1 The hydrogen emission spectrum in the visible region. The wavelengths of the lines from left to right are 410 nm, 434 nm, 486 nm, and 656 nm.

Balmer had at his disposal four frequencies from the hydrogen spectrum, which matched his formula for $M = 3, 4, 5, 6$ in turn (Figure 18.1).

He reported his findings to the professor of physics at Basel, who told him about another 12 hydrogen lines that had been discovered and measured by astronomers. Miraculously these lines fitted his formula for $M = 7 \rightarrow 18$ inclusive.

In 1888, Johannes Rydberg generalized Balmer's formula. Rydberg started with a simple rearrangement of Balmer's work:

$$\frac{1}{\lambda} = \frac{4}{B}\left(\frac{1}{2^2} - \frac{1}{M^2}\right)$$

and then replaced the number 2 with another free variable N giving:

$$\frac{1}{\lambda_{NM}} = R\left(\frac{1}{N^2} - \frac{1}{M^2}\right)$$

with N being a whole number from the sequence $\{1, 2, 3, \ldots\}$, M a whole number from the sequence $\{2, 3, 4, \ldots(N+1)\}$, λ_{NM} the line wavelength produced for the pair of numbers N and M and R the *Rydberg constant.*

As more and more hydrogen lines were discovered, they instantly fitted Rydberg's formula for different N and M values, yet nobody knew why. Balmer had produced his original formula by playing with the numbers until he found something that worked. There was no theoretical understanding of the pattern at all.

18.3.2 Bohr's Atom

While at Manchester with Rutherford, Bohr completed a sketch memo on the structure of atoms and molecules, without mentioning spectra. Indeed, it seems that he didn't hear of Rydberg's formula until February 1913, when it possibly acted as a catalyst for his thinking. By March of that year, his atomic theory was complete.

Bohr had long accepted that it was necessary to go beyond classical physics to understand the atom. The problem is that once you decide to set out in a new direction, you can never be sure which direction is going to work out. You have to take whichever clues you can. Planck's successful quantization of energy was one clue and, as Bohr later admitted, it was "in the air to use Planck's ideas in connection with such things."

In his 1913 article, Bohr set out a series of assumptions that flew in the face of known physics. First, he claimed that the single electron in hydrogen could only circle the nucleus in one of a set of fixed orbits each one of which corresponding to a *stationary state of the atom.* Historically, this was the first use of this term, which we came across in Section 15.1.5 in the context of the particle in a box.

Classical physics would allow orbits of any radius, provided the electron moved at the appropriate speed for each one. In Bohr's atom, the orbits had fixed energy values E_N, with N being the number of the orbit; $N = 1$ is what we now call *the ground state*, or the lowest energy state. We saw in Chapter 15 that a constrained particle has quantized energy, and hence a ground state.

Second, Bohr suggested that hydrogen atoms emit light when their electrons changed from a high-energy orbit to a lower-energy one. The electron could not radiate while the atom was sitting in a stationary state, contrary to what might be expected in standard theory.

Third, he stated that the ground state was not capable of emitting any light (there being no lower-energy state to change into). This at a stroke cured the problem of the electron crashing into the nucleus, but simply by the blunt assertion that it didn't.

Finally, he claimed that when the electron changed from an orbit with energy E_M to one with lower energy E_N, it emitted a single light quantum (the name photon had not yet been coined) with frequency given by:

$$E_M - E_N = hf_{NM}$$

establishing a link with the work done by Planck and Einstein.

Bohr's article also contained a specific proposal for calculating the energy in one of his stationary states. According to Bohr, the kinetic energy was quantized with values given by:

$$\text{KE of electron in orbit } N = \frac{1}{2} NhF_N$$

where F_N is the number of orbits carried out per second by the electron and N a whole number. Consequently:

$$\frac{1}{2}mv^2 = \frac{1}{2}NhF_N = \frac{1}{2}Nh\frac{v}{2\pi r}$$

or:

$$mvr = N\frac{h}{2\pi} = N\hbar$$

We recognize mvr as the angular momentum of a point particle of mass m moving at v in a circular orbit of radius r, so Bohr's statement regarding the KE is equivalent to *quantizing the angular momentum*:

$$\text{Angular momentum} = mvr = N\hbar$$

The term *quantum number* is used to denote N.

For a stable orbit, the required centripetal force $F = -mv^2/r$ must be provided by the electrostatic attraction between the charges, so:

$$\frac{-Ze^2}{4\pi\varepsilon_o r^2} = -\frac{mv^2}{r}$$

with Z being the number of protons in the nucleus. This yields:

$$\frac{Ze^2 r}{4\pi\varepsilon_o} = mv^2 r^2 = \frac{(mvr)^2}{m}$$

The quantity mvr is the angular momentum, which according to Bohr is restricted to $N\hbar$. Using this we obtain:

$$\frac{Ze^2 r}{4\pi\varepsilon_o} = \frac{(N\hbar)^2}{m}$$

So the radius the of orbit N, r_N is:

$$r_N = \frac{4\pi\varepsilon_o (N\hbar)^2}{Zme^2} = \frac{N^2}{Z}\left(\frac{4\pi\varepsilon_o \hbar^2}{me^2}\right) = \frac{N^2}{Z} a_0$$

where a_0 is the *Bohr radius* (Section 15.2.2).

The total energy, E_N is the sum of the kinetic energy and electrostatic potential energy:

$$E_N = KE + V = \frac{1}{2} mv^2 - \frac{Ze^2}{4\pi\varepsilon_o r_N}$$

Once again, as:

$$\frac{-Ze^2}{4\pi\varepsilon_o r^2} = -\frac{mv^2}{r}$$

we have:

$$\frac{Ze^2}{4\pi\varepsilon_o r_N} = mv^2$$

Using this we get:

$$E_N = \frac{Ze^2}{8\pi\varepsilon_o r_N} - \frac{Ze^2}{4\pi\varepsilon_o r_N} = -\frac{Ze^2}{8\pi\varepsilon_o r_N}$$

Inserting the formula for the Bohr radius finally brings us to:

$$E_N = -\frac{Ze^2}{8\pi\varepsilon_o r_N} = -\frac{Ze^2}{8\pi\varepsilon_o} \times \frac{Zme^2}{4\pi\varepsilon_o (N\hbar)^2}$$

$$E_N = \frac{Z^2 me^4}{32\pi^2 \varepsilon_0^2 \hbar^2} \times \frac{1}{N^2}$$

Identifying $hR_Z c = \dfrac{Z^2 me^4}{32\pi^2 \varepsilon_0^2 \hbar^2}$ implies that:

$$E_N = -\frac{hR_Z c}{N^2}$$

and:

$$R_Z = \frac{Z^2 me^4}{8\varepsilon_0^2 h^3 c}$$

Rydberg's formula then follows quite simply, setting $Z = 1$ for Hydrogen. If M is the quantum number of an electron's initial orbit, and N the orbit that it arrives in, the energy difference is:

$$E_N - E_M = \left(-\frac{hR_H c}{N^2}\right) - \left(-\frac{hR_H c}{M^2}\right) = hR_H c\left(\frac{1}{M^2} - \frac{1}{N^2}\right)$$

As $M > N$, $E_N - E_M < 0$. This is the energy *lost* by the atom, so the energy *gained* by the emitted quantum is:

$$f_{MN} = \frac{E_N - E_M}{h} = R_H c\left(\frac{1}{N^2} - \frac{1}{M^2}\right)$$

The wavelength λ_{MN} is related to the frequency f_{MN} by $c = f_{MN} \times \lambda_{MN}$, so we get:

$$\frac{c}{\lambda_{MN}} = R_H c\left(\frac{1}{N^2} - \frac{1}{M^2}\right)$$

or:

$$\frac{1}{\lambda_{MN}} = R_H\left(\frac{1}{N^2} - \frac{1}{M^2}\right)$$

which is the same as before. This calculation also produces a value for R_H as:

$$R_H = \frac{me^4}{4\varepsilon_0^2 h^3 c}$$

which gives the same value as that obtained numerically by analysis of spectra.

Bohr was quite aware that his work applied only to the simplest of atoms, hydrogen, but turned this into an advantage. At that time the community knew about a series of lines found by astronomers that couldn't be definitively linked to a known element. Bohr pointed out that an ionized helium atom with one electron removed acted rather like hydrogen, just with a heavier and more electrically charged nucleus. He calculated the value of R_{He} and was able to explain the mysterious lines, ultimately producing an agreement to five significant figures.

The atomic theory proposed by Bohr was essentially a piece of classical physics 'patched up' with some new 'quantum' ideas. Notions of quantization introduced in this 'old quantum theory' (energy, etc.) were to last, but the classical ideas such as electron orbits had to be rejected in the full light of quantum mechanics. In the end, Bohr's atom was fatally flawed, but the value of the work remained as a far-reaching step on the way to a correct theory and as a stimulus to further work.

18.3.3 Developments

Bohr's theory of the atom gained relatively rapid widespread acceptance, albeit with some reservations. Rutherford, for one, was concerned about a lack of detail in the theory. In his view, there should be a way of predicting what happens when an electron changes energy state and emits a light quantum. Clearly, an electron in orbit 4 would have three other possible lower energy orbits open to it, and so three different light quanta that it could potentially emit. Given information about the state the electron starts in, Rutherford was hoping for a method of determining the final state. Bohr's theory did not include any mechanism for calculating what was going to happen, which in retrospect was an early indication of the role that probability was to play in the full quantum theory.

Einstein picked up the stationary state idea and used it in the 1917 article where he introduced the concept of stimulated emission. This article also made use of probability and statistical arguments, which Einstein hoped could be replaced in time by a more detailed theory. He never succeeded in finding one.

A related matter is centred on the physical cause for the electron energy change and the corresponding emission of a light quantum. In classical theory, the radiation from an accelerating charge takes place continuously, until all the energy available has been radiated away. In Bohr's model, the change takes place instantaneously with no obvious cause. Bohr would come to regard strict causes as another piece of classical physics that had to be abandoned in the quantum world.

An obvious limitation to Bohr's article was the restriction to circular paths for the electrons. In 1916, Arnold Sommerfeld extended the basic idea to include elliptical orbits and introduced another quantum number k that together with Bohr's N described the shape of the ellipse. Sommerfeld went on to show that the various hydrogen orbits with the same N values but different k, had the same energy under most circumstances, which explained why Bohr's original theory had worked so well. Sommerfeld's extended theory worked for more detailed spectra and was more readily applied to multielectron atoms. As a later development, Sommerfeld brought in another quantum number to help explain atomic spectra. Having quantized the shape of the elliptical orbits using N and k, Sommerfeld thought that the *plane* of the orbit might be quantized as well. He called this *space quantization*. Interestingly, the initial use of the Stern–Gerlach experiment was to try and verify Sommerfeld's idea, as the plane of the orbit should affect the way in which an atom interacted with a nonuniform magnetic field. When they found double splitting of the atomic beam, this was initially interpreted as a confirmation of space quantization and Bohr's theory of stationary states. It was not long before the experiment was re-evaluated as evidence for spin.

The idea that electrons in atoms could occupy only fixed energy levels received decisive confirmation in 1914 when James Franck and Gustav Ludwig Hertz[9] carried out an experiment revealing the atomic energy levels in mercury vapour. They constructed an apparatus, not unlike a fluorescent tube but containing low-pressure mercury vapour. The experiment was set up to allow a current to pass through the tube when a voltage was applied across it.

As the voltage across the tube increased, so did the size of the current, showing that the electrons flowing through the tube were being accelerated. When the voltage reached 4.9 V, the current dropped sharply—almost to zero. As the voltage increased beyond 4.9 V, the current picked up again. Following that, dips in the current were observed at reasonably regular 4.9 V intervals (Figure 18.2).

These observations are elegantly explained via Bohr's theory by suggesting that the electrons flowing through the tube are involved in multiple collisions with atoms.

Let's say the accelerating voltage is set to 16 V, which is distributed along the length of the tube. An electron will accelerate through the tube until it has reached the section where the voltage is 4.9 V. Any collisions with atoms up to that point will have been harmless. The atom will not have been able to absorb any energy from the collision, as the current electrons are not carrying enough energy to bridge the gap between the ground state and the next energy level up. At 4.9 V there is enough energy in the current electrons. Consequently, collisions between electrons in the current and those in the atoms will cause the atoms to jump to a state with higher energy, pinching most of the energy from the electron in the current (Figure 18.3). This is why the current drops almost to zero. However, any recently drained electron will not stand still forever. The voltage in the tube will accelerate it again. Now, when it reaches the 9.8 V part of the tube it will have accelerated through 4.9 V again and so will have enough energy to excite another atom. The same happens again at the 14.7 V region.[10]

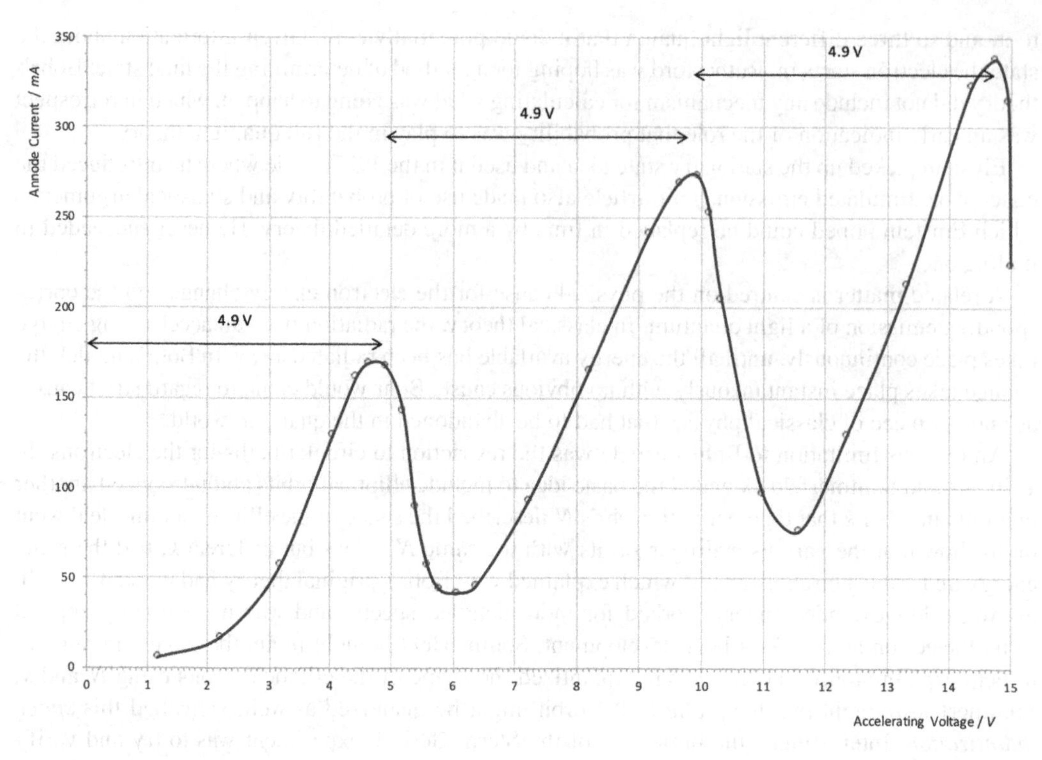

FIGURE 18.2 A representation of the results from the Franck and Hertz experiment confirming the existence of energy levels in mercury atoms.

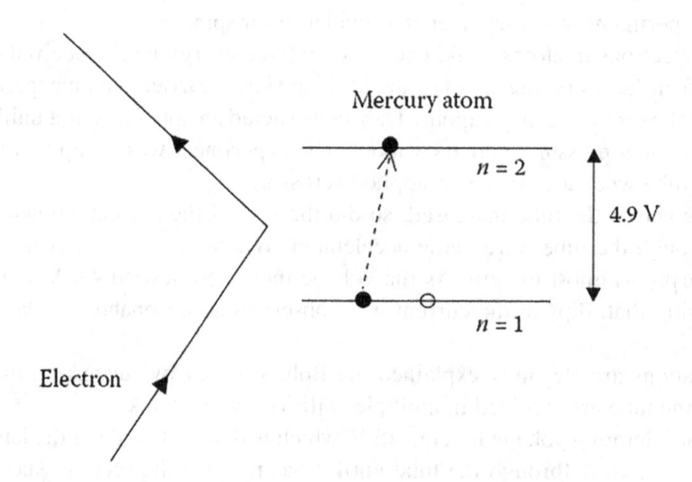

FIGURE 18.3 The principle of the Franck and Hertz experiment. If an electron with energy less than the energy difference between the $n = 1$ and $n = 2$ levels strikes an atom of mercury, no energy is absorbed from the electron. If the incoming electron's energy is greater than the energy difference, then the electron in the atom jumps to the state with higher energy absorbing just the right amount of energy and leaving the recoiling electron with the rest. The recoiled electron can then pick up more energy and go on to collide with other atoms.

As a further vindication of Bohr's ideas, Franck and Hertz also observed that the mercury vapour emitted light quanta of just the energy you would expect to be produced when the atoms returned to the ground state again.

18.4 COMPLEMENTARITY

> We are suspended in language in such a way that we cannot say what is up and what is down. The word "reality" is also a word, a word which we must learn to use correctly.
>
> *Niels Bohr*[11]

During his time as the director of the Institute of Theoretical Physics in Copenhagen, Bohr acted as a magnet, attracting some of the brightest minds in physics to visit or stay and work with him. He became a father confessor figure, with scores of young physicists bringing their ideas to him for approval. Werner Heisenberg was a particular disciple and friend. The importance of the Copenhagen institute to the development of quantum theory can be seen in the following excerpt from Heisenberg's memoir:

> During the months following these discussions an intensive study of all questions concerning the interpretation of quantum theory in Copenhagen finally led to a complete and, as many physicists believe, satisfactory clarification of the situation. But it was not a solution which one could easily accept. I remember discussions with Bohr which went through many hours till very late at night and ended almost in despair; and when at the end of the discussion I went alone for a walk in the neighbouring park I repeated to myself again and again the question: Can nature possibly be as absurd as it seemed to us in these atomic experiments?[12]

Bohr had a unique, deeply thought-out, and very subtle view of the nature of science and the interpretation of quantum theory. As such, it defies an easy and compact summary. Some of his writings read rather like an instrumentalist viewpoint, but this would be an oversimplification.

In essence, Bohr was convinced that humans could never picture the inner mechanisms of the atom. This follows from the nature of our language. As a result of our daily experience, the sort of ideas and concepts that we deal with comfortably and easily express in language are restricted to the classical world. Consequently, there is no guarantee that these ideas will automatically extend into the quantum level, which lies so far outside our common experience. Hence, when faced with the need to interpret the experiments and theory applicable to quantum reality, it is likely that classical thinking will have to be abandoned and new concepts developed.

However, there's a snag. The experiments that we carry out rely on gathering together instruments and equipment that are clearly very much bigger than atomic scales and are, consequently, described by our everyday language (including the obvious need to add technical terms for brevity). All the various measurements that we make on atoms and particles in the end come down to numbers read from dials (or similar) on a measuring apparatus. It's essential to the success of science that the equipment we use can be described in relatively ordinary terms, which makes it possible for others to reproduce the experiment.

So, we are caught in a dilemma. The experiments we carry out should not only be capable of being described in, essentially, everyday language, but they *must* also be so to make science possible. Yet, when we try to gather the results of our experiments together to make a description of the atomic world, we find that the same everyday language and ideas start to fail. Photons seem to act as particles in some circumstances and as waves in others. We can find appropriate mathematics to describe the situation, but that doesn't help us visualize or speak about photons.

Bohr hoped to find a way out of this trap by developing the *principle of complementarity.*[13] He accepted that there were two completely contradictory views of the photon that couldn't be tied together into one picture, but that didn't matter, *as you never did an experiment that required the use of both pictures at the same time.* It was acceptable to employ a form of 'double think' using one from the pair of complementary views at any one time, depending on what was most appropriate. The actual 'true' nature of the photon would never be expressible in language, which is essentially an instrument to describe the classical world, but you could shadow what was going on by keeping complementary pictures in mind.

The wave/particle views of the nature of light are different interpretations of the experimental evidence with the limitation of the classical concepts being expressed in a complementary way. In fact, we are not dealing with *contradictory* (i.e., they can't both be true at the same time) but with *complementary* (i.e., each one is true in the correct context) pictures of the phenomena, which only together offer a natural generalization of the classical modes of description.[14]

Heisenberg expressed a similar view:

> Light and matter are both single entities, and the apparent duality arises in the limitations of our language. It is not surprising that our language should be incapable of describing the processes occurring within the atoms, for, as has been remarked, it was invented to describe the experiences of daily life, and these consist only of processes involving exceedingly large numbers of atoms. Furthermore, it is very difficult to modify our language so that it will be able to describe these atomic processes, for words can only describe things of which we can form mental pictures, and this ability, too, is a result of daily experience. Fortunately, mathematics is not subject to this limitation, and it has been possible to invent a mathematical scheme—the quantum theory—which seems entirely adequate for the treatment of atomic processes; for visualisation, however, we must content ourselves with two incomplete analogies—the wave picture and the corpuscular picture.[15]

When it comes to concepts such as position, momentum, frequency, energy, and so on, they can only be defined in the context of the experiment used to measure them. If you want to talk about the frequency of light (wave picture) then you need to do an experiment where the wave picture is appropriate. If you want to talk about the momentum of a photon (particle picture) then you have to do a different sort of experiment. By insisting on this, Bohr ensured that concepts such as momentum, energy, frequency, and so on, remained rooted in the classical world, while at the same time acknowledging that they would have a limited application in the quantum realm. In his book *Atomic Theory and the Description of Nature*,[16] Bohr writes:

> [quantum theory] forces us to adopt a new mode of description designated as complimentary in the sense that any given application of classical concepts precludes the simultaneous use of other classical concepts which in a different connection are equally necessary for the elucidation of the phenomena.

This emphasis on relating the meaning of theoretical terms to the experiments in which they are measured is what gives Bohr's thinking an instrumental tinge. However, a careful reading of his writings confirms that Bohr retained the ambition of a realist, tempered by his conviction that the atomic world could not be described by limited human experience and language.[17]

These two key aspects of Bohr's thinking, that experiments in the end had to be described by classical physics and that complementary classical ideas may have to be used in tandem, formed the cornerstones of what became known as the *Copenhagen interpretation* of quantum mechanics. We will come to the Copenhagen interpretation, and with it some more aspects of Bohr's thinking, in Chapter 27.

18.4.1 Extensions

Bohr came to feel that the principle of complementarity would have a useful application in other branches of thought. Examples that Bohr considered were mind/matter and living/non-living.

Let's take the second one. Bohr was convinced that organisms comprised the same forms of matter as non-living things. There is no extra 'vital spark' present in life. Equally living creatures are subject to the same laws of nature as the rest of the material world (Bohr, however, did believe in freedom of the will) and can be studied in the same way. Of course, though, there is a snag. How do you try to find out what it is about life that makes it alive? The standard scientific approach would be to study organisms in terms of the fundamental chemical processes at work within them, in the hope of finding out what it was that made them special. In other words, one would rely on the normal scientific process of *reductionism*.

Reductionism works like this. Imagine that you wanted to find out how an old-fashioned mechanical clock worked. You would take it apart until you had a collection of pieces that could not be broken down any further, springs, pendulums, gears, etc. By carefully studying each piece you could find out what it was and what it did. Then you would know exactly how the bits could interconnect with one another and gradually trace how the whole mechanism worked. Such is the way that science tackles complicated wholes: by analysing their parts.

For this process to work, the properties of the parts have to be just the same outside the mechanism as they are when part of the mechanism. The whole does not affect the part.

Generally, this is taken as read, and undoubtedly part of the reason for the great success of science is the application of reductionism. By and large, it works. However, it pays to remember that it is an assumption. Some physicists have certainly speculated that the contextuality of quantum theory and the nature of entangled states (which we will discuss in Chapter 23) might signal a failure of reductionism at a fundamental level.

So, coming back to what Bohr had to say about life, if we were to carry out the procedure of dismantling an organism, we would undoubtedly kill it in the process. If reductionism is right, then this shouldn't matter. But, if there is some way that the context of being in a living organism affects the way in which the parts work, then the process of dismantling will not show that up. Given what the founding fathers were finding out about the contextuality of quantum theory, Bohr was sounding a justifiable warning. He felt that living/non-living might be complementary concepts in the same way that wave/particle ideas are. Similar thinking might be applied to mind/matter.

Here we are now touching on an area in which quantum-hype thinking can lead to a degree of abuse. Complementarity, and with it the uncertainty principle, does not provide a license for lazy thinking. There is a great temptation to take a couple of ideas that we don't understand, bolt them together, call them complementary, and justify pairing them, as quantum physicists say that you can do things like that.

The ideas that Bohr linked in a complementary fashion were forced on him by *experiments*. He always stressed that complementarity arises when two classical concepts, which seem opposite to one another and can't be reconciled, are nevertheless *required together* to get a full grasp of what's going on. By full grasp, of course, I mean as full as we can get, given the necessarily classically limited nature of our language and thinking.

It may be that complementarity is a helpful way forward in other branches of science, but if that is going to be the case *it will arise from a scientist's struggle to reconcile what seems to be impossible to reconcile yet is clearly indicated by experiment.*

Bohr and Heisenberg were probably the two founding fathers who thought most deeply about the philosophical consequences of quantum theory. However, Bohr's writings on philosophical matters are not very clear. His style is not easy to read. This may be one reason why pundits have cited Bohr as supporting a range of different philosophical positions: you can read some of his pronouncements in various ways.

18.5 LATER LIFE

Bohr continued his work on atoms and molecules long after the original 1913 article and made many contributions to our eventual understanding of how electrons are distributed in atoms across the periodic table. He also branched out into working on nuclear physics and became one of the foremost experts on nuclear fission—the process by which nuclei of heavy elements such as uranium can be split to release energy.

In 1922, Bohr was awarded the Nobel Prize in physics "for his services in the investigation of the structure of atoms and of the radiation emanating from them."

Bohr had built a deep friendship and collaboration with Werner Heisenberg who also made a profound series of contributions to the development of quantum mechanics (discussed in Chapter 19). However, when the Nazis occupied Denmark in 1940, the two were caught on opposite sides. The extent to which Heisenberg can properly be regarded as a Nazi collaborator, as distinct from being a patriotic man caught in an impossible situation, has been the cause of much historical debate. What can be said is that in 1941, Heisenberg came to Copenhagen to visit Bohr. Whatever happened at that meeting seriously and permanently damaged their friendship. These events have also been subject to some debate and form the setting for the brilliant play *Copenhagen* by Michael Frayn. In essence, the decisive conversation revolved around the construction of an atomic bomb, Heisenberg being in charge of the atomic program in Germany.

At the height of the war, in 1943, Bohr escaped from Denmark to Sweden from where he was subsequently smuggled to England in the bomb bay of a Mosquito fighter-bomber. Eventually, Bohr made his way to America where he became an important figure in the Manhattan Project.

In 1947, Bohr was awarded the *Order of the Elephant* from the Danish government: a considerable honour generally granted only to royalty and famous generals. The winner's family coat of arms is carved into a wall of fame. At the centre of Bohr's crest is the ancient Taoist representation of yin/yang, in which the birth of one phase is depicted in the decline of another. The inscription reads *CONTRARI SUNT COMPLEMENTA* or *Opposites Are Complements.*

After the war, Bohr returned to Denmark and campaigned for the peaceful use of atomic energy. He retained an active interest in physics and molecular biology until he died in 1962.

NOTES

1 Quoted in "The philosophy of Niels Bohr", Aage Petersen. Bulletin of the Atomic Scientists, vol 19, issue 7, 1963.
2 Thompson was awarded the Nobel Prize in 1906 for discovering the electron.
3 Bohr felt later that this impression might have been reinforced by his rather halting English.
4 Planets assuredly contain electrical charges, but due to an exquisite balance between the numbers of positive and negative charges, they are overall neutral to high precision.
5 The vertical line is actually an image of the slit through which you pass the light to direct it at the prism. In complex spectra, there are many images of different colours that have been split up by the prism.
6 The key thing here is that the light is coming from the vapor, rather than from the solid. In the hot vapor, the atoms are acting individually to produce the line spectrum. In a solid, the atoms are bonded together, which distorts and complicates the spectrum produced.
7 Rubidium and caesium.
8 Somewhat later, a famous astronomer, Edwin Hubble, was able to show that some nebulae were actually distant galaxies and not part of our own galaxy at all.
9 They were awarded the Nobel Prize in 1925 for this work.
10 Franck and Hertz's experiment is sometimes misinterpreted as showing the existence of multiple energy levels in an atom, whereas it is actually one transition between the same two levels that is repeatedly being hit.
11 Quoted in Bulletin of the Atomic Scientists, 1946.
12 Werner Heisenberg, Physics and Philosophy, Penguin, 1928, p. 30.
13 First stated on September 16, 1927, in an address to the International Congress of Physics at Como.
14 Niels Bohr, Atomic Theory and the Description of Nature, Cambridge University Press, Cambridge, 1934.
15 See note 12, p. 10.
16 See note 13.
17 Another perspective on Bohr's thinking traces the roots of complementarity to existential philosophy. Bohr was interested in the writings of Kiekegard, who was one of the founders of existentialism. Selleri has pointed out an interesting connection: "the impossibility of overcoming the conflict between thesis and antithesis—with a consequent existential pessimism—[which] was one of the cardinal features of existentialist philosophy," viewing the complementary aspects of a physical description in the same light as the thesis/antithesis dichotomy in philosophy. One of the hallmarks of existential philosophy is certain pessimism about human ability to overcome paradoxes in thinking. (F. Selleri, Quantum Paradoxes and Physical Reality, Kluwer, Dordrecht, 1990.)

19 Heisenberg

The history of physics is not only a sequence of experimental discoveries and observations, followed by their mathematical description; it is also a history of concepts. For an understanding of the phenomena, the first condition is the introduction of adequate concepts; only with the help of the correct concepts can we really know what has been observed.

Werner Heisenberg[1]

19.1 EARLY DAYS

Heisenberg was born on December 5, 1901, in Wurzburg, Germany. His maternal grandfather was the headmaster of the prestigious Maximilian Gymnasium in Munich (which Heisenberg attended), and his father, a scholar of ancient Greek philology and modern Greek literature.

Heisenberg studied physics under Arnold Sommerfeld at the University of Munich where he met his long-time friend and collaborator, Wolfgang Pauli. In the winter of 1922–1923, Heisenberg was in Gottingen working with Max Born. He returned there after gaining his PhD at Munich (1923) and became an assistant to Born.[2]

In the autumn of 1924, he went to the Institute for Theoretical Physics in Copenhagen to study under Bohr (whom he had met and impressed at Göttingen), in the process starting a deep friendship and collaboration, which was to last until the war. In 1926 he was made Lecturer in Theoretical Physics at the University of Copenhagen, and the following year, aged 26, he was Professor of Theoretical Physics at Leipzig.

Heisenberg's carrier however is regarded as being somewhat controversial due to debate over his activities during World War II. He was undoubtedly persecuted for a period; his teaching of relativity (Jewish Science, in the view of the Nazi hierarchy) brought him a great deal of disfavour. However, he was placed in charge of the Nazi atomic project. As to his ultimate role in the Nazi failure to develop an atomic bomb, Heisenberg's own accounts are not wholly consistent. The positive view of his involvement is that he continually stressed the difficulty of constructing an atomic weapon and deliberately and consistently asked for barely enough money to make it seem like his team was trying to succeed in constructing a reactor to provide power.

A more negative version suggests that Heisenberg made a mistake in one of the calculations crucial to the design of a weapon and convinced himself (wrongly) that far too much material was required to make a viable bomb.

Heisenberg repeatedly insisted that he chose to remain behind in Germany, despite the emigration of many friends and colleagues, to help rebuild German science after the war and (hopefully) the fall of the Nazi regime.

His standing as a physicist, however, is not so historically ambiguous. He was undoubtedly a giant in the development of the subject. He was the first to outline a mathematical formalism for quantum theory that was subsequently refined and developed by others, especially Heisenberg's close colleagues in Göttingen: Born and Jordan. Undoubtedly though, the contribution that he is most famous for was the development of the uncertainty principle that carries his name.

19.2 THE DEVELOPMENT OF QUANTUM THEORY

Although Heisenberg was a gifted mathematician (he had started university wishing to read mathematics and had drifted into theoretical physics under the influence of his teacher, Arnold

DOI: 10.1201/9781003225997-22

Sommerfeld), his approach to physics was undoubtedly coloured by being one of the most philosophically inclined of the founding fathers:

> I still had the feeling that the greatest difficulties did not lie in the mathematics, but at the point where the mathematics had to be linked to nature. In the end, after all, we wanted to describe nature, and not just do mathematics.[3]

This idea is borne out by an example from Heisenberg's later recollections, where he draws attention to a key conceptual development in the early history of quantum theory: the correct way to deal with the classical idea of a particle's path.

The notion that an object follows a specific path as it moves through space is something so deeply ingrained into us, that it almost seems evolutionary.[4] Consequently, we should not be surprized that the initial formulations of quantum theory included some notion of particle paths, especially when considering orbital electrons in atoms. Put simply, nobody imagined that there could be anything wrong with such an idea.

Bohr's quantized hydrogen atom had an electron in orbit around the nucleus—a closed circular path with quantum restrictions imposed on the possible radii. Sommerfeld had developed the model by including elliptical orbits, but despite these successes, one nagging issue irritated Bohr and others. If an electron circulates the nucleus in a closed path, it will execute a given number of orbits every second, which is effectively a frequency. These frequencies were easily calculated but did not correspond to the frequencies of light emitted by the atom.

In Heisenberg's first major paper on quantum theory, he resolved to get rid of any consideration of electron orbits. This radical approach was justified, in his view, by the inability of physicists to apply Bohr and Sommerfeld's quantum theory of hydrogen to more complex multielectron atoms, as well as some other technical problems (such as the meaning of orbital frequencies as mentioned earlier), which had proven to be insolvable:

> From this state of affairs, it seems more advisable to give up completely on any hope of an observation of the hitherto-unobservable quantities (such as the position and orbital period of the electron), and thus, at the same time, to grant that the partial agreement between the stated quantum rules and experiment is more-or-less coincidental and to attempt to construct a quantum-theoretical mechanics that would be analagoues to classical mechanics in which only relations between observable quantities would be present..[5]

Heisenberg decided that the classical approach to quantities such as position, momentum, and so on, would have to change and that the clues to a quantum treatment could only be found by a study of the *measurable properties of atoms*.

As a starting point, Heisenberg had some data from atomic line spectra and some calculational rules linking them together. He decided to mimic the connection between frequency and brightness expressed in these rules and apply it to position and momentum. He then set out to use his new rules for the manipulation of these quantities to solve a few simple physical situations.

The results he obtained provided some new insights into atomic phenomena. However, Heisenberg himself understood that this was only an opening skirmish in the development of a full quantum mechanics (once again we see the emphasis on *principle*):

> Whether a method for the determination of quantum-theoretical data from relations between observable quantities like the one that was proposed here can already be regarded as satisfactory, in principle, or whether that method represents only one more much-too-bold attack on the problem of formulating a quantum-theoretical mechanics (which is clearly quite physically difficult, from the outset) will first be clarified by a deeper mathematical examination of the method that was employed very superficially here..[6]

This deeper mathematical investigation was to come very soon after Heisenberg published his seminal paper. Max Born realized that Heisenberg was manipulating position and momentum as if they were *matrices*, i.e., that they were operators expressed in matrix form. Using this as a starting point, Born and Pascual Jordan developed a complete form of quantum mechanics based on matrices (*matrix mechanics*). Their paper, *On quantum mechanics,* was published in September 1925—a mere 2 months after Heisenberg's work.

Between 1925 and 1927, quantum mechanics developed as a mathematical scheme, but a consistent physical picture lagged behind. By 1926 Erwin Schrödinger had developed a rival mathematical formalism to matrix mechanics, based on matter waves (section 12.3, chapters 14 and 20). Many physicists rallied around Schrödinger's approach, both because waves seemed more 'physical' than the rather abstract matrices and because wave mathematics was much more familiar and traditional than the matrix manipulations used in the Heisenberg–Born–Jordan formulation. Heisenberg, perhaps understandably, wasn't in favour of the Schrödinger view:

> The more I think about the physical portion of Schrödinger's theory, the more repulsive I find it .... What Schrödinger writes about the visualizability of his theory 'is probably not quite right,' in other words it's crap.[7]

Heisenberg's instincts still led him to mistrust physical pictures.

In their early conversations, Heisenberg had been impressed by Bohr's insistence that "human language is obviously inadequate to describe processes within the atom since we are dealing there with a realm of experience totally shut off from direct inspection."[8] Such considerations had persuaded him to abandon thinking about paths inside the atom, an approach that had born fruit in matrix mechanics. However, there was still a conspicuous problem—experiments carried out using cloud chambers seemed to show direct physical evidence for particle paths.

19.2.1 Cloud Chamber Tracks

A cloud chamber is a very simple (at least in outline) device for studying particles. They were first used, highly successfully, in the detection of cosmic rays. Basically, cloud chambers consist of a vessel filled with gas that has been cooled and pressurized to the point at which it should condense into a liquid. The process of condensation is generally helped along by dust motes, or similar, present in the gas for the liquid to condense around. The inside of a cloud chamber is designed so that dust and other condensation points are scarce, hence the gas can be kept for an extended period, ready to condense, but without forming any liquid.

If the chamber held in this state is exposed to a flood of particles, any electrically charged ones passing through will ionize some of the gas molecules. These ions then act as condensation points and so a trail of water drops forms along the 'path' of the particle. Visually it's somewhat like contrails after an aeroplane, although their production is rather different. A photograph is then taken of the chamber's interior, revealing the particle tracks (Figure 19.1).

Given such an apparent physical representation of a particle track or path, there seemed no reason why similar paths or orbits shouldn't exist inside an atom. Yet, since the initial success of the Bohr–Sommerfeld atom, the notion of atomic orbits hadn't led to any particular insights. Furthermore, for all their apparent physical appeal, the Schrödinger matter waves didn't suggest any fixed orbital paths in atoms either. Heisenberg successfully dispensed with the idea in his paper, and the resulting mathematical machinery didn't lend itself to the description of a path in any situation.

So, on the one hand, the matrix and Schrödinger's formulations of quantum mechanics struggled to describe paths, yet on the other, they could easily be 'seen' in cloud chamber photographs. To Heisenberg, this remained the central issue preventing a fully consistent set of concepts being in place for the new quantum theory.

19.2.2 The Uncertainty Principle

In February 1927, according to his own account, Heisenberg focussed all his efforts on trying to find a quantum mechanical representation of a particle's path. He quickly realized that the problems he faced were insurmountable:

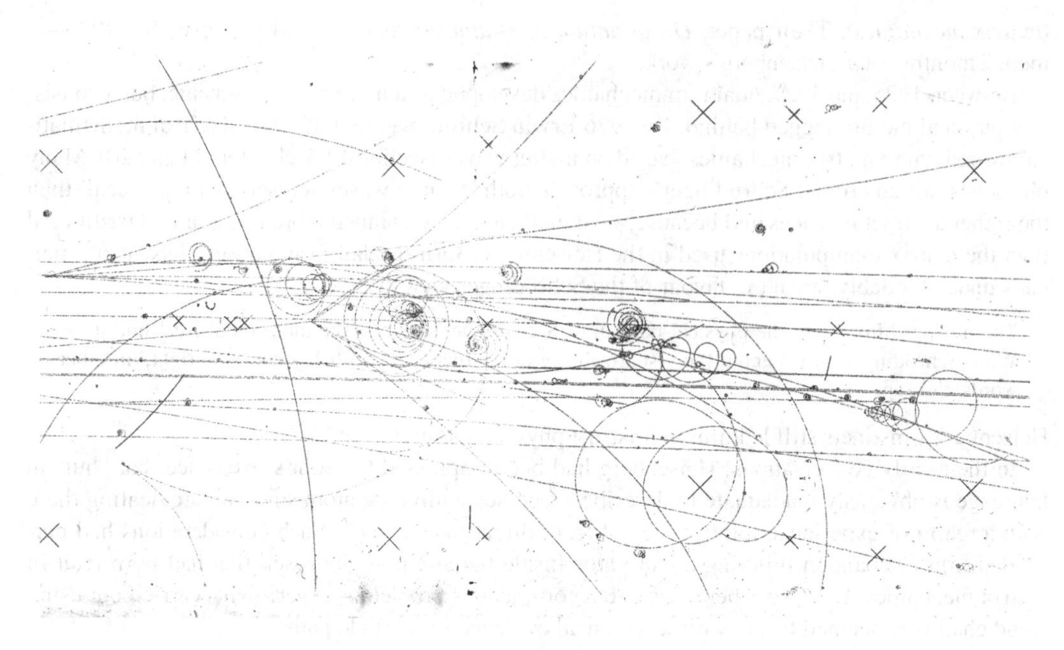

FIGURE 19.1 Cloud chambers were later replaced by bubble chambers, using essentially the same principle. Rather than having a gas cooled to its condensation temperature, bubble chambers used a liquid (hydrogen) raised to its boiling temperature. In this image, 24 GeV protons are passing through the a bubble chamber and colliding with particles in the liquid to generate reactions. The trails of ionization left by charged particles form nucleation centres for localized boiling, so tiny bubble of gas form in the liquid. {image credit CERN, © 1972 – 2022, OPEN-PHO-EXP-1972-001-1}

> I began to wonder whether we might not have been asking the wrong sort of question all along. But where had we gone wrong? The path of the electron through the cloud chamber obviously existed; one could easily observe it. The mathematical framework of quantum mechanics existed as well, and was much too convincing to allow for any changes. Hence it ought to be possible to establish a connection between the two, hard though it appeared to be.
>
> It must have been one evening after midnight when I suddenly remembered my conversation with Einstein and particularly his statement, 'It is the theory which decides what we can observe.' I was immediately convinced that the key to the gate that had been closed for so long must be sought right here.[9]

Heisenberg turned the problem around. Rather than asking how the path of a particle could be described by quantum theory, he wondered to what extent a path could be extracted from the formalism. The first question assumes that there is such a thing as a physically real path and that the problem lies with the quantum mechanical description. Heisenberg reversed the question and by showing faith in the formalism, sought to find something in quantum mechanics that would *impersonate* a continuous path in the right situation. Following the logic inherent in the quote at the start of this chapter, we think that we have seen a path in a cloud chamber image, but until we have the correct quantum concepts, we can't be sure of what we have seen and are *assuming* that it represents the path of a particle in the classical sense:

> I decided to go on a nocturnal walk through Faelled Park and to think further about the matter.
>
> We had always said so glibly that the path of the electron in the cloud chamber could be observed. But perhaps what we really observed was something much less. Perhaps we merely saw a series of discrete and ill-defined spots through which the electron had passed. In fact, all we do see in the cloud chamber are individual water droplets which must certainly be much larger than the electron. The right question should therefore be: Can quantum mechanics represent the fact that an electron finds itself approximately in a given place and that it moves approximately with a given velocity, and can we make these approximation so close that they do not cause experimental difficulties?

A brief calculation after my return to the Institute showed that one could indeed represent such situations mathematically and that the approximations are governed by what would later be called the uncertainty principle...[10]

Heisenberg realized that each droplet of water in the cloud chamber path was effectively a position measurement made on the electron. Consequently, the first would cause the electron's state to collapse into one representing a localized position of roughly the size of the liquid droplet. Such a localized position state can be expanded in terms of momentum states covering a range of momenta. This new state would then evolve, spreading out in space according to the momentum expansion. The next liquid droplet would be at a subsequent position measurement, which would once again collapse the state into a position localized in the vicinity of the second drop. Each localization of position would mean a scrambling of momentum information. The drops are not 'pearls' strung out on the thread of an electron's path. They are anchor points pulling in the state of the electron at various moments. No information about the position of the electron between droplets can be established. Heisenberg's calculations that night showed him that quantum theory could represent the situation in a cloud chamber and that states of the electron were governed by the uncertainty principle, $\Delta x \Delta p \geq \hbar/2$, where a state localized in space to within Δx can be represented by an expansion over momentum states covering a range of momenta Δp wide, where $\Delta p \geq \hbar/2\Delta x$.

Over the next few days, Heisenberg satisfied himself that any experimental situation would naturally fall within the parameters of the uncertainty principle[11] and after some prolonged discussions with Bohr (who by then was championing his own views on complementarity) published the ideas in a paper in 1927. This was the last link in the chain that established a consistent way of working within quantum theory.

19.2.3 Quantum Concepts

The derivation of the uncertainty principle was not the only important aspect of this 1927 paper. Heisenberg also developed his own views on the quantum concepts required to complete the theory, especially when it came to the meaning of quantities such as position and momentum. Following Bohr, he assumed that terms such as 'the position of a particle or 'the momentum of a particle' were defined entirely by the experiments conducted to measure them:

If one wants to be clear about what is meant by 'position of an object,' for example of an electron ..., then one has to specify definite experiments by which the 'position of an electron' can be measured; otherwise this term has no meaning at all.[12]

The uncertainty principle tells us the extent to which an experiment can be designed to measure both position and momentum with arbitrary precision. In that sense, it plugs the last hole in the interpretation.

Heisenberg's link of concept to experiment *can* be interpreted as a convenient definition of practicality, without implying anything about the properties that a particle *really* has. Viewed this way, the uncertainty relationship elucidates the *in-principal* limitations of our experimental techniques—it describes the permanently distorting lens we are forced to view the world. We may not be able to do anything about the view, but we should be clear that things do not really look like that.

We can be reasonably sure, however, that this was not the view that Heisenberg took. The use of the term 'concept' in his 1927 paper makes it clear that for him, position and momentum no longer had any validity outside that defined by the experiment. An experiment *created* the position of a particle, not measuring something that the particle had before the measurement took place (and of course the same would be true of momentum as well):

I believe that one can formulate the emergence of the classical 'path' of a particle pregnantly as follows: the 'path' comes into being only because we observe it.[13]

In some of his later writings, Heisenberg developed this view, believing that he had found a philosophical precedent for this aspect of quantum theory in Aristotle's texts. In particular, the notion of *potentia*[14]—possibilities of being—appealed to him as a way of thinking about the state of a particle:

> The probability function combines objective and subjective elements. It contains statements about possibilities or better tendencies ("potentia" in Aristotelian philosophy), and these statements are completely objective, they do not depend on any observer; and it contains statements about our knowledge of the system, which of course are subjective in so far as they may be different for different observers.[15]

With regard to paths, orbits and the statistical nature of quantum theory:

> One might perhaps call it an objective tendency or possibility, a "potentia" in the sense of Aristotelian philosophy. In fact, I believe that the language actually used by physicists when they speak about atomic events produces in their minds similar notions as the concept "potentia." So the physicists have gradually become accustomed to considering the electronic orbits, etc., not as reality but rather as a kind of "potentia."[16]

Perhaps the clearest statement of Heisenberg's thinking came in his Gifford Lectures of 1958 where he regards the quantum state as:

> a quantitative formulation of the concept of 'dynamis', possibility, or in the later Latin version, 'potentia', in Aristotle's philosophy. The concept of events not determined in a peremptory manner, but that the possibility or 'tendency' for an event to take place has a kind of reality—a certain intermediate layer of reality, halfway between the massive reality of matter and the intellectual reality of the idea or the image—this concept plays a decisive role in Aristotle's philosophy. In modern quantum theory this concept takes on a new form; it is formulated quantitatively as probability and subjected to mathematically expressible laws of nature. [17]

Heisenberg's use of potentia to describe the quantum state can probably be summarized by the term 'objective possibility'. Heisenberg is clearly inching toward a realistic interpretation of the quantum state, although he acknowledges that it has to extend the notion of reality into a sort of 'intermediate' level between the fully abstract forms of mathematics and the fully concrete form of the classical world.

'Objective possibility' is an intriguing notion that I personally find appealing. It is a serious attempt to treat the quantum state in a realistic manner, *but on its own terms*—not dictating to 'it' any preconceptions we may hold regarding its reality. In addition, it challenges us to think carefully about the meaning of 'real', which we naively tend to restrict to 'objects' that we can see and touch.

Crucially, however, if this is to be more than a philosophical label attached to a piece of mathematics, we need to find a fully worked-out theory to explain how the *potentia* of the state becomes the *actuality* of experimental results.

19.3 LATER LIFE

Einstein, for one, was deeply unhappy with the uncertainty principle, and when he met with Bohr and Heisenberg at the Solvay Conference in 1927, he did everything within his power to demonstrate that it didn't work. He imagined various experimental scenarios (*Gedankenexperiments*) which apparently violated uncertainty in creative ways. In each case, Bohr and Heisenberg were able to rebut Einstein's claims, in one case by referring to his own theory of General Relativity.

Their debates during the week of that conference are on-record and represent an extraordinary period of intellectual jousting among three of the greatest minds of that century. In the end, Einstein had to admit that quantum theory was consistent and complete, but never gave up the hope that it would be replaced by a more fundamental theory.

Heisenberg's contributions to physics did not end with his seminal papers on quantum mechanics and uncertainty. From 1927 to 1941 he was a professor at the University of Leipzig. For the next four years, he was a director of the Kaiser Wilhelm Institute for Physics in Berlin. In 1932 he was awarded the Nobel Prize in physics for his work on quantum theory.[18]

Heisenberg married in 1937 and had seven children. After the war, he organized and became director of the Max Planck Institute for Physics and Astrophysics at Göttingen, moving with the Institute, in 1958, to Munich. In 1954, he acted as the German representative on the organizing committee setting up the European nuclear physics research facility now known as CERN. On the theoretical side, he worked with Pauli, Dirac, and others on quantum field theory and made many contributions to the development of nuclear and particle physics. (It was Heisenberg who proposed the neutron–proton structure of the atomic nucleus.) Interestingly he was not a supporter of the quark model, seeing the fundamental aspects of nature in terms of mathematical relationships (symmetries) rather than fundamental particles.

Heisenberg died on February 1, 1976. The next evening a group of his collaborators, friends, and assistants walked in a candle-lit procession from the Institute to Heisenberg's home where the candles were left on the doorstep as a mark of respect.

NOTES

1 W. Heisenberg, *Encounters with Einstein, and Other Essays on People, Places, and Particles*, Princeton Science Library, Princeton, 1983, p. 19.
2 Born was the first to suggest the connection between squaring of complex amplitudes and probability. He also, with his student Pascual Jordan, refined Heisenberg's early ideas on quantum calculations into a fully developed form using matrix mathematics.
3 W. Heisenberg, *Encounters with Einstein, and Other Essays on People, Places, and Particles*, Princeton Science Library, Princeton, 1983, p. 48.
4 Some biologists have speculated that our ability to carry out mathematical calculations evolved from caveman gaining a selective advantage by being able to estimate how to throw a rock to hunt more effectively. The gulf between rock throwing and the ability to do group theory seems to indicate that evolution went way over the top on this one.
5 W. Heisenberg, *On quantum mechanical reinterpretation of kinematic and mechanical relations*, *Zeit. f Phys.* 1925, **33.**
6 See note 5.
7 Heisenberg, writing to Pauli in 1926.
8 W. Heisenberg, *Encounters with Einstein, and Other Essays on People, Places, and Particles*, Princeton Science Library, Princeton, 1983, p. 40.
9 W. Heisenberg, *Physics and Beyond,* George Allen and Unwin, London, 1971, p. 77.
10 See note 9, pp. 77–78.
11 It is thought that the first person to use the term *principle* in regard to this relationship was the English astrophysicist Eddington in his Gifford lectures of 1928, in which he referred to Heisenberg's *Principle of Indeterminacy.* Heisenberg's own preferred expression seems to have been *inaccuracy relations (Ungenauigkeitsrelationen)* or *indeterminacy relations* (*Unbestimmtheitsrelationen).*
12 W. Heisenberg, *On the physical content of quantum theoretical kinematics and mechanics*, *Zeitschrift für Physik* 1927, **43** (interestingly there is some ambiguity in the translation of the title from the original German—the term *physical* could be replaced with perceptible, perceptual, intelligible, or intuitive).
13 See note 12.
14 "The terms *actus* and *potentia* were used by the scholastics to translate Aristotle's use of the terms *energeia* or *entelecheia,* and *dynamis.* There is no single word in English that would be an exact rendering of either. Act, action, actuality, perfection, and determination express the various meanings of *actus;* potency, potentiality, power, and capacity, those of *potentia.*

In general, *potentia* means an aptitude to change, to act or to be acted upon, to give or to receive some new determination. *Potentia* is more than a mere statement of futurity, which has reference to time only; it implies a positive aptitude to be realized in the future." From Wikipedia.
15 W. Heisenberg, *Physics and Philosophy*, George Allen and Unwin, London, 1959.
16 See note 15.
17 See note 15.
18 The citation reads: For the creation of quantum mechanics, the application of which has, *inter alia,* led to the discovery of the allotropic forms of hydrogen.

20 De Broglie & Schrödinger

On principle, there is nothing new in the postulate that in the end exact science should aim at nothing more than the description of what can really be observed. The question is only whether from now on we shall have to refrain from tying description to a clear hypothesis about the real nature of the world. There are many who wish to pronounce such abdication even today. But I believe that this means making things a little too easy for oneself.

Erwin Schrödinger[1]

20.1 BEGINNINGS

Thus to describe the properties of matter as well as those of light, waves and corpuscles have to be referred to at one and the same time. The electron can no longer be conceived as a single, small granule of electricity; it must be associated with a wave and this wave is no myth; its wavelength can be measured and its interferences predicted.

Louis De Broglie[2]

In 1924, a young PhD student submitted a thesis containing the germ of an idea that would revolutionize our understanding of the material world.[3] The student was Louis De Broglie[4] (pronounced De Broy), and for a while, he had been meditating on the wave/particle nature of light. In a moment of insight, he suddenly realized that the odd dual nature of light might extend to other aspects of nature as well. Perhaps particles, such as electrons, could have a wave-like nature to balance the particle-like nature of light, which had previously been described as a wave. Echoing the Planck–Einstein formulas for photons, De Broglie proposed that electrons should have a 'pilot wave' with frequency, f and wavelength λ given by:

$$\lambda = \frac{h}{mv} \qquad f = \frac{E}{h}$$

Building on this idea, De Borglie's thesis also contains an explanation for Bohr's quantized orbits in hydrogen. As the wave associated with the circulating electron has to close on itself around the circular arc of the orbit (Figure 20.1), the circumference of an acceptable orbit can only be a whole number of wavelengths:

$$n\lambda = 2\pi r$$

Inserting the De Broglie relationship for wavelength gives:

$$n\frac{h}{mv} = 2\pi r$$

when rearranged this is equivalent to:

$$n\frac{h}{2\pi} = n\hbar = mvr$$

i.e., the quantized angular momentum that Bohr employed.

DOI: 10.1201/9781003225997-23

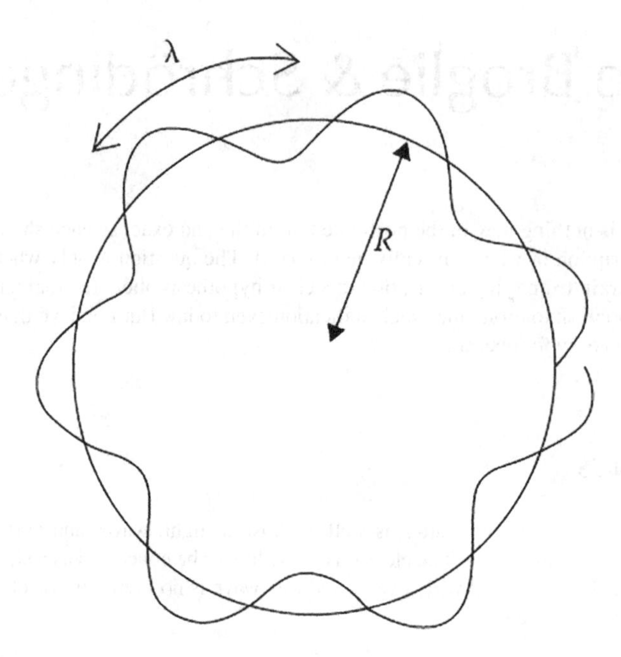

FIGURE 20.1 De Broglie's explanation for Bohr's quantized orbits. This figure shows what happens if the wavelength does not fit into the circumference of an orbit. It ends up with two different values at the same place, which can't happen. The waves must close in a neat fashion so that $n\lambda = 2\pi r$.

Between them, De Broglie, Bohr and Sommerfeld were producing the first rough sketches of the full quantum theory of the atom that ultimately developed. The limitations of these initial ideas are fully apparent. For one thing, electrons are found throughout a three-dimensional volume surrounding the nucleus of an atom, not in the two-dimensional circles used by this model.

De Broglie's ideas were greeted rather sceptically at first. His PhD supervisor even asked for a second copy so he could send it to Einstein for his opinion. Einstein gave the work his seal of approval.

News of De Broglie's thesis spread around the physics community from 1924 to 1925. In autumn 1925, Erwin Schrödinger was visiting colleagues in Switzerland and was asked to present a summary of De Broglie's work. Having studied De Broglie's thesis in some detail he became an enthusiastic supporter. In the discussions following Schrödinger's positive presentation of matter waves, it was suggested that an equation was needed to help calculate the form of the wave in different circumstances, then the theory would gain ground. This sowed the seed for the eventual development of the Schrödinger wave equation.

De Broglie was awarded the Nobel Prize in 1929. In later life, as with Einstein, he was not content with the probabilistic nature of quantum mechanics and continued to work on his 'pilot wave' idea as a way of reintroducing specific predictions.

20.1.1 Electron Diffraction

> It is rather as if one were to see a rabbit climbing a tree, and were to say, "Well, that is rather a strange thing for a rabbit to be doing, but after all there is really nothing to get excited about. Cats climb trees – so that, if the rabbit were only a cat, we would understand its behavior perfectly".
>
> *Clinton Davisson*[5]

In April 1925, Clinton Davisson and Lester H. Germer were investigating ways of improving electronic valves.[6] Consequently, they were interested in seeing how electrons fired from an electron gun scattered off a flat nickel surface in a vacuum. Unfortunately, a bottle of liquid air exploded near the apparatus and as a result, the hot nickel target became oxidized. This made the target useless for further experiments, until the oxide layer had been removed. Davisson and Germer decided to clean the nickel by heating it in a hydrogen atmosphere, then again in a vacuum, and finally by removing the top layer. After this process, the target looked to be good enough and the experiment resumed.

Remarkably the pattern of electron scattering from the newly cleaned target was completely different from that before the accident. Now the scattered electrons showed a preference for coming off at distinct angles determined by their speed.

When they examined the target closely, they saw that the previous tiny crystals bonded together within the structure had been converted into a few large crystals by the heating process. In essence, the new target was eminently suitable for demonstrating the wave nature of electrons. Having large crystals, meant that the atoms of nickel would be arranged in regular arrays across whole regions of the target. These regular atomic arrays acted somewhat like the slits in Young's experiment. Electron waves scattered off each atom and interfered with one another like the waves from the slits.[7] Davisson and Germer remarked in their 1927 paper:

> a description … in terms of an equivalent wave radiation … is not only possible, but most simple and natural. This involves the association of a wavelength with the incident electron beam, and this wavelength turns out to be in acceptable agreement with the value h/mv of the undulatory mechanics.[8]

The work of Davisson and Germer had been anticipated by an earlier experiment performed by Davisson and Charles Kunsman, who fired electron beams at metal surfaces and measured how likely the electrons were to bounce off at various angles. In 1925 a student, Walter Elsasser, heard of their results and wondered if the strange pattern of maxima and minima could be some form of interference-effect. Having checked the wavelength that De Broglie waves would have in that experiment, Elsasser realized that Davisson and Kunsman's results could be a confirmation of the existence of matter waves. He published a short note to this effect and tried to produce more definite experimental results himself. After a few fruitless months, he gave up.

20.2 ENTER THE WAVE EQUATION

Erwin Schrödinger was born in Vienna in 1887. His father, Rudolf Schrödinger, ran a linoleum factory which he had inherited from his own father. His mother, Emily Bauer, was Viennese on her father's side (he was a professor of chemistry) and English on her mother's. Consequently, Schrödinger learned both English and German at a young age. He received lessons at home from a private tutor up to the age of 10 and then entered the Akademisches Gymnasium in the autumn of 1898. After graduating from the Gymnasium, he moved to the University of Vienna to study theoretical physics. He received his PhD in 1910 and his Habitation[9] in 1914.

Schrödinger's research interests were widespread. At various times he contributed to General Relativity, statistical mechanics, radiation theory, colour vision, solid-state physics, and atomic spectroscopy. However, it was his brilliant outpouring in 1926 that has given him a lasting name in physics.

During the Christmas and New Year of 1925/1926, Schrödinger holidayed in the Alps with his mistress. On his return, he had a functioning wave equation worked out. In January 1926 Schrödinger published a paper containing the first derivation of his wave equation for De Broglie's 'pilot waves'. His derivation was restricted to cases where the system under consideration did not change with time. However, using this equation he was able to calculate the distribution of the electron wave through the volume of a hydrogen atom. At the same time, he demonstarted that the electron occupied the correct Bohr energy levels.

In the light of modern understanding, we recognize that De Broglie's 'pilot wave' has been replaced by the amplitude function $\Psi(x, y, z, t)$. However, in this first paper, there is no comment regarding the physical interpretation of Ψ. It is simply treated as an abstract concept.

A second paper followed four weeks later with a new derivation of the equation and treatment of three important physical situations.[10] By May 1926, a third paper appeared in which Schrödinger sketched out a demonstration of the mathematical equivalence of his 'wave mechanics' and the Heisenberg–Born–Jordan matrix mechanics.[11] His fourth paper of that year extended the work to time-varying systems and laid down the foundations of a very useful approximation technique.

The physics community greeted Schrödinger's work with huge enthusiasm. His wave equation was a more familiar form of mathematics than the rather austere matrices that Heisenberg, Born and Jordan had been working with. Suddenly, physicists were on ground that they found more comfortable; using techniques that they were familiar with from classical wave mechanics, albeit in a radically different context.

The successful validation that wave mechanics and matrix mechanics were essentially the same theory, just in different mathematical forms, convinced the community that they were on the right lines and that a definitive version of quantum theory was in sight.

20.2.1 Matter Waves

During the writing of his four papers, Schrödinger struggled to find a physical interpretation of the 'mechanical field scalar', Ψ. By the time the final paper had been prepared, he had settled on the notion that $e \times \Psi\Psi^*$ gave the charge density found in a small volume of space at a particular time (e being the charge of an electron).

Schrödinger had come to feel that physicists should abandon the notion of particles altogether, in favour of matter waves. In this picture, the electrons in an atom are extended vibrations in space, like the vibrations of a plucked guitar string or the surface of a drum skin. He was also able to show that his matter waves could combine to form a localized 'packet' where the wave intensity was especially strong. He was convinced that, in experimental situations, this wave packet would look like a lump of matter or particle. In Schrödinger's view, it was only a matter of time before wave mechanical calculations demonstrated how particle-like effects could be produced by waves in all relevant situations.

Another target for Schrödinger's reformulation of old quantum theory was the notion of quantum jumps, introduced by Bohr and capitalized on by Einstein. Schrödinger disliked the thought that spontaneous changes of state took place inside atoms and tried to use his wave theory to explain line spectra without the need for the sharp transitions between Bohr's stationary states. In his formulation, the different energy levels corresponded to different patterns of vibration in the electron wave. Two similar patterns could then set up 'beats'[12] between them. Schrödinger aimed to explain the radiation emitted by the atom by the beat frequency between two energy levels.

These ideas were not greeted with any degree of enthusiasm in Copenhagen where Bohr and Heisenberg were inching their way toward their interpretation of quantum mechanics. Bohr was convinced that it was necessary to abandon any attempt at a physical picture of the inner workings of the atom so there would not be a strict causal explanation for photon emission.

Matters came to a head when Schrödinger was invited to visit Bohr at his institute. According to Heisenberg's recollection of those few days, the conversations between Bohr and Schrödinger began at the train station and didn't end when Schrödinger fell ill and was confined to bed. Apparently, Bohr sat at Schrödinger's bedside haranguing him about the interpretation of quantum theory.

Schrödinger stuck to his views, succinctly expressed in the following remark, as recalled by Heisenberg: "if all this damned quantum jumping were really here to stay, I should be sorry I ever got involved with quantum theory."[13] Bohr and Heisenberg, however, were not convinced.

20.2.2 So What Is Ψ?

Any attempt to construct a direct physical meaning for Ψ must deal with several problems:

- To give a value to Ψ, in appropriate units, we need to use complex numbers and they seem to have no 'direct' physical application in the world we experience.
- For a single particle, Ψ depends on the position coordinates (x, y, z) as well as time, t. For a two-particle system, two sets of position coordinates are required, one for each particle. Then $\Psi(x_1, y, z_1; x_2, y_2, z_2;t) \times \Psi^*(x_1, y, z_1; x_2, y_2, z_2;t)$ is the probability of finding one particle at (x_1, y, z_1) and the other at (x_2, y_2, z_2) at the same time t. It is difficult to see what to make of this, physically. It does not appear to be the same value of Ψ at two different places, nor is it clear how moving one particle can affect the value of Ψ for the other. Generally, physicists talk about one value of Ψ expressed in the six-dimensional 'space' built from the possible position coordinates of the two particles. Of course, this is a purely abstract mathematical space, not something that is part of the real world.
- The collapse of the wave function, Ψ, is hard to explain, based on the notion of a spontaneous and instantaneous change across a broad span of space in a physically real wave.
- Ψ changes depending on the basis set of observables chosen, which once again is quite unlike a physical wave.
- While the wave packet that Schrödinger had employed helps create the impression of a particle of matter, it does not remain localized into a tight little lump, except under very specific circumstances. Generally, the wave packet spreads out, blurring what might be regarded as a physical position for the particle (section 15.1.6).

Although Schrödinger and others (like De Broglie) persevered in attempting to construct an interpretation in terms of matter waves, the decisive step for the orthodox view came shortly after the publication of Schrödinger's final 1926 paper. Max Born applied wave mechanics to the problem of particle scattering.[14] He considered a beam of particles aimed at a fixed object, such as an atom, and calculated how they would rebound in various directions. In a footnote, Born suggests that $\Psi\Psi^*\Delta V$ is the probability that a particle can be found in a small region of volume ΔV at a given time, t. Next, in October 1926, Born produced another paper generalising his thinking. In our slightly more modern notation, Born wrote down the expansion:

$$|\Psi\rangle = \sum_n a_n |n\rangle$$

and interpreted the complex square of a_n as the probability of a transition from state $|\Psi\rangle$ to state $|n\rangle$. However, he had not quite reached the full probabilistic view of Ψ, as he was still thinking that particles were localized lumps of matter and that the probability of finding them in different places was governed by $\Psi\Psi^*$ (a view that is called into question by the results of interference experiments with single particles).

In various writings, Born cited Einstein's work as partial inspiration for his interpretation of Schrödinger's Ψ. It was, after all, Einstein who introduced the idea of probability into quantum mechanics with his work on black body spectra and stimulated emission. However, as we already know, Einstein was never able to accept that quantum theory had to remain in principle an area where God had the freedom to "play dice".

Schrödinger always held out against the orthodox Copenhagen interpretation of quantum mechanics. As late as 1955 he was writing:

> Let me say at the outset, that in this discourse, I am opposing not a few special statements of quantum mechanics held today (1950s), I am opposing as it were the whole of it, I am opposing its basic views that have been shaped 25 years ago, when Max Born put forward his probability interpretation, which was accepted by almost everybody.[15]

20.2.3 Nobel Prizes

In 1928, Einstein nominated Heisenberg, Born, and Jordan for the Nobel Prize, but the award was not made as the committee did not think that any of the nominees met the criteria for the prize. Owen Richardson was awarded the 1928 prize in 1929.

Similarly, the announcement of the 1932 prize was delayed until 1933 when it was confirmed that Heisenberg was to be the recipient. Heisenberg wrote to Born and admitted to "bad conscience" at having been the only one to receive a prize "for work done in Göttingen in collaboration – you, Jordan and I." Born eventually received the honour in 1954.

The 1933 prize went to Schrödinger and Dirac.

20.3 SCHRÖDINGER'S PHILOSOPHY

Schrödinger maintained an interest in philosophical matters throughout his life. This spread beyond the narrow confines of the philosophy of science into deeper and broader matters. Unlike some modern advocates, Schrödinger was convinced that science was only a limited window on the world and that a complete understanding of existence and our place in it could only be obtained by combining insights obtained from a range of areas of knowledge.

In 1940, Schrödinger received an invitation to help establish an Institute for Advanced Studies in Dublin. He remained there for 17 years as director of the School of Theoretical Physics.[16] During this period he wrote on various topics. His excellent and influential book *What Is Life*? was published in 1944[17]. This remarkable volume contains many speculations about the nature of life, including the suggestion that an organism's genetic information could be passed on via a complex molecule. Both Crick and Watson, the discoverers of the shape of the DNA molecule, cited this book as an inspiration to their research.

The modern edition of What Is Life is bound together with another pamphlet, *Mind and Matter*. Both were part of the pre-university reading list sent to me before my own degree course started. Schrödinger's take on the relationship between mind and matter was heavily influenced by his wide reading of the Vedantic philosophy of Hinduism. He believed that each person's individual consciousness was an illusion and that we all form part of one undivided consciousness. Schrödinger became a great student of Indian philosophy, and there is no doubt that his work was influenced by a mystical view of the world.

Several authors have since picked up on certain apparent similarities between the statements of mystical philosophers and aspects of the quantum world. Popular books have been written to outline these connections.[18] There's no doubt that many of the founding fathers were aware of the various mystical philosophies of the Indian and Oriental cultures. It also seems clear that some of them felt a degree of sympathy with these philosophies. Niels Bohr wrote:

> For a parallel to the lesson of atomic theory regarding the limited applicability of such customary idealisations, we must in fact turn to quite other branches of science, such as psychology, or even to that kind of epistemological problems with which already thinkers like Buddha and Lao Tzu have been confronted, when trying to harmonize our position as spectators and actors in the great drama of existence.[19]

Heisenberg was also apparently influenced by mystical philosophy, although his own writings are confined to discussing aspects of Western philosophy:

> In 1929 Heisenberg spent some time in India as the guest of the celebrated Indian poet Rabindranath Tagore, with whom he had long conversations about science and Indian philosophy. This introduction to Indian thought brought Heisenberg great comfort, he told me. He began to see that the recognition of relativity, interconnectedness, and impermanence as fundamental aspects of physical reality, which had been so difficult for him and his fellow physicists, was the very basis of the Indian spiritual traditions. 'After these conversations with Tagore' he said, 'some of the ideas that had seemed so crazy suddenly made much more sense.'[20]

However, in some quarters the point has been pressed too far, transforming into an attempt to justify mystical philosophy via quantum theory.

Quantum mechanics stands or falls on its ability to describe the world, something that is open to experimental tests. The meaning of quantum theory is a different matter. It is certainly interesting to see that some modern quantum philosophical ideas are reflected in the writings of sages, but quantum theory is a scientific theory and can never be defended on the grounds that some people have seen in it a view of reality that is congruent to mystical insight.

Equally, mysticism must stand or fall on its own criteria. Many scientists would reject it out of hand. Quantum theory should never be wielded as a defence of mysticism. If nothing else, this is a dangerous approach for those interested in defending mysticism. Like any scientific theory, quantum mechanics may well be replaced in the future, potentially undermining the defence in the process.

Although some aspects of the quantum state seem to be mirrored in mystical writing (or the other way round), there is nothing in the ancient tradition that corresponds to the sharp collapse of possibilities because of measurement. Equally, the way that the familiar world arises from the underlying flux of being is not well worked out in either quantum theory or mysticism, which tends to treat the everyday world as an illusion.

I am sure that Schrödinger would not have been happy with some of the books and articles written in this area. He would have regarded any attempt to use science to justify mystical philosophy as being nonsensical, which would not have been a criticism of mysticism. Schrödinger thought that all the various branches of human thought had their contributions to make. The problem with science is that it excludes things:

> The scientific world-picture vouchsafes a very complete understanding of all that happens — it makes it just a little too understandable. It allows you to imagine the total display as that of a mechanical clockwork which, for all that science knows, could go on just the same as it does, without there being consciousness, will, endeavor, pain and delight and responsibility connected with it — though they actually are. And the reason for this disconcerting situation is just this: that for the purpose of constructing the picture of the external world, we have used the greatly simplifying device of cutting our own personality out, removing it; hence it is gone, it has evaporated, it is ostensibly not needed.... I am very astonished that the scientific picture of the real world around me is deficient. It gives a lot of factual information, puts all our experience in a magnificently consistent order, but it is ghastly silent about all and sundry that is really near to our heart, that really matters to us. It cannot tell us a word about red and blue, bitter and sweet, physical pain and physical delight; it knows nothing of beautiful and ugly, good or bad, God and eternity. Science sometimes pretends to answer questions in these domains, but the answers are very often so silly that we are not inclined to take them seriously.[21]

According to Schrödinger, the scientific worldview has had all aspects of mind removed from it, an abstraction that cuts it off from the very core of mystical insight.

NOTES

1. E. Schrödinger, *The Fundamental Idea of Wave Mechanics*, Nobel lecture, 1933, December 12.
2. De Broglie, Nobel Prize lecture, 1929, December 12.
3. His ideas were first published in a note to the journal *Comptes rendus* in 1923. (*Comptes rendus*, 1923, 177: 507–510.)
4. On the death of his elder brother in 1960, De Broglie assumed the following title: Louis-Victor-Pierre-Raymond, 7th duc De Broglie.
5. Clinton Davisson, *Franklin Inst. J.*, 1928, 205: 597.
6. Electronic valves were the forerunners of transistors and are still found in some expensive HiFi equipment.
7. Before the cleaning process, the multiple small crystals in the target would have their arrays of atoms pointing in different directions, preventing clear interference from being seen.
8. *The Phys. Rev.*, December 1927.
9. A post-PhD qualification required for entry into academic life in German-speaking universities.
10. The harmonic oscillator, the rigid rotor, and the diatomic molecule.

11 Dirac and Jordan independently provided a more rigorous proof shortly after.
12 If you set up two sounds of very similar, but not the same, frequency then you can hear a pulsing noise at a frequency between the two. This is called beats. This effect is often used to tune instruments, such as a guitar, by playing a fixed note and adjusting the tune until the beats disappear.
13 W. Heisenberg, *Physics and Beyond*, George Allen and Unwin, London, 1971, p. 75.
14 M. Born, *Zur Quantenmechanik der Stoßvorgänge*, Zeitschrift für Physik, 1926, 37: 863–867.
15 E. Schrödinger, *The Interpretation of Quantum Physics*, Ox Bow Press, Woodbridge, CN, 1995.
16 This phase of Schrödinger's career was not without controversy. Various involvements with students occurred and he fathered children from two different women (in addition to those he had with other mistresses at other times).
17 *What Is Life?* Erwin Schrödinger, Cambridge University Press, 1944, ISBN: 0-521-42708-8
18 The first successful book of this nature *The Tao of Physics* by Fritjof Capra was an influence on me as a teenager when I was thinking about becoming a physicist.
19 N. Bohr, Speech on quantum theory at Celebrazione del Secondo Centenario della Nascita di Luigi Galvani, Bologna, Italy, October 1937.
20 A conversation with Fritjof Capra, quoted in *Uncommon Wisdom*, Flamingo, 1989.
21 E. Schrödinger, *Nature and the Greeks*, now bound with *Science and Humanism*, Cambridge University Press, Cambridge, 1996.

21 Dirac

It is more important to have beauty in one's equations than to have them fit experiment... because the discrepancy may be due to minor features which are not properly taken into account and that will get cleared up with further developments of the theory... It seems that if one is working from the point of view of getting beauty in one's equations and if one has a really sound instinct, one is on a sure line of progress.

Paul Dirac[1]

21.1 DIRAC'S INFLUENCE ON QUANTUM PHYSICS

Between 1925 and 1933, Paul Adrian Maurice Dirac made a series of crucial contributions to the development of quantum mechanics and set much of the agenda for fundamental research through the following decades. Any one of his insights would have been sufficient to ensure his lasting recognition in physics, but together they amount to the work of a genius[2].

In 1925 Dirac published a paper building on Heisenberg's outline of a mathematical structure for quantum theory. He followed this up in a series of papers and his 1926 PhD thesis, demonstrating how the Schrödinger and Heisenberg approaches were different ways of looking at the same basic mathematical picture. In essence, Dirac showed how basis states could be set up, how a state can be expanded over this basis, and how one basis can be converted into another equally valid description using a different basis (the work he was apparently most proud of[3]).

Dirac's next two papers came out in 1927 and developed the theory to describe the interaction between electromagnetic waves and atoms. This work set the foundations of quantum field theory and developed the technique called 'second quantization' (an idea that we will talk about in Chapter 32). Dirac's ideas on this topic have been hugely significant. All modern theories of elementary particles are quantum field theories.

Despite all this early success, the advance that Dirac is most well-known for was published in 1928.[4] Working from first principles and following his mathematical 'nose' for beauty, Dirac produced a new equation to describe the physics of electrons. The revolutionary aspect of Dirac's equation (as it is now known) lay in how it knitted together quantum ideas and Einstein's picture of space and time (the theory of relativity) and hence overcome an evident limitation of the Schrödinger equation[5]. Unfortunately, unpicking the physics and mathematics of this striking discovery would require more page space than a book of this nature can afford.

In a starkly compact form, the equation reads:

$$i\hbar\gamma^{\mu}\,\partial_{\mu}(\psi) - mc\psi = 0$$

which looks innocent enough, but each of the four γ^{μ} s is itself a four-by-four matrix[6], and the four ∂_{μ} form parts of a vector operator acting on the four-component state vector, ψ ...

Dirac's equation is both spectacularly beautiful and incredibly fruitful, leading to many further developments and discoveries. For example, in the process of evolving mathematics that was consistent with both quantum theory and relativity, Dirac found that a natural role for spin automatically arose. In essence, the alignment between quantum theory and relativity is not possible without spin being in the mix. In some ways, Dirac's equation 'explains' spin (see Chapter 10).

However, the solutions to Dirac's equation represented an immediate puzzle. As required, the solutions represented possible electron states in different circumstances. However, they include

DOI: 10.1201/9781003225997-24

some that, mathematically, have *negative energy*. Cases with energy $+E$ and $-E$ pop out of the equation equally readily[7]. Having no clear interpretation for such negative energy states would be bad enough, but more worryingly, the natural tendency of systems to reduce their energy, when possible, would mean that all positive energy electrons in the universe should steadily convert into negative energy ones. In essence, we ought to be surrounded by negative energy electrons, whatever that might mean. Equally, no apparent conversion of this nature has been observed.

In 1929 Dirac hit on an ingenious solution. Perhaps positive energy electrons don't turn into negative energy versions *as all the negative energy states are already fully occupied*. After all, the fundamental antisymmetry of fermion states prevents any two identical fermions from being in the same state, so positive energy electrons would not be able to convert into identical negative energy states.

With all their possible states filled, this 'sea' of negative energy electrons would effectively form an unobservable background to all the stuff going on in the world, with one exception. Although positive energy electrons couldn't convert into negative energy ones, the reverse process would still be possible as not all positive energy states are filled. In principle, a negative energy electron could absorb a wandering photon, if the photon had sufficient energy to lift the electron out of its negative energy state and leave it with positive energy. The negative energy electron would then become a normal, observable, electron.

Promoting a negative energy electron in this way leaves behind a gap, or *hole*, in the negative energy states; a vacancy into which a positive energy electron could fall, radiating a photon in the process. This was precisely the effect that Dirac was trying to avoid happening on-mass. However, individual, and occasional, occurrences are not an issue, provided they can be observed in nature. Crucially, while it still existed, the hole would be capable of wandering about *as if it were a fully realized particle on its own*. As a *gap* in *negative energy states*, it would *appear to be a positive energy, positively charged particle of the same mass as the electron*. As no such particle was 'on the books' at the time, Dirac's first thought was that the holes ought to be describing protons. However, despite some effort, he was not able to make the mass come out right.

In 1931 he rejected the notion that the holes could be protons and wrote a new paper predicting the existence of a positively charged particle with the same mass as the electron and that this would be the 'hole' in the negative energy sea. One year later, Carl David Anderson published evidence for the existence of a particle exactly matching Dirac's description, which is now known as the *positron*. The first antiparticle had been discovered[8].

Eventually, physicists figured out how to develop Dirac's equation into a quantum field theory naturally incorporating both electrons and positrons without having to use the idea of a sea of negative energy electrons. This is a topic that we will touch upon in Chapter 32.

Between 1932 and 1933, Dirac wrote three more papers on quantum field theory, managing to cast it in a fashion that made its consistency with relativity apparent. The last of these papers has some particular historical interest, as it formed the basis for further work conducted by Richard Feynman (among others)[9] in 1949. Feynman's techniques for simplifying complex calculations in particle physics and his whole approach to quantum theory have been of considerable importance in the development of the subject. Feynman himself acknowledged the clues he followed from Dirac's 1926 and 1933 papers.

In 1930 Dirac was elected as a fellow of the Royal Society and was appointed as Lucasian Professor of Mathematics at the University of Cambridge[10] in 1932, a post he held for 37 years.

In 1933 Dirac shared the Nobel Prize in physics with Erwin Schrödinger, their joint citation being "for the discovery of new productive forms of atomic theory." Dirac always disliked publicity and was inclined to turn down the offer of a Nobel Prize. He eventually changed his mind when a colleague pointed out to him that declining a Nobel Prize was more likely to gain him publicity than accepting it.[11]

Dirac retired from Cambridge in 1969 and after various academic appointments took up a professorship at Florida State University, Tallahassee, in 1971, where he continued to be productive in physics, right up to his death in 1984.

Alongside lasting contributions to the development of the subject, Dirac was also responsible for writing one of the standard textbooks, *The Principles of Quantum Mechanics,* which has been used by generations of physicists. This book was first published in 1930, but it was not until the third edition in 1947 that the bra and ket notation was used.

Students who attended Dirac's lectures, especially at Cambridge where he was a fellow of St. John's College, noted how modest he was and how he never explicitly referred to his own role in the development of quantum theory. However, one such student has written:

> one did gain the impression from a slight smile that played around his features when he introduced bras and kets that this invention (and the small and harmless joke enshrined in the nomenclature) had given him great satisfaction.[12]

21.2 DIRAC, THE PERSON

> In science one tries to tell people, in such a way as to be understood by everyone, something that no one ever knew before. But in poetry, it's the exact opposite.
>
> *Paul Dirac*[13]

Dirac's father, Charles Adrien Ladislas Dirac, was born in Monthey, in the Valais Canton of Switzerland, whereas his mother, Florence Hannah Holten, came from Cornwall in England. Charles had been educated at the University of Geneva, then came to England in 1888 and taught French at a college in Bristol. Florence's father had moved to Bristol as master mariner on a Bristol ship, and she was working in the library there. Charles and Florence married in 1899, and Paul was their second child.

Dirac joined the Bristol University Electrical Engineering Department in 1918 (aged 16) and graduated three years later. However, he was unable to find a job in the post-war economic depression gripping England. In 1921 Dirac sat the Cambridge scholarship examination and was awarded a financial scholarship to study mathematics at St. John's College. Unfortunately, the local government grant that would normally help support students was not available to him, as his father had not been a British citizen for long enough to qualify. Consequently, Dirac was unable to accept the Cambridge place. Bristol University came to his rescue by offering him the chance to study mathematics for two years without paying fees. He was subsequently given a grant to undertake research at Cambridge, and so eventually arrived at St. John's.

During the academic year 1934–1935 Dirac visited the Institute for Advanced Study at Princeton (where Einstein worked for many years). There he re-established contact with another famous physicist, Eugene Wigner[14]. A chance lunchtime meeting introduced Dirac to Wigner's sister, Margit, and they were married in 1937. Margit already had two children, Judith and Gabriel Andrew, from her first marriage. They both adopted Dirac's name and Gabriel went on to become a famous pure mathematician in his own right. Margit and Paul subsequently had two daughters of their own.

There are many stories relating to Dirac's legendary taciturn nature. Dirac himself put this reserve down to his school life, where he was taught never to start a sentence until he knew how to finish it. On one occasion, he had completed a lecture and was accepting questions. A member of the audience stood up and confessed that he hadn't understood one of the equations written on the blackboard. When Dirac made no response to this remark, the chairman prompted him: "that was a statement, not a question," was Dirac's only comment.

Like many of us, Dirac occasionally fell asleep during other people's lectures; or at least he closed his eyes. Despite appearing to be asleep, Dirac would suddenly open his eyes and make some penetrating remark about the content of the lecture. On one occasion Dirac was in the audience, with his eyes closed, when the speaker stopped, scratched his head, and declared, "Here is a minus where there should be a plus. I seem to have made an error of sign." Dirac opened one eye and said, "Or an odd number of them."

Like many physicists, Dirac was fond of precision in the use of language and sceptical about ghosts and other paranormal phenomena. While attending a meeting in a castle, Dirac heard another guest discussing one of the castle's rooms where a ghost was supposed to appear at midnight. Dirac countered, "Is that midnight Greenwich time, or daylight-saving time?"

21.3 DIRAC'S VIEWS ON THE MEANING OF QUANTUM THEORY

Despite his extensive work in the field, and his writing of one of the most widely respected books on the subject, Dirac's views on the interpretation of quantum theory are not especially evident. It's possible that he held a broadly instrumental stance as in the fourth edition of *The Principles of Quantum Mechanics,* Dirac comments, "only questions about the results of experiments have real significance and it is only such questions that theoretical physics has to consider." However, this could also be a nod to the standard Copenhagen interpretation approach.

We do know that Dirac believed *state superposition* to be the most significant radical aspect of quantum theory. Again, in *The Principles of Quantum Mechanics,* he wrote:

> any [quantum] state may be considered as the result of a superposition of two or more other states .... The nonclassical nature of the superposition process is brought out clearly if we consider the superposition of two states, A and B [when an observation is made on A, result a occurs and when made on B, result b is produced] .... What will be the result of the observation when made on the system in the superposed state? The answer is that the result will be sometimes a and sometimes b, according to a probability law depending on the relative weights of A and B in the superposition process. It will never be different from both a and b.[15]

[my additions]

In this passage, Dirac is contrasting the behaviour of a quantum superposition of states with the closest classical equivalent. Generally speaking, if you have two states of a classical system, you can always construct a third possible state by merging the two. Take a simple example. A pendulum can swing back and forth, but given the right sort of support at the top, it can also swing from side to side, i.e., at 90° to the back-and-forth motion. If it can do that, then it can do both at the same time, in which case the bob of the pendulum will move round in a circle. The circular motion is *a classical superposition of the two swinging motions.* The key point here is that the *properties of a system in a classically merged state are the merged properties of the separate states.*

Things are rather different in a quantum superposition. As Dirac says, if a measurement of state $|A\rangle$ brings result a, and a measurement of state $|B\rangle$ brings result b, then a measurement of the state $\alpha|A\rangle + \beta|B\rangle$ brings a result of *either a* or b. Quantum theory allows us to calculate the probabilities of the two results, given the amplitudes in the superposed (mixed) state. Classically we would expect the measurement of $A + B$ to give us some weighted average value between a and b.

Dirac is very clear about how significant he felt this difference to be. However, he is not so forthcoming over the ontological nature of the quantum state.

Dirac was much more open on the subject of mathematical beauty. One of his repeated themes was the importance of beautiful equations in the development of theory. The quote at the start of this chapter is typical of Dirac's impassioned defences of this view.

In 1955 Dirac was a visiting professor at Moscow University, where he lectured through the autumn. During a discussion, he was asked to write a concise expression of his philosophy of physics. Dirac merely chalked up "physical laws should have mathematical beauty," an epigram that is apparently still preserved on the blackboard.

Philosophers have argued for centuries about the nature of beauty in broad terms, but at least we all have some level of access to that experience: whether it be in the face of a partner, a painting, a scenic view, or a snatch of music. Mathematical beauty is rather more austere and inaccessible without a certain fluency in mathematics.

At a rudimentary level, mathematical symbols on a page can have a rather satisfying elegance to them, rather like extraordinarily delicate Chinese calligraphy (Figure 21.1).

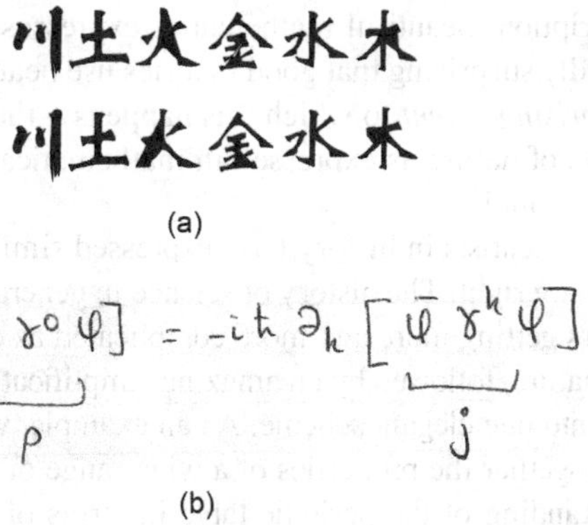

川土火金水木
川土火金水木

(a)

$$i\hbar \frac{\partial}{\partial t}\underbrace{\left[\bar{\psi}\gamma^0\psi\right]}_{\rho} = -i\hbar\,\partial_k \underbrace{\left[\bar{\psi}\gamma^k\psi\right]}_{j}$$

(b)

FIGURE 21.1 Some Chinese calligraphy (a). Part of the author's notes on the Dirac equation (b).

However, this is not what is really meant by mathematical beauty. Steven Weinberg[16] has suggested that mathematical beauty should really be called *conceptual beauty;* an appreciation of ideas that are powerful, elegant, and which have the ring of inevitability to them. Mathematics is the perfect tool for expressing abstract ideas in the simplest and most concise form. This conciseness and simplicity being a vital part of any mathematics that can be called beautiful.

My personal candidate for the most beautiful equation in mathematics is:

$$e^{i\pi} = -1$$

Part of the attraction of this little formula is the combination of three numbers from different parts of mathematics (two of which require an infinite number of decimal places for their full expression and one which has no numerical expression at all) coming together in this odd fashion to simply equal −1. There is an amazing compression of information in this pretty juxta position.

The grouping together of many ideas into a simple scheme is part of the beauty of mathematics. Something like this can be seen in a sketch drawing, where one or two deft lines have managed to catch the expression in a face.

Some of the most powerfully beautiful mathematical discoveries derive a measure of their appeal from the way they extend ideas into new regions. Such a breakthrough often links parts of mathematics that nobody had suspected a connection between; as a result, puzzles that could not be solved suddenly become tractable.

No doubt, mathematical beauty is a very abstract and intellectual sort of beauty, but it can still carry tremendous emotional power. For people gifted in the discovery of mathematics the moment of realization, the "aha!" moment, can be extraordinarily moving.

All of us have had a glimpse of this at one time or another. Perhaps it came from the satisfaction of solving a puzzle or succeeding to master a challenge. This could be a crossword, a sudoku puzzle, a piece of DIY, or the perfect cover drive[17]. Mathematicians can experience this as part and parcel of their professional life. Of course, scientists do as well. In their case, the "aha" moment relates to uncovering part of the blueprint for nature's assembly. This is often described as a joyous moment (particularly when it comes at the end of a long period of intense intellectual struggle) coupled with a profound sense of the 'rightness' or inevitability of the solution. Words such as 'beautiful', 'elegant', 'simple', and so on, are often used.

When Dirac insisted that scientific theories should contain beautiful mathematics, he was not simply voicing an aesthetic preference. His thinking was much more important than that. Dirac believed that the search for beautiful mathematics was *a valid technique to use in the process of discovery.* In part, this is obvious, if not self-fulfilling. If the aim of a scientific theory is to bring together and explain certain experimental facts, then a successful theory will do that well.

According to our description, beautiful mathematics expresses ideas in a particularly simple and concise way, so it is hardly surprising that good theories use beautiful mathematics. However, Dirac is pointing out the *surprising extent* to which this happens—the *overwhelming* beauty that is seen in the fundamental laws of nature as expressed in mathematical physics, his own equation for the electron being a prime example.

Many of the greatest scientists in history have expressed similar views. Objectively, there is even some evidence that they are right. The history of science in general, and physics in particular, shows a repeated cycle of matters getting more and more complicated as experiments probe more deeply and across wider ranges of nature, followed by an amazing simplification as some key ideas are developed to pull things together into one elegant scheme. As an example, we can point to the periodic table of the elements grouping together the properties of a wide range of different chemical elements closely followed by the understanding of the periodic table in terms of the structure of atoms. During the 1950s–1970s, the roster of elementary particles being confirmed by high energy physics experiments was appearing to grow without limit. The discovery of quarks allowed large numbers of these particles to be understood as simple combinations of three quarks[18]. There are many other similar situations.

Of course, we are still left to ask the meta question of why the laws of nature should be so profoundly beautiful. Some believe that the human brain has evolved to be especially good at working with mathematics, a survival advantage dating back to throwing rocks and spears while hunting animals for food. From this point of view, it is hardly surprising that our aesthetic sentiments, which evolved as part of this mathematical brain, should gel with our experience of the world. However, such a simplistic view once again does not do justice to the profound beauty that we are talking about.

Others see beauty at the root of nature as being a natural consequence of the universe having been created by a God who is rational, imbued with love, and the source of our aesthetic experience.

Many find no deeper significance in these thoughts at all and simply get on with the business of doing science.

Dirac was not a religious man and did not think much of philosophical reflection. It is fitting, however, that his most beautiful creation, the quantum equation for relativistic electrons[19], is inscribed on his commemorative stone in Westminster Abbey (Figure 21.2).

Given Dirac's thinking on mathematical beauty and the laws of nature, it is tempting to suppose that he must have held a realistic view of quantum theory.

FIGURE 21.2 Dirac's commemorative stone in Westminster Abbey. {Image credit: Copyright:Dean and Chapter of Westminster}

NOTES

1 P.A.M. Dirac, *Sci. Am.,* May 1963.
2 "There are two types of genius. Ordinary geniuses do great things, but they leave you room to believe that you could do the same if only you worked hard enough. Then there are magicians, and you can have no idea how they do it." Hans Bethe, 1906–2005, Nobel Prize in Physics 1967. Bethe was specifically referring to Richard Feynman in this quote, but it is arguably relevant to Dirac as well.
3 T. W. B. Kibble 1998 *Eur. J. Phys.* **19,** 017.
4 Most theoreticians regard his equation as his greatest achievement, mainly because he was indisputably first to write it down and explain its origin, at a time when several other great theoreticians (Wigner et al.) were on the trail.
5 No implied criticism here. Schrödinger's work was a brilliant and crucial development, but glancing at the equation shows, to informed eyes, that it is not consistent with relativity. It is second-order in space and only first-order in time. Relativity demands that space and time be treated on an equal footing. Physics develops in an incremental work, and you can only deal with the issues that are tractable at the time.
6 $\gamma^0 = \begin{pmatrix} I_2 & 0 \\ 0 & -I_2 \end{pmatrix}, \gamma^1 = \begin{pmatrix} 0 & \sigma_x \\ -\sigma_x & 0 \end{pmatrix}, \gamma^2 = \begin{pmatrix} 0 & \sigma_y \\ -\sigma_y & 0 \end{pmatrix}, \gamma^3 = \begin{pmatrix} 0 & \sigma_z \\ -\sigma_z & 0 \end{pmatrix}$ with I_2 being the 2×2 identity matrix and the σ the Pauli spin matrices.
7 Negative energies are not in-themselves an issue. Gravitational potential energy is negative, for example. Also, the total energy of a bound state is also negative. However, an isolated electron away from any field would only have kinetic energy + the intrinsic energy giving it rest mass. The Dirac equations generate negative solutions in this case as well.
8 As the class of antiparticles was not appreciated at the time, it was named the positron rather than the anti-electron.
9 Feynman, Schwinger, and Tomonaga were awarded the Nobel Prize in 1965 for this work.
10 This chair is currently (2022) held by Michael Cates. Previous Lucasian professors include Sir Isaac Newton, Stephen Hawking and Michael Green.
11 It is thought to be Rutherford who gave this advice.
12 J.C. Polkinghorne, in *Tributes to Paul Dirac,* J.G. Taylor, ed., Institute of Physics Publishing, Bristol, UK, 1987.
13 Quoted in H. Eves *Mathematical Circles Adieu* (Boston, MA, 1977).
14 E. Wigner, 1902–1995, Nobel Prize in 1963.
15 P.A.M. Dirac, *The Principles of Quantum Mechanics,* 4th ed., Oxford Science Publications, 2004, pp. 12–13. Square bracket inserts are my comments.
16 S. Weinberg, 1933–2021, Nobel Prize in Physics, 1979.
17 The most elegant of cricket shots.
18 If we open the date range out a bit further, then five and then six quarks have to be included to cover everything.
19 Writing the equation in the form $i\gamma\,\partial\psi = m\psi$ uses even more compact notation than that quoted in the text.

22 Conclusions

Having concluded a brief tour through the early history of quantum theory, we need to stop and consider some important themes.

From a historical point of view, a distinction can be drawn between 'old quantum theory' and the full quantum theory that we now use. Old quantum theory was in essence a patchwork, cobbled together by incorporating some quantum ideas into essentially classical physics. The work of Planck, Einstein, Bohr, and Sommerfeld can be bracketed as old quantum theory. This classification is not meant to undermine the tremendous achievement that old quantum theory represents, but simply to underline its standing as a stage on the way to a fully formed quantum theory, which emerged from the work of Heisenberg, Schrödinger, Born, and Dirac.

The gradual shift between old- and full quantum theories came as people put aside more and more classical ideas and accepted the radical revisions that were being forced on them by a thorough analysis of the experiments. The idea of a particle's path is a specific example of a 'common sense' notion that was eventually abandoned.

Equally, as the limitations of old quantum theory became more obvious, a deeper understanding replaced the rather 'ad hoc' rules that had been introduced to make progress. Planck quantized the energies of the oscillators inside black bodies as a spurious assumption. When Bohr quantized the energy inside atoms, by assuming angular momentum quantization, it became clear that these oscillators were atoms and molecules in the substance of the black body. De Broglie explained the Bohr rules by introducing matter waves, which formed specific patterns in circular orbits. Then Schrödinger brought in his wave equation and after applying it to the hydrogen atom was able to demonstrate how the full 3-D pattern adopted by the 'wave function' explained energy quantization and a whole host of other ideas as well.

Meanwhile, Heisenberg, Born, and Jordan developed matrix mechanics, from a more abstract method for manipulating amplitudes which in turn grew from a focus on direct measurement. It produced the same results when applied to the hydrogen atom, but with rather more mathematical work and with much less for physicists to get hold of and 'picture.' Mind you, this was entirely deliberate, as Heisenberg and Bohr had grown to mistrust the human ability to visualize what was going on.

Once Schrödinger, Dirac, and Jordan had shown the two approaches to be mathematically identical, physicists learned to adopt whichever set of tools was more conveniently applied to a given problem. The quantum description of spin, for example, is easily constructed in terms of matrices.[1]

The development of the photon concept is a key example of a radically new idea gradually gaining acceptance. The process works via a growing realization that more and more things can be explained in the context of the photon model. In the end, it's this ability to use a new concept to explain a wide variety of experimental results that convinces us that we are on to something. The realist would be very happy with this. Indeed, they would argue that the precise signal for a concept being part of a 'true' picture of reality is its explanatory facility.

Given the widespread success of the wave theory of light, dating back to Young in 1801, it took a great deal to convince physicists to take a particle of light seriously. Remember that Einstein himself thought that his 1905 article on black body spectra and the photoelectric effect was the only truly radical piece of work that he did. Like the Spanish inquisition, nobody could have expected photons.[2] The realist might not be happy with having to keep both wave and particle ideas going at the same time but would accept that this is what is required to do justice to experiment until a deeper understanding (quantum field theory) comes along.

The instrumentalist would simply argue that with the photon, we finally hit upon the correct key to unlock as many experimental doors as possible. If the photon concept gets us from the results of

DOI: 10.1201/9781003225997-25

experiment A to predicting what will happen in experiment B, then that's fine. Questions about the reality of the photon are unhelpful.

One historical thread that we have not mentioned is the study of radioactivity by Rutherford and others. Arguably, this is where probability first entered quantum theory. Rutherford's account of the radioactive decay of atomic nuclei relied on the moment of decay being random. Einstein, who was obviously aware of Rutherford's work, had his photons being emitted and absorbed by black body oscillators in an essentially random way. However, Einstein always felt that there should be a deeper theory explaining what appeared to be random. After all, in 1905 he had shown how the apparently random motion of pollen grains (Brownian motion) was due to the battering they received by invisibly small atoms and molecules. Bohr had photons getting emitted and absorbed by atoms when electrons changed energy states, but he gave no explanation of how this happened and no way of predicting when it would occur. Heisenberg's matrices dealt with likelihoods for different alternatives, and finally Born made the connection between $\Psi\Psi^*$ and probability.

Historically, another key idea, the amplitude, can be traced to Heisenberg's work. It started life in the analysis of electron paths in atoms (which ultimately got nowhere, leading Heisenberg to suspect that the whole path notion was a blind alley) and the patterns linking line spectra, wavelengths, and brightness. Born wrote down the state expansion in terms of amplitudes and identified the complex square of the amplitude with the probability of a transition between one state and another.

A clear common thread that runs through all these developments is the use of mathematics, which has always been the language of physics. No physicist can be happy with a theory until they can use it to calculate something. Dirac flagged the search for beautiful mathematics as being a key indication that you were on the right track. What was new and different about quantum theory was the order in which things happened. Any student of physics would be able to come up with various examples from history in which the physical concepts came first and the mathematics had to be developed (or learnt from the mathematicians) to express those ideas. Newton needed to invent calculus to do his mechanics.[3] Maxwell developed vector calculus so that he could link electric and magnetic fields in an elegant way. Einstein had to learn tensor calculus so that his physical intuitions about gravity could bear fruit.

In quantum theory, the mathematical ideas developed in a relatively orderly way. Heisenberg and Bohr continually stressed the need to develop adequate concepts, but this always lagged behind the formal part of the theory. Perhaps even now we still don't have an adequate set of concepts to understand (in the sense of picture) what's going on.

One possible reason for the success of the Copenhagen interpretation, with its expression of complementarity, is that it encouraged people to stop worrying by placing the ultimate picture beyond human understanding. However, there is something frustrating and unsatisfactory about this answer. It's like getting to the end of a murder mystery and finding the last page missing. In his later writings, Heisenberg tried to go beyond complementarity and grasp a realistic view of amplitudes by bringing in Aristotle's potentia. There have been other approaches, as we will see in Chapters 26-31.

NOTES

1 One of the virtues of Dirac's bra-ket notation is that it blurs the distinction between matrix and wave versions of the theory.
2 This is a reference to a Monty Python sketch. As I first wrote this section at 5 a.m., cut me some slack.
3 And ended up in a bitter dispute with Leibenitz about who got there first.

Part 3

23 Quantum Correlations

23.1 TWO THREADS

Before we embark on our exploration of the various interpretations of quantum theory, we need to outline one more technical matter, which we will do in Chapter 25, and take a detailed look at one of the weirder aspects of quantum reality. That is the subject of this chapter. Our discussion will disclose some of the most profound consequences of quantum theory and set stringent challenges for any interpretation.

There are two threads running through this chapter.

The first stems from Einstein's desire to discredit quantum physics as a fundamental theory of reality. The second derives from Schrödinger, who introduced the term *entanglement* to describe how interacting systems knit together so that their physical properties become correlated in surprising ways, even beyond the span of their interaction.

Einstein, working with others, posed a question to the quantum community based on a definition of reality. Applying a specially constructed entangled state, their argument seemed to contradict aspects of quantum theory and hence reveal its incompleteness. Bohr countered by exposing a weakness in their definition of reality. Many years later, Bell recast the argument by showing how entangled states demonstrate measurement correlations that cannot be reproduced by hidden variable theories unless they violate the theory of relativity. Bell moved the game on by making the discussion, in part, accessible to experimental resolution. It became a question of establishing whether these correlations, predicted by quantum theory, happened in practice. Coming right up to date, increasingly impressive experimental tests show that Bell's correlations are confirmed and that any hidden variable theory can only work in contradiction to relativity. The consequences are profound, as they cast new light on the nature of quantum reality.

23.2 IS QUANTUM THEORY COMPLETE?

Despite being responsible for a number of its key developments, Einstein was never happy with the final form of quantum theory. At first, he objected to the uncertainty principle and tried to devise some thought experiments designed to get around its limitations. In each case, Bohr and Heisenberg found a counterargument and the uncertainty principle held fast (Section 19.3).

Einstein then switched tack from trying to show that quantum theory was *inconsistent* (did not always give the same answer in similar cases) to proving that it was *incomplete* (that there were some situations not covered by the theory).

In 1935 Einstein, together with Nathan Rosen and Boris Podolsky, published a famous paper entitled "Can a quantum mechanical description of physical reality be considered complete?" In the EPR paper[1], as it is now known, the authors constructed an argument to demonstrate the need for a deeper theory to replace quantum mechanics.

The paper starts by setting the criterion for a *complete physical theory* as "every element of the physical reality must have a counterpart in the physical theory", which seems entirely reasonable, provided you can decide what the "element of the physical reality" are. That's not straightforward. According to the authors, they must be "found by an appeal to results of experiments and measurements." The workings of the world are too subtle and surprising for us to guess reliably. We need experiments to tap us on the head and say: 'it must be like this'.

Even then, it's not quite as simple as it sounds. Experiments produce quantities of data. Some aspects are important, and some are random 'noise'. We need some way of separating the key

DOI: 10.1201/9781003225997-27

components from the random fluctuations. The EPR paper contains the following suggestion, which ties theory with experiment:

> if, without in any way disturbing the system, we can predict with certainty the value of a physical quantity{theory}, then there exists an element of physical reality {experiment} corresponding to this physical quantity.

{my additions}

The bit about "disturbing the system" is there to set a trap for quantum theory.

In the next part of their short paper, the writers apply their condition for reality to a specific case. They remind us that quantum theory employs a state $|\psi\rangle$ (actually, they use the wave function ψ) to represent a system and operators, e.g., $\hat{P}$ to represent physical variables[2]. If $|\psi\rangle$ happens to be an eigenstate of $\hat{P}$, then $\hat{P}|\psi\rangle = a|\psi\rangle$ and we can say *with certainty* that a measurement of the physical variable linked to $\hat{P}$ will result in the value a. By the EPR condition, the physical variable represented by $\hat{P}$ must be an element of reality, for that system.

However, if we choose to measure some other physical variable, linked to an operator $\hat{Q}$, for example, which does not have $|\psi\rangle$ as an eigenstate, then we can't predict with certainty what a measurement will produce. We would have to expand $|\psi\rangle$ over the eigenstates of $\hat{Q}$ and calculate the expectation value for a set of results, or the probability of a given result. The only way to predict with 100% certainty the outcome of a measurement of $\hat{Q}$ is to "disturb the system" by ensuring $|\psi\rangle$ collapses into one of the eigenstates of $\hat{Q}$, via some prior measurement or preparation, before making the required measurement. So, by the EPR condition $\hat{Q}$ can't be an element of physical reality, or at least not at the same time as $\hat{P}$.[3]

The EPR argument is based on a fundamental philosophical objection to quantum theory. According to the authors, we should be able to find a set of physical variables along with a theory that allows us to predict their values with certainty, as reality must be 'out there' for us to find, not something that we can mould.

Having set up their condition for the reality of a physical variable and illustrated it for a simple quantum mechanical case, the EPR team then spring their trap by describing a situation in which it appears that quantum theory fails to describe every element of a system's reality.

23.2.1 The EPR Argument

> From this follows that either (1) the quantum-mechanical description of reality given by the wave function is not complete or (2) when the operators corresponding to two physical quantities do not commute the quantities cannot have simultaneous reality.
>
> *Einstein, Podalsky, Rosen*

Consider two particles, A and B, which interact with one another and then separate in opposite directions. After a while, their separation exceeds the range over which they can interact, and they become (classically) independent of each other. In the EPR paper the authors suggest an example state function to describe this situation:

$$\Psi(x_A, x_B) = \int_{-\infty}^{\infty} e^{\frac{i}{\hbar}(x_A - x_B + x_0)p}\, dp \tag{23.1}$$

where x_A is the position of particle A, x_B the position of particle B, x_0 is a constant and p is the momentum of the separating particles. In this scenario, the net momentum in the x direction is zero, and hence the particles have the same magnitude of momentum, but in opposite directions. There is an integral over momentum as the magnitude of the particles' momenta is not known, and so by Important Rule 3, we ought to sum over all possibilities.

This state function is somewhat curious, and the choice has been criticized by later commentators. It looks somewhat like a sum over momentum state functions as if we were expanding over a basis, but it is not quite that due to the two different position co-ordinates.

In order to understand how this state function would respond to a momentum measurement of one particle, say particle A, the authors expand the state function over momentum eigenstates. In state function terms the expansion they use is:

$$\Psi(x_A,x_B)=\int_{-\infty}^{\infty}\psi_p(x_B)u_p(x_A)dp$$

with the $u_p(x_A)=e^{\frac{i}{\hbar}px_A}$ being momentum eigenstate functions (without normalization constants) and the expansion coefficients $\psi_p(x_B)$, are interpreted as state functions for the other particle. Comparing this expansion with the original state function Eq. 23.1:

$$\Psi(x_A,x_B)=\int_{-\infty}^{\infty}e^{\frac{i}{\hbar}(x_A-x_B+x_0)p}dp=\int_{-\infty}^{\infty}\psi_p(x_B)u_p(x_A)dp=\int_{-\infty}^{\infty}\psi_p(x_B)e^{\frac{i}{\hbar}px_A}dp$$

shows that $\psi_p(x_B)=e^{-\frac{i}{\hbar}(x_B-x_0)p}$. In other words, the $\psi_p(x_B)$ are eigenstate functions of momentum for particle B, but with opposite momentum directions.

If we then measure the momentum of A, resulting in the value $\mathcal{P}$, this state function expansion collapses so only one specific momentum eigenstate of A, $u_{\mathcal{P}}(x_A)$, remains. Hence the overall state function becomes $\psi_{\mathcal{P}}(x_B)e^{\frac{i}{\hbar}\mathcal{P}x_A}=e^{-\frac{i}{\hbar}(x_B-x_0)\mathcal{P}}e^{\frac{i}{\hbar}\mathcal{P}x_A}$, i.e., we pick out the eigenstate of B with momentum $-\mathcal{P}$ at the same time.

So, measuring the momentum of A *causes an overall state collapse which allows us to predict with certainty that the momentum of B, if measured, would be* $-\mathcal{P}$. However, measuring A *does* not *"disturb the system" for B*, so we have fulfilled the criteria for the momentum being a physically real state variable for B.

The next step is to think what might happen if we choose to measure the *position* of particle A instead of its momentum. To explore this, we need to produce an expansion of the state function using position eigenstates of A.

In Section 12.2.3 we introduced Dirac's delta function as a means of describing an eigenstate of position:

$$|x=\alpha\rangle=\frac{1}{2\pi\hbar}\int_{-\infty}^{\infty}e^{\frac{i}{\hbar}p(x-\alpha)}\,|\,p\rangle dp$$

In state function terms, as used in the EPR paper this is[4]:

$$\delta(x-\alpha)=\int_{-\infty}^{\infty}e^{\frac{i}{\hbar}p(x-\alpha)}dp$$

once again, suppressing normalization considerations for notational convenience. This prompts the authors to write:

$$\Psi(x_A,x_B)=\int_{-\infty}^{\infty}\varphi_\alpha(x_B)\delta(x_A-\alpha)d\alpha$$

as their expansion over position eigenstate functions for A. They make the further identification:

$$\varphi_\alpha(x_B)=\int_{-\infty}^{\infty}e^{\frac{i}{\hbar}p(\alpha-x_B+x_0)}dp$$

so that:

$$\Psi(x_A, x_B) = \int_{-\infty}^{\infty}\left[\int_{-\infty}^{\infty} e^{\frac{i}{\hbar}p(\alpha - x_B + x_0)} dp\right]\delta(x_A - \alpha)d\alpha$$

Just to pause to explain this for a moment, $\delta(x_A - \alpha)$ is defined such that, within the integral:

$$\delta(x_A - \alpha) = 0 \qquad x_A \neq \alpha$$

$$\delta(x_A - \alpha) = 1 \qquad x_A = \alpha$$

so that the outer integral over α, which is effectively the expansion over position eigenstate functions, vanishes for all $x_A \neq \alpha$ within the range $-\infty \leq x_A \leq \infty$. When $x_A = \alpha$, the integral reduces to:

$$\Psi(x_A, x_B) = \int_{-\infty}^{\infty} e^{\frac{i}{\hbar}p(x_A - x_B + x_0)} dp$$

which is our starting state function. This whole approach is somewhat unconventional.

If we measure the position of A and obtain the value X, this picks out $\alpha = X$ from our state expansion, collapsing it into:

$$\int_{-\infty}^{\infty} e^{\frac{i}{\hbar}p(X - x_B + x_0)} dp = \int_{-\infty}^{\infty} e^{\frac{i}{\hbar}p(X - (x_B - x_0))} dp$$

which the authors point out is $\delta\big(X - (x_B - x_0)\big)$. Consequently, *particle B is also in a position eigenstate* with $X = x_B - x_0$, or $x_B = X + x_0$.

Now the trap closes.

Note what we have done. Measuring the position of particle A has "disturbed the state" of A, because we have interacted with it. We have *not* interacted with B, so we can't have disturbed its state. However, *collapsing the state of A has allowed us to predict B's position with certainty, without disturbing B, so that exact position must be a part of B's physical reality.*

Hang on a moment though. *Previously we argued that the momentum must be part of B's physical reality*, now we are saying that position is as well. The uncertainty principle prevents us from being certain of both position and momentum at the same time. We have violated the uncertainty principle.

To summarize, this is the EPR argument:

- If we choose to measure A's momentum, we can be sure of B's momentum without disturbing B, so the momentum must be part of B's reality.
- If we measure A's position instead, we can be sure of B's position, without disturbing B, so that must be part of B's reality as well.
- Whichever of A's physical variables we measure, we have not "disturbed the state" of B, so we must be finding out true things about B that have not changed due to the measurement of A.
- Yet this breaks one of the fundamental rules of quantum theory; the uncertainty principle.
- The theory doesn't allow us to construct a state with a fixed value of position and momentum at the same time.
- The theory must be incomplete. It doesn't cater for this situation.

The authors were certainly aware of one potential criticism of their argument, which they attempt to counter towards the end of the paper:

> One could object to this conclusion on the grounds that our criterion of reality is not sufficiently restrictive. Indeed, one would not arrive at our conclusion if one insisted that two or more physical quantities can be regarded as simultaneous elements of reality only when they can be simultaneously measured or predicted. On this point of view, since either one or the other, but not both simultaneously, of the quantities P and Q can be predicted, they are not simultaneously real. This makes the reality of P and Q depend upon the process of measurement carried out on the first system, which does not disturb the second system in any way. No reasonable definition of reality could be expected to permit this.

Or could it …

23.2.2 Follow-Up by David Bohm

There is an air of 'sleight of hand' about the EPR argument, as it depends on the very specific initial state function that was chosen. As mentioned earlier, modern commentators find this somewhat contentious. However, there is another version of the argument due to David Bohm[5] which uses absolutely conventional particle spin states instead.

To see this argument at work, we need to start by thinking about the possible combined states of two spin 1/2 particles, for example, two electrons. To describe the combined system, we need a collection of basis states:

$$|U\rangle|U\rangle;|D\rangle|D\rangle;|U\rangle|D\rangle;|D\rangle|U\rangle$$

where I am assuming that the first ket refers to the first particle in each case.

When you combine two spin 1/2 particles, the quantum rules governing how you add together angular momenta dictate two possible results: total spin 1 or total spin 0. Unfortunately, demonstrating this and the correctness of the states used in each case would require a deep dive unto the theory of angular momentum combinations, which nobody wants at this stage…

The total spin 0 case is straightforward; the overall state of the two electrons will be:

$$|\text{singlet}\rangle = \frac{1}{\sqrt{2}}\big(|U\rangle|D\rangle - |D\rangle|U\rangle\big)$$

which is called a *singlet state* as this is the only combination that adds up to spin 0. As the total spin is zero, $s = 0$, the spin z component must be zero as well: $m_s = 0$. Hence, we can also write the singlet state as:

$$|\text{singlet}\rangle = |s=0,\, m_s=0\rangle = \frac{1}{\sqrt{2}}\big(|U\rangle|D\rangle - |D\rangle|U\rangle\big)$$

The total spin 1 combination is slightly trickier, as if $s = 1$ there are three possible z components $m_s = -1, 0, +1$ (Section 10.2.4). Hence, we have three state functions[6]:

$$|\text{triplet 1}\rangle = |s=1,\, m_s=1\rangle = |U\rangle|U\rangle$$

$$|\text{triplet 2}\rangle = |s=1,\, m_s=0\rangle = \frac{1}{\sqrt{2}}\big(|U\rangle|D\rangle + |D\rangle|U\rangle\big)$$

$$|\text{triplet 3}\rangle = |s=1,\, m_s=-1\rangle = |D\rangle|D\rangle$$

Bohm's EPR argument starts with a pair of electrons that have been prepared in the singlet state, but which are now separating in different directions. One of the electrons reaches a Stern–Gerlach

(SG) device, set to measure its vertical spin component, which has the effect of collapsing the singlet state.

If this first electron is found to be spin UP, the singlet state collapses as follows:

$$|\text{singlet}\rangle = \frac{1}{\sqrt{2}}\left(|U\rangle|D\rangle - |D\rangle|U\rangle\right) \quad \xrightarrow{\text{collapses into}} \quad |U\rangle\,|D\rangle$$

so the other electron must be spin DOWN.

If the measurement of the first electron results in DOWN, that picks out the second pairing, and the other electron is spin UP:

$$|\text{singlet}\rangle = \frac{1}{\sqrt{2}}\left(|U\rangle|D\rangle - |D\rangle|U\rangle\right) \quad \xrightarrow{\text{collapses into}} \quad |D\rangle\,|U\rangle$$

Clearly, measuring the spin of one electron allows us to predict with certainty the spin of the other without disturbing it. By the EPR condition of reality, the vertical spin component of the second electron must be a 'real' physical property with two possible values (UP, DOWN).

The next step is to consider what might happen if the SG apparatus is set to measure the horizontal component of spin, rather than the vertical. To see what ensues we must switch bases in the singlet state by using $|L\rangle$ and $|R\rangle$ instead. Taking the relationships $|U\rangle = \frac{1}{\sqrt{2}}\left(|L\rangle + |R\rangle\right)$ and $|D\rangle = \frac{1}{\sqrt{2}}\left(|L\rangle - |R\rangle\right)$ and plugging them into our singlet state, we get:

$$|\text{singlet}\rangle = \frac{1}{\sqrt{2}}\left(|U\rangle|D\rangle - |D\rangle|U\rangle\right)$$

$$= \frac{1}{\sqrt{2}}\left[\left(\frac{1}{\sqrt{2}}\left(|L\rangle+|R\rangle\right)\right)\left(\frac{1}{\sqrt{2}}\left(|L\rangle-|R\rangle\right)\right) - \left(\frac{1}{\sqrt{2}}\left(|L\rangle-|R\rangle\right)\right)\left(\frac{1}{\sqrt{2}}\left(|L\rangle+|R\rangle\right)\right)\right]$$

$$= \frac{1}{\sqrt{2}}\left[\frac{1}{2}\left(\cancel{|L\rangle|L\rangle} - |L\rangle|R\rangle + |R\rangle|L\rangle - \cancel{|R\rangle|R\rangle}\right) - \frac{1}{2}\left(\cancel{|L\rangle|L\rangle} + |L\rangle|R\rangle - |R\rangle|L\rangle - \cancel{|R\rangle|R\rangle}\right)\right]$$

$$= \frac{1}{\sqrt{2}}\left[\frac{1}{2}\left(2|R\rangle|L\rangle - 2|L\rangle|R\rangle\right)\right]$$

$$= \frac{1}{\sqrt{2}}\left[|R\rangle|L\rangle - |L\rangle|R\rangle\right]$$

which is also a singlet state, but in the (LEFT, RIGHT) basis.

Measuring the *horizontal* spin component of electron 1 will consequently collapse the overall state so that an exact prediction can be made of electron 2's spin (if 1 is LEFT, 2 must be RIGHT, and vice versa). As this has happened without disturbing electron 2, the EPR criterion tells us that the horizontal spin is a real property of electron 2.

So…

- A measurement of the vertical spin state of particle 1 allows us to predict the vertical spin state of particle 2 with certainty, without disturbing 2. Hence by the EPR criterion, the vertical spin state is a real physical variable of 2.
- A measurement of the horizontal spin state of particle 1 allows us to predict the horizontal spin state of particle 2 with certainty, without disturbing 2. Hence by the EPR criterion, the horizontal spin state is a real physical variable of 2.

- As setting up a measurement of the horizontal state of particle 1 cannot erase or otherwise modify the reality of the vertical physical property, so, particle 2 must have *simultaneously*, real physical properties for both horizontal and vertical spins.
- Quantum theory, however, does not allow a state with precise values of *both* horizontal *and* vertical spins, as those variables are conjugate (the spin operators do not commute).
- Hence, quantum theory must be incomplete.

23.2.3 Bohr's Reply to the EPR Argument

> From our point of view we now see that the wording of the above-mentioned criterion of physical reality proposed by Einstein, Podolsky and Rosen contains an ambiguity as regards the meaning of the expression "without in any way disturbing a system." Of course there is in a case like that just considered no question of a mechanical disturbance of the system under investigation during the last critical stage of the measuring procedure. But even at this stage there is essentially the question of an influence on the very conditions which define the possible types of predictions regarding the future behavior of the system.
>
> *N Bohr*[7]

Bohr's reply appeared within two months of the EPR paper's publication. His counterargument uses the contextuality of quantum theory and the principle of complementarity (see Chapter 18) to punch a hole in the EPR definition of reality.

According to Bohr, a measurement made on particle *A* may not *physically disturb* the state of particle *B*, *but it does set up the context for any usable information about B.*

To measure the momentum of *A* we must build a specific device. In essence, we have to let *A* come to rest by colliding with something that has a pre-determined momentum and see what change it makes. The transfer of momentum from *A* to the device measures *A*'s momentum. However, if we want to measure the momentum of the *device,* we find that the uncertainty principle gets in the way. Accurately measuring the momentum of the device destroys any information about *where* it is. In which case, *the device can't act as a position reference point as well.* This applies to *both* particles. A measurement of position needs a fixed reference point to work from. When we set up the equipment to measure the momentum of particle *A*, we automatically prevent that equipment from forming a precise reference point for a position measurement of *A* or *B*.

Let's say we then set out to measure the position of particle *B*. We can't compare the position of *B* with anything to do with particle *A as we can't fix the position of the equipment at B's end relative to the equipment measuring A*. We have effectively blurred the position of particle *B* by measuring the momentum of particle *A*. As Bohr puts it:

> If we choose to measure the momentum of one of the particles, we lose through the uncontrollable displacement inevitable in such a measurement any possibility of deducing from the behaviour of this particle the position of the … apparatus, and have thus no basis whatever for predictions regarding the location of the other particle.

A similar argument applies if we choose to measure the position of *A* instead of its momentum. Accurate position measurements rely on fixed objects to act as reference points. When particle *A* collides with a device designed to record its position, there is bound to be an exchange of momentum between the device and the particle. With the device fixed in place we can't measure that momentum exchange. We need an accurate fix of the position of the device to measure the position of *A*, so we can't be sure of the device's momentum. This ruins any potential momentum measurement we might try to make of *B*. Momentum can only be known relative to some object. A cup of tea on a table in front of us may be stationary relative to us, but as we are on a train passing through a station, the tea will be in rapid motion relative to people standing on the platform.

Bohr explains this in the following way:

By allowing an essentially uncontrollable momentum to pass from the first particle into the mentioned support, however, we have by this procedure cut ourselves off from any future possibility of applying the law of conservation of momentum to the system consisting of the … two particles, and therefore have lost our only basis for an unambiguous application of the idea of momentum in predictions regarding the behaviour of the second particle.

The key difference between the philosophical positions expressed by EPR and Bohr hinges on interpreting the phrase "without disturbing the system".

To Bohr a property, such as momentum or position, has a meaning *only in the context of the experimental equipment used to measure it.* Particle *B* may not encounter the equipment at all. In which case, any information we have about *B* is useless *as there is nothing we can do with it.* If we would like to use this information, e.g., to predict where *B* will be after a while, and to check that prediction experimentally, then *B* must come in contact with the measuring equipment used for *A*, or some other apparatus has been linked to that equipment.

As far as complementarity is concerned, Bohr insisted that the concepts of momentum and position work only within a specific experimental context. The two particles can either live in the same context, in which case limited comparisons of the same properties can be made, or not, in which case *we have two separate bundles of information that can't be related to one another.*

We have to imagine what Bohr's response to the Bohmian version of the EPR argument might be, but it would doubtless hinge on denying the EPR assertion that setting up to measure a horizontal spin component has no impact on the reality of a vertical spin component. To Bohr, and other members of the Copenhagen school, a physical variable only has 'reality' within the context of a measurement tuned to reveal that variable.

23.2.4 Einstein and Bohr

These two great physicists were never able to reach a common ground; their philosophical viewpoints were too different.

For Einstein, the probabilities in quantum mechanics were unacceptable in a fundamental theory. He had himself pioneered the use of statistics in calculating the behaviour of large numbers of particles (e.g., in Brownian motion, see Chapter 17), but this was different. In that context, the need for statistics was the inevitable consequence of having too many particles, with their associated positions and momenta, to deal with in a calculation. In principle, a sufficiently powerful calculating device could cope (aside from the issues to do with the chaos in the system) where a human could not and remove the need for statistical calculations.

Bohr and his followers were convinced that the probabilities in quantum theory are *fundamental*, not a consequence of our inability to measure or calculate.

Einstein believed in a layer of reality underneath quantum theory that contains 'hidden variables'—properties of particles that we haven't yet discovered. Think of it like this. When you look at an old-fashioned mechanical clock face, you can see the hands going around, but you may not be able to see the hidden mechanisms driving them. A careful study of the hands as they move might enable us to figure out the properties of the clock, but we would have a much better understanding if we could get inside to see the workings. In Einstein's view, quantum mechanics is a theory of the hands; we need another theory that looks inside the mechanism. At the moment we can't see the 'mechanism' inside quantum systems, as we've not discovered the 'hidden variables' involved.

Without any technique for measuring these hidden variables (if they exist), we can't make definite predictions about the outcome of experiments. When we repeat experiments, we are effectively averaging over the various (unknown) possible values of these variables, which is why we end up using probability in quantum theory.

As Einstein expressed it in a letter to Schrodinger[8]:

But it seems certain to me that the fundamentally statistical character of the theory is simply a consequence of the incompleteness of the description … It is rather rough to see that we are still in the stage of our swaddling clothes, and it is not surprising that the fellows struggle against admitting it (even to themselves).

23.3 SCHRÖDINGER INTRODUCES ENTANGLEMENT

When two systems, of which we know the states by their respective representatives, enter into temporary physical interaction due to known forces between them, and when after a time of mutual influence the systems separate again, then they can no longer be described in the same way as before, viz. by endowing each of them with a representative of its own. I would not call that one but rather the characteristic trait of quantum mechanics, the one that enforces its entire departure from classical lines of thought. By the interaction the two representatives (or ψ -functions) have become entangled. To disentangle them we must gather further information by experiment, although we knew as much as anybody could possibly know about all that happened.

Erwin Schrödinger[9]

The term 'entanglement' probably entered the physics vocabulary as Schrödinger read an essay to the Cambridge Philosophical Society in October 1935. At the start of his article, Schrödinger argues that once two particles have interacted, the combined state of the two-particle system can no longer be written as a product of individual particle states:

$$\left|\text{Particle } A \text{ having interacted with } B\right\rangle \neq \left|A\right\rangle\left|B\right\rangle$$

The only way to write the combined state would be as an *entanglement* of individual states, preventing you from saying, with any certainty, which particle is in which state.

A state such as:

$$\left|\text{singlet}\right\rangle = \frac{1}{\sqrt{2}}\left(|U\rangle|D\rangle - |D\rangle|U\rangle\right)$$

is an entangled state as the first particle could be in either $|U\rangle$ or $|D\rangle$. In an entangled state, there is no way to describe the behaviour of one particle without referring to the other. The spin singlet state is a relatively simple example of an entangled state. Something more complex would be:

$$|\Psi\rangle = \sum_{n,m}\psi(x_1, x_2)|\phi_n\rangle|\varphi_m\rangle$$

In essence, any state which is not factorizable into distinct particle states is an entangled state.

Dirac saw the defining characteristic of quantum theory, that which marked it out as being totally different to classical physics, as the existence of *superposition*, i.e., that a quantum state can be constructed from a combination of states which would be mutually exclusive in classical physics:

$$|\psi\rangle = \sum_n a_n \mid n\rangle$$

Schrödinger saw *entanglement* as the defining characteristic of quantum mechanics. He felt it was of *sinister* importance as it is involved in every act of measurement.

23.3.1 ENTANGLEMENT AND MEASUREMENT

By the interaction the two representatives (or ψ -functions) have become entangled. To disentangle them we must gather further information by experiment, although we know as much as anyone could know about all that happened. Of either system, taken separately, all previous knowledge may be entirely lost, leaving us but one privilege: to restrict experiments to one only of the two systems. After reestablishing one representative by observation, the other one can be inferred simultaneously. In what follows the whole of this procedure will be called disentanglement. Its sinister importance is due to its being involved in every measuring process and therefore forming the basis of the quantum theory of measurement, threatening us thereby with at least a regressus in infinitum, since it will be noticed that the procedure itself involves measurement.

Erwin Schrodinger[10]

Any measuring apparatus is a material object, and so composed of atoms with their constituent protons, neutrons and electrons. In principle, therefore, the measuring device should fall under the purview of the same laws of physics as those constituent atoms. Consequently, the apparatus itself should be describable from within quantum theory. It might be that the details are too mathematically complex to deal with in practice, but that is not undermining the principle.

As an example, consider an SG apparatus set to measure the horizontal spin state of an electron. In broad terms, this apparatus can be in one of three quantum states:

$|\varphi_n\rangle$ n – neutral, the idling state before a measurement, or after having been reset

$|\varphi_L\rangle$ L – the aparatus is displaying a result, which is LEFT

$|\varphi_R\rangle$ R – the aparatus is displaying a result, which is RIGHT

If we know that an electron approaching the SG device is in the state $|L\rangle$ (as we grabbed it from the LEFT channel of a previous magnet), then the combined quantum state of the electron and apparatus prior to the measurement is $|\varphi_n\rangle|L\rangle$. After the electron interacts with the device, if it is any good at taking the measurement, the state must have evolved into $|\varphi_L\rangle|L\rangle$. This process is governed by the Schrödinger equation, or equivalently the $\hat{U}(t)$ operator:

$$\hat{U}(t)\left(|\varphi_n\rangle|L\rangle\right) = |\varphi_L\rangle|L\rangle$$

No states have collapsed in this process. For the first time, we are treating the measuring device as something that can be described by quantum theory as well, rather than the bold statements made in previous chapters, that a measurement collapses the state. This is just a normal evolution, governed by linear quantum equations.

Similarly, if the arriving electron is in the state $|R\rangle$, we must have:

$$\hat{U}(t)\left(|\varphi_n\rangle|R\rangle\right) = |\varphi_R\rangle|R\rangle$$

If we do not know the initial spin state of the electron, we need to write:

$$|\psi\rangle = a|L\rangle + b|R\rangle$$

with $a^2 + b^2 = 1$, and then the interaction with the measuring magnet becomes:

$$\hat{U}(t)\left(\{a|L\rangle + b|R\rangle\}|\varphi_n\rangle\right) = a|\varphi_L\rangle|L\rangle + b|\varphi_R\rangle|R\rangle$$

The device and the electron have become entangled with each other.

There is the possibility, indeed likelihood, that the measuring apparatus is not completely efficient (some electrons pass through without being detected) nor completely accurate (some results are wrongly registered). This would make the states after measurement:

$$|\chi_1\rangle = p|\varphi_N\rangle|L\rangle + q|\varphi_R\rangle|L\rangle + r|\varphi_L\rangle|L\rangle$$

$$|\chi_2\rangle = P|\varphi_N\rangle|R + Q|\varphi_R\rangle|R + R|\varphi_L\rangle|R\rangle$$

so that:

$$\hat{U}(t)\left(\left\{a\left|L\right\rangle + b\left|R\right\rangle\right\}\left|\varphi_n\right\rangle\right)$$

$$= a\left\{p\left|\varphi_N\right\rangle\left|L\right\rangle + q\left|\varphi_R\right\rangle\left|L\right\rangle + r\left|\varphi_L\right\rangle\left|L\right\rangle\right\} + b\left\{P\left|\varphi_N\right\rangle\left|R\right\rangle + Q\left|\varphi_R\right\rangle\left|R\right\rangle + R\left|\varphi_L\right\rangle\left|R\right\rangle\right\}$$

$$= \left|\varphi_N\right\rangle\left\{ap\left|L\right\rangle + bP\left|R\right\rangle\right\} + \left|\varphi_R\right\rangle\left\{aq\left|L\right\rangle + bQ\left|R\right\rangle\right\} + \left|\varphi_L\right\rangle\left\{ar\left|L\right\rangle + bR\left|R\right\rangle\right\}$$

We now come to a puzzling and rather worrying conclusion. If we take the seeming natural step of describing a measuring apparatus as a quantum system, then we can never, within quantum mechanics, see the result of a measurement. No state collapse takes place, just entanglement between quantum states. This is why state collapse has to be added in as a non-linear (and so not quantum, at least in our current understanding) additional assumption. The process of state collapse turns potential (*potentia* in Heisenberg's term) into an actual instance. It would appear that quantum theory requires something outside of itself to render reality from the potential of the quantum state.

This is one of the major philosophical challenges to the interpretation of quantum theory. As we shall see, some approaches try to work within quantum theory to describe state collapse, while some deny that it ever happens!

One, reasonably obvious, step is to add a further mechanism that records the value displayed on the measuring device, in the hope that might produce state collapse. However, if this device is part of our natural world, then it would also be quantum describable and all that happens is that the state of this apparatus becomes part of the entanglement as well.

There is, however, a further intriguing variation to mention now, for further discussion in later chapters. At some point in the chain of observation, apparatus to apparatus, the results of the measurement have to be observed by a human being, if any science is to happen. Now, being material objects ourselves, we are also presumably subject to quantum laws, in which case the observer might have states $\left|H_n\right\rangle$, $\left|H_L\right\rangle$, $\left|H_R\right\rangle$ corresponding to not having seen the result, seeing the result to be LEFT and seeing the result to be RIGHT respectively. Having interacted with the measuring apparatus, either by exchange of photons for visual recording or having the device interact audibly with the observer etc., those states become part of the entanglement:

$$a\left|\varphi_L\right\rangle\left|L\right\rangle\left|H_L\right\rangle + b\left|\varphi_R\right\rangle\left|R\right\rangle\left|H\right\rangle_R$$

However, arguably this is still not a recorded observation. There are still the philosophically and neurologically, extremely deep waters of how sensory impulses get translated into the qualia of conscious awareness. Some physicists and philosophers have seen this as a marker for where the quantum chain of entanglement ends, and state collapse happens; when the observation enters human consciousness.

23.3.2 The Sorry Tail of Schrödinger's Cat

> One can even set up quite ridiculous cases. A cat is penned up in a steel chamber, along with the following device (which must be secured against direct interference by the cat): in a Geiger counter there is a tiny bit of radioactive substance, so small, that perhaps in the course of the hour one of the atoms decays, but also, with equal probability, perhaps none; if it happens, the counter tube discharges and through a relay releases a hammer which shatters a small flask of hydrocyanic acid. If one has left this entire system to itself for an hour, one would say that the cat still lives if meanwhile no atom has decayed. The psi-function of the entire system would express this by having in it the living and dead cat (pardon the expression) mixed or smeared out in equal parts.
>
> *Erwin Schrodinger*[11]

In fact, the mere act of opening the box will determine the state of the cat, although in this case there were three determinate states the cat could be in: these being Alive, Dead, and Bloody Furious.

T Pratchett[12]

The story of Schrödinger's cat has assumed the status of a quantum mechanical fairy tale. It's amazing how thoroughly this cat has entered the quantum zeitgeist, as there is only one reference to the unfortunate creature in the whole of Schrödinger's written output.

Schrödinger constructed this gedankenexperiment in order to illustrate how strange quantum states are (Figure 23.1) and also the 'sinister' nature of entanglement, as it spreads along the observation chain. The point is to apply standard quantum mechanical procedures to produce an outcome that would be visible at a classical level without recourse to measurement instruments.

The radioactive source exists in a mixed state (superposition) $|\psi\rangle = a|\text{decayed}\rangle + b|\text{not decayed}\rangle$ while it is unobserved. The hammer and flask combination acts as a measuring device observing the radioactive source, so the state of the flask is entangled with that of the source:

$$|\Psi\rangle = a|\text{broken}\rangle|\text{decayed}\rangle + b|\text{intact}\rangle|\text{not decayed}\rangle$$

As breaking the flask releases poison gas, which would kill the cat, the existential state of the cat is also ultimately entangled into the situation:

$$|\Phi\rangle = a|\text{dead}\rangle|\text{broken}\rangle|\text{decayed}\rangle + b|\text{alive}\rangle|\text{intact}\rangle|\text{not decayed}\rangle$$

Having a human being lift the lid on the steel case, to observe whether the cat is alive or dead, does not improve the situation, as the human's state entangles into $|\Phi\rangle$. State collapse does not happen at any stage in the observation chain. That being the case, we are forced to conclude that the cat is in a quantum state of being alive/dead, while unobserved.

FIGURE 23.1 In Schrödinger's cat experiment, a small radioactive device is linked to a trigger controlling the release of poison gas. After a period of time, the quantum state of the radioactive device is a superposition of states ($= a|\text{decayed}\rangle + b|\text{not decayed}\rangle$). Both the state of the gas and ultimately the state of the cat are entangled with this, so it is not possible to tell, in the quantum description, if the cat is alive or dead.

It is one thing to say that an electron is LEFT/RIGHT or that a photon is traversing two paths through an apparatus, it is quite something else to say that a cat, or any animal, is alive/dead.

Bohr's view on this, developed with Heisenberg and others in the Copenhagen school, was to assume some macroscopic scale at which classical physics replaces quantum theory. Hence, we have to describe measurement devices at this level as being classical, which stops the entanglement and triggers state collapse.

Modern physicists are starting to take a different view, as successful experimental attempts to view quantum effects are reaching ever larger sizes. It is starting to look as if there is no 'cut-off' at which quantum theory is replaced by classical mechanics.

23.4 JOHN BELL AND BOHM'S EPR

> In a theory in which parameters {hidden variables} are added to quantum mechanics to determine the results of individual measurements, without changing the statistical predictions, there must be a mechanism whereby the setting of one measuring device can influence the reading of another instrument, however remote. Moreover, the signal involved must propagate instantaneously, so that such a theory could not be Lorentz invariant {locally causal}.
>
> *John Bell*[13]
>
> *[my additions]*

While Einstein was chastising the other founding fathers for, in his view, surrendering too soon and accepting quantum randomness as fundamental, rather than continuing to seek an underlying deterministic 'hidden variable' theory, the debate was largely philosophical. There was no experimental clue that could potentially lead to the hidden variables, and quantum theory was steadily gaining traction in wider and wider areas of physics. Then in 1964, John Bell developed a theorem which placed severe constraints on the nature of any putative deterministic hidden variable theory (hereafter just *hidden variable theory*). In essence, if such a theory were to successfully reproduce the results of quantum theory, it would have to violate another of Einstein's cherished assumptions about the way the world works: *local causality*.

According to Einstein's theory of relativity, no material object can move faster than the speed of light. Consequently, any event has a radius of influence surrounding it, which widens over time. Say event A happens at $(x, t) = (x_A, t_A)$. Any other event, B, happening at $(x, t) = (x_B, t_B)$ lies outside the radius of influence of A if $(x_B - x_A) > c(t_B - t_A)$. In the terminology of relativity, two such events have *spatial separation*. If a photon emitted from A could reach the location of B prior to that event happening, then B could *in principle* be influenced by A. In which case, $(x_B - x_A) \leq c(t_B - t_A)$ and the two events have *temporal separation*. Local causality restricts the causal consequences of an event to within its radius of influence, i.e., it is the assumption that causality is constrained by the speed of light.

Bell wondered what would happen in the context of an entangled EPR-type state, e.g., the singlet state used in the Bohm argument, if the measurement made on particle B was spatially separated from the measurement made on A. Quantum theory would have the state collapsing at the moment the measurement was made on A. Not only that, but the theory also demands that the collapse happens *everywhere across space at the same moment*. While this appears, at first glance, to be a direct violation of relativity, *there is no material object carrying the information of the state collapse*, so in principle, it can happen everywhere instantaneously.

However, if there is a hidden variable theory underneath quantum mechanics, and these variables influence the outcome of experiments, then to get EPR correlations between measurements made on entangled particles, Bell could see that there would be some contexts in which data had to flow from A to B faster than the speed of light.

23.4.1 Bell's Argument

Suppose we prepare two particles, A and B in a singlet state:

$$|\psi\rangle = \frac{1}{\sqrt{2}}\left(|\Theta+\rangle_A|\Theta-\rangle_B - |\Theta-\rangle_A|\Theta+\rangle_B\right)$$

where $|\Theta+\rangle$ is spin UP along some arbitrary SG angle, Θ, and $|\Theta-\rangle$ is spin DOWN along the same angle. Such a state, $|\psi\rangle$, is rotationally invariant, which means that is a singlet for *any* angle chosen (see the derivation in 23.2.2 for an illustration of this).

The particles are allowed to separate in linearly opposite directions. We place a Stern-Gerlach device in the path of A (SG1) and a separate device (SG2) in the path of B. Also, we arrange for SG2 to be much further from the origin of the particles than SG1, ensuring that the measurement of A taking place at SG1 is temporally prior and spatially separate to the measurement of B at SG2 (Figure 23.2). Importantly, we are *not* specifying that the two devices are aligned to the same angle, nor that they are aligned with the state angle, Θ. In fact, we will want them at different angles.

Particle A now reaches SG1, which has been set to some arbitrary angle ϑ_1. As a result, A will emerge from the device either along the UP channel for this angle or along DOWN for that angle. Over a lot of A B singlet pairs, the distribution UP/DOWN at SG1 will be 50:50, as a singlet state retains its nature when rotated into any angle.

When particle B arrives at SG2 we will find a perfect *anti-correlation* between the results, if the angle of SG2 is also set to ϑ_1, i.e., if A is UP then B will be DOWN and vice versa.

If SG2 is instead set at 90° to SG1, then the two sets of results will be *statistically independent*; whatever the result at SG1(ϑ_1) the UP/DOWN at SG2$\left(\vartheta_1 + \frac{\pi}{2}\right)$ is unpredictable, with 50:50 probability (compare this to LEFT/RIGHT and UP/DOWN cases for vertical and horizontal).

The interest lies in the intermediate angles.

As a result of the measurement SG1(ϑ_1), particle A will be either UP or DOWN along that angle. This makes particle B DOWN or UP along the same angle. However, if SG2 is set to a different angle ϑ_2 then the emergence of B along UP/DOWN for that angle is random, *but statistically correlated to* ϑ_1.

If particle A emerges from SG1(ϑ_1) in the state $|\vartheta_1+\rangle_A$, then particle B must be $|\vartheta_1+\rangle_B$. To find out what happens to B when it arrives at SG2(ϑ_2) we need to expand $|\vartheta_1-\rangle_B$ using the basis $\left\{|\vartheta_2+\rangle_B, |\vartheta_2-\rangle_B\right\}$:

$$|\vartheta_1-\rangle_B = \cos\left(\frac{\vartheta_2-\vartheta_1}{2}\right)|\vartheta_2+\rangle_B + \sin\left(\frac{\vartheta_2-\vartheta_1}{2}\right)|\vartheta_2-\rangle_B$$

(from Section 3.3.6) showing that the probability of UP along ϑ_2 is $\cos^2\left(\frac{\vartheta_2-\vartheta_1}{2}\right)$ and that for DOWN is $\sin^2\left(\frac{\vartheta_2-\vartheta_1}{2}\right)$, i.e., the probabilities are not equal and depend on $\vartheta_2 - \vartheta_1$.

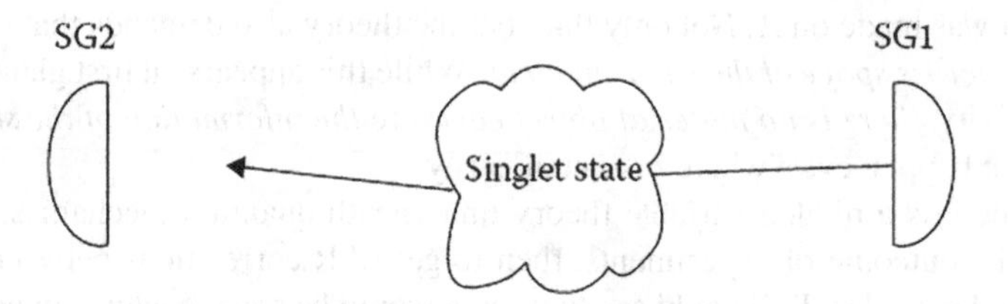

FIGURE 23.2 In this experiment two particles, created in a singlet state, are heading toward a pair of S–G magnets. The right-hand side S–G magnet is rather closer to the place where the particles were created.

The same expansion for the case $|\vartheta_1 -\rangle_A, |\vartheta_1 +\rangle_B$ is:

$$|\vartheta_1 +\rangle_B = \sin\left(\frac{\vartheta_2 - \vartheta_1}{2}\right)|\vartheta_2 +\rangle_B - \cos\left(\frac{\vartheta_2 - \vartheta_1}{2}\right)|\vartheta_2 -\rangle_B$$

(for the case $\vartheta_2 - \vartheta_1 = \pi/2$, then we have $\sin\left(\frac{\pi}{4}\right) = \cos\left(\frac{\pi}{4}\right) = 1/\sqrt{2}$ giving the 50:50 distribution alluded to earlier).

The challenge for a locally causal hidden variable theory is to reproduce the probabilities, or the expectation values, for the intermediate angle cases.

23.4.2 A Toy Model

Let us assume the existence of some hidden variable, λ, which, together with the angles of the SG magnets, determines the measured spin states of the particles. The hidden variable is assumed to be a state variable of a two-particle system, the value of which is set at some point prior to the particles taking part in the experiment. If we had knowledge of this variable's value, we could predict the outcome of a spin measurement at any angle ϑ, via some currently unknown physics relating λ and ϑ. We also assume that the value of λ is unchanged during the experiment.

To encode the imagined physics, we assume two functions $A(\vartheta_1, \lambda)$ and $B(\vartheta_2, \lambda)$, one for each particle. These functions take the values ± 1 to indicate spin UP/DOWN for particle A or B according to the value of λ and the SG angle ϑ. Local causality is enshrined in these functions, as they are only dependent on the hidden variable *and the local angle of the SG magnet their particle will interact with.*

Particle A now interacts with $SG1(\vartheta_1)$ producing the value +1 or −1.

$$A(\vartheta_1, \lambda) = \pm 1$$

Sometime later, particle B encounters SG2 set to the same angle, ϑ_1. If the theory is to reproduce the quantum mechanical result, then:

$$B(\vartheta_1, \lambda) = \mp 1$$

in exact anti-correlation to the result at SG1, which must be true no matter what angle ϑ_1 is chosen for both magnets. This does not present a problem. It is easy to imagine two functions which produce opposite signs given the same values of input variables.

Now let us change the situation. Particle A still encounters $SG1(\vartheta_1)$, but particle B arrives at SG2 with the angle set to $\vartheta_1 + \frac{\pi}{2}$. Now the *quantum mechanical* prediction is that B will be UP/DOWN at this angle with a 50:50 distribution, uncorrelated to the result at SG1. Again, this does not present a problem. We could imagine that λ works with the angle settings to produce anticorrelated results when the angles are equal but are effectively randomized when the second angle is different.

However, these are not the only two possible experiments.

If particle A is measured by $SG1(\vartheta_1)$ and B by $SG2(\vartheta_2)$ where $\vartheta_1 \neq \vartheta_2$ and $\vartheta_1 \neq \vartheta_1 + \frac{\pi}{2}$, then the function $B(\vartheta_2, \lambda)$ has to produce results which *are statistically correlated with A* and *dependent on the value of ϑ_1, even though this is not an input variable of the function.*

We could extend the functions to $A(\vartheta_1, \vartheta_2, \lambda)$ and $B(\vartheta_1, \vartheta_2, \lambda)$, breaking locality, but this would require some collusion, by which SG1 would be 'aware' of the angle setting of SG2 and vice versa. Furthermore, with modern equipment, it is possible to construct an experiment where the magnets are set to random angles while the particles are in flight. In this scenario, if the quantum mechanical results hold true, the hidden variable theory would have to incorporate collusion between the magnets in a dynamic way, i.e., not that they had settled into knowing about each other's settings over time, before the start of the experiment. That dynamic mechanism *would also have to act faster than the speed of light, given the spatial separation of the measurement events.*

In essence, this is Bell's argument. By looking at a range of angle settings, it is difficult for a local hidden variable theory to keep up with the quantum predictions. That's not to say that the quantum predictions are correct. That must be tested by experiment. For that to be possible, this qualitative argument must be quantified.

23.4.3 Bell's Formula

Working with a toy local hidden variable theory, like the one outlined in the previous section, Bell was able to calculate a statistical limit on the correlations that such a theory could produce between spatially separated measurements (Figure 23.3).

He also demonstrated that quantum mechanics predicted correlations that lay outside those limits.

It was then possible to construct experiments to measure the *actual correlations* to see if they matched the quantum predictions or adhered to the limits of a local hidden variable theory.

Following Bell's argument, the first step is to define the nature of the correlations that are to be measured and calculated using hidden variable theory and, separately, quantum mechanics.

Experimental Correlations, S_e

We imagine an experiment that takes place as before, with two particles A and B and detectors SG1 and SG2. Each detector records spin UP (+) or spin DOWN (−) at a specific angular setting. We release a pair of particles, prepared in a singlet state, to arrive at their respective detectors. Each release and subsequent measurement at SG1 and SGE we call an *event*.

FIGURE 23.3 John Stuart Bell of CERN. The CHSH inequality and the outline of an experiment to test it can be seen on the board. (Image credit: CERN.)

N_{++}	Number of times SG1 is + and SG2 is −
N_{--}	Number of times SG1 is − and SG2 is −
N_{-+}	Number of times SG1 is − and SG2 is +
N_{+-}	Number of times SG1 is + and SG2 is −

Sometimes both detectors will record UP in the same event, sometimes both DOWN and occasionally an UP/DOWN combination. The number of occasions each of these occurrences takes place is recorded, as below:

The *experimental measure* of the *statistical correlation between the detectors* at angles ϑ_1, ϑ_2 is defined as:

$$E_e(\vartheta_1, \vartheta_2) = \frac{\Sigma \text{product of spin values} \times \text{number of occurrences}}{\text{Total number of events}}$$

$$= \frac{(+1)(+1)(N_{++}) + (-1)(-1)(N_{--}) + (-1)(+1)(N_{-+}) + (+1)(-1)(N_{+-})}{N_{++} + N_{--} + N_{-+} + N_{+-}}$$

$$= \frac{(N_{++}) + (N_{--}) - (N_{-+}) - (N_{+-})}{N_{++} + N_{--} + N_{-+} + N_{+-}}$$

The subscript 'e' denotes that this is a measure derived from experimental data, rather than theoretical calculation.

Clearly, if each detector was doing its own thing, with no relationship between their results, then $E_e(\vartheta_1, \vartheta_2) = 0$.

To quantify things further, a standard experimental protocol is defined, where each magnet can adopt one of two possible angles ϑ_1, ϑ_1' for SG1 and ϑ_2, ϑ_2' for SG2. After a run of events, with the detectors at each possible setting, we will have $E_e(\vartheta_1, \vartheta_2)$, $E_e(\vartheta_1, \vartheta_2')$, $E_e(\vartheta_1', \vartheta_2)$, $E_e(\vartheta_1', \vartheta_2')$, each of which would be zero if there was no connective physics. In practice, either quantum mechanics or its hidden variable successor is correlating the results at SG1 with those at SG2, either via the collapse of entangled states (quantum theory) or pre-set values of λ (hidden variables). The extent to which these correlations are present is not revealed by the separate E_e, but via a parameter derived from them:

$$S_e = \left| E_e(\vartheta_1, \vartheta_2) - E_e(\vartheta_1, \vartheta_2') \right| + \left| E_e(\vartheta_1', \vartheta_2) + E_e(\vartheta_1', \vartheta_2') \right|$$

The reason why this combination of the E_e is chosen will become apparent shortly.

The quantity, S_e, is the experimentally determined measure of the correlations between measurements at SG1 and SG2, given singlet state particles and a range of angular settings. As we will see, various experiments have measured S_e. Our task now is to see what a hidden variable theory would predict for these correlations, given that it must reproduce quantum theory. As we will see, dealing with a set of possible angles using a fixed λ per event limits the correlations that can be provided.

Local Hidden Variable Correlations, S_H

To calculate the correlations in a local hidden variable theory, we assume that the variable λ is randomly set for the two-particle system involved in each event. The likelihood of finding a given value of λ is determined by the probability density $\rho(\lambda)$. If this were a fully worked out theory, then having some function for $\rho(\lambda)$ would be crucial[14]. However, the argument at hand is to determine the limits of *any* hidden variable theory, so we do not need to know $\rho(\lambda)$ in any explicit mathematical form. However, like any probability density, we must have:

$$\int_{\{\text{range of possible values for } \lambda\}} \rho(\lambda)d\lambda = 1$$

Then:

$$E_H(\vartheta_1,\vartheta_2) = \int_{\{\lambda\}} A(\vartheta_1, \lambda)B(\vartheta_2, \lambda)\rho(\lambda)d\lambda$$

is the predicted correlation between SG1 and SG2 at a given set of angles across all possible values of λ. Note that I have abbreviated 'range of possible values of λ" to $\{\lambda\}$ in the integration limits. Also, the subscript 'H' denotes a measure derived from theoretical calculations based on a hidden variable theory.

Next, on a different tack, we write:

$$a = A(\vartheta_1,\lambda) \qquad a' = A(\vartheta_1',\lambda)$$
$$b = B(\vartheta_2,\lambda) \qquad b' = A(\vartheta_2',\lambda)$$

simply for notational compactness. Now we tabulate all the possibilities for b and b', assuming the angles to be fixed, but the values of λ to be different in each row.

	$b = B(\vartheta_2,\lambda)$	$b' = A(\vartheta_2',\lambda)$	$b+b'$	$b-b'$
λ = 1st value	+1	+1	+2	0
λ = 2nd value	−1	−1	−2	0
λ = 3rd value	+1	−1	0	2
λ = 4th value	−1	+1	0	−2

Hence, for any value of λ one of $(b+b')$ or $(b-b')$ is ±2, while the other is zero. This impacts the following construction:

$$|ab - ab'| + |a'b + a'b'| = |a(b-b')| + |a'(b+b')|$$

which then can only be: $2a$ or $2a'$.
(our earlier choice for the combination is now revealed).

Finally, we build a combination of correlations to predict S_H, *to compare with* S_e.

$$S_H = |E_H(\vartheta_1,\vartheta_2) - E_H(\vartheta_1,\vartheta_2')| + |E_H(\vartheta_1',\vartheta_2) + E_H(\vartheta_1',\vartheta_2')|$$

$$= \left|\int_{\{\lambda\}} ab\rho(\lambda)d\lambda - \int_{\{\lambda\}} ab'\rho(\lambda)d\lambda\right| + \left|\int_{\{\lambda\}} a'b\rho(\lambda)d\lambda + \int_{\{\lambda\}} a'b'\rho(\lambda)d\lambda\right|$$

$$= \int_{\{\lambda\}} \{|ab - ab'| + |a'b + a'b'|\}\rho(\lambda)d\lambda$$

$$= \int_{\{\lambda\}} 2a\rho(\lambda)d\lambda \quad \text{or} \quad \int_{\{\lambda\}} 2a'\rho(\lambda)d\lambda$$

$$\leq \int_{\{\lambda\}} 2\rho(\lambda)d\lambda$$

$$\le 2\int_{\{\lambda\}} \rho(\lambda)d\lambda$$

$$\le 2$$

We now have the Clauser, Horne, Shimony and Holt (CHSH) parameter[15], which is a generalization of Bell's original result:

$$S_H = \left|E_H(\vartheta_1,\vartheta_2) - E_H(\vartheta_1,\vartheta_2')\right| + \left|E_H(\vartheta_1',\vartheta_2) + E_H(\vartheta_1',\vartheta_2')\right| \le 2$$

for a local hidden variable theory.

If a fixed hidden variable, λ, albeit randomly distributed event-by-event, is to account for the measurement results at SG1 and SG2 (along with their angular settings) then $S_e = S_H \le 2$.

It is worth recalling the fundamental limitation placed on the possible physics within our toy theory. The functions that determine the results at SG1 and SG2 are $A(\vartheta_1,\lambda)$ and $B(\vartheta_2,\lambda)$ and hence *locally dependent* on the fixed (for each event) value of λ and the angular setting of the SG apparatuses.

Quantum Mechanical Correlations, S_q

The quantum mechanical prediction is a little trickier to obtain. We start with a singlet state:

$$|\psi\rangle = \frac{1}{\sqrt{2}}\left(|+\rangle|-\rangle - |-\rangle|+\rangle\right)$$

and calculate the expectation value of the operator $\hat{\sigma}_n\hat{\sigma}_m$, which is an appropriately rotated combination of Pauli spin matrices:

$$E_q(\vartheta_1,\vartheta_2) = \langle\psi|\hat{\sigma}_m\hat{\sigma}_n|\psi\rangle$$

In this construction, $\boldsymbol{n}$ and $\boldsymbol{m}$ are (physical) unit-length vectors pointing along the measurement axis of SG1 and SG2 respectively, so $\boldsymbol{n} = (n_x, n_y, n_z)$ with $n_x^2 + n_y^2 + n_z^2 = 1$ and $\boldsymbol{m} = (m_x, m_y, m_z)$ with $m_x^2 + m_y^2 + m_z^2 = 1$. We then build $\hat{\sigma}_n$ using the vector dot product of $\boldsymbol{n}$ into $\boldsymbol{\sigma} = (\sigma_x, \sigma_y, \sigma_z)$. In other words:

$$\hat{\sigma}_n = n_x\sigma_x + n_y\sigma_y + n_z\sigma_z = \begin{pmatrix} n_z & n_x - in_y \\ n_x + in_y & -n_z \end{pmatrix}$$

Similarly

$$\hat{\sigma}_m = \begin{pmatrix} m_z & m_x - im_y \\ m_x + im_y & -m_z \end{pmatrix}$$

The expectation value is then:

$$\frac{1}{\sqrt{2}}\left\{\begin{pmatrix} 1 & 0 \end{pmatrix}_n \begin{pmatrix} 0 & 1 \end{pmatrix}_m - \begin{pmatrix} 0 & 1 \end{pmatrix}_n \begin{pmatrix} 1 & 0 \end{pmatrix}_m\right\} \begin{pmatrix} m_z & m_x - im_y \\ m_x + im_y & -m_z \end{pmatrix}_m$$

$$\times \begin{pmatrix} n_z & n_x - in_y \\ n_x + in_y & -n_z \end{pmatrix}_n \frac{1}{\sqrt{2}}\left\{\begin{pmatrix} 1 \\ 0 \end{pmatrix}_n \begin{pmatrix} 0 \\ 1 \end{pmatrix}_m - \begin{pmatrix} 0 \\ 1 \end{pmatrix}_n \begin{pmatrix} 1 \\ 0 \end{pmatrix}_m\right\}$$

where you have to be very careful about which matrix acts on which vector (I have used subscripts to indicate the pairings). If you allow the n matrix to act on its column vectors to its right, you get:

$$\frac{1}{\sqrt{2}}\left\{\begin{pmatrix} n_z \\ n_x + in_y \end{pmatrix}\begin{pmatrix} 0 \\ 1 \end{pmatrix} - \begin{pmatrix} n_x + in_y \\ -n_z \end{pmatrix}\begin{pmatrix} 1 \\ 0 \end{pmatrix}\right\}$$

Allowing the m row vectors to act on the m matrix to their right produces:

$$\frac{1}{\sqrt{2}}\left\{\begin{pmatrix} 1 & 0 \end{pmatrix}\begin{pmatrix} m_x + im_y & -m_z \end{pmatrix} - \begin{pmatrix} 0 & 1 \end{pmatrix}\begin{pmatrix} m_z & m_x - im_{yz} \end{pmatrix}\right\}$$

Multiplying the two together gives[16]:

$$E_q(\vartheta_1, \vartheta_2) = -(n_x m_x + n_y m_y + n_z m_z) = -\cos(\vartheta_1 - \vartheta_2)$$

Hence, the quantum parameter, S_q, (subscript 'q' to indicate a result based on quantum calculations) is:

$$S_q = \left|E_q(\vartheta_1, \vartheta_2) - E_q(\vartheta_1, \vartheta_2')\right| + \left|E_q(\vartheta_1', \vartheta_2) + E_q(\vartheta_1', \vartheta_2')\right|$$
$$= \left|-\cos(\vartheta_1 - \vartheta_2) + \cos(\vartheta_1 - \vartheta_2')\right| + \left|-\cos(\vartheta_1' - \vartheta_2) - \cos(\vartheta_1' - \vartheta_2')\right|$$

Now, if we happen to pick the combination $\vartheta_1 = \pi/2$, $\vartheta_1' = 0$, $\vartheta_2 = \pi/4$, $\vartheta_2' = -\pi/4$, the quantum prediction is[17]:

$$S_q = \left|-\cos(\pi/2 - \pi/4) + \cos(\pi/2 + \pi/4)\right| + \left|-\cos(0 - \pi/4) - \cos(0 + \pi/4)\right|$$
$$= \left|-\cos(\pi/4) + \cos(3\pi/4)\right| + \left|-\cos(-\pi/4) - \cos(\pi/4)\right|$$
$$= \left|-\frac{1}{\sqrt{2}} - \frac{1}{\sqrt{2}}\right| + \left|-\frac{1}{\sqrt{2}} - \frac{1}{\sqrt{2}}\right| = 2\sqrt{2}$$

clearly greater than the upper bound on S_H.

So, provided the quantum prediction is backed up by experiment, i.e., that $S_e = S_q \geq 2$, quantum theory cannot be replaced by a local hidden variable theory.

Its now all down to experiment to find S_e and see how it compares to S_H and S_q.

23.4.4 Aspect's Experiment

In 1982 a team led by Alain Aspect performed the first full test of Bell's inequality. With this experiment, for the first time, the detector angles could be changed while the particles were in flight, a key limitation of earlier tests. The experimental configuration is shown in Figure 23.4.

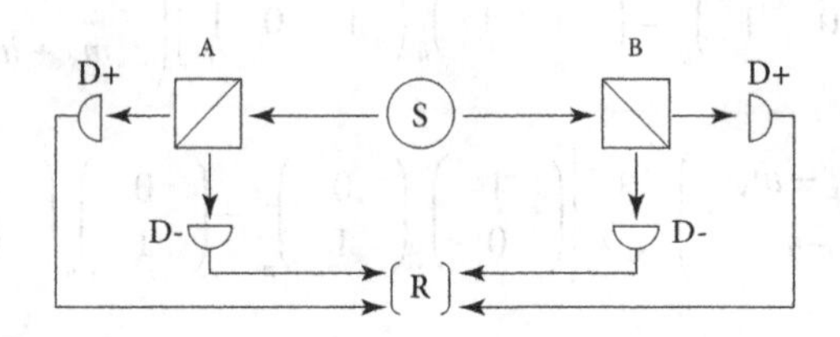

FIGURE 23.4 Aspect's experiment to test Bell's inequality.

The source, S, produced pairs of photons in an entangled state of polarization. The photons moved away from the source in opposite but co-linear directions. Detectors A and B were placed on either side of S and separated by 13 m. Arriving at the detectors, the equivalent of an SG magnet measurement was performed on each photon, separating their polarization states. The devices could be set to different angles: the left-hand one to either angle a or a', and the right-hand one to b or b'. As with an SG detector, the photon then travelled in one direction to be detected at $D+$ or along another to be registered at $D-$. The results from these detectors on either side of the source were then passed to R where coincidences accumulated to build the N_{++}, N_{--}, N_{+-}, N_{-+} counts needed to calculate S_e.

During an experimental run the angles at A and B were switched between $a = 0°$, $a' = 45°$ and $b = 22.5°$, and $b' = 67.5°$, a combination[18] giving $S_q = 2\sqrt{2}$. This value must be 'massaged' slightly to account for detector efficiency (sometimes they miss things), which reduces S_q to 2.70. Aspect's team measured $S_e = 2.697 \pm 0.015$ in agreement with the quantum-theoretical S_q and exceeding S_H.

Since Aspect's pioneering experiment, the basic protocol has been refined by further investigations which have closed more and more loopholes.

- **Detection loophole**
  Detectors cannot be 100% efficient, they miss particles from time to time. It is possible to compensate for this by adjusting the value of S_H, but that does not address the 'conspiracy' problem. In principle, all the events that are missed could be of a different statistical nature which, if included in the results, agree with the upper limit on S_H. Generally, the 'fair sample' hypothesis justifies the assumption that missed events are of the same nature as those that are seen. In any case, as technology has improved so detector efficiency for photons, for example, has increased from 20%–30% up to 93%.
- **Locality loophole**
  There needs to be a balance between the speed at which the particles travel, the distance between the detectors and the rate at which the detectors can take their measurements. The ideal is that the measurement at B lies outside the causal radius of A. Again, as technology has improved so has the distance between A and B without significant particle loss during flight. This loophole has effectively been closed by an experiment[19] performed with an AB separation of 18 km.
- **Memory loophole**
  If A and B remain in fixed locations then the so-called memory loophole is a potential issue, i.e., that conditions local to the two detectors could have an influence on successive measurements. There is also the possibility that the physical parameters in the locations might be varying with time. However, it can be shown that provided each successive measurement uses a different random setting, this should have no effect on the outcome. The first experiments[20] to close the detection, locality and memory loopholes in the same protocol were published in 2015.
- **Free choice loophole**
  One of the assumptions made in deriving S_H is that the detector settings a, a', b, b' are independent of λ (technically that $\rho(\lambda \mid ab) = \rho(\lambda)$). Conceivably, some event or sequence of events, at some point in the history of the universe could influence both the detector settings and the hidden variable and hence appear to violate the limits of a local hidden variable theory. Attempts to overcome this loophole by using quantum systems to generate random numbers that determine the detector settings are invalid on the grounds that we do not know if these systems are similarly determined. A modern (2018) experiment[21] used the light from two separate quasars to determine the detector settings during the experiment. Given the distance between the two quasars and their distance from Earth, an event that mutually determines both detector settings would have to have happened more than 7.9×10^9 years ago. While this is clearly not impossible, it does stretch credulity.

23.5 IMPLICATIONS

Einstein could not have seen where the EPR argument would lead, from Bohm's simpler and more practical version to Bell's analysis in terms of local hidden variables to the Aspect's experiment and the results that have been produced since. Quantum theory has survived every test thrown at it and thanks to Bell's argument, it now seems impossible that it will ever be replaced by a local hidden variable theory. Of course, the option of a hidden variable theory that doesn't obey local causality is still on the table.

Quantum theory survives Bell's test because entangled states collapse at the first measurement. Remarkably this collapse affects both particles in the entangled state no matter how far apart they may be. Bohr simply shrugged this off as a consequence of having to use the same experimental context for both particles, an argument that is logically compelling but doesn't seem to do justice to the remarkable properties of entangled states.

Bell's formula is broken by entangled quantum states, so if we want to take a realistic view of what's happening, state collapse must be a real physical change happening everywhere at once. Perhaps we ought to stop thinking of our two entangled particles as being separate objects at all. At the very least, *the experimental evidence (not philosophical musing) points to an underlying non-local reality.*

It does not take long for a rather disquieting thought to arise: since the start of the universe in the Big Bang, it is very likely that many, if not all, of the particles in the universe have interacted at some point in their history. It is entirely possible, then, that there is a level of reality that is completely entangled. This raises serious questions for standard scientific assumptions, such as reductionism[22] as well as challenging realism and the assumption that the world is composed of objects with an independent reality.

NOTES

1 Available at https://journals.aps.org/pr/pdf/10.1103/PhysRev.47.777.
2 I use P and Q here to match the terminology of the paper, making some of the references in the quoted sections more intelligible for the reader.
3 For ease of expression, I am blurring the distinction between a physical quantity and the operator that represents its measurement in quantum theory.
4 As mentioned previously, the delta function is not a 'function' in the true mathematical sense. It is more properly defied within the theory of distributions, its fundamental property being: $\int_{-\infty}^{\infty} f(x)\delta(x-a)dx = f(a)$. We will be following a more heuristic approach, as did physicists at the time.
5 David Bohm 1917–1992, theoretical physicists and philosopher. Professor of theoretical physics, Birkbeck College London. We will be discussing more of his work in later chapters.
6 If you would like a bit of practice in manipulating states, then try showing that all three triplet states are orthogonal to one another and to the singlet state.
7 N. Bohr, *Phys. Rev.*, 1935, **48**: 696–702.
8 Letter to Erwin Schrödinger, 1950, from Letters on Wave Mechanics: Correspondence with H. A. Lorentz, Max Planck, and Erwin Schrödinger, Philosophical Library/Open Road, ISBN: 1453204687.
9 E. Schrödinger, Discussion of probability relations between separate systems, *Proc. Cambridge Phil. Soc.*, **31**, 1935.
10 See note 9.
11 E. Schrödinger, The present situation in quantum mechanics, *Natur-wissenschaften,* 1935, **23**: 807–812, 823–828, 844–849.
12 T Pratchett, Lords and Ladies, Corgi, 2013.
13 J. S. Bell, On the Einstein Podolsky Rosen Paradox, *Physics* **1**(3): 195–200, 1964.
14 Note that the probabilities here are not an inherent quantum randomness in nature. They are covering our ignorance of the actual physics going on. Part of the hidden variable theory would be the ability to know that value in any given case.

15 J.F. Clauser, M.A. Horne, A. Shimony, R.A. Holt, Proposed experiment to test local hidden-variable theories, *Phys. Rev. Lett.*, **23**(15): 880–884, 1969.
16 This is another vector product relationship. The dot product between two unit-vectors is $\cos(\text{angle between the vectors})$.
17 This choice of angles happening to be the set that gives the largest value of S_q.
18 NB these are different angles to the ones used in the quantum prediction in the previous section as the aspect experiment is dealing with spin 1 photons, rather than spin ½ electrons, but they do lead to the same S_q value.
19 D. Salart, A. Baas, J.A.W. van Houwelingen, N. Gisin, H. Zbinden, Spacelike Separation in a Bell Test Assuming Gravitationally Induced Collapses. *Phys. Rev. Lett.*, **100**(22): 220404, 2008.
20 E.G. Hensen, et al., Loophole-free Bell inequality violation using electron spins separated by 1.3 kilometres, *Nature*, **526** (7575): 682–686, 2015.
21 Rauch et al., Cosmic Bell Test Using Random Measurement Settings from High-Redshift Quasars, *Phys. Rev. Lett.,* **121**, 080403, 2018 https://journals.aps.org/prl/pdf/10.1103/physrevlett.121.080403.
22 The assumption that a complex system can be completely understood by examining the properties of its component parts. Being part of a whole, does not influence the properties of the parts. The properties of entangled states challenge the notion at singlet state can still be regarded as a combination of separate particles.

24 Quantum Computing

24.1 HISTORICAL PERSPECTIVE

In some ways, the roots of interest in quantum computing date back to a 1965 observation by Gordon Moore[1], the co-founder of Intel. Having noted that the trend in manufacturing development had seen the number of transistors per square inch that could be fabricated on a chip doubling roughly every 18 months, Moore proposed that this would continue into the future[2]. *Moore's law*, as it came to be known, is a popular rule-of-thumb for computing development[3]. In 1971, the typical scale of a chip-based transistor was ~ 10 μm. As of 2020, that had reduced to ~ 5 nm. If that were to continue unabated, then eventually the individual components would be at an atomic scale, and hence quantum theory would be directly applicable to their function. It then becomes natural to explore the notion that quantum physics might enable a new way for computations to take place.

Early developments include Paul Benioff's publication of a quantum mechanical version of a Turing machine[4] (see Section 24.2) in 1980. In 1982 Richard Feynman produced an abstract model demonstrating how a quantum system could be used for computations[5]. A few years later, in 1985, David Deutsch[6] proposed the first universal quantum Turing machine[7] and in 1989 laid the foundations of the quantum circuit model[8]. Work on quantum algorithms accelerated in the 1990s with discoveries made by Deutsch, Josza and Simon. In particular, the 1994 publication of Shor's algorithm[9] prompted a huge uptake of interest. Using Shor's process, a quantum computer would be able to decompose any integer into its prime factors much more rapidly than a classical computer. While this does not sound like an especially significant mathematical process, it lies at the basis of all modern forms of encryption. The security of encoded information lies with the practical impossibility of using computers to crack the code, in essence because the crucial process, prime factoring, would take too long. With the prospect that quantum computers could remove the time logjam, their practical and theoretical development started to receive significant funding.

Since 2000, developments in this area have seen remarkable growth. New forms of algorithms have been developed to run on quantum computers and progress has been made towards their practical realization. A variety of different systems are being explored including cold ion traps, photon optics, along with solid state and condensed matter versions.

Despite the tremendous growth in the field, we await any form of general theoretical demonstration that quantum computers would outperform their classical counterparts in a range of tasks, rather than one or two specific ones. Also, there is, as yet, no guarantee that the practical development of a working quantum computer can be scaled up to the point where the algorithms under development would run in a real system.

24.2 THE FUNDAMENTALS OF DIGITAL COMPUTING

In 1900, the great mathematician David Hilbert[10] published a set of 23 outstanding issues or problems in mathematics, which did much to set the agenda for mathematical research in the 20th century[11].

Hilbert believed that mathematics could and should be set up as a *formal system*, starting from a set of agreed axioms and using logical methods to develop the proofs of mathematical conjectures from those axioms. In that sense, mathematics was potentially 'mechanical'; a proof could be derived by a set of logical moves performed in a defined sequence starting from an input, the axioms, and ending up with a proposition, which was the statement to be proven[12]. Each particular set of moves in sequence, appropriate to the problem at hand, was an *algorithm*. In its original

DOI: 10.1201/9781003225997-28

formulation, Hilbert's 10th problem asked for an algorithm that would decide if an example of a specific type of equation could be solved. However, it developed into something with broader aims; is there a general algorithmic process that will decide if any given mathematical proposition is true or false?

In 1936, Alan Turing[13] published a paper[14] in which he structured the problem by imagining a collection of literal mechanisms, *Turing machines*, each one being built to execute a specific algorithm. He envisioned these machines as all following the same basic construction:

- The machine would have a finite number of distinct internal states
- The machine could only be in one of these states at any one time
- The machine had access to a limitless supply of 'tape' divided into discrete 'cells' which can be blank or contain a character. The machine can read the character from a cell and write characters into cells.
- The machine was able to move the tape back and forth so that it could read and write to and from different sections of the tape[15].
- The machine would have the options to change state, write a character and move the tape in response to reading a character off the tape.

Based on this specification, Turing was able to demonstrate the existence of a *Universal Turning machine*, a mechanism capable of acting in lieu of any of the hard-wired machines. The Universal machine would read its instructions from the tape, know which algorithm it was to perform and access the data from the tape as well. In the Universal machine, we see the mathematical and structural foundations of the modern digital computer.

From here, Turing was able to compose an argument, involving the Universal machine, that ended in a paradox. The only way of avoiding the paradox was to deny the existence of one key algorithm that acted as a keystone in the argument. In this way, Turing was able to answer Hilbert's 10th problem – there is at least one properly framed mathematical problem that cannot be solved algorithmically.

In the process of addressing one of Hilbert's problems, Turing laid the foundations for digital computing. He went on to construct a range of devices, first for cracking Nazi codes during The War, and then later as research projects.

24.2.1 A Bit More Information

As is well known, conventional computers work, at root, in binary, with the basic unit of information or 'bit' being a state that can be set to 1 or 0. A *bitstring* is a sequence of such bits transmitted or operated upon in sequence or simultaneously (in parallel). Whenever we use a computer to email, surf the internet, do complex calculations in quantum chromodynamic field theory, or write the occasional book, underneath the high-level interface between us and the machine, it is working away crunching bits.

In the mathematical-scientific discipline of *Information Theory*, the bit is also taken as the basic unit, or atom if you like, of information.

24.2.2 Logic Gates

In classical computing, the process of 'crunching bits' is carried out by *logic gates*. These are physically realized electronic circuits within chips. They act on voltage levels, with +5 V representing a bit value of 1 and 0 V a bit value of 0. A logic circuit is then strung together from combinations of gates.

There are a variety of different types of logic gate, but all of them can be constructed out of combinations of NAND (Not-AND) gates. The function of an idealized gate is summarized in a *Truth*

Input		Output
A	B	Q
0	0	1
1	0	1
0	1	1
1	1	0

FIGURE 24.1 The circuit symbol and Truth Table for a two-input NAND gate.

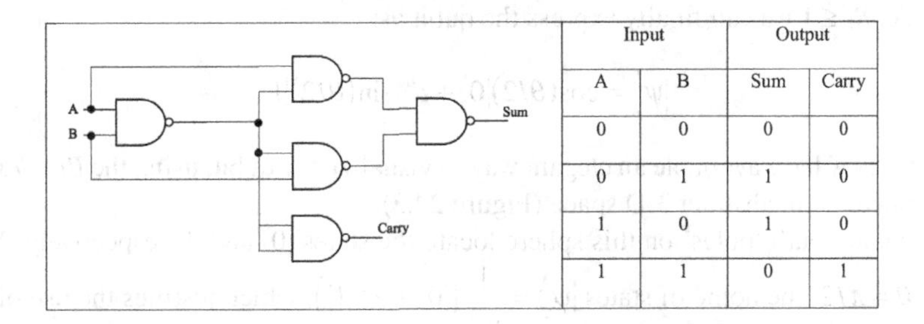

Input		Output	
A	B	Sum	Carry
0	0	0	0
0	1	1	0
1	0	1	0
1	1	0	1

FIGURE 24.2 A half adder circuit from NAND gates.

Table, which shows the gate's output as a function of the different possible inputs. Figure 24.1 shows the circuit symbol and Truth Table for a two-input NAND gate.

NAND gates can be strung together to carry out logical operations. For example, the *half adder* circuit is shown in Figure 24.2 along with its Truth Table

24.3 QUANTUM ANALOGUES

Unsurprisingly, some of the foundational concepts of quantum computing have been set up in analogy with counterparts in the digital regime. This allows us to view quantum theory from a different, more informational, perspective, without altering the formalism in any extensive way. Aside from the intrinsic value of the theoretical and practical developments of possible computing technologies, spin-offs are applicable to scientific research into quantum foundations. For example, it will be necessary to hold environmental decoherence at bay for quantum information processing. The ability to hold a quantum system in a superposition state without environmental effects getting in the way would be very useful in pure research.

24.3.1 Qubits

In his 1995 paper[16], Benjamin Schumacher suggested the term *qubit* to refer to two-state systems (such as $|U\rangle/|D\rangle$ or $|L\rangle/|R\rangle$ for spin ½ systems) in an obvious analogy with bits used in classical computing.

A qubit $|\psi\rangle$ is then:

$$|\psi\rangle = \alpha|0\rangle + \beta|1\rangle$$

using $|0\rangle$ and $|1\rangle$ as a convenient orthonormal basis for the two-state system. Note that in quantum computing, it is conventional to have $|0\rangle = \begin{pmatrix} 1 \\ 0 \end{pmatrix}$ and $|1\rangle = \begin{pmatrix} 0 \\ 1 \end{pmatrix}$, so $|\psi\rangle = \begin{pmatrix} \alpha \\ \beta \end{pmatrix}$.

Given that α and β are both complex numbers, and hence can be written $Re^{i\vartheta}$, it would appear at first glance that a qubit has 4 independent degrees of freedom embedded within it $(R_1, R_2, \theta_1, \theta_2)$:

$$|\psi\rangle = \alpha|0\rangle + \beta|1\rangle = R_1 e^{i\theta_1}|0\rangle + R_2 e^{i\theta_2}|1\rangle$$

As the absolute phase is of no physical relevance, we can rotate the state by $e^{-i\theta_1}$ without any loss of generality:

$$|\psi\rangle = R_1|0\rangle + R_2 e^{i(\theta_2-\theta_1)}|1\rangle$$

The normalization constraint forces $R_1^2 + R_2^2 = 1$, so $R_2 = \sqrt{1-R_1^2}$, and we are down to two degrees of freedom. As $R_1 \leq 1$ we can finally express the qubit as:

$$|\psi\rangle = \cos(\theta/2)|0\rangle + e^{i\varphi}\sin(\theta/2)|1\rangle$$

As a bonus, we now have available an elegant way to visualize the qubit, using the *Bloch sphere*[17] – a unit radius sphere in an abstract 3-D space (Figure 24.3).

The North and South 'poles' on this sphere locate the states $|0\rangle$ and $|1\rangle$ respectively. That makes the equator, $\theta = \pi/2$, the home of states $|\psi\rangle = \frac{1}{\sqrt{2}}\left(|0\rangle + e^{i\varphi}|1\rangle\right)$, which justifies the use of $\theta/2$ in the trigonometric representation earlier. Every pure state of the qubit is then visualizable as a point on the surface of the Bloch sphere. A little thought also shows that a mixed state density matrix $\hat{D}_m = P_0|0\rangle\langle 0| + P_1|1\rangle\langle 1|$ can be found among the points inside the body of the sphere.

Physically, a qubit can be realized in any system context where there are (just) two orthogonal states available. We have already noted that the spin states of spin ½ particles are one example. The horizontal vs vertical polarization states of a photon are another well-used context. However, there are many other possibilities which are in active technological development.

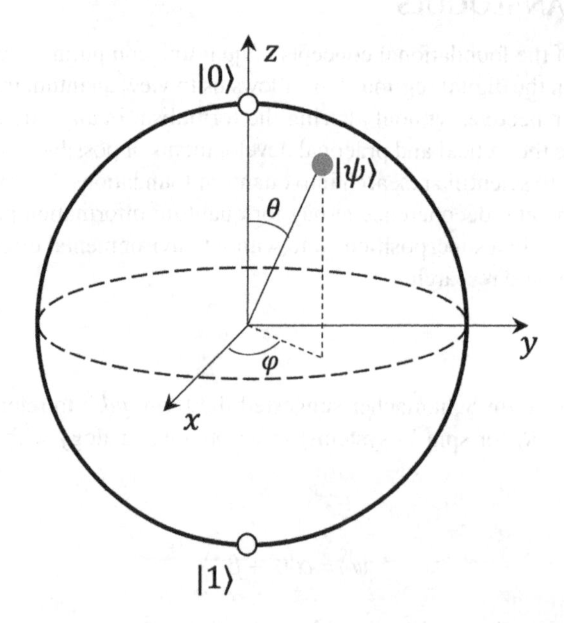

FIGURE 24.3 The Bloch sphere is a convenient way to visualise a qubit. All pure qubit states are points on the surface of the sphere.

In principle, the information-carrying potential in a qubit is very high. For example, data that would normally take n bits to specify could be encoded in a single qubit as the relative phase, φ, to n bits of precision. However, data extraction is problematic as any measurement of the state would only reveal $|0\rangle$ or $|1\rangle$, governed by the probabilities.

24.3.2 Quantum Gates

One approach to quantum computing, the quantum circuit model, is based on *quantum gates* which act on qubits as logic circuits act on bits. There are a range of quantum gates, each of which can be represented by a unitary matrix. For example:

Pauli Gates

These three gates X, Y, Z act on a single qubit and their functions are represented by the Pauli matrices σ_1, σ_2, σ_3 respectively.

Gate	Matrix Representation	Action	Bloch Sphere
X	$\begin{pmatrix} 0 & 1 \\ 1 & 0 \end{pmatrix}$	$\begin{pmatrix} 0 & 1 \\ 1 & 0 \end{pmatrix}\begin{pmatrix} 1 \\ 0 \end{pmatrix}=\begin{pmatrix} 0 \\ 1 \end{pmatrix}$ $\begin{pmatrix} 0 & 1 \\ 1 & 0 \end{pmatrix}\begin{pmatrix} 0 \\ 1 \end{pmatrix}=\begin{pmatrix} 1 \\ 0 \end{pmatrix}$	Rotate by π about the x -axis
Y	$\begin{pmatrix} 0 & -i \\ i & 0 \end{pmatrix}$	$\begin{pmatrix} 0 & -i \\ i & 0 \end{pmatrix}\begin{pmatrix} 1 \\ 0 \end{pmatrix}=\begin{pmatrix} 0 \\ i \end{pmatrix}$ $\begin{pmatrix} 0 & -i \\ i & 0 \end{pmatrix}\begin{pmatrix} 0 \\ 1 \end{pmatrix}=\begin{pmatrix} -i \\ 0 \end{pmatrix}$	Rotate by π about the y -axis
Z	$\begin{pmatrix} 1 & 0 \\ 0 & -1 \end{pmatrix}$	$\begin{pmatrix} 1 & 0 \\ 0 & -1 \end{pmatrix}\begin{pmatrix} 1 \\ 0 \end{pmatrix}=\begin{pmatrix} 1 \\ 0 \end{pmatrix}$ $\begin{pmatrix} 1 & 0 \\ 0 & -1 \end{pmatrix}\begin{pmatrix} 0 \\ 1 \end{pmatrix}=\begin{pmatrix} 0 \\ -1 \end{pmatrix}$	Rotate by π about the z -axis

Hadamard Gate

This is another single qubit gate which generates superpositions.

Gate	Matrix Representation	Action	Bloch Sphere
H	$\frac{1}{\sqrt{2}}\begin{pmatrix} 1 & 1 \\ 1 & -1 \end{pmatrix}$	$\frac{1}{\sqrt{2}}\begin{pmatrix} 1 & 1 \\ 1 & -1 \end{pmatrix}\begin{pmatrix} 1 \\ 0 \end{pmatrix}=\frac{1}{\sqrt{2}}\begin{pmatrix} 1 \\ 1 \end{pmatrix}$ $\frac{1}{\sqrt{2}}\begin{pmatrix} 1 & 1 \\ 1 & -1 \end{pmatrix}\begin{pmatrix} 0 \\ 1 \end{pmatrix}=\frac{1}{\sqrt{2}}\begin{pmatrix} 1 \\ -1 \end{pmatrix}$	Rotate by π about the axis vector $(1, 0, 1)$

CNOT Gate

The Controlled Not takes two qubits and then applies the X gate to the *input qubit*, $|\psi\rangle$, if the *control qubit*, $|\phi\rangle$, is $|1\rangle$, otherwise it leaves it alone. To employ a matrix representation, we must combine the two qubits, using a tensor product, e.g.:

$$|1\rangle \otimes |\psi\rangle = \begin{pmatrix} 0 \\ 1 \end{pmatrix} \otimes \begin{pmatrix} \alpha \\ \beta \end{pmatrix} = \begin{pmatrix} 0\times & \begin{matrix} \alpha \\ \beta \end{matrix} \\ 1\times & \begin{matrix} \alpha \\ \beta \end{matrix} \end{pmatrix} = \begin{pmatrix} 0 \\ 0 \\ \alpha \\ \beta \end{pmatrix}$$

$$|0\rangle \otimes |\psi\rangle = \begin{pmatrix} 1 \\ 0 \end{pmatrix} \otimes \begin{pmatrix} \alpha \\ \beta \end{pmatrix} = \begin{pmatrix} 1\times & \begin{matrix} \alpha \\ \beta \end{matrix} \\ 0\times & \begin{matrix} \alpha \\ \beta \end{matrix} \end{pmatrix} = \begin{pmatrix} \alpha \\ \beta \\ 0 \\ 0 \end{pmatrix}$$

Gate	Matrix Representation	Action	Bloch Sphere
$CNOT_{\phi\psi}$	$\begin{pmatrix} 1 & 0 & 0 & 0 \\ 0 & 1 & 0 & 0 \\ 0 & 0 & 0 & 1 \\ 0 & 0 & 1 & 0 \end{pmatrix}$	$\begin{pmatrix} 1 & 0 & 0 & 0 \\ 0 & 1 & 0 & 0 \\ 0 & 0 & 0 & 1 \\ 0 & 0 & 1 & 0 \end{pmatrix}\begin{pmatrix} 0 \\ 0 \\ \alpha \\ \beta \end{pmatrix} = \begin{pmatrix} 0 \\ 0 \\ \alpha \\ \beta \end{pmatrix}$ $\begin{pmatrix} 1 & 0 & 0 & 0 \\ 0 & 1 & 0 & 0 \\ 0 & 0 & 0 & 1 \\ 0 & 0 & 1 & 0 \end{pmatrix}\begin{pmatrix} \alpha \\ \beta \\ 0 \\ 0 \end{pmatrix} = \begin{pmatrix} \alpha \\ \beta \\ 0 \\ 0 \end{pmatrix}$	Rotate by π about the x -axis if control qubit is $|1\rangle$

In summary, and this will be useful later, the CNOT gate switches the amplitudes of the input qubit if the control qubit is $|1\rangle$ otherwise, the amplitudes are unchanged.

Gates like these can be linked together to form logic circuits, just as their digital counterparts. For example, if we apply a Hadamard gate to a qubit set to $|0\rangle_A$, we transform it into $\frac{1}{\sqrt{2}}\left(|0\rangle_A + |1\rangle_A\right)$. This is illustrated by the simple diagram shown in Figure 24.4.

If we now introduce a second qubit, $q_B = |0\rangle_B$, the overall state is:

$$H(q_A)q_B = \frac{1}{\sqrt{2}}\left(|0\rangle_A|0\rangle_B + |1\rangle_A|0\rangle_B\right)$$

Using $H(q_A)$ to act as the control bit for a CNOT gate applied to q_B (Figure 24.5):
we end up with the state:

$$\text{CNOT}\{H(q_A)q_B\} = \frac{1}{\sqrt{2}}\left(|0\rangle_A|0\rangle_B + |1\rangle_A|1\rangle_B\right)$$

q_A — H — $\frac{1}{\sqrt{2}}(|0\rangle_A + |1\rangle_A)$

$|0\rangle_A$

FIGURE 24.4 Using a Hadamard gate to transform q_A.

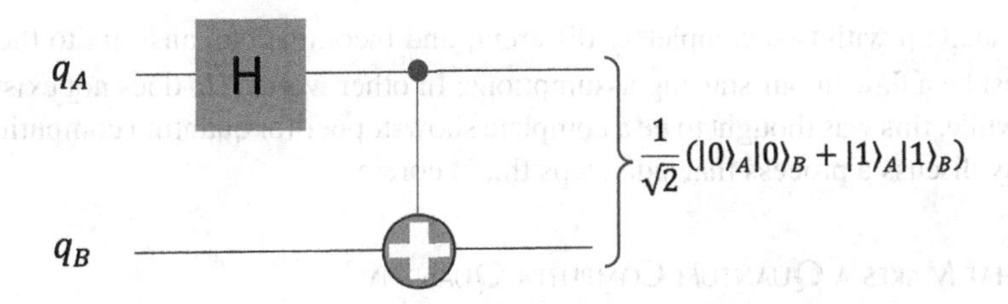

FIGURE 24.5 The output of the Hadamard gate acting on q_A is now used as the control bit for a CNOT gate acting on q_B.

In other words, we have entangled the states together. While a very minor example, this serves to illustrate how gates can be linked into circuits in order to manipulate qubits.

However, there is one important restriction on manipulating qubits that does not apply to digital bits and which has profound implications for the design of a quantum computer.

24.3.3 The No-Cloning Theorem

Making a copy of a bitstring is a common process in digital computing. If nothing else, having a stored copy of a bitstring before processing is useful in error correction processes. However, the no-cloning theorem applies to qubits, preventing copies from being made[18].

In essence, there are two ways in which a qubit can be manipulated, *measurement* or *unitary transformation*. A measurement collapses the qubit into one of its component states, thus erasing the information content in the amplitudes. So, measurements can't be used in any process intending to copy a qubit. Any other valid quantum process must be unitary, so we assume a unitary operator, $\hat{U}_C$, which serves to copy a qubit, $|\psi\rangle_A$, from one system to another:

$$\hat{U}_C\left(|\psi\rangle_A |e\rangle_B\right) = |\psi\rangle_A |\psi\rangle_B$$

where $|e\rangle_B$ is some arbitrary initial state of system B. Assuming the original qubit to be $|\psi\rangle_A = \alpha|0\rangle_A + \beta|1\rangle_A$, the result of this process must be:

$$\begin{aligned}\hat{U}_C\left(|\psi\rangle_A |e\rangle_B\right) &= \left(\alpha|0\rangle_A + \beta|1\rangle_A\right)\left(\alpha|0\rangle_B + \beta|1\rangle_B\right) \\ &= \alpha^2|0\rangle_A|0\rangle_B + \alpha\beta|0\rangle_A|1\rangle_B + \alpha\beta|1\rangle_A|0\rangle_B + \beta^2|1\rangle_A|1\rangle_B\end{aligned}$$

On the other hand, if $|e\rangle_B = A|0\rangle_B + B|1\rangle_B$, then:

$$\begin{aligned}|\psi\rangle_A |e\rangle_B &= \left(\alpha|0\rangle_A + \beta|1\rangle_A\right)\left(A|0\rangle_B + B|1\rangle_B\right) \\ &= \alpha A|0\rangle_A|0\rangle_B + \alpha B|0\rangle_A|1\rangle_B + \beta A|1\rangle_A|0\rangle_B + \beta B|1\rangle_A|1\rangle_B\end{aligned}$$

Followed by:

$$\begin{aligned}\hat{U}_C\left(|\psi\rangle_A |e\rangle_B\right) &= \hat{U}_C\left(\alpha A|0\rangle_A|0\rangle_B\right) + \hat{U}_C\left(\alpha B|0\rangle_A|1\rangle_B\right) + \hat{U}_C\left(\beta A|1\rangle_A|0\rangle_B\right) + \hat{U}_C\left(\beta B|1\rangle_A|1\rangle_B\right) \\ &= \alpha A|0\rangle_A|0\rangle_B + \alpha B|0\rangle_A|0\rangle_B + \beta A|1\rangle_A|1\rangle_B + \beta B|1\rangle_A|1\rangle_B \\ &= \alpha(A+B)|0\rangle_A|0\rangle_B + \beta(A+B)|1\rangle_A|1\rangle_B\end{aligned}$$

As we have ended up with two completely different, and incompatible, answers to the same problem, there must be a flaw in our starting assumptions. In other words, $\hat{U}_C$ does not exist.

For some while, this was thought to be a complete showstopper for quantum computing. However, we will shortly discuss a process that side-steps this theorem.

24.3.4 What Makes a Quantum Computer Quantum?

Quantum computers, hypothetical or real, manipulate systems whose behaviour is governed by quantum theory. However, that does not in itself make such a device a quantum computing mechanism. There is, presumably, some aspect of quantum behaviour that is being exploited in order to leverage computing power.

Surprisingly, there is a lack of consensus on this issue, which may be one reason why more quantum algorithms have not been developed[19]. Interference, entanglement and contextuality have all been proposed, indeed David Deutsch, who has made significant contributions to the field, suggests that quantum computers work as they operate simultaneous across branching realities (see Chapter 28 for a treatment of the Many Worlds Interpretation). One key question is the interpretation of the process:

$$\left(\sum_k |k\rangle\right)|0\rangle \rightarrow_f\rightarrow \sum_k |k\rangle|f(k)\rangle \tag{24.1}$$

which, at face value, suggests that the quantum computer has taken the original qubit which is in the superposition $\sum_k |k\rangle$, and used it to calculate all the possible values $f(k)$ simultaneously. Deutsch, and others, accept this interpretation and propose that the only way of making sense of this is to support a Many Worlds view. However, this approach is not without its critics[20]. Even if we accept that EQ24.1 indicates simultaneous processing, it is unclear how to extract the results. Measurement of the output results in its collapse into one state, $|K\rangle|f(K)\rangle$ destroying information about the other values. Were it not for the no-cloning theorem, multiple copies could be made of the 'answer state' and each subjected to a measurement in order to (randomly) extract more information.

Looking at this from another perspective, consider a *satisfiability problem* from Boolean logic[21]. We are presented with a proposition involving a number of variables and need to see if any assignment of True or False to the variables results in the overall statement being True. For a simple example, take the statement "a AND NOT b". The truth table is:

a	b	NOT b	a AND NOT b
F	F	T	F
T	F	T	T
F	T	F	F
T	T	F	F

showing that there is one combination (row 2) that does satisfy the overall statement. In general, with n variables in the statement, 2^n combinations have to be checked. As suggested by Pitowsky, a quantum computer would seem well placed to deal with this[22]. Each variable is assigned to a qubit, with $|0\rangle$ being False and $|1\rangle$ being True. One row in the truth table would then be $|\varepsilon_1\rangle_1|\varepsilon_2\rangle_2|\varepsilon_3\rangle_3\ldots|\varepsilon_n\rangle_n$ with $\varepsilon_i = (0, 1)$. We also encode the statement into some sequence $|\ldots\phi\ldots\rangle$, which is the same length as the number of bits that would be needed to encode the statement into a conventional Turing machine. Finally, we need another qubit which keeps track of the overall truth value, $\{|Y\rangle, |N\rangle\}$. So, we start with the state:

$$|\ldots\phi\ldots\rangle|\varepsilon_1\rangle_1|\varepsilon_2\rangle_2|\varepsilon_3\rangle_3\ldots|\varepsilon_n\rangle_n|N\rangle$$

The action of a conventional program can be represented by a unitary transformation $\hat{U}_C$:

$$\hat{U}_C\left(|\ldots\phi\ldots\rangle|\varepsilon_1\rangle_1|\varepsilon_2\rangle_2|\varepsilon_3\rangle_3\ldots|\varepsilon_n\rangle_n|N\rangle\right)=|\ldots\phi\ldots\rangle|\varepsilon_1\rangle_1|\varepsilon_2\rangle_2|\varepsilon_3\rangle_3\ldots|\varepsilon_n\rangle_n|T(\phi;\varepsilon_1,\varepsilon_2,\varepsilon_3,\ldots,\varepsilon_n)\rangle$$

with T being the truth value dependent on the specific values of the variables. The code simply cycles through all the possibilities checking them one by one.

If instead we start with $|\ldots\phi\ldots\rangle|0\rangle_1|0\rangle_2|0\rangle_3\ldots|0\rangle_n|N\rangle$ and apply a Hadamard gate to the first qubit we get:

$$\frac{1}{\sqrt{2}}|\ldots\phi\ldots\rangle\left(|0\rangle+|1\rangle\right)_1|0\rangle_2|0\rangle_3\ldots|0\rangle_n|N\rangle$$

Continuing to apply H to each qubit in turn produces:

$$|\Phi\rangle=\frac{1}{\sqrt{2^n}}\sum_{\varepsilon_1,\varepsilon_2\ldots}|\ldots\phi\ldots\rangle|\varepsilon_1\rangle_1|\varepsilon_2\rangle_2|\varepsilon_3\rangle_3\ldots|\varepsilon_n\rangle_n|N\rangle$$

i.e., the state is a superposition of each possible truth assignment to the variables. Now we use $\hat{U}_C$ on this state, which effectively evaluates all possibilities in one hit:

$$\hat{U}_C|\Phi\rangle=\frac{1}{\sqrt{2^n}}\sum_{\varepsilon_1,\varepsilon_2\ldots}|\ldots\phi\ldots\rangle|\varepsilon_1\rangle_1|\varepsilon_2\rangle_2|\varepsilon_3\rangle_3\ldots|\varepsilon_n\rangle_n\,T(\phi;\varepsilon_1,\varepsilon_2,\varepsilon_3,\ldots\,\varepsilon_n)\rangle$$

The problem will be trying to extract the result. Any measurement will collapse $\hat{U}_C|\Phi\rangle$ into one of the terms in the sum, revealing T. However, if there is only one assignment which works for the overall statement, we have a chance of $1/2^n$ of finding it. A Turning machine programmed to try each one in turn would do just as well. However, Pitowsky believes that this argument does suggest a fruitful way of thinking about quantum computation and the advantage that it might offer. The trick is to create clever superpositions which enhance the probability of a result, compared with 'guesswork'. As Scott Aaronson puts it[23]:

> The goal in devising an algorithm for a quantum computer is to choreograph a pattern of constructive and destructive interference so that for each wrong answer the contributions to its amplitude cancel each other out, whereas for the right answer the contributions reinforce each other. If, and only if, you can arrange that, you'll see the right answer with a large probability when you look.

This is not to say that quantum computers are not going to represent a huge advance in certain areas. However, it does suggest that they may not be the revolution in all areas that they are sometimes heralded to be.

24.4 QUANTUM TELEPORTATION

Physicists are often tempted to use whimsical nomenclature or references to Sci-Fi in their work (Figure 24.6). We are human after all. Unfortunately, while the community understand this and can extract the real meaning from the whimsey, the press often uses headlines that pick up on the apparent meaning, which can lead to some confusion or misunderstanding for the public. Such is the case with *quantum teleportation*. This is a process which uses entangled Bell-type states, along

FIGURE 24.6 If only it were true… (Image credit: XKCD https://xkcd.com/465/.)

with a more traditional classical communication channel, to enable the state of a system to be passed from one place to another. Crucially, it is the *state* that is passed, not any material substance. The term *teleportation*, with all its Star Trek connotations, is at best only indirectly relevant. However, the process is important as a technical advance relevant to quantum computing as well of being of interest in fundamental theory[24].

As per an EPR type experiment, we start with an entangled pair of systems, *A* and *B*, which are then separated and *A* sent to Alice, who is our operator, and *B* to Bob, the receiver. In addition, Alice has the qubit $|\psi\rangle_C = \alpha|0\rangle_C + \beta|1\rangle_C$ realized on a third system, *C*. Now imagine that Alice wishes to let Bob know the nature of her extra qubit. The obvious way of doing that would be for Alice to determine the values of α and β and then pick up the telephone to Bob. However, this plan has two serious snags. Firstly, determining the amplitudes would require a measurement, destroying the qubit in the process. Also, this would not be possible without an ensemble of the qubits, a single measurement only resulting in either $|1\rangle$ or $|0\rangle$. Even then, Alice can only send the values of α and β to a finite level of precision, so Bob can never have exact knowledge.

The trick is to employ the EPR pair. It turns out that Alice can send just two bits of information to Bob, which will enable him to rotate the state of his system, *B*, so that it matches Alice's qubit exactly (Figure 24.7).

To explain how this works, we start with the 4 possible Bell-type entanglements of a pair of qubits expressed in the two-state systems, *A* and *B*:

$$|\Phi^+\rangle_{AB} = \frac{1}{\sqrt{2}}\{|0\rangle_A|0\rangle_B + |1\rangle_A|1\rangle_B\} \qquad |\Phi^-\rangle_{AB} = \frac{1}{\sqrt{2}}\{|0\rangle_A|0\rangle_B - |1\rangle_A|1\rangle_B\}$$

$$|\Psi^+\rangle_{AB} = \frac{1}{\sqrt{2}}\{|0\rangle_A|1\rangle_B + |1\rangle_A|0\rangle_B\} \qquad |\Psi^-\rangle_{AB} = \frac{1}{\sqrt{2}}\{|0\rangle_A|1\rangle_B - |1\rangle_A|0\rangle_B\}$$

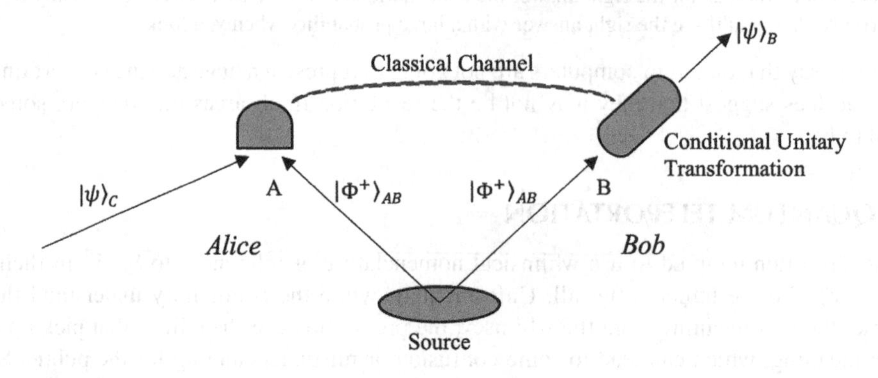

FIGURE 24.7 Using entanglement to teleport a qubit.

We arbitrarily select and prepare one of these states, say $\left|\Phi^+\right\rangle$, and separate the systems, without breaking the entanglement, sending A to Alice and B to Bob[25].

At Alice's end, she then has a total state which is the tensor product of her qubit and $\left|\Phi^+\right\rangle_{AB}$:

$$|\mathbb{Q}\rangle_{ABC} = |\psi\rangle_C \otimes \left|\Phi^+\right\rangle_{AB} = \left(\alpha|0\rangle_C + \beta|1\rangle_C\right) \otimes \frac{1}{\sqrt{2}}\left\{|0\rangle_A|0\rangle_B + |1\rangle_A|1\rangle_B\right\}$$

$$= \frac{1}{\sqrt{2}}\left\{\alpha|0\rangle_C|0\rangle_A|0\rangle_B + \alpha|0\rangle_C|1\rangle_A|1\rangle_B + \beta|1\rangle_C|0\rangle_A|0\rangle_B + \beta|1\rangle_C|1\rangle_A|1\rangle_B\right\}$$

Remembering, of course, that she only has systems A and C at her physical location.

Alice now follows a pre-defined procedure, the first step of which is to use the $|\psi\rangle_C$ qubit as the control of a CNOT gate (see Section 24.3.2) which acts on A.

$$\text{CNOT}_{CA}|\mathbb{Q}\rangle_{ABC} = \frac{1}{\sqrt{2}}\left\{\alpha|0\rangle_C|0\rangle_A|0\rangle_B + \alpha|0\rangle_C|1\rangle_A|1\rangle_B + \beta|1\rangle_C|1\rangle_A|0\rangle_B + \beta|1\rangle_C|0\rangle_A|1\rangle_B\right\}$$

Next, she applies a Hadamard gate to $|\psi\rangle_C$:

$$\begin{aligned}
H_C\left(\text{CNOT}_{CA}|\mathbb{Q}\rangle_{ABC}\right) &= \frac{1}{\sqrt{2}}\left\{\alpha|0\rangle_C|0\rangle_A|0\rangle_B + \alpha|0\rangle_C|1\rangle_A|1\rangle_B + \beta|1\rangle_C|1\rangle_A|0\rangle_B + \beta|1\rangle_C|0\rangle_A|1\rangle_B\right\} \\
&= \frac{1}{2}\alpha\left(|0\rangle_C|0\rangle_A|0\rangle_B + |1\rangle_C|0\rangle_A|0\rangle_B\right) + \frac{1}{2}\alpha\left(|0\rangle_C|1\rangle_A|1\rangle_B + |1\rangle_C|1\rangle_A|1\rangle_B\right) \\
&\quad + \frac{1}{2}\beta\left(|0\rangle_C|1\rangle_A|0\rangle_B - |1\rangle_C|1\rangle_A|0\rangle_B\right) + \frac{1}{2}\beta\left(|0\rangle_C|0\rangle_A|1\rangle_B - |1\rangle_C|0\rangle_A|1\rangle_B\right) \\
&= \frac{1}{2}|0\rangle_C|0\rangle_A\left(\alpha|0\rangle_B + \beta|1\rangle_B\right) + \frac{1}{2}|1\rangle_C|0\rangle_A\left(\alpha|0\rangle_B - \beta|1\rangle_B\right) \\
&\quad + \frac{1}{2}|0\rangle_C|1\rangle_A\left(\beta|0\rangle_B + \alpha|1\rangle_B\right) + \frac{1}{2}|1\rangle_C|1\rangle_A\left(-\beta|0\rangle_B + \alpha|1\rangle_B\right)
\end{aligned} \tag{24.2}$$

If Alice now measures her two qubits A and C, the overall state will collapse into one of the four terms in Eq. 24.2.

Finally, Alice communicates to Bob the results of her measurement using a classical channel of some form. As there are only four possibilities for Alice's results, two bits of data are sufficient to tell Bob all he needs to know. He can then follow the rules in Table 24.1 to convert the state of B into the required echo of $|\psi\rangle_C$.

Bob's choice of which unitary transformation to apply is dictated entirely by the communication from Alice. He does not have to know the state of his system to make the choice. He does not even need to know the original qubit that Alice is trying to transfer to him.

In quantum circuit terms, the process is illustrated in Figure 24.8.

The teleportation process sidesteps the strictures of the no-cloning theorem as the original qubit, q_C, is destroyed in the process:

$$|\psi\rangle_C|e\rangle_B \rightarrow |\psi\rangle_C|\psi\rangle_B \qquad \text{cloning}$$

$$|\psi\rangle_C|e\rangle_B \rightarrow |\phi\rangle_C|\psi\rangle_B \qquad \text{teleportation}$$

TABLE 24.1
Converting the State of Qubit *B* Depending on the Result of Alice's Measurement

Alice's Measurement Result	Bits Transferred	Bob's State	Conversion Required
$\lvert 0\rangle_C \lvert 0\rangle_A$	00	$\alpha\lvert 0\rangle_B + \beta\lvert 1\rangle_B$	None
$\lvert 0\rangle_C \lvert 1\rangle_A$	01	$\beta\lvert 0\rangle_B + \alpha\lvert 1\rangle_B$	Apply X gate
$\lvert 1\rangle_C \lvert 0\rangle_A$	10	$\alpha\lvert 0\rangle_B - \beta\lvert 1\rangle_B$	Apply Z gate
$\lvert 1\rangle_C \lvert 1\rangle_A$	11	$-\beta\lvert 0\rangle_B + \alpha\lvert 1\rangle_B$	Apply X gate then Z gate

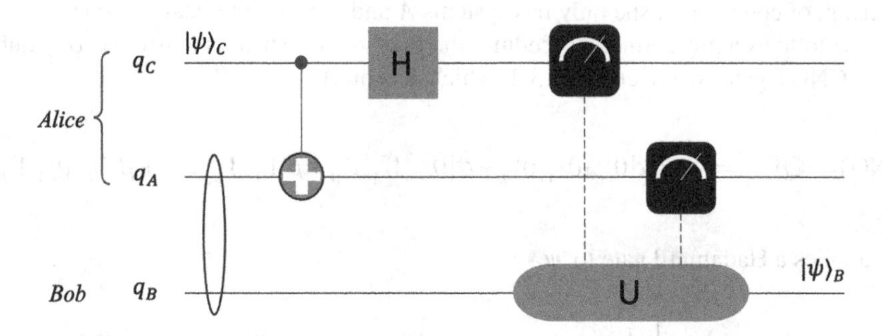

FIGURE 24.8 A quantum circuit for teleporting the state $\lvert\psi\rangle$ from system *C* to *B*.

In short, cloning is a 'copy and paste' operation, whereas teleportation is 'cut and paste'.

24.4.1 Experimental Implementation

As of 2021, the record for quantum teleportation over distance stands at 1400 km set by a Chinese team in 2017[26]. Their experiment used a combination of a ground station (*Ngari*, in Tibet) and an orbiting satellite, *Micius*, at an altitude ~ 500 km. The ground station prepared a pair of entangled photons, using the two-state vertical/horizontal polarization basis. A third photon on the ground carried the qubit to be teleported in its polarization state. This photon was transmitted via telescope to the receiver in orbit and subsequently measured. By preselecting and recording one of two states on the ground, the team was able to avoid transmitting data classically to the receiving station and instead applying a phase shift or not to the data in the analysis stage to confirm teleportation had taken place. Tibet was chosen due to its high altitude, making the air thinner, reducing its impact on photon transmission. During a satellite transit over Ngari, the distance from ground to receiver varied from ~ 500 km to ~ 1400 km. The team recorded an 80% successful transfer rate[27].

Ground-based experiments have achieved a greater successful transfer rate, albeit over much shorter distances. In 2020, a group reported > 90% success rate of over 44 km of fibre optic.

24.5 PRACTICAL QUANTUM COMPUTERS

As of 2022, Google is working with its *Sycamore* 53 qubit quantum computer[28]. This device is superconductor based, requiring the chip to be cooled to 53 mK and housed in a vacuum chamber. The qubits on the chip are 0.2 mm across and constructed from a superconducting aluminium strips with two accessible energy levels. Each qubit is linked to four others, via tuneable junctions that can vary their interaction strength. Qubits can be manipulated via microwave pulses at different frequencies, causing the qubit states to flip.

In 2019, Google reported that Sycamore performed a task in 200 seconds that they estimated would take the world's most powerful classical computer, the IBM *Summit*, over 10,000 years to carry out[29]. IBM subsequently disputed this claim and suggested that a properly configured version of Summit could cut the time to 2.5 days. In any case, Google's calculation was effectively checking the output of a quantum random number generator, which has limited practical application. However, since then (2020) Google has published the results of a wave function approximation calculation in quantum chemistry using Sycamore[30]. The calculation resulted in the binding energy of a hydrogen chain up to H_{12}.

A team from the University of Science and Technology of China (USTC) has developed *Jiŭzhāng 2.0*, a photon-based quantum computer with 113 qubits. Their original 76 qubit system was estimated to perform a task 10^{14} times faster than Summit. Once again, the level of computational advantage has been disputed by other researchers.

In November 2021, IBM announced *Eagle*[31], another superconductor-based quantum chip with 127 qubits and has a roadmap[32] to *Osprey* in 2022 with 433 qubits and then 1000+ qubits in *Condor* in 2023. While the sheer number of qubits available in a system is important, gate accuracy also needs to be considered when evaluating a system. To make real quantum computations viable, a gate error rate $< 1\%$ needs to be achieved.

At this point, it is clear that a multi-national race is in progress to develop ever better quantum computers, with no particular system basis showing a decisive advantage. Specific calculations have been performed that demonstrate some promise in terms of quantum speed advantage, but any claim of decisive quantum supremacy is hotly disputed.

NOTES

1 Gordon Earle Moore (1929–).
2 In reality, Moore's projected doublings started to taper off around 2010, although new processes were in 2018 that have the prospect of following the trend for a few more years.
3 Moore, Gordon E Cramming more components onto integrated circuits intel.com. Electronics Magazine, 1965-04-19.
4 Benioff, Paul. The computer as a physical system: A microscopic quantum mechanical Hamiltonian model of computers as represented by Turing machines. *Journal of Statistical Physics* 22 (5): 563–591, 1980.
5 Feynman, Richard. Simulating Physics with Computers. *International Journal of Theoretical Physics.* 21 (6/7): 467–488, 1982.
6 D. Deutsch (1953–) Visiting Professor in the Department of Atomic and Laser Physics at the Centre for Quantum Computation, Oxford University.
7 Deutsch, D. Quantum Theory, the Church-Turing Principle and the Universal Quantum Computer. *Proceedings of the Royal Society of London*, 400: 97–117, 1985.
8 Deutsch, D. Quantum Computational Networks. *Proceedings of the Royal Society of London*, 425: 73–90, 1989.
9 Shor, P. W. Algorithms for Quantum Computation: Discrete Logarithms and Factoring. *Proceedings of the 35th Annual Symposium on Foundations of Computer Science (SFCS '94)*, 124–34, Washington, DC: IEEE Computer Society, 1994.
10 David Hilbert (1862–1943), Professor of Mathematics, Göttingen University.
11 Hilbert, D. Mathematical Problems, Lecture delivered before the International Congress of Mathematicians at Paris in 1900.
12 Any mathematical conjecture was assumed to be either true (i.e., could be proven from the axioms) or false, (i.e., the negation of the conjecture could be proven from the axioms). Gödel's incompleteness theorems wrecked that approach.
13 Alan Mathison Turing (1912–1954), University of Manchester.
14 Turing, A.M. On Computable Numbers, with an Application to the Entscheidungsproblem. *Proceedings of the London Mathematical Society*, 2, 42(1): 230–65, 1937.
15 Roger Penrose has pointed out that moving an infinitely long tape back and forth would pose an issue, so suggested instead that the machine would be able to move itself along the tape like a train along a track.
16 Schumacher, B. Quantum Coding. *Physical Review A*, 51: 2738–2747.
17 Bloch, F. (1905–1983), first Director-General of the particle physics research institution, CERN and Stanford University.

18 Wootters, W.K. and Zurek, W.H. A Single Quantum Cannot Be Cloned. *Nature*, 299: 802–803, 1982. https://doi.org/10.1038/299802a0. Also, Dieks, D. Communication by EPR Devices. *Physics Letters A*, 92: 271–272, 1982. https://doi.org/10.1016/0375-9601(82)90084-6.
19 For a discussion of this issue, see https://plato.stanford.edu/entries/qt-quantcomp/.
20 One line of argument against will make more sense when we have discussed Many Worlds. However, in brief one aspect of this interpretation is that the different worlds branch from each other as environmental decoherence takes hold. However, quantum computers work by holding decoherence at bay so that delicate quantum superpositions are preserved. It is unclear how to reconcile these opposing requirements.
21 Boolean logic was introduced by George Boole in The Mathematical Analysis of Logic (1847). Essentially it is a form of algebra based on the operations AND, OR, NOT using algebraic symbols that have the values True and False.
22 Pitowsky, I. Quantum Speed-up of Computations. *Philosophy of Science* 69: S168–S177, 2002.
23 https://www.quantamagazine.org/why-is-quantum-computing-so-hard-to-explain-20210608/.
24 Bennett, C.H., Brassard, G., Grepeau, C. and Jozsa, R. Teleporting an Unknown Quantum State via Dual Classical and Einstein-Podolsky-Rosen Channels. *Physical Review Letters* 70, 1895–1899, 1993. https://doi.org/10.1103/PhysRevLett.70.1895. Also, Bouwmeester, D., et al. Experimental Quantum Teleportation. *Nature* 390: 575–579, 1997. https://doi.org/10.1038/37539 and Boschi, D., et al. Experimental Realization of Teleporting an Unknown Pure Quantum State via Dual Classical and Einstein-Podolsky-Rosen Channels. *Physical Review Letters* 80: 1121–1125, 1998. https://doi.org/10.1103/PhysRevLett.80.1121.
25 We can prepare this state conveniently using the process in Section 24.2.2.
26 Ren, J.G., Xu, P., Yong, H.L., et al. Ground-to-satellite quantum teleportation. *Nature* 549: 70–73, 2017. https://doi.org/10.1038/nature23675.
27 The actual measure is the quantum teleport fidelity. By choosing a specific state to be teleported and making a series of measurements using an ensemble of received states, the fidelity can be calculated as the overlap between the measured density of states and the original qubit.
28 54 qubits were fabricated, but one failed.
29 Arute, F., Arya, K., Babbush, R. et al. Quantum supremacy using a programmable superconducting processor. *Nature* 574: 505–510, 2019.
30 https://arxiv.org/abs/2004.04174.
31 https://research.ibm.com/blog/127-qubit-quantum-processor-eagle.
32 https://research.ibm.com/blog/ibm-quantum-roadmap.

25 Density Operators

25.1 GREAT EXPECTATIONS

Before we can effectively discuss the various interpretations of quantum mechanics, there is one more piece of technical apparatus from the theory that we need to describe.

In Chapter 5, I introduced the *expectation value* - the average of a set of measurement results taken from a collection of systems in the same state. A straightforward calculation of the expectation value of operator $\hat{O}$ applied to state $|\psi\rangle$ takes the form:

$$\langle\hat{O}\rangle = \langle\psi|\hat{O}|\psi\rangle$$

Expanding $|\psi\rangle$ and $\langle\psi|$ over the basis $\{|\,i\rangle\}$:

$$|\psi\rangle = \sum_c a_c|c\rangle \qquad \langle\psi| = \sum_r a_r^*\langle r|$$

the expectation value becomes:

$$\langle\hat{O}\rangle = \sum_c\sum_r a_r^* a_c\langle r|\hat{O}|c\rangle$$

(The reason for the rather unconventional choice of r and c to represent the integer summation indices will become evident shortly.)

Remembering that $a_c = \langle c|\psi\rangle$ and $a_r^* = \langle\psi|r\rangle$ and tinkering with the ordering of terms, we get:

$$\langle\hat{O}\rangle = \sum_c\sum_r \langle\psi|r\rangle\langle r|\hat{O}|c\rangle\langle c|\psi\rangle$$

In the middle of this construction is the *matrix element* $\hat{O}_{rc} = \langle r|\hat{O}|c\rangle$ which is the general form of an operator rendered into a matrix representation. The $\langle c|\psi\rangle$ factors are then the elements of a *column vector* and $\langle\psi|r\rangle$ those of a *row vector* (clearly r is the *row index* in the matrix and c the *column index*[1]):

$$\langle\hat{O}\rangle = \begin{pmatrix} \langle\psi|1\rangle & \langle\psi|2\rangle & \langle\psi|3\rangle & \dots & \langle\psi|n\rangle \end{pmatrix} \begin{pmatrix} \langle 1|\hat{O}|1\rangle & \langle 1|\hat{O}|2\rangle & \langle 1|\hat{O}|3\rangle & \dots & \langle 1|\hat{O}|n\rangle \\ \langle 2|\hat{O}|1\rangle & \langle 2|\hat{O}|2\rangle & \langle 2|\hat{O}|3\rangle & \dots & \langle 2|\hat{O}|n\rangle \\ \langle 3|\hat{O}|1\rangle & \langle 3|\hat{O}|2\rangle & \langle 3|\hat{O}|3\rangle & \dots & \langle 3|\hat{O}|n\rangle \\ \dots & \dots & \dots & \dots & \dots \\ \langle n|\hat{O}|1\rangle & \langle n|\hat{O}|2\rangle & \langle n|\hat{O}|3\rangle & \dots & \langle n|\hat{O}|n\rangle \end{pmatrix} \begin{pmatrix} \langle 1|\psi\rangle \\ \langle 2|\psi\rangle \\ \langle 3|\psi\rangle \\ \dots \\ \langle n|\psi\rangle \end{pmatrix}$$

DOI: 10.1201/9781003225997-29

This construction can be extended to cater for an infinite number of elements, as would be required for dealing with a continuous basis. The matrix element would then be, for example:

$$\hat{O}_{rc} = \int a_r^*(x)\hat{O}a_c(x)dx$$

However, there is another way of viewing this, by moving the order of terms around:

$$\langle \hat{O} \rangle = \sum_c \sum_r \langle r|\hat{O}|c\rangle \langle c|\psi\rangle \langle \psi|r\rangle$$

which is mathematically the same but notice now that the term $\langle c|\psi\rangle\langle\psi|r\rangle$ *could be viewed as a matrix element,* provided we consider $|\psi\rangle\langle\psi|$ to be an operator.

This is reminiscent of the *projection operator*:

$$\hat{P}_i = |i\rangle\langle i|$$

that we met in Section 12.2.4, but different; this is the *density operator*:

$$\hat{D} = |\psi\rangle\langle\psi|$$

with matrix elements $\hat{D}_{cr} = \langle c|\psi\rangle\langle\psi|r\rangle$

(Note that now, due to the way things have worked out, *c* is the *row index* and *r* the *column index* for *this* matrix. Any matrix element takes the form $\hat{M}_{\text{row, column}} = \langle \text{row}|\hat{M}|\text{column}\rangle$.)

This makes the expectation value:

$$\langle \hat{O} \rangle = \sum_c \sum_r \langle c|\hat{D}|r\rangle\langle r|\hat{O}|c\rangle$$

At the core of this expression, is the combination $\langle c|\hat{D}|r\rangle\langle r|\hat{O}|c\rangle$ which is the product of two matrix elements. Consequently, this product must *represent an element of a single matrix formed from multiplying the other two together.* To deconstruct this, let's examine the process of multiplying two matrices:

$$A \times B = \begin{pmatrix} a_{11} & a_{12} & a_{13} & \cdots \\ a_{21} & a_{22} & a_{23} & \cdots \\ a_{31} & a_{32} & a_{33} & \cdots \\ \cdots & \cdots & \cdots & \cdots \end{pmatrix} \times \begin{pmatrix} b_{11} & b_{12} & b_{13} & \cdots \\ b_{21} & b_{22} & b_{23} & \cdots \\ b_{31} & b_{32} & b_{33} & \cdots \\ \cdots & \cdots & \cdots & \cdots \end{pmatrix}$$

$$= \begin{pmatrix} (a_{11}b_{11} + a_{12}b_{21} + a_{13}b_{31} + \cdots) & (a_{11}b_{12} + a_{12}b_{22} + a_{13}b_{32} + \cdots) & (a_{11}b_{13} + a_{12}b_{23} + a_{13}b_{33} + \cdots) & \cdots \\ (a_{21}b_{11} + a_{22}b_{21} + a_{23}b_{31} + \cdots) & (a_{21}b_{12} + a_{22}b_{22} + a_{23}b_{32} + \cdots) & (a_{21}b_{13} + a_{22}b_{23} + a_{23}b_{33} + \cdots) & \cdots \\ (a_{31}b_{11} + a_{32}b_{21} + a_{33}b_{31} + \cdots) & (a_{31}b_{12} + a_{32}b_{22} + a_{33}b_{32} + \cdots) & (a_{31}b_{13} + a_{32}b_{23} + a_{33}b_{33} + \cdots) & \cdots \\ \cdots & \cdots & \cdots & \cdots \end{pmatrix}$$

Each term has the structure:

$$(A \times B)_{RC} = \sum_k a_{Rk} b_{kC} \tag{25.1}$$

which is the general form of a matrix product. If we are only interested in the terms lying on the diagonal arm of the product (those in grey above), then we are picking out the elements with $R = C$. In other words:

$$\text{diagonal element } N = \sum_k a_{Nk} b_{kN} \tag{25.2}$$

where N specifies the intersection between row & column involved.

Moving back to our expectation value, and adding in a bracket to guide the eye:

$$\langle \hat{O} \rangle = \sum_c \sum_r \langle c|\hat{D}|r\rangle\langle r|\hat{O}|c\rangle = \sum_c \langle c|\hat{D}|r\rangle\langle r|\hat{O}|c\rangle = \sum_c \left(\sum_r \hat{D}_{cr}\hat{O}_{rc} \right)$$

we can see that the piece inside the bracket is the diagonal term from the product's row/column c, i.e., $N = c$ in Eq. 25.2. As we are summing over c, *the expectation value is a sum along the diagonal of the product matrix, i.e., from our previous example*:

$$(a_{11}b_{11} + a_{12}b_{21} + a_{13}b_{31} + \cdots) + (a_{21}b_{12} + a_{22}b_{22} + a_{23}b_{32} + \cdots) + (a_{31}b_{13} + a_{32}b_{23} + a_{33}b_{33} + \cdots) + \cdots$$

Mathematicians have a name for adding up the diagonal elements of a matrix (mathematicians have a name for most things) - it's called taking the *trace* of the matrix. Using this terminology, we can write:

$$\langle \hat{O} \rangle = \text{Trace}(\hat{D} \times \hat{O}) = \text{Tr}(\hat{D} \times \hat{O}) = \sum_c \left(\sum_r \hat{D}_{cr}\hat{O}_{rc} \right) \tag{25.3}$$

25.2 WHY BOTHER?

The real power of the density operator approach comes to the fore when dealing with a situation where we do not have enough information to be sure what state the system is in.

Imagine that we have a whole collection of identical systems, some of which are in state $|\psi_1\rangle$, some in state $|\psi_2\rangle$, etc. We might not know which system is in which state, and we might not even know how many systems are in any one given state. As a practical example, think about a beam of electrons that have not passed through any Stern–Gerlach magnet. Chances are that the spin states of the electrons are completely random. Perhaps all we know is the probability of finding an electron in each state:

$$P_1 = \text{Prob}(|\psi_1\rangle) \qquad P_2 = \text{Prob}(|\psi_2\rangle) \qquad P_3 = \text{Prob}(|\psi_3\rangle) \qquad \text{etc.}$$

To be clear, these probabilities *have nothing to do with quantum effects.* They simply represent *our ignorance of the details of what is happening in this case.* They are not related to quantum amplitudes, and hence are not *fundamental* probabilities[2].

In an instance like this, the overall expectation value of a series of measurements made on our collection of systems (one measurement per system) would be:

$$\langle \hat{O} \rangle = P_1 \langle \psi_1|\hat{O}|\psi_1\rangle + P_2 \langle \psi_2|\hat{O}|\psi_2\rangle + P_3 \langle \psi_3|\hat{O}|\psi_3\rangle + \cdots = \sum_m P_m \langle \psi_m|\hat{O}|\psi_m\rangle \tag{25.4}$$

which is simply the weighted average of the expectation values for each separate state in the collection. Now, if I build a density operator:

$$\hat{D} = \sum_m P_m |\psi_m\rangle\langle\psi_m|$$

its matrix elements will be (using the same indices as before):

$$\hat{D}_{cr} = \left\langle c \middle| \hat{D} \middle| r \right\rangle = \sum_m P_m \left\langle c \middle| \psi_m \right\rangle \left\langle \psi_m \middle| r \right\rangle$$

Multiplying this matrix by the matrix representing the operator $\hat{O}$, results in the product elements (using Eq. 25.1 with an obvious notion for the indices):

$$\left(\hat{D} \times \hat{O}\right)_{RC} = \sum_k \hat{D}_{Rk} \hat{O}_{kC}$$

$$= \sum_k \left(\sum_m P_m \left\langle R \middle| \psi_m \right\rangle \left\langle \psi_m \middle| k \right\rangle \right) \left\langle k \middle| \hat{O} \middle| C \right\rangle$$

$$= \sum_m P_m \left\langle R \middle| \psi_m \right\rangle \sum_k \left\langle \psi_m \middle| k \right\rangle \left\langle k \middle| \hat{O} \middle| C \right\rangle$$

Noting that $\sum_k |k\rangle\langle k| = \hat{I}$ allows us to dispose of the sum over k :

$$\left(\hat{D} \times \hat{O}\right)_{RC} = \sum_m P_m \left\langle R \middle| \psi_m \right\rangle \left\langle \psi_m \middle| \hat{O} \middle| C \right\rangle$$

If we now take the trace of this product, we mandate $R = C = N$ and sum over N. This gives:

$$\mathrm{Tr}\left(\hat{D} \times \hat{O}\right) = \sum_N \left(\hat{D} \times \hat{O}\right)_{NN} = \sum_N \left(\sum_m P_m \left\langle N \middle| \psi_m \right\rangle \left\langle \psi_m \middle| \hat{O} \middle| N \right\rangle \right)$$

A little re-ordering of the terms:

$$\mathrm{Tr}\left(\hat{D} \times \hat{O}\right) = \sum_m \sum_N P_m \left\langle \psi_m \middle| \hat{O} \middle| N \right\rangle \left\langle N \middle| \psi_m \right\rangle$$

gives the opportunity to use $\sum_N |N\rangle\langle N| = \hat{I}$, to produce:

$$\mathrm{Tr}\left(\hat{D} \times \hat{O}\right) = \sum_m P_m \left\langle \psi_m \middle| \hat{O} \middle| \psi_m \right\rangle$$

i.e., (Eq. 25.4) the expectation value. That's rather neat…

A density operator of the form:

$$\hat{D}_p = |\psi\rangle\langle\psi|$$

is often called a *pure state*, whereas:

$$\hat{D}_M = \sum_m P_m |\psi_m\rangle\langle\psi_m|$$

is a *mixed state*.

25.3 THE DENSITY OPERATOR AND EPR/BOHM-TYPE EXPERIMENTS

To see how all of this works in practice, we need to take a specific example.

A team of scientists get funding to carry out a very large EPR experiment of the type suggested by Bohm & Bell. They place one detector (Alice) on the Moon and another (Bob) on Earth. A space station placed rather nearer to the Moon than the Earth, is a source of particles prepared in a spin 0 singlet state and then released to fly off in opposite directions. Alice is set up to measure the vertical spin component. As the space station is nearer to the Moon, particles will arrive at Alice, causing the state to collapse before their counterparts arrive at Bob. However, there isn't enough time for the results of the Alice experiment to be sent to Earth before the other particle arrives at Bob.

As we know that the particles start in the singlet state, it's clear that if Alice measures $|U\rangle$ then the particle flying toward Earth must be in state $|D\rangle$, and vice-versa. Of course, in truth sitting on Earth, we do not know any of this, as we don't get the results of Alice's measurement before our particle arrives.

From our point of view, *we are forced to build a mixed state density operator to represent any particles heading our way.* Again, as the combination started out in the singlet state, we know that Alice has a 50:50 chance of measuring $|U\rangle$ or $|D\rangle$, so our arriving particle has a 50:50 chance of being $|D\rangle$ or $|U\rangle$. The density operator must then be:

$$\hat{D}_M = P_1|\psi_1\rangle\langle\psi_1| + P_2|\psi_2\rangle\langle\psi_2|$$

$$= \frac{1}{2}|D\rangle\langle D| + \frac{1}{2}|U\rangle\langle U|$$

Using the column vector and row vector representations:

$$|U\rangle = \begin{pmatrix} 1 \\ 0 \end{pmatrix} \qquad \langle U| = \begin{pmatrix} 1 & 0 \end{pmatrix}$$

$$|D\rangle = \begin{pmatrix} 0 \\ 1 \end{pmatrix} \qquad \langle D| = \begin{pmatrix} 0 & 1 \end{pmatrix}$$

The matrix form of the density operator (the *density matrix*) is then:

$$\hat{D}_M = \frac{1}{2} \times \begin{pmatrix} 0 \\ 1 \end{pmatrix} \otimes \begin{pmatrix} 0 & 1 \end{pmatrix} + \frac{1}{2} \times \begin{pmatrix} 1 \\ 0 \end{pmatrix} \otimes \begin{pmatrix} 1 & 0 \end{pmatrix}$$

where the ⊗ symbol reminds us that we are not involved in normal matrix multiplication, as the factors are in the wrong order for that. This is called a *tensor product*, and it works by embedding the column vector into the slots in the row vector, multiplied by the elements in the row. Hence:

$$\hat{D}_M = \frac{1}{2}\left(\begin{pmatrix} 0 \\ 1 \end{pmatrix} \times 0 \quad \begin{pmatrix} 0 \\ 1 \end{pmatrix} \times 1 \right) + \frac{1}{2}\left(\begin{pmatrix} 1 \\ 0 \end{pmatrix} \times 1 \quad \begin{pmatrix} 1 \\ 0 \end{pmatrix} \times 0 \right)$$

$$= \frac{1}{2}\begin{pmatrix} 0 & 0 \\ 0 & 1 \end{pmatrix} + \frac{1}{2}\begin{pmatrix} 1 & 0 \\ 0 & 0 \end{pmatrix}$$

$$= \begin{pmatrix} 1/2 & 0 \\ 0 & 1/2 \end{pmatrix}$$

Note that the 'off-diagonal' elements in *this* matrix are zero, which is certainly not always the case. In general density matrices, these elements will not be zero, but that doesn't matter as the expectation value is *always* given by tracing over the density matrix multiplied by the relevant operator, a process that only picks out the diagonal terms.

For example, as the z component spin operator $\hat{S}_z$ can be written as the matrix:

$$\hat{S}_z = \begin{pmatrix} 1 & 0 \\ 0 & -1 \end{pmatrix}$$

the expectation value for a vertical spin measurement here on Earth is:

$$\hat{S}_z = \text{Tr}\left(\hat{D}_M \times \hat{S}_z\right) = \text{Tr}\left(\begin{pmatrix} 1/2 & 0 \\ 0 & 1/2 \end{pmatrix} \times \begin{pmatrix} 1 & 0 \\ 0 & -1 \end{pmatrix}\right) = \text{Tr}\left(\begin{pmatrix} 1/2 & 0 \\ 0 & -1/2 \end{pmatrix}\right)$$
$$= (1/2) + (-1/2) = 0$$

which is exactly what you would expect given a 50:50 chance of UP/DOWN at Bob.

25.3.1 Representing a State

Next, let's look at a slightly different situation. Alice does not measure vertical spin; the SG magnet on the Moon has been set to some other angle, say horizontal for convenience.

However, we are expecting vertical eigenstates to arrive at Bob on Earth, as we assume that's how the Alice experiment has been set up.

Half of the time, Alice will measure $|R\rangle$ and so we get $|L\rangle$ on Earth. As we don't know this, *we carry on making vertical spin measurements* of the $|L\rangle$ or $|R\rangle$ states, resulting in either $|U\rangle$ and $|D\rangle$ with equal chance.

Overall, we still get 50:50 $|U\rangle/|D\rangle$ *no matter what Alice comes up with.* In other words, our density operator is still the same. For any angle that Alice chooses, we get the right expectation value using the *original density operator.*

In a fit of pure rebellion, the Alice team decides to point their magnet along some angle α between vertical and horizontal. Half the time, they will get spin UP along this angle, $|\alpha +\rangle$, and half the time spin DOWN, $|\alpha -\rangle$. Expanding these states in terms of the vertical basis, I am going to suppose that for angle α we get:

$$|\alpha +\rangle = \frac{1}{\sqrt{3}}\left(|U\rangle + \sqrt{2}|D\rangle\right)$$

$$|\alpha -\rangle = \frac{1}{\sqrt{3}}\left(\sqrt{2}|U\rangle - |D\rangle\right)$$

In this context, half the time Alice will measure $|\alpha +\rangle$, which means that we measure $|\alpha -\rangle$. On one-third of those occasions, our measurement of the vertical spin axis will give us $|D\rangle$ and for the other two-thirds of the time we get $|U\rangle$. However, if Alice measures $|\alpha -\rangle$, we receive $|\alpha +\rangle$. On one-third of those occasions we measure $|U\rangle$ and the rest of the time $|D\rangle$. *Overall*, we get $|U\rangle$. and $|D\rangle$ 50% of the time each. The density operator is correct once again.

Effectively there is no physically measurable difference between $|\alpha +\rangle$ and $|\alpha -\rangle$ as far as Bob on Earth is concerned. Indeed, even if team Bob decided to rebel as well and set their SG magnet to angle α by pure luck, they will still only see 50% spin UP along this angle and 50% spin DOWN, just as they would with any angle.

If you stop to think about it, *no measurement made on Earth can tell us what angle the Alice team has chosen.* This fits, as otherwise, *we would have invented a way of sending information faster than the speed of light.* It would be like the post office on the space station sending out two letters, one to Alice and the other to Bob. When the scientists of the Alice team read their letter, they find a 'speeling mistook' that they correct. This collapses the state and corrects the spelling of our letter while it is still on the way. That would be faster than light signalling, something that is apparently forbidden in our universe.

25.3.2 The Density Operator and Entangled States

Starting with a pure state density operator $\hat{D}_p = |\psi\rangle\langle\psi|$, we can expand it out over a basis $\{|i\rangle\}$ resulting in:

$$\hat{D}_p = \sum_j \sum_i a_j^* a_i |j\rangle\langle i|$$

The matrix elements of this operator are then:

$$\left(\hat{D}_p\right)_{rc} = \langle r\left|\left(\sum_j \sum_i a_j^* a_i |j\rangle\langle i|\right)\right|c\rangle = \sum_j \sum_i a_j^* a_i \langle r|j\rangle\langle i|c\rangle$$

If our basis is properly orthonormal, then $\langle i|c\rangle = 0$ for $i \neq c$ and $\langle i|c\rangle = 1$ for $i = c$ (the same being true for $\langle r|j\rangle$. Applying this to our matrix element forces it into the shape:

$$\left(\hat{D}_p\right)_{rc} = \sum_r \sum_c a_r^* a_c \langle r|r\rangle\langle c|c\rangle = \sum_r \sum_c a_r^* a_c$$

So now we see that:

$$\hat{D}_p = \begin{pmatrix} a_1^* a_1 & a_1^* a_2 & \dots & a_1^* a_m \\ a_2^* a_1 & a_2^* a_2 & \dots & a_2^* a_m \\ \dots & \dots & \dots & \dots \\ a_m^* a_1 & a_m^* a_2 & \dots & a_m^* a_m \end{pmatrix}$$

the diagonal elements being the probabilities for $|\psi\rangle$ to collapse into one of the basis states.

If our original state happened to be one of the bases, say $|k\rangle$, so that the density operator is $\hat{D}_p = |k\rangle\langle k|$, then every off-diagonal term in the matrix will vanish, and all the diagonal terms aside from the one at $r = k$, $c = k$ will vanish as well:

$$\hat{D}_p = |k\rangle\langle k| = \begin{pmatrix} 0 & 0 & \dots & 0 \\ 0 & 0 & \dots & 0 \\ 0 & 0 & a_k^* a_k & 0 \\ 0 & 0 & \dots & 0 \end{pmatrix}$$

However, this is clearly quite a specialist case. It is more likely that we have a mixed density operator, where we know that the state is in one of the bases, but we are not sure which one. We may know the relative probabilities; in which case we are able to construct a mixed density operator of the form:

$$\hat{D}_M = P_1|1\rangle\langle 1| + P_2|2\rangle\langle 2| + \cdots + P_n|n\rangle\langle n|$$

giving a matrix:

$$\hat{D}_M = \begin{pmatrix} P_1 & 0 & 0 & \cdots & 0 \\ 0 & P_2 & 0 & \cdots & 0 \\ \cdots & \cdots & \cdots & \cdots & \cdots \\ 0 & 0 & 0 & \cdots & P_n \end{pmatrix}$$

As we generally choose a basis made from the eigenstates of some physical variable, the diagonal density matrix must represent a mixture of eigenstates with the diagonal terms giving the probability of being in each one. We can write:

$$\Pr(I) = \langle I|\hat{D}_M|I\rangle \tag{25.5}$$

where I is a specific eigenvalue from the set $\{i\}$ with eigenstates $\{|i\rangle\}$

This is rather like the situation you find after a run of repeated measurements of the physical variable, which is a hint that a diagonal density matrix may have something to do with the measurement problem.

25.4 THE DENSITY MATRIX AND THE MEASUREMENT PROBLEM

The idea that the density matrix gives a different perspective on the measurement problem can be further developed by considering the density matrix for an entangled state, such as our standard singlet state:

$$|\psi_s\rangle = \frac{1}{\sqrt{2}}\left(|U\rangle|D\rangle - |D\rangle|U\rangle\right)$$

Constructing the density matrix for this state involves a slightly tedious calculation, but it will lead to some useful insight, so here we go:

$$\begin{aligned}
\hat{D}_p &= |\psi_s\rangle\langle\psi_s| \\
&= \frac{1}{2}\left(|U\rangle|D\rangle - |D\rangle|U\rangle\right) \times \left(\langle U|\langle D| - \langle D|\langle U|\right) \\
&= \frac{1}{2}\left\{|U\rangle|D\rangle\langle U|\langle D| - |U\rangle|D\rangle\langle D|\langle U| - |D\rangle|U\rangle\langle U|\langle D| + |D\rangle|U\rangle\langle D|\langle U|\right\} \\
&= \frac{1}{2}\left\{\left(|U\rangle\otimes\langle U|\right)\otimes\left(|D\rangle\otimes\langle D|\right) - \left(|U\rangle\otimes\langle D|\right)\otimes\left(|D\rangle\otimes\langle U|\right)\right. \\
&\quad \left. - \left(|D\rangle\otimes\langle U|\right)\otimes\left(|U\rangle\otimes\langle D|\right) + \left(|D\rangle\otimes\langle D|\right)\otimes\left(|U\rangle\otimes\langle U|\right)\right\}
\end{aligned}$$

Looking at each of the tensor products involved within the ( ) brackets:

$$|U\rangle \otimes \langle U| = \begin{pmatrix} 1 \\ 0 \end{pmatrix} \otimes \begin{pmatrix} 1 & 0 \end{pmatrix} = \begin{pmatrix} 1 & 0 \\ 0 & 0 \end{pmatrix}$$

$$|D\rangle \otimes \langle D| = \begin{pmatrix} 0 \\ 1 \end{pmatrix} \otimes \begin{pmatrix} 0 & 1 \end{pmatrix} = \begin{pmatrix} 0 & 0 \\ 0 & 1 \end{pmatrix}$$

$$|D\rangle \otimes \langle U| = \begin{pmatrix} 0 \\ 1 \end{pmatrix} \otimes \begin{pmatrix} 1 & 0 \end{pmatrix} = \begin{pmatrix} 0 & 0 \\ 1 & 0 \end{pmatrix}$$

$$|U\rangle \otimes \langle D| = \begin{pmatrix} 1 \\ 0 \end{pmatrix} \otimes \begin{pmatrix} 0 & 1 \end{pmatrix} = \begin{pmatrix} 0 & 1 \\ 0 & 0 \end{pmatrix}$$

Carrying this out for each term in the density matrix gives:

$$\hat{D}_p = \frac{1}{2}\left\{\begin{pmatrix} 1 & 0 \\ 0 & 0 \end{pmatrix}_A \otimes \begin{pmatrix} 0 & 0 \\ 0 & 1 \end{pmatrix}_B - \begin{pmatrix} 0 & 1 \\ 0 & 0 \end{pmatrix}_A \otimes \begin{pmatrix} 0 & 0 \\ 1 & 0 \end{pmatrix}_B - \begin{pmatrix} 0 & 0 \\ 1 & 0 \end{pmatrix}_A \otimes \begin{pmatrix} 0 & 1 \\ 0 & 0 \end{pmatrix}_B + \begin{pmatrix} 0 & 0 \\ 0 & 1 \end{pmatrix}_A \otimes \begin{pmatrix} 1 & 0 \\ 0 & 0 \end{pmatrix}_B\right\}$$

I have added subscripts to the various matrices to indicate if they apply to particle A or B.

Suppose we were given particle B to experiment on, the entanglement having been broken by a measurement on particle A, but we have not been told what the result of the measurement of A was. The way to deal with that situation is to average over all the possible outcomes of the measurement A. In other words, *we trace over the matrices related to A*. This produces the *reduced density matrix*.

$$\left(\hat{D}_p\right)_{\text{reduced}} = \frac{1}{2}\left\{\mathrm{Tr}\begin{pmatrix} 1 & 0 \\ 0 & 0 \end{pmatrix}_A \otimes \begin{pmatrix} 0 & 0 \\ 0 & 1 \end{pmatrix}_B - \mathrm{Tr}\begin{pmatrix} 0 & 1 \\ 0 & 0 \end{pmatrix}_A \otimes \begin{pmatrix} 0 & 0 \\ 1 & 0 \end{pmatrix}_B - \mathrm{Tr}\begin{pmatrix} 0 & 0 \\ 1 & 0 \end{pmatrix}_A \otimes \begin{pmatrix} 1 & 0 \\ 0 & 0 \end{pmatrix}_B + \mathrm{Tr}\begin{pmatrix} 0 & 0 \\ 0 & 1 \end{pmatrix}_A \otimes \begin{pmatrix} 1 & 0 \\ 0 & 0 \end{pmatrix}_B\right\}$$

$$= \frac{1}{2}\left\{\begin{pmatrix} 0 & 0 \\ 0 & 1 \end{pmatrix}_B + \begin{pmatrix} 1 & 0 \\ 0 & 0 \end{pmatrix}_B\right\}$$

$$= \begin{pmatrix} 1/2 & 0 \\ 0 & 1/2 \end{pmatrix}_B$$

Amazingly *the reduced density matrix formed by tracing over everything to do with particle A is exactly the same as the density matrix we constructed in the previous section to describe what was going to happen at the Bob detector.* Tracing over A has reduced the density matrix so that it now just contains every bit of information that we could extract about B.

This becomes especially intriguing when we cycle back to the measurement problem. As we have seen, the process of measurement produces an entanglement between the device and the system being measured. Equally, as it is impossible to totally isolate any apparatus from its environment, there will be an (incredibly complex) entanglement between the device and the environment as well. To extract the information about the measurement, we would have to trace over all the environmental factors. In practice, this is a fiercely complex mathematical problem, so the work that has been done is in the form of approximations. However, those results have been interesting as it would appear that *tracing over the environment diagonalises the density matrix*, i.e., all the off-diagonal elements vanish (or approximate to zero). One of our earlier results showed that a diagonal density matrix represents a collection of systems in eigenstates of the measurement device. On that basis, it looks as if e*nvironmental decoherence*, as this effect is called, reduces the density matrix to the measurement results. The process does not pick out any single element from the diagonal, so a realist would still be left with the mystery of instantiation on measurement. However, it is a promising line of attack if you have a more instrumentalist stance, or if you wish to regard quantum descriptions as only being applicable to a group of identical systems, rather than each individual within the collection. From this perspective, the density matrix is the most appropriate entity to represent general collections of systems.

However, if you prefer to seek out a realistic interpretation of amplitudes and quantum states, then the density matrix is not adding anything to the ontological stakes. It is just a very handy mathematical tool for dealing with practical situations.

NOTES

1 It is helpful to remember the ordering of terms and indices to make reading future equations easier.
2 In Chapter 26, we will distinguish between classical and quantum probabilities in a more formal sense.

26 Interpretations

By now we have so many interpretations, that it must be clear to all that there is some basic ambiguity as to what the formalism is telling us about the nature of quantum processes and their detailed relation to those occurring in the classical domain. The fact that we have very little idea of how to quantise gravity should provide a salutary warning to anyone claiming that we have fully understood the nature of quantum processes. I believe the difficulties the formalism presents is an indication that we need a very deep revision of our physical concepts and that the difficulties will certainly not be removed by a naive return to classical concepts.

B J Hiley[1]

26.1 WHAT IS AN INTERPRETATION?

Every physical theory needs a form of interpretation. When we write down a string of mathematical formulas, there has to be some link between the symbols on the page and the measurements that we make if the formulas are to be physically relevant. This is one of the jobs an interpretation is designed for.

While we were doing classical mechanics, there was no real problem involved; the intuitive association between mathematics and measurement was so close that a formal interpretation wasn't really needed. The theory of fields (gravity & electromagnetism) was more challenging to conceptualize, but even then physicists converged on an interpretive approach relatively quickly.

With quantum theory, things have changed. The intimate link between mathematics and measurement has been broken. We have a successful set of mathematical rules allowing us to manipulate symbols such as $|\varphi\rangle$, but that still leaves us free to argue over what $|\varphi\rangle$ means.

Not every mathematical symbol used in a theory needs a direct link to measurement. We associate the operator $\hat{p} = -i\hbar \dfrac{d}{dx}$ (in its position representation) with momentum and calculate $\langle\varphi|\hat{p}|\varphi\rangle$ to predict the average value of momentum observed. However, $\hat{p}$ itself is not in any sense 'measured' and it could well be (as some physicists think) that $|\varphi\rangle$ is nothing more than a mathematical convenience used to get us from one set of results to another, without representing anything in the real world. After all, we divide the surface of the Earth up with lines of latitude and longitude to help us find our way about, but we don't expect to find the lines drawn on the ground.

Arguably, though, there is something more to an interpretation than a simple set of rules to match symbols to physical properties. Many physicists and philosophers look to an interpretation to enshrine some physical picture of the world. Not a literal picture like a drawing, but something that we can glimpse in our minds as to the 'shape' of what's out there. This is partly done by the links between symbol and property, but as in the poem,[2] it is good to see what the whole elephant looks like…

Judged against that requirement, any instrumentalist view is not really an interpretation. Instrumentalists either do not require science to produce a picture of the world or are not fixed to one picture rather than any other. The appeal is the practicality; you can get on with doing science this way. It's an example of "shut up and calculate." However, as an approach, it does not seem likely to attract people into doing science in the first place. More plausibly, we enter the scientific establishment with realistic ambitions, or at least realistic assumptions. Coming across quantum theory for the first time can certainly undermine this thinking.

DOI: 10.1201/9781003225997-30

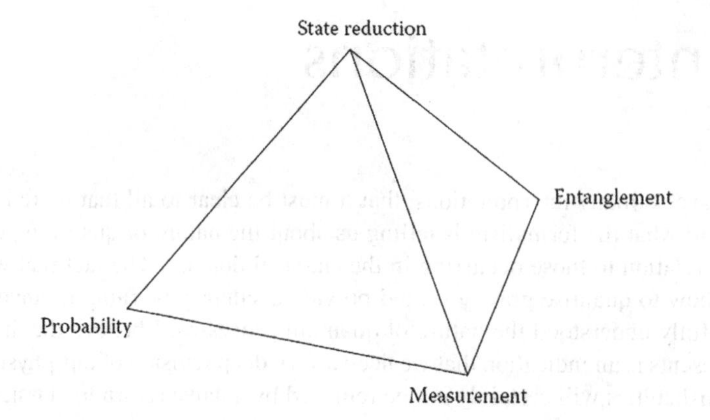

FIGURE 26.1 Chris Isham's "quaternity of problems" in the interpretation of quantum theory. They are drawn at the four corners of a tetrahedron to emphasize their interconnectedness.

26.2 A COLLECTION OF PROBLEMS

In his influential book on quantum theory, Chris Isham[3] draws attention to four key issues (Figure 26.1) that any interpretation of quantum theory must face:

1. The nature of probability
2. The reduction of the state vector
3. Entanglement
4. The role of measurement

One way of distinguishing between the various interpretations is to categorize how they approach these four issues.

26.2.1 The Nature of Probability

Fundamental to probability theory is the notion of a *sample space*. This is an *exhaustive* list of *exclusive* events that may come about in a given context. For example, with a simple coin toss the possible outcomes are Heads (H) and Tails (T), in which case the sample space is merely $S = \{H, T\}$. With a six-sided die, the sample space is $S = \{1, 2, 3, 4, 5, 6\}$. As the list is exhaustive, one event, s_I, in the list $\{s_i\}$ must happen. However, only one of the $\{s_i\}$ can happen at any one time, hence the events are exclusive.

Derived from the sample space is the *event algebra*, ε. This is the collection of all possible *compound events* based on the sample space. Taking our die as an example, the event $s_I = 4$ is a simple event from the event algebra. However, "s is odd", "$s<5$" etc., are all members of the event algebra. If the sample space contains a finite number of events, n, and if we count the entire sample space as a single event, I (something happened) along with an event called $\varnothing$ (nothing happened), then the total number of compound events in the event algebra is 2^n. To see this, consider building a compound event by saying yes (Y) or no (N) to each event in the sample space. Thus, the compound event "$s<5$" is the choice sequence $(1_Y, 2_Y, 3_Y, 4_Y, 5_N, 6_N)$. The presence, or otherwise, of each event in the sequence is a binary choice, so the total number of ways of building compound events out of a sequence of n events is 2^n.

The elements of the event algebra follow the rules of Boolean logic. For example, if we have $A =$ "$s<5$" and $B =$ "s is odd" then A AND $B = B$ AND $A = (1_Y, 2_N, 3_Y, 4_N, 5_N, 6_N)$, whereas A OR $B = B$ OR $A = (1_Y, 2_Y, 3_Y, 4_Y, 5_Y, 6_N)$.

A *probability distribution* assigns a positive number, $0 \le p \le 1$, *the probability* $p = \Pr(s_i)$ to each event, s_i, in the sample space, subject to $\sum_i \Pr(s_i) = 1$. Elements of the event algebra are given probabilities according to the event selections, so "$s<5$" has probability $\Pr(1)+\Pr(2)+\Pr(3)+\Pr(4)$. The event probabilities may well be equal, if the die is fair for example, but the scheme does not require that.

So much for the theory. A much thornier issue is the *meaning* of the probability that is assigned. One way to get a toehold on this is to consider a collection consisting of a very large number, N, of nominally identical systems. Such a collection is called an *ensemble*. Across the ensemble, we would expect each of the possible events within the sample space to be displayed by one or more of the systems. For example, that the event s_I had occurred in a number $\mathrm{N}(s_I)$ of the systems in the ensemble. In saying this, we are assuming that when we observe the ensemble all the systems have had the chance to 'select' one of the events and have not been able to change since then. Consequently:

$$\frac{\mathrm{N}(s_I)}{N} \approx \Pr(s_I)$$

where $\Pr(s_i)$ is the probability that a single system, chosen at random from the ensemble, will be in state s_I. One of the basic assumptions of probability theory is that:

$$\Pr(s_I) = \lim_{N \to \infty} \{\mathrm{N}(s_I)/N\} \tag{26.1}$$

which indicates how the probability can be obtained from an estimation that is increasingly reliable, as more and more examples are added to the ensemble. In practical terms, the working ensemble can be constructed from a set of systems of the same type (identical in every way that matters to the outcomes) or from a sequence of experiments carried out over time using the same system. If you take precautions so that the system is re-set before each experiment, that the experiments are performed in a consistent manner and that the outcomes of a previous experiment do not affect any of the later ones, then the procedure leads to a perfectly acceptable ensemble.

Having a robust process for estimating the probabilities of the events in a sample space still does not entirely get to the root of the matter. At the very least *it is unclear what we are actually estimating.*

In the case of a coin or a die, their physical nature goes some way in determining the probabilities in the sample space. A coin has two sides; a (conventional) die has six faces.

The way the coin is manufactured, the extent to which the metal is distributed across its width, etc., will all help to influence how the coin lands. To experimentally check the coins fairness, we would have to fix, as well as we can, other factors to do with the forces applied during the toss, air resistance etc., during a run of tosses. Small variations from toss to toss mean that we can never fully see the coin's nature. It just becomes progressively clearer as we make more tosses (build up the ensemble) or at least that's the assumption. By trying a large number of coin tosses, we're hoping that these fluctuations balance out from toss to toss: there will be as many occasions when an uneven toss tips the coin one way as the other, so overall these effects will smear out.

Let's say that we toss the coin 100 times and get 57 heads. Our estimation of the probability of a head is 57/100 or 0.57. This could mean one of several things:

- the value is a first approximation to some objective quantity which in some sense 'exists' outside to any attempt to measure it. The probability is *objective*.
- The value is a measure of some properties physically expressed in the structure of the coin. This would make the probability objective, but it would mean something slightly different in each scenario.

- The value specifically refers to that collection of 100 coin tosses, with limited use or validity outside of that collection. The probability is then a good way to summarize what happened in a given ensemble.

We generally assume that our probability estimation is pertinent outside the ensemble used to obtain it. We use such numbers to make further predictions; an approach that at least tacitly assumes some level of objectivity to the quantity we are estimating. Essentially, we would expect a second identical ensemble to display the same probabilities, at least to some level of accuracy.

If we take the same coin and toss it 1000 more times, so that 578 heads come out, the probability estimation is now 0.578. This might be a more accurate estimation of the 'underlying' probability, or it might just refer to that collection of 1000 tosses. After all, it would be possible to do another 1000 tosses with the same coin and end up with 643 heads. That could be a pointer that our original estimate was wrong, or it might just be one of those statistical things that happens from time to time.

The probability formula, Eq. 26.1 $\Pr(s_I) = \lim_{N\to\infty}\{N(s_I)/N\}$, appears at first glance to be mathematically similar to a concept from classical mechanics, where we define the *average speed* as:

$$\text{average speed} = \frac{\text{distance travelled}}{\text{elapsed time}}$$

When we watch an object move, we believe that it has a speed from moment to moment. But in an instant, it cannot travel any distance. Equally, over a more extended duration, it is likely that the speed will change, at least by small amounts, from moment to moment. Hence the estimation distance/time can only ever reveal the *average speed* during that time. We are led to define an *instantaneous speed*, v, via:

$$v = \lim_{\delta t\to 0}\{\delta x/\delta t\}$$

allowing the mathematics to gel with our intuition.

At first glance, similarity between this definition and Eq. 26.1, tricks us into thinking that the improving estimation, $N(s_i)/N$, as $N \to \infty$, converges on some objective value in the same way that average speed converges on instantaneous speed. We might have a *feeling* that there is some 'objective probability' we are estimating, but we can't 'see' it in the same way as speed.

We then have a dilemma: it is difficult to justify the use of probabilities to project outside of the ensemble used to estimate their value without some recourse to objectivity, yet we have no evident physical peg to hang this notion on.

In the case of the coin or die, we may have a glimpse of objectivity in their shape and construction which directly influences the probability.

On the other hand, probabilities sometimes derive from the application of physical laws. Take the example of a gas. The huge number of particles involved in even the smallest volume of gas makes trying to calculate exactly what is happening, based on the motion of every single particle, a hopeless task. However, the collection of particles is a natural ensemble, suggesting that a probabilistic approach would be fruitful.

Figure 26.2 shows the Maxwell–Boltzmann distribution for molecular speeds in a gas at three different temperatures. The distributions have been calculated for one million particles at each temperature, based on a calculation of the relative probabilities for each speed, founded on the appropriate laws of nature and standard statistical techniques.

If we pick a single particle out of the gas, the distribution tells us the probability that the chosen particle will have a selected speed. Nonetheless, it does not follow that the probability is a *property* of the particle. After all, each particle has a speed, whereas the probability distribution refers to *every speed* that a randomly chosen particle *might* have.

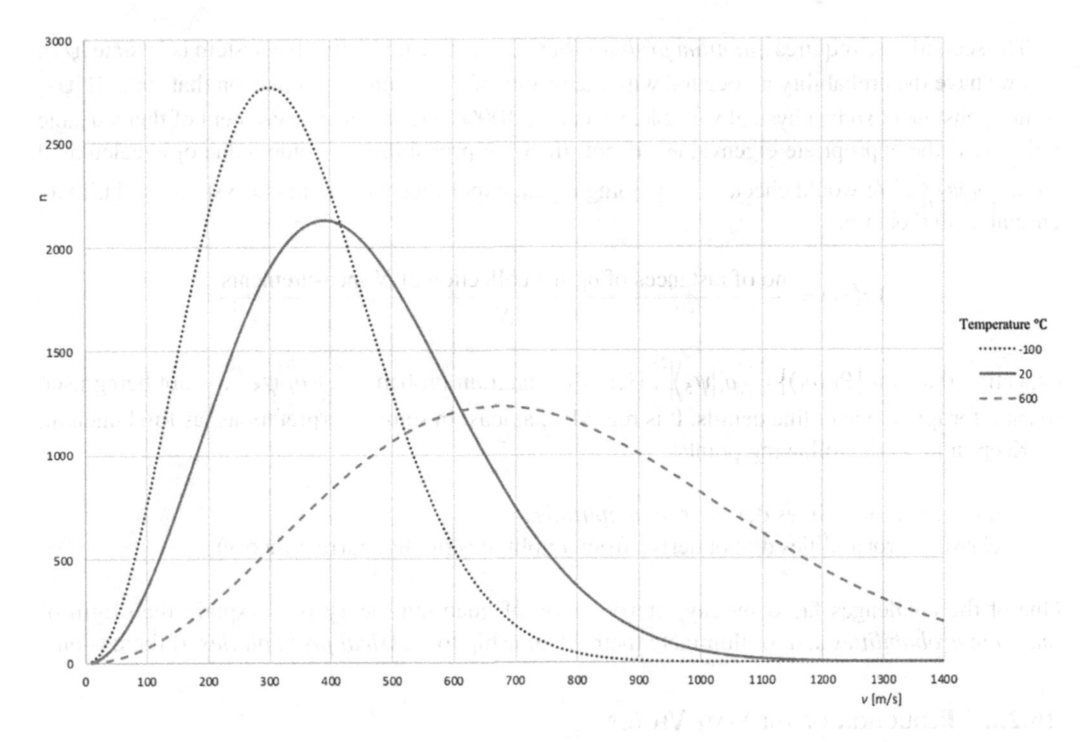

FIGURE 26.2 The Maxwell–Boltzman distribution of speeds within a gas at three different temperatures: –100°C, 20°C, and 600°C. The horizontal axis is speed in meters per second, and the vertical axis shows the number of particles at a given speed. Each distribution contains a total of one million particles.

Effectively we are dividing the gas into sub-ensembles, each of which contains just those particles with a given speed (or within a small range of a given speed). When we say that p is the probability of a selected particle having a specific speed, we mean that the chances of it having come from a given group is p. Then we experiment on a real gas to see how well it corresponds to the theory.

Here we have an example of a situation where probability is being used *to cover up our ignorance about the fine details of what's going on.* However, *we talk as if this probability exists in some objective way and that all the real gases measure up to this standard, to some extent.*

In quantum theory, probability pops up in two different ways.

In the first case, we deploy probability techniques when we are unsure which state applies to a quantum system. To be clear, this refers to an ensemble of systems, each of which is in a particular state $|\psi_I\rangle$ from a collection $\{|\psi_i\rangle\}$. However, in the case of any specific system, we are not aware of its state[4]. Alternatively, we may have one system and not be sure what state it is in. In either case, we construct a density matrix to summarize our information about the system:

$$\hat{D} = \sum_n P_n |\psi_n\rangle\langle\psi_n|$$

where the P_n are the probabilities of finding a system in state $|\psi_n\rangle$. We either estimate P_n by counting systems in sub-ensembles, one for each possible state, or calculate it based on some knowledge of how the systems are prepared. In either case, the probability is being used in the same fashion as per our gas particle, i.e., as a means of dealing with incomplete or inexact knowledge. We will refer to probabilities of this sort as *classical probabilities.*

The second case requires *quantum probabilities*. Even if we know that the system is in state $|\psi_5\rangle$, say, we have the probability associated with the results of a measurement made on that state. If $|\psi_5\rangle$ is an eigenstate of some physical variable, we can be 100% sure that a measurement of that variable will reveal the appropriate eigenvalue[5]. If not, then the probability of some value o_I is calculated from $|\langle o_I|\psi_5\rangle|^2$. We would check this by doing repeat experiments on identical systems, to build an ensemble, and obtain:

$$\Pr(o_I) = \frac{\text{no of instances of } o_I \text{ in a collection of } N \text{ measurements}}{N}$$

expecting that $\lim_{N\to\infty}\{\Pr(o_I)\} = |\langle o_I|\psi_5\rangle|^2$. Here the quantum probability, $|\langle o_I|\psi_5\rangle|^2$, is not being used to cater for ignorance of fine details. It is regarded, at least in some interpretations, as fundamental.

Keep in mind the following points:

- quantum probabilities derive from *amplitudes*
- classical probabilities do not derive from amplitudes (within current theory).

One of the challenges faced by any interpretation of quantum theory is to explain the origin of *quantum probabilities* and to illuminate their relationship to *classical probabilities*, if there is one.

26.2.2 Reduction of the State Vector

When we use an expression such as "the state of a system" or "the system is in state $|\psi\rangle$", our language carries inherently realistic overtones. Such statements are convenient linguistic forms but should not be taken to imply that the user believes in the objective reality of quantum states.

Some physicists deny that quantum states refer to individual examples of a system. They regard the state as a description of the ensemble, not the individuals within it. After all, to measure the expectation value of any physical property we have to make a series of measurements on identical systems. If the expectation value belongs to the collection, then it is a simple extension to think of the *state* in the same way.

However, taking this ensemble view leaves an open question regarding the physical properties in the systems and whether they have fixed values.

Let's imagine that we have an ensemble where the quantum states $\{|i\rangle\}$ with eigenvalues $\{o_i\}$ are distributed. Perhaps these values are distributed randomly among the members of the ensemble, to be revealed and averaged during a set of measurements. Or perhaps each measurement causes an individual system to manifest a value out of the spectrum of possibilities. A realist would ascribe a state to each separate system and hence divide the ensemble into several sub-ensembles, one for each $|i\rangle$. Someone with a less-realistic viewpoint would prefer to say that the ensemble is divided into sub-ensembles and the state refers to each sub-ensemble, but not the systems within.

In either case, we would describe the ensemble of states by the density operator:

$$\hat{D} = \sum_i P_i|i\rangle\langle i|$$

If we now perform a measurement, or collection of measurements, on the ensemble, we can use the results to divide the systems into their respective sub-ensembles. We can pick out all the members in the sub-ensemble of systems that ended up in state $|3\rangle$, for example. This sub-ensemble of $|3\rangle$ systems is described by $\hat{D}_3 = |3\rangle\langle 3|$. We have 'reduced the density matrix' from $\hat{D}$ to $\hat{D}_3$:

$$\hat{D} = \sum_i P_i|i\rangle\langle i| = \sum_i P_i\hat{D}_i \quad \Rightarrow \quad \hat{D}_3 = |3\rangle\langle 3|$$

a process that is not especially mysterious; it just reflects a change in our detailed knowledge of the systems.

As a next step, let's imagine making a further set of measurements on the $|3\rangle$ sub-ensemble. The physical property of choice is represented by an operator $\hat{O}$ with eigenstates $\{|\varphi_k\rangle\}$ and $|3\rangle$ is not a mutual eigenstate with any of them. After the measurement, the collection of states $|3\rangle$ has been divided into a collection of even smaller sub-sub-ensembles, one for each $|\varphi_k\rangle$ in $\{|\varphi_k\rangle\}$. The number of systems in each of these smaller piles is:

$$\text{Number of systems in sub-sub-ensemble } |\varphi_K\rangle = |\langle\varphi_K|3\rangle|^2 \times \text{number of systems in sub-ensemble } |3\rangle$$

where $|\langle\varphi_K|3\rangle|^2$ is a quantum probability.

When we describe this process, we talk in terms of the state $|3\rangle$ having reduced or collapsed into one of states in $\{|\varphi_k\rangle\}$:

$$|3\rangle \Rightarrow |\varphi_K\rangle$$

$$|3\rangle\langle 3| \Rightarrow |\varphi_K\rangle\langle\varphi_K|$$

for example. A realist views this as a process taking place in each individual system. In other words, some physical change has taken place, system by system. Consequently, the quantum probability must be a measure of the likelihood of this physical change, and so presumably related to the *extent* of the change, i.e., how different $|\varphi_K\rangle$ is from $|3\rangle$.

For those who are more instrumentalist in thinking, there is no physical change in any system. We are, once again, focussing on a smaller number of systems that are described as a set by $|\varphi_K\rangle$.

However, there is a catch. Imagine that we have performed this measurement on each system in the collection $|3\rangle$ so that the systems have been subdivided into piles of $\{|\varphi_k\rangle\}$, and that we then get some colleagues to make a further series of measurements, *without telling them exactly what we have done.* They are allowed to know that we've made a measurement of $\hat{O}$, so they know that they're dealing with a collection of states $\{|\varphi_k\rangle\}$, but won't know which system is which. From their point of view, *this is the first case that we discussed earlier* – ignorance of which system the state is in. In which case, they must describe the collective by the density operator:

$$\hat{D}_O = \sum_k p_k |\varphi_k\rangle\langle\varphi_k|$$

but *we* can go a little further than this. As we are aware of the full details, we know that the systems started off in state $|3\rangle$ and hence the various probabilities p_k in their density operator are actually $p_k = |\langle\varphi_k|3\rangle|^2$, making $\hat{D}'_o$ from our point of view:

$$\hat{D}'_o = \sum_k |\langle\varphi_k|3\rangle|^2 |\varphi_k\rangle\langle\varphi_k|$$

However, as they are describing the same thing, the two density operators must surely be the same, $\hat{D}_O = \hat{D}'_o$. In which case, the difference between a classical and quantum probability, in this scenario, is just a matter of perspective. From their point of view, they are treating the p_k as classical probabilities, but we know them to be quantum probabilities.

Several questions consequently arise:

- Is there any difference really between quantum and classical probabilities?
- Do they convert from one to the other under different circumstances?

- Is the distinction a matter of perspective and knowledge about the background?
- Can we maintain a sensible distinction between:

$$\hat{D} = \sum_i P_i |i\rangle\langle i| \Rightarrow \hat{D}_3 = |3\rangle\langle 3| \text{ and } |3\rangle \Rightarrow |\varphi_K\rangle$$

if the probabilities are of the same form?

These are all questions that a realistic interpretation must attempt to resolve.

26.2.3 ENTANGLEMENT

An entanglement is a state of more than one system that can't be factorized into a combination of unique states for each system, $|\Psi\rangle \neq |\psi\rangle|\varphi\rangle$. The existence of entangled states in quantum theory is remarkable, but not just because of their amazing physical properties; they also potentially cut across one of the basic assumptions behind much of Western philosophy and science.

A particularly Western approach to studying the world is the use of *reductionism*: the notion that any system can be completely understood by examining its constituent parts. For this to work, we need to assume that the parts behave in the same fashion when they are members of the whole as they do when they are isolated. One problem with entangled states is that they blur the distinction between *part* and *whole*.

At one level of approximation, two electrons can be treated separately and the behaviour of one has no influence on the other. However, in detail, every electron is in an entangled state with all the other electrons in the universe, just from the fact that they are identical fermions and hence have to be in antisymmetric states. Allow them to take part in an EPR-type experiment and they become even more deeply entangled. With a combined state such as $|\Psi\rangle = \frac{1}{\sqrt{2}}\left[|\varphi_A\rangle|\psi_B\rangle - |\psi_A\rangle|\varphi_B\rangle\right]$, we can no longer say with any certainty that electron A is in state $|\varphi\rangle$ while electron B is in state $|\psi\rangle$. This is tantamount to having two parts, but not being able to tell which is which or indeed who is doing what, undermining the notion of parts. Also, the correlated behaviour of the two electrons in this state indicates that they are certainly affected by being part of the whole. The erosion of this view is, if anything, further promoted by quantum field theory.

Entanglement presents us with a philosophical challenge, one that threatens to pick away at our notion of what a 'thing' is, but at another level, it also presents us with a practical problem: why does anything happen at all?

As we have seen, the interaction between a particle and a measuring device necessarily entangles particle state with device state (provided we accept that the measuring device can be described by quantum theory):

$$|\Psi\rangle|\varphi_n\rangle = \left(a|U\rangle + b|D\rangle\right)|\varphi_n\rangle \quad \Rightarrow \quad a|U\rangle|\varphi_U\rangle + b|D\rangle|\varphi_D\rangle$$

where $|\varphi_n\rangle$, $|\varphi_U\rangle$, and $|\varphi_D\rangle$ represent the neutral, recording UP, and recording DOWN states of the measuring device, respectively. While the entanglement exists, neither the particle nor the measuring device can be said to be in one state or another. If we take a literal interpretation, the measuring device (like Schrödinger's cat) is in an intermediate form of existence that we certainly don't observe in practice.

It is not just a measurement interaction that will entangle states. Any interaction between two quantum systems will entangle their states. Consequently, the entanglement spreads like an infectious disease.

Entanglement is governed by the $\hat{U}(t)$ operator constructed from the Hamiltonian specifying the energy of interaction between device and particle. Being a linear operator, $\hat{U}(t)$ cannot produce state collapse, so the manner by which a possibility becomes an actuality is an open question of physics, as well as a challenge for interpretation.

26.2.4 Measurement

There are two questions we must answer if the measurement problem is to be solved:

1. What makes a measurement interaction any different from an ordinary interaction?
2. What causes a state to collapse if indeed it happens at all?

Returning to our earlier discussion of probability, it should be clear that there is no measurement problem if we relate states to ensembles of systems. A state $|\varphi\rangle$ refers to an ensemble of systems and a state $|i\rangle$ to one of the smaller sub-ensembles within. No physical change has taken place, just a shift in our focus. This disposes of the measurement problem in one sense but replaces it with another.

How do we know which systems go into which pile? If we simply sift them by making a set of measurements, we are suggesting that the systems' properties have values before we make a measurement of them. This sounds somewhat like the sorting of electrons that we tried in Section 2.2, and it didn't work then (see Section 26.3.2). An ensemble-type view of state collapse follows if we accept the limitations of instrumentalism, but in turn, it raises its own fundamental problems.

A realistic view of state collapse seems part and parcel of a realistic approach to the measurement problem, as without a state collapsing in some manner, the states of a system are hopelessly entangled with those of the measuring device. If state collapse actually happens, then quantum probabilities must exist, and so amplitudes appear to have some level of objective reality.

To state the measurement problem in a more formal way, we would point to three separate assumptions:

1. the wave function is a complete description of a quantum system;
2. the wave function evolves linearly according to the equations of quantum theory;
3. a measurement produces a value of a physical variable (subject to normal experimental constraints)

and assert that these are not compatible with each other. A hidden variable approach can then be seen as addressing the issue through denying assumption 1. The Many Worlds interpretation tackles the problem by denying assumption 3. Objective collapse theories seek to replace 2. Proponents of the consistent histories approach assert that state collapse happens on the page during a calculation and is not a physical process.

26.3 IMPORTANT THEOREMS

When evaluating interpretations, it is worth considering two important theorems, alongside Isham's tetrahedron of issues: Bell's inequality and the Kochen-Specker theorem. Both serve to constrain any hidden variable theories that can reproduce the results of quantum theory. In the process, they illuminate some of the more subtle aspects of the quantum world.

26.3.1 Bell's Inequality

As we have already seen, Bell's inequality is violated by the statistical correlations present between quantum mechanical results. Experiments are telling us, subject to the steady closure of increasingly implausible loopholes, that quantum reality is fundamentally non-local[6]. Taken to its logical conclusions, non-locality undermines any attempt to define particles as independent wholes, and consequently the ultimate application of reductionism.

However, the violation of Bell's inequality does not rule out hidden variables. It just means that the variables would have to communicate at faster than light speeds. Any interpretation of quantum theory must make these correlations plausible without violating relativity. A realistic view of an entangled state would regard the particles as part of one whole, so that the correlations exist within the system, not between two (or more) systems.

26.3.2 The Kochen-Specker Theorem

If we merely wanted to reproduce the randomness inherent in quantum processes, we could simply specify a randomly distributed hidden variable for each physical quantity. Arguably, that would not get us very far, as it would be no advance on what we already have. Also, measurement outcomes are not simply *random* (at least when not using a prepared eigenstate), they are also *correlated* with other measurements, a feature that would not be reproduced by a set of purely random hidden variables.

Bell's inequality already deals with the inability to reproduce quantum correlations using local hidden variables. The Kochen-Specker theorem speaks to the *contextuality* of quantum theory[7].

The whole point of proposing hidden variables is to underpin aspects of quantum physics with something a little more 'classical'. One of the weirdest characteristics of quantum theory is the way that it deals with the values of physical variables. We have repeatedly argued that physical variables do not have definite values prior to any measurement. The process of measurement actualizes potentials present within the state. This contrasts sharply with classical physics, where systems are presumed to have values of physical variables irrespective of any measurement that may happen. Hidden variable theories seek to revert to definite values, possibly distributed according to some other variable, such as λ (which we used in our proof of Bell's inequality), that is currently not known to us. Hence, one of the foundational assumptions of a hidden variable theory would be:

Value Definiteness (V): all the physical variables of a system have definite values at all times.

There is also a connected premise: experiments reveal values of physical properties that exist independently of our decision to measure them. Hopefully, the measured values faithfully correspond to the actual values of the physical properties, but this is not a *necessary* aspect to the assumption.

Equally, one would wish to construct a theory based on the related assumption of:

Noncontextuality (N): the value of an observable belonging to a system is independent of the measurement context, i.e., the value is independent of any specific technique used to measure it and *any other measurements that could potentially take place at the same time.*

Finally, to reproduce quantum effects, we must grant the existence of:

Operators (O): we can allocate a specific mathematical operator and corresponting orthonormal basis, to each physical property of the system.

The Kochen-Specker theorem proves that *the combination of* V+N+O *cannot be applied consistently.*

26.3.3 Proving the Kochen-Specker Theorem

Opening Moves

Whenever you first try and come to terms with a proof purporting to deal with a fundamental aspect of theory, one of the challenges is to temporarily unlearn what you already know. It is hard to avoid bringing in ideas 'too soon' and hence generate confusion. Please keep this in mind over the next few pages.

The key issue that we are trying to address is the process by which values are ascribed to physical variables. Let us discount what we have already learned in that context and try and build something up from scratch that incorporates assumption V from the previous section.

We assume that we can relate any physical variable, Q, to an operator $\hat{Q}$ (assumption O from before) and real number q which is the *value* of the variable when the system is in some arbitrary state $|\psi\rangle$. Quite how this relationship works is left as an open question. However, we will write $q = v(\hat{Q})$ to propose some *value function*, v, which is doing the work for us. Of course, we are tempted to say that if $|\psi\rangle$ is an eigenstate of $\hat{Q}$ then $v(\hat{Q})$ will be the eigenvalue, but we should avoid specifying a specific way that v works for the purposes of the proof[8].

This whole procedure is embodying assumption V, as measurement is not coming into this. The physical variable Q *has* the value $q = v(\hat{Q})$ when the system is in state $|\psi\rangle$ (I'll drop the reference to the state from now on; just carry it as an aspect in the background).

If these values are to be of any use to us and work at all like classical values, then for any two physical variables A & B:

$$v(\hat{A}\hat{B}) = v(\hat{A}) \times v(\hat{B})$$

$$v(\hat{A} + \hat{B}) = v(\hat{A}) + v(\hat{B})$$

However, we might bear in mind that these relationships are only going to work if the operators commute.

Back in Section 12.2.4, we introduced *projection operators* with the following properties:

PROJECTION OPERATORS:

Given a basis set $\{|i\rangle\}$ of eigenstates of an observable $\hat{O}$ with eigenvalues o_i then

$$\hat{P}_i = |i\rangle\langle i| \qquad \sum_i \hat{P}_i = \hat{I} \qquad \hat{P}_i\hat{P}_j = \delta_{ij}\hat{P}_j \qquad \hat{O} = \sum_i o_i\hat{P}_i$$

These will be important to the Kochen-Specker theorem and form an increasing part of our consideration of various interpretations in the chapters to come.

To start our proof, we pick a basis set $\{|i\rangle\}$ off the shelf with associated $\{\hat{P}_i\}$. Ascribing a value to the operators $\hat{P}_i$ via $v(\hat{P}_i)$, we can reason as follows:

$$v(\hat{P}_i^2) = v(\hat{P}_i\hat{P}_i) = v(\hat{P}_i)v(\hat{P}_i) = \left[v(\hat{P}_i)\right]^2$$

However, as $\hat{P}_i^2 = \hat{P}_i$, it must be the case that:

$$v(\hat{P}_i^2) = \left[v(\hat{P}_i)\right]^2 = v(\hat{P}_i\hat{P}_i) = v(\hat{P}_i)$$

Hence $v(\hat{P}_i) = 0$ or $v(\hat{P}_i) = 1$. This will be crucial.

Now we assume that at least one other physical variable Q has a non-zero value, q, for state $|\psi\rangle$. In which case:

$$q = v(\hat{Q}) = v(\hat{I}\hat{Q}) = v(\hat{I})v(\hat{Q}) = v(\hat{I})q$$

So, $v(\hat{I}) = 1$.

Finally, as $\sum_i \hat{P}_i = \hat{I}$:

$$\sum_i v(\hat{P}_i) = v(\hat{I}) = 1$$

Given that each $v(\hat{P}_i) = 0$ or 1, it follows that in any case (as applied to a state $|\psi\rangle$) *one of the* $\hat{P}_i$ *in the set has value 1 and all the others must have value* 0. How we would establish which of the $\hat{P}_i$ would take the value 1 would be a question for additional physics relevant to the state under consideration.

Development

A formal proof of the Kochen-Specker theorem would require a certain degree of preliminary work to establish that:

1. it is sufficient to prove the theorem using operators with only three eigenstates and
2. it is sufficient to prove the theorem using linear combinations of basis states with real coefficients.

which is unfortunately beyond what we can achieve here. However, if we take such establishment work for granted, we can proceed as follows[9]:

Consider a basis set of three orthonormal states $\{|1\rangle, |2\rangle, |3\rangle\}$ with associated projection operators $\hat{P}_1 = |1\rangle\langle 1|$, $\hat{P}_2 = |2\rangle\langle 2|$ and $\hat{P}_3 = |3\rangle\langle 3|$. With this combination:

$$\sum_i \hat{P}_i = \hat{P}_1 + \hat{P}_2 + \hat{P}_3 = \hat{I} \qquad \sum_i v(\hat{P}_i) = v(\hat{P}_1) + v(\hat{P}_2) + v(\hat{P}_3) = 1$$

One of the $\hat{P}_i$ has the value 1 with the others 0, and the value of the identity operator is 1.

The assumption of value definiteness, V, would insist that these values pre-exist in an arbitrary state, ready to be discovered in the process of measurement. Arguably, we have already seen in earlier chapters that this view cannot be consistently applied, but we grant it licence in this context to see how things play out in the proof.

The three states can be represented in a column vector format as:

$$|1\rangle = \begin{pmatrix} 1 \\ 0 \\ 0 \end{pmatrix} \qquad |2\rangle = \begin{pmatrix} 0 \\ 1 \\ 0 \end{pmatrix} \qquad |3\rangle = \begin{pmatrix} 0 \\ 0 \\ 1 \end{pmatrix}$$

without any loss of generality. For one of these states, $v(\hat{P}) = 1$ and for the other two $v(\hat{P}) = 0$.

It is possible to construct a range of linear combinations of these states that also form orthonormal bases. Indeed, we can translate from one basis to another using Dirac's techniques. For example, the set:

$$|1'\rangle = \begin{pmatrix} 0 \\ 1 \\ 0 \end{pmatrix} \qquad |2'\rangle = \frac{1}{\sqrt{2}}\begin{pmatrix} 1 \\ 0 \\ 1 \end{pmatrix} \qquad |3'\rangle = \frac{1}{\sqrt{2}}\begin{pmatrix} -1 \\ 0 \\ 1 \end{pmatrix}$$

is an orthonormal basis, as is:

$$|1''\rangle = \frac{1}{\sqrt{2}}\begin{pmatrix} 1 \\ 1 \\ 0 \end{pmatrix} \qquad |2''\rangle = \frac{1}{2}\begin{pmatrix} -1 \\ 1 \\ \sqrt{2} \end{pmatrix} \qquad |3''\rangle = \frac{1}{2}\begin{pmatrix} 1 \\ -1 \\ \sqrt{2} \end{pmatrix}$$

Table 26.1 contains a collection of basis sets. Our initial basis is shown in column 1. The other columns are variations on that initial set, one per column, across the table.

Note that for compactness in the table, the normalization constants are not shown, and the states are displayed as row vectors rather than column vectors.

TABLE 26.1
The Basis Sets Used in the Proof of the K-S Theorem

1	2	3	4	5	6	7	8	9	10	11
$\hat{P}_i^1$	$\hat{P}_i^2$	$\hat{P}_i^3$	$\hat{P}_i^4$	$\hat{P}_i^5$	$\hat{P}_i^6$	$\hat{P}_i^7$	$\hat{P}_i^8$	$\hat{P}_i^9$	$\hat{P}_i^{10}$	$\hat{P}_i^{11}$
$(1, 0, 0)$	$(0, 1, 0)$	$(1, 0, 0)$	$(1, 1, 0)$	$(0, 1, 0)$	$(0, -1, 1)$	$(0, 1, 0)$	$(1, -1, 0)$	$(1, 0, 0)$	$(-1, 0, 1)$	$(1, 0, 0)$
$(0, 1, 0)$	$(-1, 0, 1)$	$(0, -1, 1)$	$(-1, 1, \sqrt{2})$	$(1, 0, \sqrt{2})$	$(\sqrt{2}, 1, 1)$	$(\sqrt{2}, 0, 1)$	$(1, 1, \sqrt{2})$	$(0, 1, \sqrt{2})$	$(1, \sqrt{2}, 1)$	$(0, \sqrt{2}, 1)$
$(0, 0, 1)$	$(1, 0, 1)$	$(0, 1, 1)$	$(1, -1, \sqrt{2})$	$(-\sqrt{2}, 0, 1)$	$(-\sqrt{2}, 1, 1)$	$(-1, 0, \sqrt{2})$	$(-1, -1, \sqrt{2})$	$(0, -\sqrt{2}, 1)$	$(1, -\sqrt{2}, 1)$	$(0, -1, \sqrt{2})$
$(1, 1, 0)$			$(-\sqrt{2}, 0, 1)$	$(-\sqrt{2}, 1, 1)$	$(-1, 0, \sqrt{2})$	$(-1, -1, \sqrt{2})$	$(0, -\sqrt{2}, 1)$	$(1, -\sqrt{2}, 1)$	$(0, -1, \sqrt{2})$	
$(1, -1, 0)$			$(0, \sqrt{2}, 1)$							

For the moment, focus on the top three rows in the table (those above the gap).

As each combination of the initial basis set is also a basis, each column represents a set of projection operators. Take, for example, column 11 containing the vectors (showing their normalization constants) $\begin{pmatrix}1 & 0 & 0\end{pmatrix}$, $\frac{1}{\sqrt{3}}\begin{pmatrix}0 & \sqrt{2} & 1\end{pmatrix}$ and $\frac{1}{\sqrt{3}}\begin{pmatrix}0 & -1 & \sqrt{2}\end{pmatrix}$.

These give rise, via tensor multiplication, to:

$$\hat{P}_1^{11} = \begin{pmatrix}1\\0\\0\end{pmatrix}\otimes\begin{pmatrix}1 & 0 & 0\end{pmatrix} = \left(\begin{pmatrix}1\\0\\0\end{pmatrix}\times 1 \quad \begin{pmatrix}1\\0\\0\end{pmatrix}\times 0 \quad \begin{pmatrix}1\\0\\0\end{pmatrix}\times 0\right) = \begin{pmatrix}1 & 0 & 0\\0 & 0 & 0\\0 & 0 & 0\end{pmatrix}$$

$$\hat{P}_2^{11} = \frac{1}{3}\begin{pmatrix}0\\\sqrt{2}\\1\end{pmatrix}\begin{pmatrix}0 & \sqrt{2} & 1\end{pmatrix} = \frac{1}{3}\begin{pmatrix}0 & 0 & 0\\0 & 2 & \sqrt{2}\\0 & \sqrt{2} & 1\end{pmatrix}$$

$$\hat{P}_3^{11} = \frac{1}{3}\begin{pmatrix}0\\-1\\\sqrt{2}\end{pmatrix}\begin{pmatrix}0 & -1 & \sqrt{2}\end{pmatrix} = \frac{1}{3}\begin{pmatrix}0 & 0 & 0\\0 & 1 & -\sqrt{2}\\0 & -\sqrt{2} & 2\end{pmatrix}$$

(using the superscript to label the column number, not a power of the operator) so that:

$$\hat{P}_1^{11} + \hat{P}_2^{11} + \hat{P}_3^{11} = \begin{pmatrix}1 & 0 & 0\\0 & 0 & 0\\0 & 0 & 0\end{pmatrix} + \frac{1}{3}\begin{pmatrix}0 & 0 & 0\\0 & 2 & \sqrt{2}\\0 & \sqrt{2} & 1\end{pmatrix} + \frac{1}{3}\begin{pmatrix}0 & 0 & 0\\0 & 1 & -\sqrt{2}\\0 & -\sqrt{2} & 2\end{pmatrix} = \hat{I}$$

Hence, for this set, $\sum_i \hat{P}_i^{11} = \hat{I}$ and $v\left(\hat{P}_1^{11}\right) + v\left(\hat{P}_2^{11}\right) + v\left(\hat{P}_3^{11}\right) = 1$.

As before, and *as for every column in the table*, one of the projection operators in the set has value 1 and the others in the set have value 0.

The 11 variations shown here are part of a full set of 33 generated in a systematic and symmetrical manner from the initial basis. For detailed reasons, the full set of 33 must to exist for the proof to work, but the actual argument only needs the 11 shown (or indeed any 11 taken from the 33).

The extra two rows (below the gap) that show states in some of the columns will be explained in the next section.

Endgame

Given that we have not specified the state $|\psi\rangle$ for the system in question and given that we have not made any specific assertion of how $v\left(\hat{P}\right)$ works, I am at liberty to select which of the projectors in each column has the value 1. At least I can do that initially. We will see how the consequences play out.

We now turn to Table 26.2. In column 1, I have arbitrarily selected $|3\rangle$ to be the state with $v\left(\hat{P}_3^1\right) = 1$ and shaded it appropriately. The other states are now forced to have value $v\left(\hat{P}_1^1\right) = v\left(\hat{P}_2^1\right) = 0$. These states have been shaded differently to indicate $v\left(\hat{P}\right) = 0$ for them. This is implementing *value definiteness*. All other instances of $|2\rangle$ and $|3\rangle$ in the table must also have $v\left(\hat{P}\right) = 0$.

TABLE 26.2
Opening Moves in the Final Stages of the Kochen-Specker Proof

1	2	3	4	5	6	7	8	9	10	11
$\hat{P}_i^1$	$\hat{P}_i^2$	$\hat{P}_i^3$	$\hat{P}_i^4$	$\hat{P}_i^5$	$\hat{P}_i^6$	$\hat{P}_i^7$	$\hat{P}_i^8$	$\hat{P}_i^9$	$\hat{P}_i^{10}$	$\hat{P}_i^{11}$
(1, 0, 0)	(0, 1, 0)	(1, 0, 0)	(1, 1, 0)	(0, 1, 0)	(0, −1, 1)	(0, 1, 0)	(1, −1, 0)	(1, 0, 0)	(−1, 0, 1)	(1, 0, 0)
(0, 1, 0)	(−1, 0, 1)	(0, −1, 1)	$(-1, 1, \sqrt{2})$	$(1, 0, \sqrt{2})$	$(\sqrt{2}, 1, 1)$	$(\sqrt{2}, 0, 1)$	$(1, 1, \sqrt{2})$	$(0, 1, \sqrt{2})$	$(1, \sqrt{2}, 1)$	$(0, \sqrt{2}, 1)$
(0, 0, 1)	(1, 0, 1)	(0, 1, 1)	$(1, -1, \sqrt{2})$	$(-\sqrt{2}, 0, 1)$	$(-\sqrt{2}, 1, 1)$	$(-1, 0, \sqrt{2})$	$(-1, -1, \sqrt{2})$	$(0, -\sqrt{2}, 1)$	$(1, -\sqrt{2}, 1)$	$(0, -1, \sqrt{2})$
(1, 1, 0)			$(-\sqrt{2}, 0, 1)$	$(-\sqrt{2}, 1, 1)$	$(-1, 0, \sqrt{2})$	$(-1, -1, \sqrt{2})$	$(0, -\sqrt{2}, 1)$	$(1, -\sqrt{2}, 1)$	$(0, -1, \sqrt{2})$	
(1, −1, 0)			$(0, \sqrt{2}, 1)$							

The state $|3\rangle$ in column 1 has been selected to have the value 1. It has been dark shaded to indicate this. As only one projector in a set can have value 1, the others in the same column must be value 0. They have been shaded in a lighter colour to indicate their different value. All other occurrences of the same states have also been shaded.

This is implementing *non-contextuality*, as the occurrence of a state, say $|2\rangle$, in another column is placing it in a different measurement context – there are a different set of projection operators acting along with it as part of that basis. In Table 26.2, I have shaded all occurrences of states as per their values determined by column 1.

Turning our attention to the last two rows in the table, the additional states contained here are orthogonal to *one* of the states listed above them in the same column. In column 1, the states $\begin{pmatrix}1 & 1 & 0\end{pmatrix}$ and $\begin{pmatrix}1 & -1 & 0\end{pmatrix}$ are both orthogonal to $\begin{pmatrix}0 & 0 & 1\end{pmatrix}$. For example:

$$\begin{pmatrix}1 & 1 & 0\end{pmatrix}\begin{pmatrix}0\\0\\1\end{pmatrix} = 0$$

These extra states do not form part of the basis, nor are they needed for the projection operators in the column to sum to the identity. Indeed, you will find them elsewhere in the table where they are part of a different basis; the extra vector $\begin{pmatrix}1 & 1 & 0\end{pmatrix}$ from column 1 also pops up as part of the basis in column 4.

While they might at first glance appear to be decorative excess baggage, these states play a vital role in the argument. Given that an extra state $|K\rangle$ is orthogonal to a specific state $|I\rangle$ from the set $\{|i\rangle\}$ in the same column, i.e., $\langle K|I\rangle = 0$, then as $\hat{P}_K^j = |K\rangle\langle K|$ we have $\hat{P}_K^j\hat{P}_I^j = |K\rangle\langle K|I\rangle\langle I| = 0$. Having picked the value of the projector in the 3rd row of the first column to be 1, $v\left(\hat{P}_3^1\right) = 1$, then the value of both extra states in the same column must be zero. After all, $v\left(\hat{P}_K^1\hat{P}_3^1\right) = v\left(\hat{P}_K^1\right)v\left(\hat{P}_3^1\right) = v\left(\hat{P}_K^1\right)\times 1$. However, as $\hat{P}_K^j\hat{P}_3^j = 0$ it follows that $v\left(\hat{P}_K^1\hat{P}_3^1\right) = 0$. So, $v\left(\hat{P}_K^1\right)\times 1 = 0$, hence $v\left(\hat{P}_K^1\right) = 0$.

As a result, all the states in the last two rows of column 1 (below the gap) must have $v\left(\hat{P}\right) = 0$. They have been shaded for $v\left(\hat{P}\right) = 0$, *along with any other occurrences of the same states across the table*. The argument, of course, extends to any of the extra states in the other columns. We will find that the extra states picked to be shown in these rows have been judiciously (and cunningly) selected from elsewhere in the table. The extra states in each column are *orthogonal to the state that we pick to have value 1*. Hence, I am being prescient regarding some of the upcoming selections. However, this is not some trickery in the proof. If you think of the proof as a progression from column to column, then once I have picked the $v = 1$ state from a column, I proceed to scour the rest of the table looking for states orthogonal to my selected state and copy them into the column below the gap. The key point for these states is that they must have value 0, given that their selected orthogonal partner has $v = 1$. They are copied below the gap in the column to make the reason why they have $v = 0$ more evident and they are shaded there, as well as their other locations in the table. The table is presented in its complete form in order to economize on space, rather than building it column by column over several tables.

If I now move to column 2, I am blocked from selecting the state in row 1 as its value has already been fixed as 0 from its occurrence in column 1. So, I will pick the state in the third row and set its value to be 1.

Proceeding with the random selection process where I can, given that some states have already been assigned values from applying the rules V and N, there are no issues until I get to column 5. There I no longer have any choice over the state with value 1. The only free state left is in the second row (Table 26.3)

From this point onwards, the selections are all forced, until we reach column 11 (Table 26.4).

Column 11 closes off the proof. In this column, we see that *all the basis states (or rather the projection operators derived from them) have their values pre-assigned by the consequences of the*

TABLE 26.3
By the Time My Arbitrary Selections Get to the Fifth Column, There Is No Free Choice Remaining

1	2	3	4	5	6	7	8	9	10	11
$\hat{P}_i^1$	$\hat{P}_i^2$	$\hat{P}_i^3$	$\hat{P}_i^4$	$\hat{P}_i^5$	$\hat{P}_i^6$	$\hat{P}_i^7$	$\hat{P}_i^8$	$\hat{P}_i^9$	$\hat{P}_i^{10}$	$\hat{P}_i^{11}$
(1, 0, 0)	(0, 1, 0)	(1, 0, 0)	(1, 1, 0)	(0, 1, 0)	(0, −1, 1)	(0, 1, 0)	(1, −1, 0)	(1, 0, 0)	(−1, 0, 1)	(1, 0, 0)
(0, 1, 0)	(−1, 0, 1)	(0, −1, 1)	$(-1, 1, \sqrt{2})$	$(1, 0, \sqrt{2})$	$(\sqrt{2}, 1, 1)$	$(\sqrt{2}, 0, 1)$	$(1, 1, \sqrt{2})$	$(0, 1, \sqrt{2})$	$(1, \sqrt{2}, 1)$	$(0, \sqrt{2}, 1)$
(0, 0, 1)	(1, 0, 1)	(0, 1, 1)	$(1, -1, \sqrt{2})$	$(-\sqrt{2}, 0, 1)$	$(-\sqrt{2}, 1, 1)$	$(-1, 0, \sqrt{2})$	$(-1, -1, \sqrt{2})$	$(0, -\sqrt{2}, 1)$	$(1, -\sqrt{2}, 1)$	$(0, -1, \sqrt{2})$
(1, 1, 0)			$(-\sqrt{2}, 0, 1)$	$(-\sqrt{2}, 1, 1)$	$(-1, 0, \sqrt{2})$	$(-1, -1, \sqrt{2})$	$(0, -\sqrt{2}, 1)$	$(1, -\sqrt{2}, 1)$	$(0, -1, \sqrt{2})$	
(1, −1, 0)			$(0, \sqrt{2}, 1)$							

TABLE 26.4
The Trap Closes

1	2	3	4	5	6	7	8	9	10	11
$\hat{P}_i^1$	$\hat{P}_i^2$	$\hat{P}_i^3$	$\hat{P}_i^4$	$\hat{P}_i^5$	$\hat{P}_i^6$	$\hat{P}_i^7$	$\hat{P}_i^8$	$\hat{P}_i^9$	$\hat{P}_i^{10}$	$\hat{P}_i^{11}$
(1, 0, 0)	(0, 1, 0)	(1, 0, 0)	(1, 1, 0)	(0, 1, 0)	(0, −1, 1)	(0, 1, 0)	(1, −1, 0)	(1, 0, 0)	(−1, 0, 1)	(1, 0, 0)
(0, 1, 0)	(−1, 0, 1)	(0, −1, 1)	$(-1, 1, \sqrt{2})$	$(1, 0, \sqrt{2})$	$(\sqrt{2}, 1, 1)$	$(\sqrt{2}, 0, 1)$	$(1, 1, \sqrt{2})$	$(0, 1, \sqrt{2})$	$(1, \sqrt{2}, 1)$	$(0, \sqrt{2}, 1)$
(0, 0, 1)	(1, 0, 1)	(0, 1, 1)	$(1, -1, \sqrt{2})$	$(-\sqrt{2}, 0, 1)$	$(-\sqrt{2}, 1, 1)$	$(-1, 0, \sqrt{2})$	$(-1, -1, \sqrt{2})$	$(0, -\sqrt{2}, 1)$	$(1, -\sqrt{2}, 1)$	$(0, -1, \sqrt{2})$
(1, 1, 0)			$(-\sqrt{2}, 0, 1)$	$(-\sqrt{2}, 1, 1)$	$(-1, 0, \sqrt{2})$	$(-1, -1, \sqrt{2})$	$(0, -\sqrt{2}, 1)$	$(1, -\sqrt{2}, 1)$	$(0, -1, \sqrt{2})$	
(1, −1, 0)			$(0, \sqrt{2}, 1)$							

In column 11, all the basis states are forced to have value 0 from their occurrences earlier in the sequence. Hence there can be no value 1 in that set. The transfer from random selection to enforced selection is indicated by the marked row now being the second row from column 5 onwards.

selections made in earlier columns. Furthermore, *all the values are zero*, in violation of our claim $\sum_i v(\hat{P}_i) = v(\hat{I}) = 1$.

26.3.4 Consequences

We have arrived at a contradiction. Our starting point was the assertion that some value could be ascribed to an observable via $v(\hat{P})$. We used sensible assumptions about how these real-numbered values would combine to prove that projection operators can only take the values 0 or 1 and that one of the projectors in a set must be 1 while the others are zero. This was applied to a collection of basis sets, one per column, keeping in mind value definiteness (V) and non-contextuality (N). We have discovered that it is not possible to consistently ascribe values across the collection. To avoid the contradiction, we must abandon either value definiteness, non-contextuality or the whole program of assigning values to observables.

While this result was demonstrated specifically for projection operators, *any* observable can be written as an expansion over projectors, $\hat{O} = \sum_i o_i \hat{P}_i$, so we are *generally*, rather than *specifically*, in deep trouble. In broader terms: *it is not possible to allocate specific values of a physical variable to a state, simply by a consideration of the state itself. The context of the measurement, i.e., the other measurements that might be made at the same time, must be considered as well.*

This is a long way from being classical physics.

26.4 CARNEGIE HALL

There's an old joke in which a New York tourist asks a cab driver how to get to Carnegie Hall (a famous concert venue); the helpful reply comes "well, I wouldn't start from where you are."[10]

If we want to be scientists with a realistic view of what we do, then quantum theory is not the best place to start. But it's not like we have a choice. Quantum theory was not invented as a way for physicists to give philosophers migraines[11]. It has come from our attempts to describe the world as revealed by experiments. That being the case, I believe that we have to fight to retain a realistic interpretation of quantum theory, albeit one that forces us to modify our views of what the world is like and potentially expand our definition of what counts as reality.

NOTES

1. B. J. Hiley, Active information and teleportation, *Epistemological and Experimental Perspectives on Quantum Physics*, eds. D. Greenberger et al., 113–126, Kluwer, Netherlands, 1999.
2. *The Blind Men and the Elephant* by John Godfrey Saxe, 1816–1887.
3. C. Isham, *Lectures on Quantum Theory: Mathematical and Structural Foundations*, Imperial College Press, London, 1995. C. Isham, 1944–, Professor of Physics, Imperial College London.
4. In the case of $|\Psi\rangle = a|\psi\rangle + b|\varphi\rangle$ is different. We are not sure if the system is in $|\psi\rangle$ or $|\varphi\rangle$, but we know that it is in $|\Psi\rangle$.
5. Subject, of course, to precision and accuracy limitations in the procedure and devices.
6. Non-locality is acknowledged in the consistent histories approach but requires a more refined view of its meaning.
7. S. Kochen and E. P. Specker, The Problem of Hidden Variables in Quantum Mechanics, *Journal of Mathematics and Mechanics*, vol 17 No 1, 1967.
8. In any case, we are trying to see if we can avoid the quantum issues surrounding physical variables having definite values irrespective of measurement. So, we should not tie the operation of the value function to what we already believe to be the case for quantum theory.
9. Based on a version of the proof presented by A Peres in Two Simple Proofs of the Kochen-Specker Theorem, *Journal of Physics A*, vol 24, L175–LI78, 1991.
10. The alternate answer runs, "You have to practice, man...."
11. That's just a useful biproduct....

27 The Copenhagen Interpretation

The soothing Heisenberg–Bohr philosophy – or religion? – is so nicely contrived that for now it offers the true believer a soft pillow from which he is not easily rousted.

Albert Einstein[1]

27.1 BOHR'S INFLUENCE

Bohr had a profound personal and scientific influence on a generation of physicists. However, his writings on the philosophy of quantum theory are notoriously unclear, which caused some debate about the correct way to interpret his views, even at the time.

With this sort of confusion among his close colleagues, it's surprising that Bohr had the range of influence that he did. Perhaps this was due to his personal magnetism and conviction. As Pauli pointed out:

> Bohr himself integrated, in lectures at international congresses and at those carefully planned conferences in Copenhagen, the diverse scientific standpoints and epistemological attitudes of the physicists, and thereby imparted to all participants in these conferences, the feeling of belonging, in spite of all their dissensions, to one large family.[2]

Bohr took less interest in the mathematical development of quantum theory, preferring to focus on the concepts and the use of language to describe atomic events. As Heisenberg once said, Bohr was "primarily a philosopher and not a physicist."[3] As a result, some physicists at the time felt that Bohr had sorted out all the philosophical aspects of the theory, although they found his ideas hard to follow, for example, the notion of complementarity. Originally, Bohr introduced complementarity (Chapter 18) to deal with the apparently contradictory experimental results that were cropping up. The term was rapidly and widely adopted, so that everyone could discuss its application, even if there was no consensus as to its exact meaning. As it had Bohr's stamp of approval, there had to be something to it, so people bought into the idea.

Unfortunately, Bohr wrote very little about state reduction and measurement. His 1927 Como lecture contains the following:

> Now, the quantum postulate implies that any observation of atomic phenomena will involve an interaction with the agency of observation not to be neglected. Accordingly, an independent reality in the ordinary physical sense can neither be ascribed to the phenomena nor to the agencies of observation. After all, the concept of observation is in so far arbitrary as it depends upon which objects are included in the system to be observed. Ultimately, every observation can of course be reduced to our sense perceptions. The circumstance, however, that in interpreting observations use has always to be made of theoretical notions, entails that for every particular case it is a question of convenience at which point the concept of observation involving the quantum postulate with its inherent "irrationality" is brought in.[4]

Apparently, Bohr later regretted the use of the term 'irrationality' in this context. He simply meant to express the idea that state collapse cannot be described by the quantum equations, and that at some stage, there had to be a 'cut'[5] between the quantum description of the microworld and the classical description of the apparatus.

As far as state reduction is concerned, Bohr seems to have treated this as the necessary change in the mathematical description when you move from one experimental context to another. He regarded such issues as being related to mathematical formalism, not the physical interpretation.

DOI: 10.1201/9781003225997-31

All the apparatus of Schrödinger equations, wave functions, states, operators, and the like, were effectively abstract rules that meant nothing unless they were anchored to classical concepts via experiments. When talking about the Schrödinger equation in 1958, Bohr wrote:

> we are here dealing with a purely symbolic procedure, the unambiguous physical interpretation of which in the last resort requires a reference to a complete experimental arrangement.[6]

Arguably Bohr had an instrumental view of the mathematics of quantum theory and a realistic view of the concepts used to describe events.

Although two of our four conceptual issues held little interest for Bohr, he was undoubtedly one of the most gifted minds of the twentieth century (Einstein once commented that Bohr's *thinking* was clear, it was only his *writing* that was obscure) and his work on other aspects of the interpretation of quantum theory needs to be reconned with, even today.

27.2 BOHR'S VIEW OF QUANTUM THEORY

Some historians of physics argue that Bohr's thinking developed markedly during his career. In particular, they suggest that his ideas had to change in some crucial ways after the Einstein, Podolsky, and Rosen (EPR) paper had been published. Personally, I don't go along with this and agree with those historians who say that Bohr's ideas were fully in place before the EPR argument was available. For one, the speed at which he produced the counterargument tends to suggest that he didn't need a lot of time to rethink things. Possibly the process of composing his counter to EPR helped to make his thinking clearer.

As to the nature of Bohr's approach, experts point to four key topics that he repeatedly emphasized:

1. Classical concepts must be used to describe the results of any experiment.
2. During a measurement it is impossible to separate a quantum object from the apparatus, as you can't control the interaction between the two.
3. The results of one experimental arrangement cannot necessarily be related to another (which could be seen as a foreshadow of the Kochen-Specker theorem).
4. An accurate description of an object in terms of its position in space and time cannot be constructed alongside an accurate description of its energy and momentum, so the classical way of explaining how the world works must be replaced by something new.

At various stages, Bohr tried to arrange these ideas into a logically compelling sequence, so that he could justify one in terms of another. Although he never entirely succeeded in doing this in a satisfactory fashion, these four points taken together form an overall picture of both the quantum world and the role of physics. They are an interlocking set of ideas that you must 'buy into' as a whole. Over the next few sections, we will take them one by one.

27.2.1 Classical Concepts Must Be Used to Describe the Results of Any Experiment

> [I]t lies in the nature of physical observation, that all experience must ultimately be expressed in terms of classical concepts[7]
>
> Ultimately, every observation can, of course, be reduced to our sense perceptions.[8]
>
> [H]owever far the phenomena transcend the scope of classical physical explanation, the account of all evidence must be expressed in classical terms. The argument is simply that by the word "experiment" we refer to a situation where we can tell others what we have done and what we have learned and that, therefore, the account of the experimental arrangement and of the results of the observations must be expressed in unambiguous language with suitable application of the terminology of classical physics.[9]

Bohr felt that the ultimate purpose of physics was to describe the world in clear language that everyday people could understand. Naturally, this language, as well as the whole of our thinking, is rooted in the classical world and the basic physics by which it operates. This makes it very hard to strip away our ingrained ways of thinking and talking, to start again in quantum theory.

Clearly, raw experimental results (computer displays, pointer readings, photographic films, detector readouts, etc.) are part of the classical world, and so describable in classical terms using everyday language (with the obvious technical additions).

However, Bohr was not simply noting the classical nature of raw results. In his view, it was not simply *practically valid* to use classical physics in this way, it was *logically necessary* as well. It's like the board on which we play the game. Classical physics is logically prior to quantum physics, not just historically earlier. Although the overall pattern of results from different experiments may not fit any of our classical theories, we still need to describe each separate experiment using classical concepts.

In this area, Bohr is possibly on shaky ground. First, many would argue that physics achieves its aims by using a mathematical description of the world, and the concepts used in mathematics often outstrip everyday language. Second, there is no evident reason why we should not attempt to develop new concepts in quantum theory and then show, in some way, how the classical concepts arise from them in the right circumstances. However, mathematics needs to be anchored into experiment, for predictions to be checked, and at that stage, classical concepts must be used, according to Bohr. Also, developing new concepts is far from an easy task, especially if your whole mode of thinking is built on other conceptual lines. As Bohr wrote to Schrödinger:

> I am scarcely in complete agreement with your stress on the necessity of developing 'new' concepts. Not only, as far as I can see, have we up to now no clues for such a re-arrangement, but the 'old' experimental concepts seem to me to be inseparably connected with the foundation of man's power of visualising.
>
> ...it continues to be the application of these [classical] concepts alone that make it possible to relate the symbolism of the quantum theory to the data of experience. [10]

It's easy to suggest that Bohr was being close-minded, but it's perfectly reasonable to start from a set of assumptions that can't be fully justified, other than via the successful theory that develops out of them. Bohr's key points link together in a mesh, and you can't tug at the threads of one without unravelling the lot. Consequently, we can only see the full nature of any one of Bohr's ideas when we see how they interact with the others.

Interestingly, although we can criticize Bohr over the use of classical concepts, the ultimate need to refer to them was widely accepted at the time.[11] Even Einstein, his most trenchant critic over quantum theory, seemed to accept this point.

27.2.2 During a Measurement It Is Impossible to Separate a Quantum Object from the Apparatus

As part of setting up an experiment to measure the position of an electron at a given time, we must make sure that the equipment is rigidly attached to the rest of the world. When the electron interacts with the experiment, it will cause the equipment to move slightly[12]. Unless this movement can be dissipated into a very high mass, e.g., the planet, it will ruin our ability to pin down exactly where the electron is.

The equipment can be very simple. A single sheet of metal with a small hole in it will do, provided the sheet is fixed to something. If an electron passes through the hole, we can get a fix on its position, in the plane of the sheet, to within the width of the hole. In principle, we can make the hole as small as we like, so we can have as good a measurement of position as required. However, in passing through the hole, the electron may well have bounced off one of the edges. We can't tell if this has happened, as we are not measuring the momentum of the electron at the same time, nor are we able to see if the sheet of metal recoiled slightly as its fixed in place. We might have measured the momentum of the electron before it arrived at the hole, but now that measurement is useless as the momentum has changed in an undetectable way.

In classical physics, we don't worry so much about this, as we can always assume that the effect of any impact between the electron and the edge of the hole is small enough not to matter. According to Bohr, though, this is no longer valid in the quantum realm. Planck had shown how energy and momentum are exchanged in finite lumps that can be quite large compared with the original energy and momentum of the electron. Bohr called this the "quantum postulate" and talked in terms of an "interaction with the agency of observation":

> Now, the quantum postulate implies that any observation of atomic phenomena will involve an interaction with the agency of observation not to be neglected. Accordingly, an independent reality in the ordinary physical sense can neither be ascribed to the phenomena nor to the agencies of observation.

or:

> Notwithstanding the difficulties which hence are involved in the formulation of the quantum theory, it seems, as we shall see, that its essence may be expressed in the so-called quantum postulate, which attributes to any atomic process an essential discontinuity, or rather individuality, completely foreign to the classical theories and symbolised by Planck's quantum of action.[13]

As Bohr never formally defined the quantum postulate, we can't be sure exactly what he meant by the term and its use in his writing is not completely consistent. In any case, there are certain circumstances where energy and momentum don't have to be exchanged in quantized lumps. However, the idea that the measuring device interacts with the object being measured in an *uncontrollable way* is important:

> Indeed the finite interaction between object and measuring agencies conditioned by the very existence of the quantum of action entails—because of <u>the impossibility of controlling the reaction of the object on the measuring instruments</u> if these are to serve their purpose—the necessity of a final renunciation of the classical ideal of causality and a radical revision of our attitude towards the problem of physical reality.[14]

[my emphasis – underlined]

Imagine that we try to detect the interaction between a particle and a measuring device. We can only do this by inserting more equipment into the system. According to Bohr, this is a completely new experiment, and hence we can't count on the results being the same as before (think back to Chapter 1 and the Pockels cell) invalidating the effort. In any case, the particle is going to interact with any device that we use to detect its exchange with the main equipment. The only way to examine that interaction would be to introduce a third device, and the chain simply develops. This is the ground of Bohr's account of the contextuality of quantum behaviour.

If we accept that the interaction between a quantum object and a measuring device is not just an 'uncontrollable disturbance' but also one that we can't directly measure, then we can follow Bohr's suggestion that "an independent reality in the ordinary sense can neither be ascribed to the phenomena nor to the agencies of observation." In classical mechanics, we could talk about the object being studied as if it were separate from the equipment being used to study it, as we could always rely on the interaction between them being small enough not to radically change the object. The quantum world cannot be chopped up in such an easy manner. The interaction between measured and measuring knits them together into a whole.

Although this sounds like the entanglement between a quantum system and an experimental apparatus, Bohr was thinking in different terms. We could say that a state $|\varphi\rangle$ belongs to the *combined system of the quantum object and measuring device linked together by an interaction that can't be minimized*. Bohr is stressing the contextuality of quantum theory and the wholeness of the object–equipment system.

Bohr went further than this. Given the need to use classical ideas such as energy, momentum, position, and so on, to describe the results of our experiments, and now accepting that we can't split an object off from the apparatus used to measure it, it follows that quantum objects will have classical properties *in the context of the measurement used*. In other words, the measurement context defines the properties. What we mean by momentum is defined by the equipment used to measure it in a particular experiment.

This does two things for us. It gives us another reason why it is impossible to get around the problem of interaction between the measuring device and the object being studied. After all, an experiment to measure the position of an electron prevents an accurate *definition* of the energy and momentum of the electron, so even if it does bounce off the edge of a hole, we can't hope to follow the details of what happened in those terms. It also sets up Bohr's reply to the EPR argument, which as we saw in Chapter 23, is based on the whole experimental arrangement providing a context for what's going on. It's not an interaction between the measuring device and the distant particle that is mucking things up; the experimental equipment *defines the concepts that can be used consistently*:

> From our point of view we now see that the wording of the above-mentioned criterion of physical reality proposed by Einstein, Podolsky and Rosen contains an ambiguity as regards the meaning of the expression "without in any way disturbing a system." Of course there is in a case like that just considered no question of a mechanical disturbance of the system under investigation during the last critical stage of the measuring procedure. But even at this stage there is essentially the question of an influence on the very conditions which define the possible types of predictions regarding the future behavior of the system. [15]

So, the properties of a quantum object must be described classically and are valid only in the context of the particular experiment that is underway.

Going back to the uncontrolled disturbance that takes place during measurement, it's easy to see how this could be related to the uncertainty principle. After all, it is the argument behind Heisenberg's gamma-ray microscope (Section 13.2.3). However, for Bohr, this was not the core reason for uncertainty. He believed that the uncertainty principle reflected the difficulty in defining the two classical concepts (say position and momentum) at the same time in the same experiment, a point that he convinced Heisenberg to accept and include in a later draft of his uncertainty paper.

Although Bohr stressed the need to use classical ideas to describe experimental equipment and results, from time to time he applied quantum ideas, such as uncertainty, to measuring devices. This was particularly the case at the Fifth Solvay Conference, held in Brussels in 1927. During the meeting, Einstein repeatedly presented arguments, using his famous thought experiments, that were intended to show the inconsistency of quantum theory. Heisenberg and Bohr collaborated in countering every single argument Einstein came up with. Their counterarguments generally worked by applying the uncertainty principle to parts of the experimental setup Einstein had devised, such as the metal sheet with a hole in it that I mentioned earlier.

Bohr was quite aware that there was a possible inconsistency here. On the one hand, he claimed that you had to resort in the end to classical ideas, and on the other hand, he was applying a quantum concept such as uncertainty to the equipment. A clarification was required, and it came in his reply to the EPR experiment.

Let's say that during an experiment to measure the position of a particle, using a metal sheet with a hole [slit] in it, you want to measure the amount of momentum the sheet gets from the impact of the electron. In which case, the sheet has to be treated:

> as regards its position relative to the rest of the apparatus ... like the particle traversing the slit, as an object of investigation, in the sense that the quantum mechanical uncertainty relations regarding its position and momentum must be taken explicitly into account.[16]

The key part here is the phrase "be treated, like the particle traversing the slit, as an object of investigation." Bohr is linking the metal sheet in with the quantum object, something that he can easily justify in terms of the uncontrolled disturbance between the two. Not only are you then able to use the uncertainty principle on the metal sheet, but consistency also requires that you must.

The position of the metal sheet must be measured relative to the rest of the equipment, so an uncontrolled disturbance takes place in that interaction. Clearly, there is a danger of building an infinite chain with no resolution, but Bohr had an answer. At some point you must use a completely classical description of the equipment:

> In the system to which the quantum mechanical formalism is applied, it is of course possible to include any intermediate auxiliary agency employed in the measuring process [but] some ultimate measuring instruments must always be described entirely on classical lines, and consequently kept outside the system subject to quantum mechanical treatment.[17]

It then becomes a question of deciding where to draw the line that prevents one device from measuring another, etc., ad infinitum. According to Bohr, the sequence can stop when you reach "a region where the quantum mechanical description of the process concerned is effectively equivalent with the classical description"[18] or in more obvious terms "the use, as measuring instruments, of rigid bodies sufficiently heavy to allow a completely classical account of their relative positions and velocities."[19]

So, the quantum object links to part of the equipment, and that part links to another part, etc. At some point, you draw a line. On one side of the line, you are going to treat everything quantum mechanically, and on the other side, everything is classical. The best place to do this is when you get to part of the equipment that is so comparatively massive, it can absorb the sort of energy and momentum involved without notably altering its motion (or lack of).

It's easy to think of this in terms of an entanglement between quantum object and equipment that is broken by a classical dividing line, which is exactly how the Copenhagen interpretation is generally publicized. However, Bohr was not thinking this way. Entanglement was part of the formal mathematical structure of the theory, and so Bohr didn't take a great deal of interest in it.

From within Bohr's system, the whole picture makes complete sense. However, some modern physicists feel that this was a 'cop out' on Bohr's part. As everything is made of atoms and molecules, including measuring devices, and quantum theory is the appropriate atomic theory, it simply will not do to draw an arbitrary line across the world and assert that two different theories must be used either side of the line, with no link between them.

In some respects, the relationship between quantum theory and classical mechanics mirrors that for other theories, e.g., Einstein's theory of General Relativity (GR) and Newtonian gravity. However, with General Relativity and Newtonian theory, the one does not depend on the other. GR is always the 'correct' theory as it applies in a wider range of circumstances and can reproduce the predictions of Newtonian gravity, in the correct limit of a weak field. While this could be regarded as a line drawn, as per quantum theory/classical mechanics, there is a profound difference. GR does not need the line to make sense of its predictions. In Bohr's terms, quantum theory has a regression of device measuring device until you trip over the classical line. While calculations can demonstrate how quantum theory reproduces classical predictions in the appropriate limit, we could not use it to replace classical mechanics in all circumstances without an alternative theory of state collapse.

The modern disquiet regarding the Heisenberg cut is further fuelled by the increasingly compelling experimental evidence for larger and larger systems showing quantum behaviour (see Section 28.1.1), making it more difficult to subscribe to the existence of a definite dividing line.

27.2.3 The Results of One Experimental Arrangement Cannot Necessarily Be Related to Another

> The extent to which renunciation of the visualisation of atomic phenomena is imposed upon us by the impossibility of their subdivision is strikingly illustrated by the following example to which Einstein very early called attention and often has reverted. If a semi-reflecting mirror is placed in the way of a photon, leaving two possibilities for its direction of propagation, the photon may either be recorded on one, and only one, of two photographic plates situated at great distances in the two directions in question, or else we may, by replacing the plates by mirrors, observe effects exhibiting an interference between the two reflected wave-trains. In any attempt of a pictorial representation of the behaviour

> of the photon we would, thus, meet with the difficulty: to be obliged to say, on the one hand, that the photon always chooses one of the two ways and, on the other hand, that it behaves as if it had passed both ways.[20]

Here Bohr is drawing attention to the Mach–Zehnder experiment that we discussed in Chapter 1. He is using it as an example to ram home another of his key ideas: *we must give up any attempt to picture exactly what a quantum object is like.*

Bohr's argument up to this point has been along the following lines:

- Experiments must be described in classical terms.
- Quantum objects cannot be isolated from the context of the experiment.
- So, quantum objects must also be described in classical terms appropriate to that experiment.

All of which leaves open an obvious question—why doesn't classical physics work? Where is the room for a quantum particle to behave in a quantum fashion? This is where *complementarity* comes in.

The properties of a quantum object are so context-dependent that *we can't compare one experiment with another.* Any attempt to knit together classical ideas to produce a theory of what's going on is bound to fail. There is no firm *experimentally ubiquitous* rock on which to build a structure. The only hope is to adopt the idea of complementarity. Stick to a particular picture in a particular context and accept that you need to use a range of different pictures, which are impossible to relate to one another in terms of a complete classical theory.

As we can't find a set of classical concepts that work equally well in every single experiment, we are forced to use probability. When we relate one set of experimental results to another, the classical concepts in one will not perfectly fit in the other. We may need to use a totally different complementary picture in the second experiment. However, we can make a link between them by using probabilities.

Bohr made this point while talking about the difference between the wave and photon pictures of light:

> In this situation, there could be no question of attempting a causal analysis of radiative phenomena, but only, by a combined use of the contrasting pictures, to estimate probabilities for the occurrence of the individual radiation processes. However, it is most important to realize that the recourse to probability laws under such circumstances is essentially different in aim from the familiar application of statistical considerations as practical means of accounting for the properties of mechanical systems of great structural complexity. In fact, in quantum physics we are presented not with intricacies of this kind, but with the inability of the classical frame of concepts to comprise the peculiar feature of indivisibility, or "individuality," characterising the elementary processes.[21]

Bohr is making it clear that these are not classical probabilities, but quantum probabilities brought about by trying to fit concepts from one experiment into another. The probabilities measure how well the ideas fit. Bohr talked about a 'rupture' in the causal description of a particle when we make a measurement. In other words, poking an object with a measuring device, designed to measure its position, means that we can no longer be sure of its energy and momentum, and so we are forced to use probability to figure out what is going to happen next, as we don't have a complete set of accurate classical values on which to base a prediction. This links in well with the final key idea in our list.

27.2.4 Classical Explanations

Most popular accounts, along with a few professional ones, talk about Bohr's complementarity solely in terms of wave/particle duality. However, Bohr didn't think of wave/particle duality as the most important application of complementarity. When he first introduced the idea, he stressed the complementary roles of a space–time description and a causal one:

> the very nature of quantum theory thus forces us to regard the space-time co-ordination and the claim of causality, the union of which characterizes the classical theories, as complementary but exclusive features of the description.[22]

Classical physics takes various measurements and uses laws of motion to predict the future. With something simple like a classical particle, you need to measure its position at a given moment and the forces acting on it. Alternatively, instead of the force/position approach, you can work entirely in terms of energy (kinetic and potential) and momentum and use their conservation laws to predict what's going to happen. In either case, you need to know where it is now *and* the energy/momentum it has now in order to figure out where it's going to be in the future.

According to Bohr, this is not going to work with quantum objects. If you try to measure the position of an electron at a given time (the "space–time coordination" in Bohr's terminology), you will necessarily muck up any information you have about the energy and momentum, because the experiment allows an 'uncontrolled disturbance' to take place and limits the extent to which energy and momentum can be defined. Without a good knowledge of energy and momentum, we can't apply the conservation laws (the causal rules) to predict an outcome. The link between the two, so crucial in classical theory, is broken.

We can talk about energies and momenta *or* positions in space and time. This is how quantum objects get the room that they need to behave in a quantum fashion. This is also why we need probability to link our experiments together. Exact knowledge of position blurs our knowledge of momentum, via the uncertainty principle (or Bohr's "uncontrolled disturbance"), so we can only predict a range of possibilities within the blur.

27.2.5 Drawing the Threads Together

When we experiment on quantum objects, such as electrons and photons, we find that certain traditional classical ideas, e.g., position, work fine but others, such as momentum, don't seem to fit. When we do a different experiment on the same quantum object, the picture can change so that momentum works, but the position doesn't.

The reason for this lies in the interaction between the quantum object and the measuring equipment, which we neither control nor know about in detail. This interaction knits the object and the equipment together into one whole system. Consequently, we can only talk about the classical properties of a quantum object within the context of an experiment that defines its relevance.

We are forced to stick with an attempted classical description, as we don't have another way of thinking. To try and develop specific quantum concepts would be like trying to teach a person who was blind from birth the difference between red and blue.

The problem is that we can't use classical physics anymore, as the properties seen in one experiment don't always transfer to another experiment with a different context. The best we can do is link experiments together using states or wave functions, which tell us how likely we are to get a certain fit between the results.

Our best hope for a physical picture of what is happening is by a complementary 'double think,' which flips between classical pictures, each of which is a partial picture of the truth.

However, we must be careful. The unbreakable link between a quantum object and a measuring device also extends the need to apply quantum principles (such as uncertainty) to the equipment. This will get us into trouble, as we can't necessarily define the properties of the apparatus consistently from one experiment to the next. In the end, we are stopped from extending this right across the universe because at some scale the quantum effects are small enough to ignore and we can use a completely classical description.

Of course, we have no idea what view Bohr's might have taken of Bell's inequality and the Kochen-Specker theorem, but it is tempting to suppose that both would simply have been regarded as inevitable consequences of the intimate link between concept and measurement or experiment and the quantum nature of systems.

27.3 HEISENBERG AND POTENTIA

It seems that the first person to use the term "Copenhagen interpretation" was Heisenberg, in his 1955 article on Bohr's contribution to physics:

> The months which followed Schrödinger's visit were a time of the most intensive work in Copenhagen, from which there finally emerged what is called the "Copenhagen interpretation of quantum theory".... From the spring of 1927, therefore, there existed a complete, unambiguous mathematical procedure for the interpretation of experiments on atoms or for predicting their results.... Since the Solvay conference of 1927, the "Copenhagen interpretation" has been fairly generally accepted, and has formed the basis of all practical applications of quantum theory.[23]

Recent analysis suggests that Heisenberg's picture of one Copenhagen interpretation, on which the dust had settled by 1927, is an oversimplification. Perhaps the publication of Bohr's reply to the EPR paper (1935) is a more accurate date for the crystallization of ideas. Be that as it may, it is clear that none of the founding fathers used the term "Copenhagen interpretation" at the time.

Heisenberg was a close friend and collaborator of Bohr's for many years and took great interest in both the mathematical and philosophical sides of the theory. He was very well read in philosophy and talked to representatives of several philosophical schools about the implications of quantum theory. Heisenberg clearly felt that he and Bohr had cracked the problem of quantum interpretation. Hence, we also need to examine what Heisenberg has to say on the subject.

In his book *Physics and Philosophy*, Heisenberg devotes an entire chapter to his version of the Copenhagen interpretation. Right from the off, he puts his cards on the table and states his key point - that the Copenhagen interpretation:

> ...starts from a paradox. Any experiment in physics, whether it refers to the phenomena of daily life or to atomic events, is to be described in the terms of classical physics. The concepts of classical physics form the language by which we describe the arrangements of our experiments and state the results. We cannot and should not replace these concepts by any others.[24]

Clearly, he is in complete agreement with Bohr on this issue.

Where he differed from Bohr was over the relationship between mathematics and language. Heisenberg felt that mathematics was primary: set up the mathematics and find the language to describe it, rather than get the language (ideas) right and then set up the mathematics. Consequently, Heisenberg took a more objective view of the quantum state and drew conclusions about the nature of quantum objects that Bohr was reluctant to accept. Using an idea that dates to Aristotle, Heisenberg viewed the quantum state as *an objective description of an object's potential* - the collection of possible or latent properties that come to be when a measurement takes place. To Heisenberg, quantum systems don't have properties, until they become manifest in a measurement. He would have taken a great interest in the Kochen-Specker result and seen it as a confirmation of this idea.

While Bohr's view of state collapse is not entirely clear, Heisenberg regarded it (at least in his later writings) as a real change in the system brought about by making the *potentia* that exist between measurements *actual* at the time of measurement. As a result, it's not possible to say what is 'really' happening between one measurement and the next.

If a quantum object 'exists' at all between measurements, then it's in a strange sort of potential existence; not at all like the existence of things that we prod in the everyday world:

> Therefore, the transition from the 'possible' to the 'actual' takes place during the act of observation. If we want to describe what happens in an atomic event, we have to realize that the word 'happens' can apply only to the observation, not to the state of affairs between two observations."[25]

Although Heisenberg certainly agreed with the notion of complementarity, there are some subtle differences between his views and those of Bohr. When he talks of the complementarity between a space–time coordinate description of a quantum object and a causal description[26], he stresses the two forms of state evolution: one governed by $\hat{U}(t)$, and the other by state collapse:

> The space–time description of the atomic events is complementary to their deterministic (causal) description. The probability function [wave function or density matrix] obeys an equation of motion … its change in the course of time is completely determined by the quantum mechanical equation $\left(\hat{U}(t)\right)$, but it does not allow a description in space and time. The observation, however, enforces the description in space and time [state collapse] but breaks the determined continuity of the probability function by changing our knowledge of the system.
>
> [brackets = my inserts]

In other words, the normal evolution of a state is determined by $\hat{U}(t)$, but when we make a measurement, this is interrupted and the state collapses into an actuality at a given point in space and time. In coming to this view, Heisenberg was undoubtedly influenced by von Neumann's work on the measurement problem (published in 1932), which is discussed in the next section.

Regarding the 'probability function':

> It should be emphasised, however, that the probability function does not in itself represent a course of events in the course of time. It represents a tendency for events and our knowledge of events. The probability function can be connected with reality only if one essential condition is fulfilled: if a new measurement is made to determine a certain property of the system. Only then does the probability function allow us to calculate the probable result of the new measurement. The result of the measurement again will be stated in terms of classical physics.

This comes back to Heisenberg's realistic view of an objectively existing quantum state, bought at the expense of a very different sort of 'reality' which we are not used to, but which Aristotle may have anticipated.

In another key passage about the probability function (in this context he means a density matrix) Heisenberg says that:

> It contains statements about possibilities or better tendencies ('potentia' in Aristotelian philosophy), and these statements are completely objective, they do not depend on any observer; and it contains statements about our knowledge of the system, which of course are subjective in so far as they may be different for different observers. In ideal cases the subjective element in the probability function may be practically negligible as compared with the objective one. The physicists then speak of a pure case.

The "subjective" statements are the classical probabilities in the density matrix, and the "objective" statements are the quantum probabilities within each state that makes up the density matrix. Clearly, Heisenberg accepts the reality of both forms of probability and the distinction between them; the quantum probabilities reflect an object's potentia.

In another interesting variation on Bohr, Heisenberg saw the contrast between Schrödinger's waves and his own matrix mechanics as a mathematical reflection of the complementarity between wave and particle. To him, the Schrödinger equation is the wave view and the matrix mechanics is the particle view. As Bohr stressed ideas more than mathematics, he felt that you needed both wave and particle pictures in complementary balance to describe a situation, whereas Heisenberg was content that either mathematical view could capture the whole truth (it was just a case of picking the form of mathematics that was most easily applied to a given problem).

It is now 100 years since Bohr and Heisenberg debated the meaning of quantum theory. As is often the case, the passage of time has tended to corrupt and amalgamate the historical picture. If asked to name the key features of the Copenhagen interpretation, most contemporary physicists would probably come up with something like Heisenberg's version, rather than pure Bohr. However, there is still one thing missing. I mentioned earlier that von Neumann had influenced Heisenberg's view of state collapse. Von Neumann's treatment of the measurement problem is the final piece in the Copenhagen interpretation jigsaw.

27.4 VON NEUMANN AND MEASUREMENT

John von Neumann[27] is regarded as one of the principal mathematicians of his age. He is famous for important contributions to logic, set theory, game theory, and computer science. He enters the story of the Copenhagen interpretation via his book *Mathematical Principles of Quantum Mechanics* published in 1932 (translated into English in 1955)[28]. In this book, von Neumann set out the standard mathematical approach to quantum mechanics (with a little added terminology from Dirac). His book is a landmark in the development of quantum theory.

In this rigorous treatise, von Neumann specifies the smallest set of assumptions needed to develop the full theory. The Dirac-von Neumann assumptions are:

1. The state of an isolated physical system at time t, is represented by a state vector $|\psi(t)\rangle$ in a mathematical space called *Hilbert space*.
2. Every measurable physical quantity O is represented by an operator $\hat{O}$ (of a certain mathematical type) and its eigenvectors $\{|O_n\rangle\}$ form a basis in the Hilbert space.
3. The result of measuring O must be one of the eigenvalues o_n of $\hat{O}$.
4. When a measurement takes place, the probability of obtaining the result o_n is $\left|\langle O_n|\psi\rangle\right|^2$.
5. If a measurement results in the value o_n, then the state of the system immediately after the result is $|O_n\rangle$.
6. The time evolution of a state vector is governed by the Schrödinger equation: $i\hbar\frac{d\psi(x,t)}{dt} = \hat{H}\psi(x,t)$, where $\psi(x,t) = \langle x, t|\psi\rangle$ and $\hat{H}$ is an operator constructed to mirror the appropriate Hamiltonian for the system.

Von Neumann's book also contains the first full coverage of measurement, treating the equipment in a quantum theoretical way. Bohr had applied the uncertainty principle to some experimental arrangements while countering Einstein's arguments against the consistency of quantum theory. He had also acknowledged that experimental equipment should have quantum physical properties, but at some point, you had to draw a line, when the quantum effects were small enough not to worry about.

Von Neumann took this further. In his account, he shows how quantum theory needs two ways to describe how a state can change over time if it's to work at all. First, we have the "continuous and causal changes in the course of time," for example, what we have called $\hat{U}(t)$ processes. Second, there are "the discontinuous, non-causal and instantaneously acting experiments or measurements"[29] i.e., state collapses. Crucially, von Neumann was able to demonstrate mathematically that $\hat{U}(t)$ processes can't lead to a state collapse of the sort required by a measurement. Consequently, state collapse is not part of the mathematical structure of the theory. For this reason, it needs to be 'pasted on' to make things work. In his own words (labelling the continuous, causal, $\hat{U}(t)$ evolution process 2, and state collapse process 1)[30]:

> quantum mechanics describes the events which occur in the observed portions of the world, so long as they do not interact with the observing portion, with the aid of the processes 2 but as soon as such an interaction occurs, that is, a measurement, it requires the application of process 1.

[By "observed portions of the world" von Neumann is clearly referring to a quantum system, with the "observing portion" being the measuring equipment. However, quantum systems are not 'observed' *between* measurements. In which case, we perhaps ought not to use a term such as "events which occur" to describe how states change during $\hat{U}(t)$ evolution. I suspect that Heisenberg would not have liked it, and neither would Bohr. Between measurements the quantum state is a collection of evolving possibilities, not a set of realized events. See also the quotation from Heisenberg at the beginning of Section 27.5.]

It is this 'grafted on' extra assumption that lies at the heart of many criticisms of the Copenhagen interpretation.

Some alternatives attempt to do without the assumption of state collapse by modifying the equations governing $\hat{U}(t)$ so that they can cause a form of collapse.

Other approaches do without state collapse at all (which leads to a Many Worlds type of interpretation) or try to use the interaction between quantum systems and the environment to bring about a reduction of the density matrix (in which case the quantum state applies to the ensemble, not the individual).

Interestingly, von Neumann himself might not have been that committed to the extra assumption he introduced. Apparently, in conversation with a colleague he accepted that it might be possible to modify quantum theory so that $\hat{U}(t)$ brings about state collapse.

Von Neumann's written output doesn't contain any statements of his views on realism vs instrumentalism. It's also not easy to find any definitive remarks about how state collapse comes about. This is another odd historical quirk, for physicists generally think that von Neumann was the first person to suggest that state collapse happens when an experimental result enters someone's conscious mind. While that view can be read into what von Neumann wrote and may certainly have been a sub-text to his ideas, he holds his cards on the close to his chest on this issue.

27.4.1 The Mind of an Observer

Arguably, Von Neumann's book might not have had the influence it now enjoys had Fritz Wolfgang London[31] and Edmond Bauer[32] not written a monograph that helped present von Neumann's ideas in a simpler form[33]. Historically, this work appears to be the first explicit written suggestion that mind is involved in state collapse.

In contrast to von Neumann's rather cautious approach, London and Bauer come right out and say that a measurement doesn't happen until the results are 'registered' in the consciousness of an observer:

> The observer has a completely different point of view: for him it is only the object and the apparatus which pertain to the external world, to that which he calls 'objective'. By contrast, he has within himself relations of a very peculiar character.... For he can immediately give an account of his own state ... namely, to cut the claim of statistical co-ordination ... by saying 'I am in state a_j' or more simply: 'I see $A = a_j$.[34]'

There is a unique feature of human observers, rather than mechanical ones: the ability to *introspect*, to know something about oneself. This, London and Bauer claim, is what is needed to break the entanglement. The measurement apparatus can't 'know' what state it's in, but when the measurement value enters the mind, we're able to say: "I can see that the value is $A = a_j$"

Eugene Wigner took up this point in an argument that is now known as 'Wigner's friend.' In simple terms, it runs as follows.

Suppose we start off with a system in a state $a_1|1\rangle + a_2|2\rangle$ and ask some friend to observe it. After the observation, the entangled state of system and friend is:

$$a_1|1\rangle|\text{friend sees 1}\rangle + a_2|2\rangle|\text{friend sees 2}\rangle$$

If we ask the friend "what did you see?" we will get the reply "I saw 1" with probability $|a_1|^2$ and "I saw 2" with probability $|a_2|^2$.

So far, there is nothing remarkable here. But what if I then ask, "what did you see before I asked you?"

Now we have a problem. The answer is going to be "I just told you, I saw 1" or "I just told you, I saw 2," which seems to imply that the state just after my friend looked at the system was *either* $|1\rangle|\text{friend sees 1}\rangle$ *or* $|2\rangle|\text{friend sees 2}\rangle$, not the entangled state.

This would be a very different situation if the 'friend' was actually some other quantum system.

In Wigner's original argument, $|1\rangle$ was the state of an illuminated light bulb and $|2\rangle$ the light bulb's off state. The friend was then replaced by an atom, which would go into an excited state if it picked up some light from the bulb or remain in its ground state if the bulb did not come on. In this case, the correct quantum description is evidently:

$$a_1|1\rangle|\text{atom excited}\rangle + a_2|2\rangle|\text{atom in ground state}\rangle$$

which has observably different properties to the mixed state:

$$a_1|1\rangle|\text{friend sees 1}\rangle + a_2|2\rangle|\text{friend sees 2}\rangle$$

In Wigner's words:

> If the atom is replaced by a conscious being the wave function appears absurd because it implies that my friend was in a state of suspended animation before he answered my question. It follows that the being with a consciousness must have a different role in quantum mechanics than the inanimate measuring device: the atom considered above.[35]

Consciousness causes state collapse.

Among modern day physicists, a few (notably Henry Stapp of the Lawrence Berkley Laboratory) are following this line of thinking, at least to the point of suggesting that quantum theory is intimately linked with nature of the mind[36]. Roger Penrose has proposed a modification to quantum theory to explain state collapse and then used this to develop some ideas of how consciousness arises in the brain.

27.5 THE DEEP END...

The use of mind as the ultimate reason for state collapse was not generally accepted by the founding fathers. Heisenberg, for one, specifically rejected this view:

> If we want to describe what happens in an atomic event, we have to realize that the word 'happens' can apply only to the observation, not to the state of affairs between two observations. It applies to the physical, not the psychical act of observation, and we may say that the transition from the 'possible' to the 'actual' takes place as soon as the interaction of the object with the measuring device, and thereby with the rest of the world, has come into play; it is not connected with the act of registration of the result by the mind of the observer.[37]

The potential incorporation of mind into physics has caught the imagination of some commentators. However, it is an approach that has many problems.

Some feel that mind, with its self-reflective nature, lies forever beyond the reach of science. In which case, the crucial role of measurement has been put out of play as far as further research is concerned[38]. This would close science off in an unsatisfactory manner.

More significantly, if mind is responsible for collapsing states, what is 'going on' when there is no mind to observe things? As Einstein once asked Bohr "Do you really think the moon isn't there if you aren't looking at it?"[39] The universe seems to have existed perfectly well before human beings came along to observe it. If mind is necessary to collapse states, then that would seem to require some 'mind' to be permanently present in the universe.

One obvious suggestion, of course, is God.

I don't want to be rude about a suggestion like this, but it seems to me that there are as many *theological* problems with it as there are physical ones. If it is God's continual 'observation' that selects quantum outcomes, then surely, we are quite right to 'blame' him for all the evils, both moral and physical, taking place in the world.

A more startling answer has come from John Wheeler. Earlier in his career, Wheeler collaborated with both Bohr and Einstein on aspects of nuclear physics and relativity. He continued to work on the deepest problems in physics right up to his death in 2008. "The time left for me on earth is

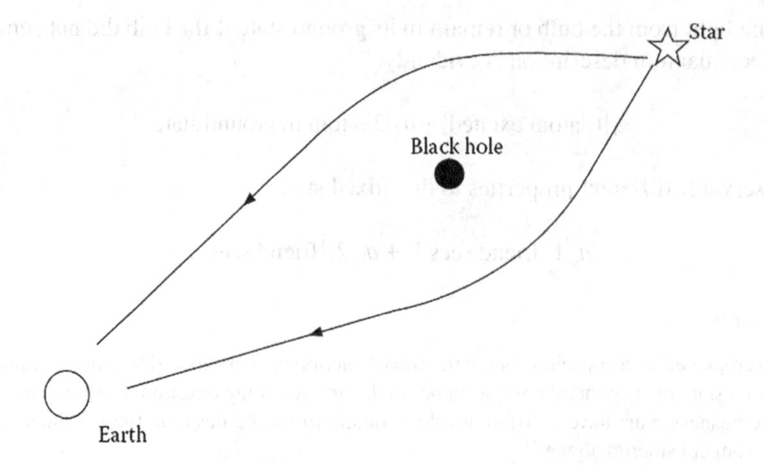

FIGURE 27.1 John Wheeler's cosmic version of the delayed choice experiment. The light from a star in a distant galaxy is deflected by the gravitational pull of a black hole. This means that a photon has two possible paths for reaching Earth. The light from this star can be collected in a telescope/mirror arrangement and used to create an interference pattern. With the star being very distant, it is likely that we will detect only one photon at a time. If we place a Pockels cell in one path, we can detect which way the photon travelled and destroy the interference. This would imply that we can influence the photon's behaviour although it left the star millions of years ago.

limited," he wrote, "and the creation question is so formidable that I can hardly hope to answer it in the time left to me. But each Tuesday and Thursday I will put down the best response that I can, imagining that I am under torture."[40]

Accepting the possibility that mind has a role in state collapse, Wheeler develops the point by considering an extension to the delayed choice experiment we discussed in Section 1.5.3. In such an experimental context, the random setting of a Pockels cell determines if we get an interference pattern or not, *even if the setting is changed after a photon interacted with the half-silvered mirror.*

In Wheeler's version, the photon comes from a distant galaxy and the half-silvered mirror is replaced by a strong gravitational source, like a black hole, which is also potentially millions of light years from Earth. The curved space-time in the vicinity of a black hole bends light ray paths, so that there are (at least) two possible routes to Earth from the distant galaxy (Figure 27.1). An individual photon leaving the galaxy must then be in a superposed state, indicative of travelling one side of the black hole or the other. With modern telescopes, the light can be detected photon-by-photon and used in an interference experiment. In principle, we would have one telescope monitoring each path around the black hole and combine their light output in an interferometer. If, however, a Pockels cell (or similar) was used to resolve the photon's path, the interference would be destroyed.

While there is an interference pattern, the photon passed the back hole along both paths. With no interference pattern, the photon clearly selected one route or the other. So, if we take the quantum theory seriously, our decision made now can influence the photon, although it passed the black hole millions of years ago.

From this example (and remember that standard delayed choice experiments have been successfully conducted in a laboratory) Wheeler concluded that we can influence the universe back in time by 'measurements' that we take now. In his view, the universe has a built-in feedback loop, allowing us to contribute to the ongoing creation of the past, present and the future. Our observations now help to 'actualize' the past. In his journal, Wheeler apparently wrote:

> No space, no time, no gravity, no electromagnetism, no particles. Nothing. We are back where Plato, Aristotle and Parmenides struggled with the great questions: How Come the Universe, How Come Us, How Come Anything? But happily also we have around the answer to these questions. That's us.[41]

Wheeler invented several catching phrases to summarise his views on different aspects of theory. In the context of quantum mechanics, he spoke of the *participatory universe*:

> We are participators in bringing into being not only the near and here but the far away and long ago. We are in this sense, participators in bringing about something of the universe in the distant past and if we have one explanation for what's happening in the distant past why should we need more?[42]

and coined the term 'it from bit', building on the notion of measurement as a binary 'bit' process that we first alluded to in Section 12.2.4.

> It from bit. Otherwise put, every it — every particle, every field of force, even the space-time continuum itself — derives its function, its meaning, its very existence entirely — even if in some contexts indirectly — from the apparatus-elicited answers to yes-or-no questions, binary choices, bits. It from bit symbolizes the idea that every item of the physical world has at bottom — a very deep bottom, in most instances — an immaterial source and explanation; that which we call reality arises in the last analysis from the posing of yes-no questions and the registering of equipment-evoked responses; in short, that all things physical are information-theoretic in origin and that this is a participatory universe.[43]

27.6 CRITICISMS OF THE COPENHAGEN VIEW

Clearly, if any interpretation is going to replace Copenhagenism, then it must address perceived shortcomings of that approach. Hence, if we are to evaluate the alternatives, we must first be clear about the improvements needed.

27.6.1 The Problem of the Cut

Nobel Prize winning physicist Steven Weinberg has clearly expressed one of the major criticisms of the Copenhagen approach:

> The Copenhagen interpretation describes what happens when an observer makes a measurement, but the observer and the act of measurement are themselves treated classically. This is surely wrong: Physicists and their apparatus must be governed by the same quantum mechanical rules that govern everything else in the universe.[44]

Physicists are used to having different theories with different ranges of application. After all, the simplifications and approximations that you make in one domain don't necessarily apply to another branch of physics. This is something different. Quantum theory deals with the microworld, and so should be the bedrock for everything else. We ought to be aiming, some would say, to build the classical world out of quantum behaviour, whereas the Copenhagen interpretation only works if we draw a sharp demarcation line between the two.

To counter Bohr's philosophical point, that we are in the end forced to use classical concepts due to the way in which our minds work, critics would argue that we are dealing with a mathematical theory. While we may have to make do with half pictures and inconsistent views as visualizations of the mathematics, as long as the calculations work, this does not matter.

All of this, though, is rather philosophical. There is another more immediate practical problem. Bohr placed 'the cut' between classical and quantum at "a region where the quantum mechanical description of the process concerned is effectively equivalent with the classical description," but this is becoming a question of how far 'up' in size we're prepared to tolerate. Seeing quantum behaviour, such as interference, with electrons and photons is one thing, but technology has now advanced to the point where the wave aspects of large molecules can be observed.

The latest generation of matter wave experiments exploits a curious phenomenon first reported in 1836 by Henry Fox Talbot (a pioneer in photography) and rediscovered with a theoretical explanation by Lord Rayleigh in 1881. While examining the light emerging from a diffraction grating[45], Talbot noticed that an exact duplicate of the slit pattern could be observed through a lens at a certain distance from the grating. Raleigh was able to use the theory of interference and diffraction to explain this effect and prove that the *Talbot length*, z_T, is related to the period of the grating, a, and the wavelength, λ, by[46]:

$$z_T = \frac{2a^2}{\lambda}$$

provided $\lambda \ll a$. Viewed from above, the interference pattern draws out a *Talbot carpet*, as shown in Figure 27.2

In December 2019, a paper was published in *Nature Physics* reporting on an experiment to demonstrate the wave aspect of molecules comprising up to 2000 atoms[47]. The experimental arrangement is outlined in Figure 27.3.

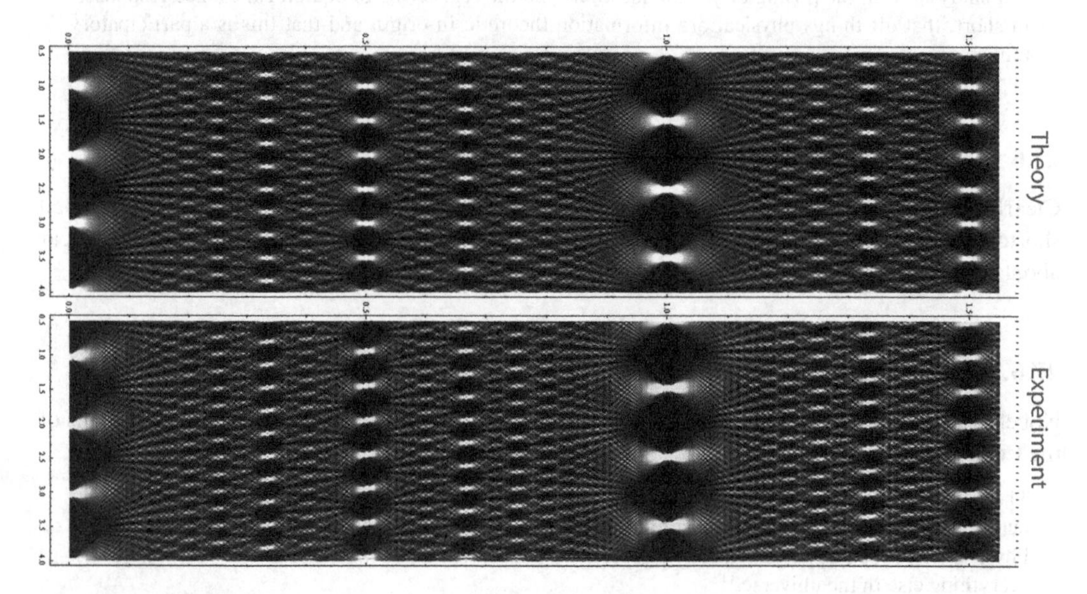

FIGURE 27.2 The optical Talbot carpet in theory (top) and experimental results (bottom). At the far left you can see diffraction through slits producing a pattern that is reproduced periodically. The horizontal scale is multiples of the Talbot length, z_T. (Image from W.B. Cast et at Realization of optical carpets in the Talbot and Talbot-law configurations, Optics Express, 9 November 2009 Vol 17 no 23, reproduced with kind permission.)

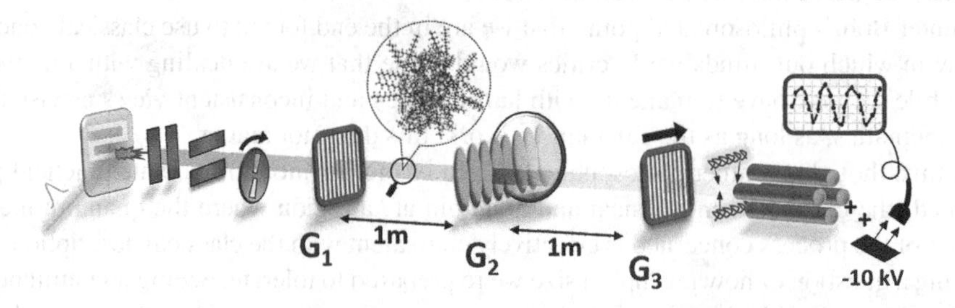

FIGURE 27.3 A matter wave Kapitza-Dirac Talbot-Lau interferometer. G1 and G3 are physical gratings, while G2 is a grating formed from an optical standing wave. At the far left of the experimental arrangement, the molecular beam is formed by a powerful laser scanning across a glass slide coated with the appropriate molecules. The beam is then collimated by the vertical combination of vertical and horizontal slits. The rotating disk with slots cut unto it acts as a velocity selector. (Image credit: Image provided by the Quantum Nanophysics Group, University of Vienna, based on Fein et al., *Nat. Phys.* 15, 1242 (2019).)

One novel aspect of this arrangement is the use of an optical standing wave formed by directing laser light through a lens towards a mirror. The resulting pattern of varying light intensity provides a non-physical diffraction grating via an effect known as the dipole force[48].

The molecular beam was formed by scanning a laser over a glass slide coated with the specially prepared, highly polarizable molecules. A combination of fixed slits and a rotating velocity selector refined the beam before it entered the instrument.

The first grating, G1, acted as a regular lateral series of sources with de Broglie waves diffracting through each slit in the grating. The optical grating, G2, was the source of the Talbot carpet, with each source in G1 producing a Talbot image after G2, shifted by multiples of the G1 period. Provided the periods of G1 and G2 match, these images will constructively interfere. The third scanning grating G3 was placed at an integer multiple of the appropriate Talbot length and moved back and forth laterally across the image. As G3 also had the same period, it progressively blocked and transmitted the molecules from the image as it scanned across the field of view. Finally, the molecules were laser ionised and detected. A sinusoidal variation in molecular intensity was produced by the movement of G3. However, the period of this variation is not dependent on the wavelength of the molecular beam. To reveal the quantum origins of the pattern, the group varied the power of the laser deployed in the optical grating. The results are shown in Figure 27.4.

The experiment was extremely delicate and appropriate corrections had to be made for the effects of gravity and the Coriolis force due to the Earth's rotation. Nevertheless, the team demonstrated

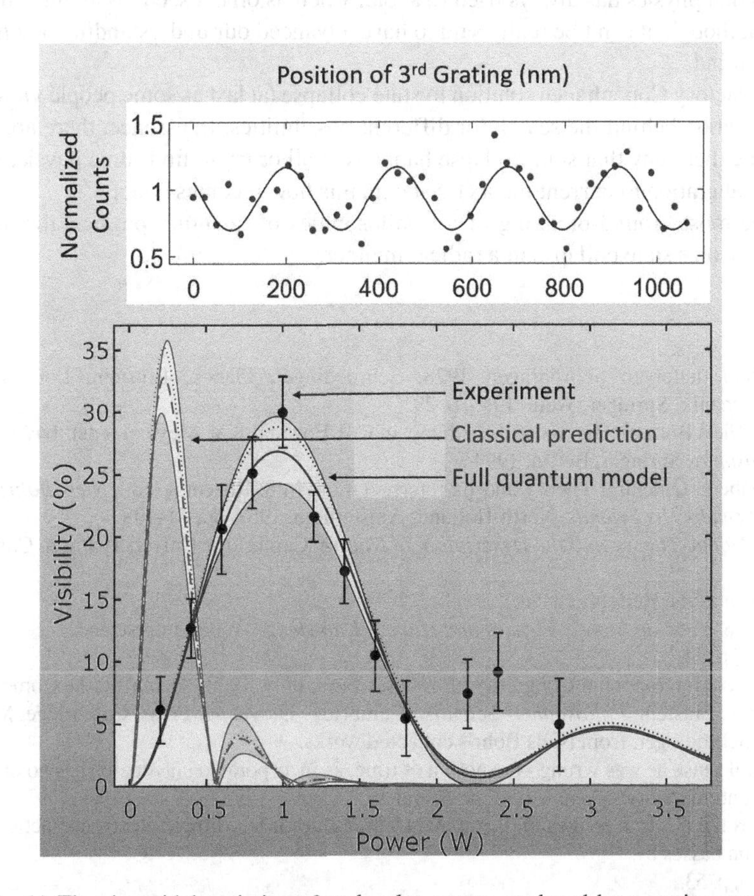

FIGURE 27.4 (a) The sinusoidal variation of molecular counts produced by scanning grating G3 across the interference pattern. (b) The variation in fringe visibility (contrast between maximum and minimum molecular counts) as a function of laser power in the optical grating G2. (Image credit: Image provided by the Quantum Nanophysics Group, University of Vienna, based on Fein et al., *Nat. Phys.* 15, 1242 (2019).)

the existence of de Broglie waves in molecules of significant size and mass. Given the velocities involved, the de Broglie wavelengths of the molecules were ~ 5.3×10^{-14} m, i.e., five orders of magnitude *smaller than the diameter of the molecules.*

As experimental techniques develop, along with the technology, the likelihood is that larger and larger molecules will be shown to have wave aspects. Indeed, the group projects that the basic techniques should be extendable to proteins and small viruses[49]. The prospect of interference in viral structures is especially intriguing, given that viruses are classed as living structures. To show interference at that scale would be powerfully symbolic.

27.6.2 Problem of Collapse

Bohr regarded wave function (or state) collapse as being a consequence of shifting the complementary focus, whereas Heisenberg saw it as potentiality coming into actuality. Others believe it to be triggered by an experimental result entering the consciousness of an observer.

Bohr's views are open to the criticism expressed by Weinberg: to call for some objects to be treated by quantum theory and others classically seems odd, when larger objects are constructed from smaller objects such as electrons and atoms.

It could be argued that Heisenberg's ideas have simply given a different name to the problem. State collapse due to consciousness polarizes opinions. Some see it as reverting to the sort of mystical vagueness that physics has always tried to avoid, whereas others see it as an intriguing possibility. As an idea, though, it can't be really said to have advanced our understanding of either quantum physics or the mind.

The unsatisfactory Copenhagen solution to state collapse (at last as some people view it) has been a driving motivation behind the search for different possibilities. In essence, there are two alternatives. You can either deny that state collapse happens at all or try to find some physical mechanism (in essence an alteration to current theory) that explains how it comes about.

Perhaps the most ground-breaking of all philosophies of quantum physics, the *Many Worlds interpretation* denies state collapse in a radical manner.

NOTES

1 A. Einstein, letter to Schrödinger, 1928, Schroedinger, Planck, Einstein, Lorentz, Briefe zur Wellenmechanik, Springer, Wein, 1963, p. 29.
2 W. Pauli, Niels Bohr on His Sixtieth Birthday, in C.P. Enz and K.V. Meyenn **(eds),** *Writings on Physics and Philosophy*, Springer, Berlin, 1994.
3 W. Heisenberg, Quantum Theory and Its Interpretation, in S. Rozental (ed.), *Niels Bohr: His Life and Work as Seen by His Friends*, North-Holland, Amsterdam, 1967, pp. 94–108.
4 N. Bohr, *Atomic Theory and the Description of Nature*, Cambridge University Press, Cambridge, MA, 1934.
5 Also known as the Heisenberg cut.
6 *Essays 1958–1962 on Atomic Physics and Human Knowledge*, Wiley Interscience.
7 See note 4, p. 94.
8 Lecture to the International Congress of Physics, Como, Italy, 1927—hereafter the Como lecture.
9 From Albert Einstein: *Philosopher–Scientist*, Cambridge University Press, Cambridge, MA, 1949.
10 Letter to Schrödinger, from Niels Bohr's collected works.
11 In a general sense he was wrong - the notion of time, as incorporated classically, has no obvious equivalent in quantum theory - there is no time operator.
12 Sometimes the electron strikes the equipment, sometimes it is a simple electromagnetic interaction as the electron passes by.
13 See note 6, p. 53.
14 N. Bohr, Can a quantum-mechanical description of physical reality be considered complete? *Phys. Rev.*, 1935, 48.
15 See note 13.
16 See note 9.

17 N. Bohr, *New Theories in Physics*, International Institute of International Co-operation, Paris, 1939.
18 See note 9.
19 N. Bohr, *Quantum Physics and Philosophy: Causality and Complementarity*, in Philosophy at Mid Century, Florence, 1958.
20 See note 5.
21 See note 5.
22 The Como lecture.
23 W. Heisenberg, The Development of the Interpretation of Quantum Theory, in W. Pauli, L. Rosenfeld, and V. Weisskopf (eds), *Niels Bohr and the Development of Physics*, McGraw Hill, New York, 1955.
24 W. Heisenberg, *Physics and Philosophy*, George Allen & Unwin, London, 1959, Chapter 3.
25 See note 24.
26 It is possible that Heisenberg mis-understood Bohr on this point. Where Bohr identified the causal description with conservation of energy, Heisenberg saw it as the deterministic evolution of Schrödinger's wave function.
27 J. von Neumann, 1903–1957, Hungarian mathematician, professor at the Institute for Advanced Study in Princeton.
28 J. von Neumann, *Mathematical Foundations of Quantum Mechanics*, Princeton University Press, Princeton, NJ, 1955. Princeton University Press; New edition (27 Feb. 2018)
29 See note 27.
30 See note 27.
31 Fritz Wolfgang London, 1900–1954, Professor Duke University.
32 Edmond Bauer, 1880–1963, Collège de France.
33 London, F., Bauer, E. (1939). La Théorie de l'Observation dans la Mécanique Quantique, issue 775 of Actualités Scientifiques et Industrielles, section Exposés de Physique Générale, directed by Paul Langevin, Hermann & Cie, Paris, translated by Shimony, A., Wheeler, J.A., Zurek, W.H., McGrath, J., McGrath, S.M. (1983), at pp. 217–259 in Wheeler, J.A., Zurek, W.H. editors (1983). Quantum Theory and Measurement, Princeton University Press, Princeton, NJ.
Note that this monograph appeared many years before the English translation of von Neumann's original work (1955).
34 See note 30. In the quotation, A refers to an observable physical property of the system and $a_j$one of the eigenvalues.
35 E.P. Wigner, Remarks on the Mind-Body Question, in I.J. Good (ed.), *The Scientist Speculates*, Heinemann, London, 1961.
36 Given the importance of such an effect, if it could be shown to be correct, it is surprising that experimental tests have not been more forthcoming. See, for example: Radin et al., Psychophysical modulation of fringe visibility in a distant double-slit optical system, *Physics Essays*, March 2016 DOI: 10.4006/0836-1398-29.1.014.
37 See note 24.
38 In my view, mind is an important feature of the world that demands being incorporated into a satisfactory account of reality. However, I do not believe that it lies beyond the reach of physics, albeit a new form of physics for the future.
39 As quoted in: Einstein and the quantum theory A. Pais *Rev. Mod. Phys.* 51, 863 – Published 1 October 1979. Bohr's reply was supposedly: "can you prove to me the opposite?"
40 As quoted by Dennis Overbye in his New York Times article: Peering Through the Gates of Time, March 12 2002.
41 See note 37.
42 From the transcript of a radio interview as part of *The Anthropic Universe*. Science Show. February 18, 2006
43 Wheeler, John Archibald (1990). *A Journey Into Gravity and Spacetime*. Scientific American Library. New York: W.H. Freeman. ISBN 0-7167-6034-7.
44 Einstein's mistakes, *Phys. Today*, November 2005, 31.
45 A regular pattern of parallel slits ruled into an otherwise opaque material.
46 In some treatments, the Talbot distance is defined as a^2 / λ i.e., half of the length used here.
47 Quantum superposition of molecules beyond 25 kDa, Fein et al., *Nature Physics*, Vol 15, December 2019, 1242–1245.
48 In essence, this comes about due to the interaction between an induced electrical dipole in the molecule and the oscillating electrical field in the optical standing wave.
49 https://physicsworld.com/a/probing-the-limits-of-the-quantum-world/.

28 The Many Worlds Interpretation

I do believe, however, that at this time the present theory is the simplest adequate interpretation. The hidden variable theories are, to me, more cumbersome and artificial—while the Copenhagen interpretation is hopelessly incomplete because of its a priori reliance on classical physics (excluding in principle any deduction of classical physics from quantum theory, or any adequate investigation of the measuring process), as well as a philosophic monstrosity with a "reality" concept for the macroscopic world and denial of the same for the microcosm.

Hugh Everett III[1]

28.1 EVERETT, WHEELER, BOHR & DEWITT

In 1953, Hugh Everett III, a talented graduate[2], with a National Science Foundation scholarship, was accepted onto the mathematics PhD programme at Princeton. In 1954, he transferred to physics and Wheeler became his supervisor. At the time, Wheeler's own research interests lay in Einstein's theory of gravity, General Relativity. However, in 1942 he had supervised Richard Feynman's thesis on the path integral approach to quantum theory and was clearly ready to accept another student who was keen to explore quantum foundations. Everett set out to develop an interpretation that had no additional postulates; it simply went where the equations lead.

The first draft of his thesis was handed over in September 1955. While Wheeler thought that the work was of great value, he was:

> ...frankly bashful about showing it to Bohr in its present form, valuable and important as I consider it to be, because of parts subject to mystical interpretation by too many unskilled readers.[3]

Wheeler and Bohr were fast friends, having worked together on the theory of nuclear fusion. As he was the living authority on the foundations of quantum theory, Wheeler sort Bohr's approval for Everett's ideas, hoping that his endorsement would lead to the work being published in the proceedings of the Royal Danish Academy of Sciences and Letters, promoting what Wheeler saw as an important contribution.

In response, Everett expanded the draft it into what is now known as 'the long thesis', presenting his interpretation, the relative state formulation, in detail alongside six alternative views, including Bohr's. This copy, although still regarded as a draft, was sent to Bohr in April 1956 as a prelude to Wheeler's own visit later in the year. By all accounts, the reception to Everett's approach was decidedly frosty. Discussions between Wheeler, Everett and Bohr proceeded via correspondence through 1956. At Wheeler's insistence, Everett reduced the long thesis from 130 pages to 30 and dampened down his criticism of Bohr's approach. This thesis was successfully defended for the PhD award in 1957 and subsequently published, after some further minor revisions, in Reviews of Modern Physics[4]. While this was a well-regarded journal, it was rather a step down from Wheeler's original hopes for the publication. In addition, it came out in the midst of a special edition devoted to the proceedings of a conference on gravitation. The nature of the journal and the specifics of the edition probably contributed to the paper's lack of attention from other physicists.

In 1959 Everett visited Copenhagen to meet with Bohr (something that Wheeler had suggested). It appears that the meeting didn't go well, and Bohr remained unimpressed by Everett's ideas.

DOI: 10.1201/9781003225997-32

Understandably disappointed, Everett returned to his hotel and started work on applying some well-known mathematical techniques to process optimization. This line of thinking eventually took him into private sector military research and made him a multimillionaire.

Things remained quiet for the relative state formulation until 1970 when Bryce DeWitt[5] wrote an article on Everett's theory[6]. DeWitt had been the editor of the journal where Everett's paper had appeared and had been in correspondence with him since 1957. Initially sceptical, DeWitt had become an enthusiastic supporter and tried to bring Everett's ideas to a wider audience. He also coined the term *Many Worlds Interpretation.* In 1973, DeWitt published a book that gathered over 500 papers on the interpretation of quantum theory[7]. Everett's long thesis was included, generating a great deal of interest, in both mainstream physics and the science fiction fringe.

By 1976, Wheeler had retired from Princeton and taken up a position at the University of Texas, alongside DeWitt. Together they organized a conference on the Many Worlds approach and persuaded Everett to attend. Everett became quite a star and was reportedly surrounded by a throng of enthusiastic student groupies. Wheeler proposed founding a new research institute at Santa Barbara where Everett could return to work on quantum theory, but Everett chose to remain with the various companies that he had founded. He died of a heart attack in 1982.

28.2 THE RELATIVE STATE FORMULATION

As a graduate student, Everett studied textbooks by von Neumann and Bohm and was struck by what he called the "magic process in which something quite drastic occurred, while in all other times systems were assumed to obey perfectly natural continuous laws," [8] in other words, state collapse. Conversations with other graduate students convinced him to try and find a way of constructing quantum theory that got rid of the 'magic process' and relied instead on the perfectly regulated and familiar $\hat{U}(t)$ evolution due to the Schrödinger equation.

As part of this new approach, Everett took a fresh look at the issue of entanglement. While he accepted that the separate systems in an entangled combination do not have well defined individual states, he proposed that the state of one system could be defined *relative to the state of the other.* In Everett's words (referring to an interaction between a measuring device and measured system, leading to an entanglement of their states):

> As a result of the interaction the state of the measuring apparatus is no longer capable of independent definition. It can be defined only relative to the state of the object system. In other words, there exists only a correlation between the states of the two systems.[9]

We can tease out Everett's meaning if we think of two systems, M (taken to be a measuring device) with a state variable A and orthonormal states $\{|m_j\rangle\}$ and S (a system ultimately measured by M) with state variable O and orthonormal states $\{|s_k\rangle\}$. Their combined general (not necessarily factorizable) state can always be written as:

$$|\Psi\rangle = \sum_k \sum_j \alpha_{kj} |s_k\rangle |m_j\rangle$$

incorporating an element, $\alpha_{kj} = \langle s_k|\Psi\rangle\langle m_j|\Psi$. While the combined system is in a definite state $|\Psi\rangle$, the two sub-systems, M and S cannot be ascribed definite states independently of each other (unless $\alpha_{kj} = 0$ for all but one choice of k, j). However, Everett showed how a *unique relative state* of one system could be assigned which was *dependent on the choice of state for the other.*

For example, if we pick out the specific state $|s_K\rangle$ of S then the relative state of M, subject to this choice is:

RELATIVE STATE

$$\left|M \text{ relative to } \left|s_K\right\rangle\right\rangle = \left|M; s_K\right\rangle = \sum_j \alpha_{Kj}\left|m_j\right\rangle \tag{28.1}$$

Hence:

$$\left|\Psi\right\rangle = \sum_k\sum_j \alpha_{kj}\left|s_k\right\rangle\left|m_j\right\rangle = \sum_k \left|s_k\right\rangle\left|M; s_k\right\rangle$$

To illustrate how this works in a simple example, take a standard measurement, where an SG device interacts with a spin ½ particle in state $\left|\psi\right\rangle = a\left|U\right\rangle + b\left|D\right\rangle$. We have an initial state $\left|\psi\right\rangle_S\left|\phi_0\right\rangle_M$ which evolves into a post-interaction state $\left|\Psi\right\rangle = a\left|U\right\rangle\left|m_U\right\rangle + b\left|D\right\rangle\left|m_D\right\rangle$. Hence, in this case, we have $\left\{\left|s_k\right\rangle\right\} = \left\{\left|s_1\right\rangle, \left|s_2\right\rangle\right\} = \left\{\left|U\right\rangle, \left|D\right\rangle\right\}$ and $\left\{\left|m_j\right\rangle\right\} = \left\{\left|m_1\right\rangle, \left|m_2\right\rangle\right\} = \left\{\left|m_U\right\rangle, \left|m_D\right\rangle\right\}$. Having written:

$$\begin{aligned}\left|\Psi\right\rangle &= \sum_k\sum_j \alpha_{kj}\left|s_k\right\rangle\left|m_j\right\rangle \\ &= \alpha_{11}\left|s_1\right\rangle\left|m_1\right\rangle + \alpha_{12}\left|s_1\right\rangle\left|m_2\right\rangle + \alpha_{21}\left|s_2\right\rangle\left|m_1\right\rangle + \alpha_{22}\left|s_2\right\rangle\left|m_2\right\rangle \\ &= \alpha_{11}\left|U\right\rangle\left|m_U\right\rangle + \alpha_{12}\left|U\right\rangle\left|m_D\right\rangle + \alpha_{21}\left|D\right\rangle\left|m_U\right\rangle + \alpha_{22}\left|D\right\rangle\left|m_D\right\rangle\end{aligned}$$

consistency forces $\alpha_{12} = \alpha_{21} = 0$, $\alpha_{11} = a$ and $\alpha_{22} = b$. In which case the relative states are:

$$\left|M;U\right\rangle = \sum_j \alpha_{1j}\left|m_j\right\rangle = \alpha_{11}\left|m_1\right\rangle + \alpha_{12}\left|m_2\right\rangle = a\left|m_U\right\rangle$$

$$\left|M;D\right\rangle = \sum_j \alpha_{2j}\left|m_j\right\rangle = \alpha_{21}\left|m_1\right\rangle + \alpha_{22}\left|m_2\right\rangle = b\left|m_D\right\rangle$$

A crucial feature of the relative state $\left|M; s_K\right\rangle$ is that it is *unique*, despite appearing to depend specifically on the choice of basis $\left\{\left|m_j\right\rangle\right\}$. Consider switching to a different basis $\left\{\left|n_g\right\rangle\right\}$ with $\left|m_j\right\rangle = \sum_g \epsilon_{jg}\left|n_g\right\rangle$. In which case:

$$\left|M; s_K\right\rangle = \sum_j \alpha_{Kj}\left|m_j\right\rangle = \sum_j \alpha_{Kj}\sum_g \epsilon_{jg}\left|n_g\right\rangle$$

Remembering that $\alpha_{Kj} = \left\langle s_K\right|\Psi\rangle\left\langle m_j\right|\Psi\rangle = \left\langle s_k\right|\Psi\rangle\left(\sum_h \epsilon^*_{hj}\left\langle n_h\right|\right)\left|\Psi\right\rangle$, we have:

$$|\mathrm{M}; s_K\rangle = \sum_j \langle s_K|\Psi\rangle \sum_h \epsilon^*_{hj} \langle n_h|\Psi\rangle \sum_g \epsilon_{jg} |n_g\rangle$$

$$= \sum_j \langle s_K|\Psi\rangle \sum_{h,g} \langle n_h|\Psi\rangle \epsilon^*_{hj}\epsilon_{jg} |n_g\rangle$$

Applying the condition $\sum_j \epsilon^*_{hj}\epsilon_{jg} = \delta_{gh}$ reduces this to:

$$|\mathrm{M}; s_K\rangle = \sum_h \langle s_K|\Psi\rangle\langle n_h|\Psi\rangle|n_h\rangle = \sum_h \alpha'_{Kh}|n_h\rangle$$

which is exactly what we would have obtained if we had started from $\{|n_g\rangle\}$ rather than $\{|m_j\rangle\}$.

Everett felt that the existence of a unique relative state in M for each choice of state in S was of fundamental significance:

> There does not, in general, exist anything like a single state for one subsystem of a composite system. That is, subsystems do not possess states independent of the states of the remainder of the system, so that the subsystem states are generally correlated. One can arbitrarily choose a state for one subsystem, and be led to the relative state for the other subsystem. Thus we are faced with a fundamental relativity of states, which is implied by the formalism of composite systems. It is meaningless to ask the absolute state of a subsystem - one can only ask the state relative to a given state of the remainder of the system.[10]

Everett is declaring the relativity of states to be a *natural consequence of entanglement*. We might say that the *absolute state* $|\Psi\rangle$ is a linear combination of *relative states* $|M; s_k\rangle$ as $|\Psi\rangle = \sum_k |s_k\rangle|M; s_k\rangle$.

28.3 MEASUREMENT RECORDS

According to Everett, a proper measurement must include a process by which the results are added to a store of previous results to form a complete sequential record. To model this, he assumed that the value of a state variable, A, of M changed to record a measurement.

Suppose that the initial state of $M + S$, prior to the measurement, is $|\Psi\rangle = |\psi\rangle_S|\phi_0\rangle_M$. The state of the apparatus and the state of the system are, in this configuration, uncorrelated. For a measurement to take place, an interaction must happen between the two systems, albeit for a brief period. If we represent this interaction as a specific evolution operator $\hat{\mathcal{U}}$, then $|\Psi'\rangle = \hat{\mathcal{U}}|\Psi\rangle$ is the combined state after the measurement has happened.

Any state of $M + S$, including $|\Psi'\rangle$, can be written as an expansion over the basis $\{|s_k\rangle|m_j\rangle\}$, so we need to understand how $\hat{\mathcal{U}}$ acts on this basis. If our memory record is built from from the eigenvalues m_j, then presumably:

$$\hat{\mathcal{U}}\left(|s_k\rangle|m_j\rangle\right) = |s_k\rangle|m_j + gs_k\rangle = |s_k\rangle|m_q\rangle$$

denoting the fact that the measurement result, finding $O = s_k$, has brought about a change in M by shifting it into a new state, with A's new eigenvalue being $m_q = m_j + gs_k$. The factor g is introduced to account for the likelihood that the values of O are 'encoded' into the memory record in some format.

Now we can use our basis $\{|s_k\rangle|m_j\rangle\}$, to write:

$$|\Psi\rangle = \sum_k\sum_j \langle s_k|\psi\rangle\langle m_j|\phi_0\rangle|s_k\rangle|m_j\rangle$$

and so:

$$|\Psi'\rangle = \hat{\mathcal{U}}|\Psi\rangle = \hat{\mathcal{U}}\left(\sum_k\sum_j \langle s_k|\psi\rangle\langle m_j|\phi_0\rangle|s_k\rangle|m_j\rangle\right) = \sum_k\sum_j \langle s_k|\psi\rangle\langle m_j|\phi_0\rangle|s_k\rangle|m_j + gs_k\rangle$$

If we make things a little less visually cluttered, by writing $\alpha_{kj} = \langle s_k|\psi\rangle\langle m_j|\phi_0\rangle$, then we have:

$$\begin{aligned}|\Psi'\rangle &= \sum_k\sum_j \alpha_{kj}|s_k\rangle|m_j + gs_k\rangle \\ &= \sum_k |s_k\rangle\sum_j \alpha_{kj}|m_j + gs_k\rangle \\ &= \sum_k |s_k\rangle|M;s_k\rangle\end{aligned}$$

This reveals the state $|\Psi'\rangle$ to be a linear superposition of products of eigenstates of S with the relative state of M for each s_k, $|s_k\rangle|M;s_k\rangle$. In this situation, we can't say that O has a particular value s_k, but we can say that the unique *relative states* of M are in a 1-1 correspondence with each value.

To ensure that these are 'good' measurements, we must engineer things so that $\delta m_j \ll g\delta s_k$, i.e., the interval between eigenvalues of A in the vicinity of m_j is very much less than g times the intervals between eigenvalues of O in the vicinity of s_k. In more prosaic terms, the measuring device must have sufficient resolution to be able to separately encode values of s_k.

It is now an opportune moment to demonstrate that the relative states are orthogonal to each other. We start with two different relative states:

$$|\mathrm{M};s_K\rangle = \sum_j \alpha_{Kj}|m_j\rangle \qquad |\mathrm{M};s_L\rangle = \sum_n \alpha_{Ln}|m_n\rangle$$

and then form the bra of the second one:

$$\langle M;s_L| = \sum_n \alpha^*_{nL}\langle m_n|$$

so that we can construct the product:

$$\begin{aligned}\langle \mathrm{M};s_L|\mathrm{M};s_K\rangle &= \sum_{j,n}\alpha^*_{nL}\alpha_{Kj}\langle m_n|m_j\rangle = \sum_n \alpha^*_{nL}\alpha_{Kn} \\ &= \sum_n \langle\Psi|s_L\rangle\langle\Psi|m_n\rangle\langle s_K|\Psi\rangle\langle m_n|\Psi\rangle = \sum_n \langle\Psi|m_n\rangle\langle m_n|\Psi\rangle\langle\Psi|s_L\rangle\langle s_K|\Psi\rangle \\ &= \langle\Psi|s_L\rangle\langle s_K|\Psi\rangle \\ &= \delta_{LK}\end{aligned}$$

In a conventional interpretation, orthogonality is crucially important as it guarantees that a system can only manifest one value of a physical property at any one time. In this context, the relative states being unique and mutual orthogonal guarantees that no experiment taking place along one 'branch' of a measurement chain can ever reveal the outcome on another branch.

The next step is to think about what would happen with a third system N capable of observing S and M.

28.3.1 And the Next One...

Our third system, N, would also have to have a memory record, capable of recording the states of *both* S and M. Hence there must be (at least) two physical variables B and C of N. If the process is to work properly, B and C must have definite values at the same time, and so have mutual eigenstates $\{|B_l,C_m\rangle\}$. We take B as the record of S states and C to record the state of M.

So, if N observes S, the operation at basis level is:

$$\hat{U}_{NS}|s_k\rangle|m_j\rangle|B_l,C_m\rangle = |s_k\rangle|m_j\rangle|B_l + Gs_k,C_m\rangle$$

Whereas if N were to observe M :

$$\hat{U}_{NM}|s_k\rangle|m_j\rangle|B_l,C_m\rangle = |s_k\rangle|m_j\rangle|B_l,C_m + Hm_j\rangle$$

We now have two rules that we can apply to an initial state $|\Psi\rangle = |\psi\rangle_S|\phi_0\rangle_M|\zeta\rangle_N$ as it evolves due to the measurement interaction. We start by deconstructing the initial state:

$$\begin{aligned}
|\Psi\rangle &= |\psi\rangle_S|\phi_0\rangle_M|\zeta\rangle_N \\
&= \left(\sum_k \langle s_k|\psi\rangle|s_k\rangle\right)\left(\sum_j \langle m_j|\phi_0\rangle|m_j\rangle\right)\left(\sum_{l,m} \langle B_l,C_m|\zeta\rangle|B_l,C_m\rangle\right) \\
&= \sum_{k,j,l,m} \langle s_k|\psi\rangle\langle m_j|\phi_0\rangle\langle B_l,C_m|\zeta\rangle|s_k\rangle|m_j\rangle|B_l,C_m\rangle \\
&= \sum_{k,j,l,m} \alpha_{kj}\beta_{lm}|s_k\rangle|m_j\rangle|B_l,C_m\rangle
\end{aligned}$$

using $\alpha_{kj} = \langle s_k|\psi\rangle\langle m_j|\phi\rangle$ and defining $\beta_{lm} = \langle B_l,C_m|\zeta\rangle$.

Allowing the two measuring systems to observe S directly we get:

$$\begin{aligned}
\hat{U}_{NS}\left(\hat{U}_{MS}\left(|\psi\rangle_S|\phi\rangle_M|\zeta\rangle_N\right)\right) &= \hat{U}_{NS}\left(\hat{U}_{MS}\sum_{k,j,l,m} \alpha_{kj}\beta_{lm}|s_k\rangle|m_j\rangle|B_l,C_m\rangle\right) \\
&= \hat{U}_{NS}\left(\sum_{k,j,l,m} \alpha_{kj}\beta_{lm}|s_k\rangle|m_j + gs_k\rangle|B_l,C_m\rangle\right) \\
&= \sum_{k,j,l,m} \alpha_{kj}\beta_{lm}|s_k\rangle|m_j + gs_k\rangle|B_l + Gs_k,C_m\rangle
\end{aligned}$$

Now we allow N to observe M:

$$\hat{U}_{NS}\left(\sum_{k,j,l,m} \alpha_{kj}\beta_{lm}|s_k\rangle|m_j+gs_k\rangle|B_l+Gs_k,C_m\rangle\right)$$

$$=\sum_{k,j,l,m}\alpha_{kj}\beta_{lm}|s_k\rangle|m_j+gs_k\rangle\left|B_l+Gs_k,C_m+H(m_j+gs_k)\right\rangle$$

$$=\sum_{k,j}\alpha_{kj}|s_k\rangle|m_j+gs_k\rangle|N;s_k,\ m_j+gs_k\rangle$$

$$=\sum_{k}|s_k\rangle|M;s_k\rangle|N;s_k,\ m_j+gs_k\rangle$$

The measurement record of N correlates with that in M and to the system S. Given an initial state of S, N observes the same value as M and furthermore *observes that M observes it.*

Once again, the relative states separate out into orthogonal collections.

This demonstration is very important. Problems would arise if N could observe a memory record of M that doesn't contain the actual observed value of S that N has checked. Each possible measurement outcome is consistently linked into one of a collection of orthogonal relative states. From the point of view of the last measurement in the chain, each sequence tells a consistent story. It's not possible to observe a result that belongs to one relative state from the vantage point of another.

28.4 THE ONTOLOGICAL STEP

So far, this has been a reasonably routine analysis of the measurement problem, with a couple of interesting twists added. The real revolutionary aspect comes from the ontological gloss that Everett placed on this mathematics.

Holding to his belief that state collapse is a "magic process" that should not be considered in quantum physics, Everett suggested that all the orthogonal relative states *are equally valid ontologically.* All of them are true *at the same time.* The state never collapses, it just continues to evolve according to the dictates of the Schrödinger equation, with each 'branch', i.e., collection of correlated relative states, accumulating new and consistent correlations.

However, we clearly do not observe systems existing in such strangely combined states. Everett's answer simply incorporates us and our observations into the analysis. After all, in Everett's view, we are just as much 'measuring devices' as a Stern–Gerlach (SG) magnet. We have our memory records as well. According to the discussion from the previous section, *we must exist in a combined state with each of our possible memory records correctly linked with the experimental results.* As Everett puts it:

> with each succeeding observation (or interaction), the observer state "branches" into a number of different states. Each branch represents a different outcome of the measurement and the corresponding eigenstate for the object-system state. All branches exist simultaneously in the superposition after any given sequence of observations. The "trajectory" of the memory configuration of an observer performing a sequence of measurements is thus not a linear sequence of memory configurations, but a branching tree, with all possible outcomes existing simultaneously in a final superposition with various coefficients in the mathematical model.

Clearly, the physicists responsible for reviewing Everett's paper raised the obvious objection: we don't experience ourselves to be in a combined state of this sort. To counter this Everett added the following footnote:

> In reply to a preprint of this article some correspondents have raised the question of the "transition from possible to actual," arguing that in "reality" there is – as our experience testifies – no such splitting of observers states, so that only one branch can ever actually exist. Since this point may occur to other readers the following is offered in explanation. The whole issue of the transition from "possible" to "actual" is taken care of in the theory in a very simple way – there is no such transition, nor is such a transition necessary for the theory to be in accord with our experience. From the viewpoint of the theory all elements of a superposition (all "branches") are "actual," none are any more "real" than the rest. It is unnecessary to suppose that all but one are somehow destroyed, since all separate elements of a superposition individually obey the wave equation with complete indifference to the presence or absence ("actuality" or not) of any other elements. This total lack of effect of one branch on another also implies that no observer will ever be aware of any "splitting" process.

This is why the consistency of the process, as we discussed in Section 28.3.1, is vitally important. Each branched version of us has a separate and fully consistent memory record and no access to the memory record existing in any other branch. Hence, *we have no way to know that we are in a superposition of states.*

As part of his defence, Everett made an interesting comparison with another situation from the history of physics. When Copernicus placed the Sun at the centre of the universe, with the Earth and other planets in orbit around it, people denied that this could be correct as they had no sensation of movement while standing on the surface of the Earth. At the time, Copernicus had no answer to this. It was only later with the work of Galileo and Newton that physicists realized that we couldn't possibly have any awareness of the Earth's movement, having to share it ourselves. The fundamental relativity of motion prevents us from being able to notice. In this way, the fully worked out Copernican theory explains why it is that we can't tell (at least without sophisticated measurements) that it's correct.

Everett claimed the same for his theory. All we need to do is take the mathematics of quantum theory completely seriously. It's telling us exactly what's happening from an ontological point of view. This includes explaining why we can't tell that the universe, and us with it, is actually a collection of mutually exclusive evolving branches.

There is a certain elegance to this, which followers of Everett cite as a very attractive feature. There is no need to add anything to quantum theory as it stands, no collapse of state, and no modifications to the equations. Everything is perfectly correct. You just have to treat it seriously.

28.5 MANY WORLDS ARRIVES

The context behind these simultaneously existing branches of states is not entirely clear in Everett's original paper. It was DeWitt who introduced the term *Many Worlds* and started talking about parallel worlds within the one universe.

Nowadays we don't quite follow the same terminology, using *universe* where DeWitt would use *world* and *multiverse* for DeWitt's *universe.* However, I prefer the original, less hyperbolic, expressions, partly as it distinguishes the context of the quantum interpretation from some developments in modern cosmology, which also talk about multiverses. The multiverse/universe terminology has also become polluted by its use in science fiction. Everett's worlds are not so closed off from one another as the parallel universes often used in sci-fi.

While multiple copies of observers in the various worlds are completely unaware of one another's existence, due to the lack of any overlap between the measurement records, that does not mean that the worlds are completely cut off from one another. After all, *this is what causes interference.* When a photon arrives at a half-silvered mirror, its state splits so that in one branch the photon travels along one path, and in another it sets off on the other path. As yet, the two branches have not completely separated (so we can't really call them different worlds), as there has been no measurement made and no memory record recorded. In which case, the relative states are not orthogonal. The amplitudes can interfere and when the photon arrives at the detector, multiple further branches are created, one for each possible position in the interference pattern.

The wave function is hence the true meta-reality spread across different worlds. We can't claim ultimate reality for any one world separately:

> By virtue of the temporal development of the dynamical variables the state vector decomposes naturally into orthogonal vectors, reflecting a continual splitting of the universe into a multitude of mutually unobservable but equally real worlds, in each of which every good measurement has yielded a definite result.[11]

Everett's own terminology of 'splitting' is possibly unhelpful as it conjures a picture like that of an amoeba undergoing division – two (or more) worlds created from one. It would be better to talk of an 'unravelling' whereby two or more branches that were capable of interference gradually separate into non-overlapping versions.

28.6 MANY WORLDS MATURES

The Everett interpretation is quite popular among quantum cosmologists, who are trying to write down a wave function for the universe. In this context, they will clearly struggle to find a measuring apparatus, or an observer, who stands outside the universe to collapse its state[12]. The results of a pole of 72 leading cosmologists[13] returned 58% agreement with the statement "yes I think that the Many Worlds Interpretation is true".

However, appealing to one fraction of the physics community is not enough. If Everett's radical interpretation is going to persuade the rest of us, it must have some telling advantages.

In this, we should not dismiss the simple elegance of the viewpoint. David Wallace[14] expresses it in the following way:

> In recent work on the Everett (Many-Worlds) interpretation of quantum mechanics, it has increasingly been recognized that any version of the interpretation worth defending will be one in which the basic formalism of quantum mechanics is left unchanged. Properties such as the interpretation of the wave-function as describing a multiverse of branching worlds, or the ascription of probabilities to the branching events, must be emergent from the unitary quantum mechanics rather than added explicitly to the mathematics.[15]

If all the challenges to a realistic theory itemized by Isham and discussed in Chapter 26 can be answered from within the mathematical formalism of quantum theory *as it stands*, rather than by adding a collapse assumption or tinkering with the equations, then that would be significant progress. We need to review each in turn.

28.6.1 THE NATURE OF PROBABILITY

There is an acute issue with probability in the context of the Many Worlds interpretation.

Whenever we arrive at a measurement process, each possibility happens. In that sense, there is an ensemble, distributed across the various worlds or branches in the universe. There is no obvious vantage point from which we could observe this ensemble, but that is not the primary issue. Also, there is clearly a sample space: the various measurement outcomes, as normal. However, *the distribution of outcomes across the ensemble of worlds does not reflect any probability calculated from the expansion amplitudes.* If we follow the progress of a state such as $|\psi\rangle = \frac{1}{\sqrt{10}}\left(|U\rangle + 3|D\rangle\right)$, we do not see nine $|D\rangle$ worlds for every $|U\rangle$ world. We get two worlds, one $|D\rangle$ and one $|U\rangle$.

Imagine running a sequence of experiments with electrons in state $|\psi\rangle$ to confirm the expectation value $\langle \hat{S}_z \rangle = -0.8$. The first experiment branches into $|D\rangle$ and $|U\rangle$ worlds. The second experiment in the $|D\rangle$ world branches into $|D\rangle$ and $|U\rangle$ again. Surely, in one branch the unfortunate observer will end up with a memory record $(-1, -1, -1, -1, \ldots)$ rendering the expectation value meaningless.

Conventionally, this issue is split into two distinct but related problems.

1. **The incoherence problem**: When we are about to carry out an experiment, we act as if we're not sure what the result will be and calculate the relative probabilities of the outcomes. How can we make sense of a process that *assigns probabilities to outcomes that are certain to occur across the branching worlds*?
2. **The quantitative problem**: Even if we can make sense of our experience of not being sure of the outcome of an experiment, why should the results be determined by the quantum probability, $\text{prob}(x) = |\langle x|\varphi\rangle|^2$ (Born Rule), rather than some other rule that assigns probabilities to branches?

As normal, I have a physicist standing by to observe the results of a measurement made on a quantum system described by state $|\varphi\rangle$. This physicist is a follower of the Everett interpretation, so she knows that at some point in the future, after the measurement has been made and observed, there will be many copies of her—at least one for each possible measurement outcome. She will also be aware that each of these copies is equally 'real' and will have a consistent memory of events up to the point of discovering which actual result is present in her world.

The state ascribed to each of her future selves has been determined by the evolution of the combined set of states, describing the physicist, the measuring device, and the quantum system, according to the Schrödinger equation (or the appropriate $\hat{U}(t)$ operator). State collapse has been sliced out of this viewpoint, and so the amplitudes involved do nothing other than determine how states interfere with one another. Relating amplitudes to probabilities via $\text{prob}(x) = |\langle x|\varphi\rangle|^2$ has no traction unless state collapse picks out one of the possibilities according to this probability rule. Yet, if we speak to our physicist before the measurement is made, she will admit that she's not sure what the outcome will be.

At first glance, such a statement doesn't make any sense in the context of the Many Worlds interpretation: everything that can happen does. In practice, the physicist will calculate the probabilities of the different possible outcomes and use them to construct an expectation value. How can she justify doing this in the Many Worlds interpretation? This is the incoherence problem. David Wallace summarizes it in the following way: "how, when every outcome actually occurs, can it even make sense to view the result of a measurement as uncertain?"[16] (Note that he is using 'uncertain' in the sense of not being sure of an outcome, rather than the technical sense connected with the uncertainty principle.)

Everett's Solution

Everett contended that an observer who monitors a sequence of measurements on a quantum system will have an apparently random sequence of results in their memory record. In a development of this argument, due to Graham and DeWitt, we start with an ensemble of N systems each of which is in the initial state $|\sigma\rangle$, with a measuring device in state $|\phi_0\rangle$. The overall state is then $|\psi_0\rangle = |\sigma\rangle_1 |\sigma\rangle_2 |\sigma\rangle_3 \ldots |\sigma\rangle_n \ldots |\sigma\rangle_N |\phi_0\rangle$. Now we carry out a sequence of measurements of O with eigenstates $\{|s_i\rangle\}$ and $\alpha_I = \langle s_I|\sigma\rangle$ one system after another. After the measurements have all taken place, the overall state is $|\psi_N\rangle$ where:

$$|\psi_N\rangle = \sum_n \left(\sum_i \langle s_i|\sigma\rangle_n |s_i\rangle_n |M; s_i\rangle_n \right)$$

Each observer then has a 'memory sequence' $\{s_1, s_2, s_3 \ldots s_n, \ldots s_N\}$ corresponding to the values observed for system 1, system 2 etc. If we use a function, f, to count the number of occurances of a particular value, s, in a selected sequence (in fractional terms):

$$f(s;\ \{s_1, s_2,\ s_3 \ldots s_n,\ \ldots s_N\ \}) = f(s;\ \{s_i\}) = \frac{1}{N}\sum_n \delta_{ss_i}$$

we can compare this count with what we would expect for a series of measurements in a 'wave function collapse' scenario. In which case, we would expect each s that we pick to occur a fraction $|\langle s|\sigma\rangle|^2$ times. The variance between our count from the memory sequence and the quantum probability prediction is, for each selected s:

$$\text{Var}(s;\{s_i\}\ ) = \left[f(s;\ \{s_i\}) - |\langle s|\sigma\rangle|^2\right]^2$$

giving an overall variance, for the whole sequence:

$$\text{Var}(\{s_i\}) = \sum_s \left[f(s;\ \{s_i\}) - |\langle s|\sigma\rangle|^2\right]^2$$

We now pick a criterion of *randomness* by specifying a number ε which is as small as we wish, but non-zero. A particular memory sequence is declared to be random if $\text{Var}(\{s_i\}) < \varepsilon$ and non-random if $\text{Var}(\{s_i\}) \geq \varepsilon$. The memory sequence of our unlucky observer who got spin down every time would assuredly be in the non-random category.

If we then take our state $|\psi_N\rangle$ and manually chuck out of the summation any branch where the memory sequence is non-random, we end up with a state $|\psi_N^\varepsilon\rangle$. It can then be shown that:

$$\lim_{N\to\infty}\left[|\psi_N\rangle - |\psi_N^\varepsilon\rangle\right] = 0$$

for any ε that we select. In other words, if we take enough measurements, the impact of the non-random sequences is negligible. The argument then goes on to suggest that the Born rule arises out of the formalism, without having to be added as an extra postulate. However, that approach has been criticized as relying on assumptions that have not been justified.

Other Approaches

Two other possible ways of dealing with the incoherence problem have been suggested. Both place the probability in the mental attitude of the observer. We might say that our physicist, waiting to carry out an experiment, knows that she will end up as one of her future copies, but can't be sure which one. The 'uncertainty' is then internal, not something to do with the physics at all[17]. Or we might allow the experiment to happen, so the branching into different worlds takes place, but then to ask the physicist which world she is in *before she sees the result*. At this point, she knows that he is in one world or another but can't be sure which one. This would be the necessary uncertainty[18].

Even if we can make some sort of sense out of the language of 'indecision' and 'uncertainty' in the Many Worlds interpretation, that still leaves the quantitative problem to solve. In a way, it makes it a bigger issue, for we are now faced with having to justify why our subjective uncertainty should be related to $\text{prob}(x) = |\langle x|\varphi\rangle|^2$.

In 1999 David Deutsch transformed the discussion by publishing a proof of the Born rule, $\text{prob}(x) = |\langle x|\varphi\rangle|^2$, based on *decision theory*[19], apparently anchoring our subjective feeling of uncertainty in a calculatable manner. The work has since been refined by Wallace[20] and Saunders[21].

Decision Theory Enters the Argument

In Deutsch's view, rather than being a weakness of the Many Worlds approach, the issue of probability is a strength. After all, as he points out in his paper, although we can calculate quantum probabilities quite happily, we're not sure what they *mean*, even in the context of standard interpretations (Section 26.2.1). Deutsch defines a quantity that can *play the part of probability* from *within the theory itself* clearing up the issue regarding the meaning of quantum probability and placing it in the realm of the observer, but in a quantitative and objective manner.

Decision theory sets out rules for people to make sensible, logical, and consistent decisions in circumstances where they might not know all the details involved.

Imagine that you're in a situation with a range of possible actions (things to do) that you might take. Each action has a specific outcome. Decision theory says that the way forward, assuming you want to make the choice on *logical* rather than *emotional* grounds, is to give each outcome a *value* (a quantity specifying how much it matters to you) and then pick the action that gives the best return.

In practice, it's more likely that each action will have a range of possible outcomes and you can't be sure which will in fact happen. In which case, each outcome should be given a *weight*, a number between 0 and 1, with the total weight of all the outcomes for each action adding up to 1. Then you calculate:

$$\text{Expectation} = \sum_{\text{outcomes}} \text{weight} \times \text{value}$$

for each action and pick the one with the greatest expectation.[22]

Clearly, if we interpret this weight as a conventional probability, then the expectation takes the same form as the expectation value calculated via quantum theory. Decision theory justifies our use of probability: something of that nature must exist if we are going to make sensible and logical decisions.

Deutsch's strategy was to introduce a few assumptions from decision theory into the Everett interpretation and then to show how they work with quantum theory to create a natural value measure that is the same as the expectation value from conventional theory. This then justifies the use of the Born rule.

As an example of one such assumption, imagine that you're given a choice of actions leading to three possible outcomes, A, B, and C. If you prefer A to B and also find B preferable to C then you must, if you're being logical, prefer A to C. Such assumptions guarantee that we can quantify our preferences by giving each outcome a value, which in the end can be written down as a number, $\mathcal{V}[A]$, which stands for the value of outcome A.

Decision theory models situations in terms of games, and Deutsch follows this line by defining a specific quantum game. Suppose we set up a situation in which we make a measurement on some quantum system in state $|\varphi\rangle$. The observable we measure is represented by an operator $\hat{X}$ and has a range of eigenvalues $\{x_i\}$. The rules of the game ask us to make a stake on the outcome of the measurement, given that we will receive a payoff equal in monetary terms to the value x_I that appears in the measurement. If we stake an amount less than the measurement result, we will make a profit. However, we'll have lost out if the measurement comes in less than the stake.

Deutsch defines the value of the game as the maximum amount that a sensible player is prepared to stake on the outcome of the measurement.

Let's say that we have a quantum system in a state $|\varphi\rangle$, which has two equally possible eigenvalues of position $|3\rangle$ and $|7\rangle$, so that $|\varphi\rangle = \frac{1}{\sqrt{2}}\left(|3\rangle + |7\rangle\right)$. Tabulating the outcomes for various stakes:

	RESULT $\|3\rangle$		RESULT $\|7\rangle$	
Stake £3	Outcome	£0	Outcome	+£4
Stake £4	Outcome	–£1	Outcome	+£3
Stake £5	Outcome	–£2	Outcome	+£2
Stake £6	Outcome	–£3	Outcome	+£1
Stake £7	Outcome	–£4	Outcome	£0

We can see that the maximum sensible amount to stake would be £5, and hence this is the value of the game as defined by Deutsch.

This is all well and good, but in a more general situation, with a more complicated state, a player must have some rule for determining the value of a game.

If we follow a conventional quantum theory approach, then it's easy to see that the value of a game is the same as the expectation value, $\langle x \rangle$ of the measurement results. Sometimes the measurement will come in with $x > \langle x \rangle$ and we make a profit, sometimes we lose out when $x < \langle x \rangle$, but on balance, we come out neutral. The game value we arrived at in the earlier table, £5, is the same as the expectation value of the state $|\varphi\rangle = 1/\sqrt{2}\left(|3\rangle + |7\rangle\right)$.

Expectation values, however, are the quantum probability weighted average over all possible measurement results. In the Many Worlds theory, a likelihood probability, as the quantum probability ultimately is, loses meaning, so the expectation value is in danger of sinking as well. Deutsch's proof rescues the expectation value by showing that it is the value that should be attached to a general game, without introducing any probabilistic ideas.

The complete argument is too intricate to follow through in detail here, but the result is of such importance that it's worth seeing an outline of how it fits together.

As a piece of terminology let's write $\mathcal{V}\left[|\varphi\rangle\right]$ as being the value of a game played using a known state $|\varphi\rangle$. I'm going to assume that $|\varphi\rangle$ can be written as a sum over some (but not necessarily all) of the eigenstates of a measurement $\hat{X}$, $\left\{|x_i\rangle\right\}$.

If we happen to play the game using one of the eigenstates of $\hat{X}$, then the value of such a game (the maximum stake that a player is prepared to make) is x_i, the eigenvalue of that state. So, $\mathcal{V}\left[|x_i\rangle\right] = x_i$, a result that will be useful later.

We now need a couple of sensible rules. First, the value of a game played as the *banker* (the person paying out on the other side) is the negative of the value of a game from the *player's* point of view:

$$\mathcal{V}\Big[\underbrace{\sum a_i|x_i\rangle}_{\text{Player}}\Big] = -\mathcal{V}\Big[\underbrace{\sum a_i|x_i\rangle}_{\text{Banker}}\Big] = \mathcal{V}\Big[\underbrace{\sum a_i|-x_i\rangle}_{\text{Player}}\Big] \qquad \text{Rule 1}$$

where in the last step, we switch player with banker and say that a banker paying out £x on a result of x is the same as a player receiving –£x on a result of $-x$. The second rule is:

$$\mathcal{V}\Big[\underbrace{\sum a_i|x_i + k\rangle}_{\text{Player}}\Big] = k + \mathcal{V}\Big[\underbrace{\sum a_i|x_i\rangle}_{\text{Player}}\Big] \qquad \text{Rule 2}$$

Hence the value of a game played with all the x values shifted along by an amount k is the same as that of a normal game $+k$. Think of it like this; if you were a player asked to choose between two different games, one in which you had a payment based on adding k to the measurement result, and another game based on paying you the measurement result and then giving an additional k, you wouldn't really care which you played.

Having got hold of these two rules, we can apply them to a simple state such as:

$$|\varphi\rangle = \frac{1}{\sqrt{2}}\left(|x_1\rangle + |x_2\rangle\right)$$

Rule 2 tells us that:

$$\mathcal{V}\left[\frac{1}{\sqrt{2}}\left(|x_1+k\rangle+|x_2+k\rangle\right)\right]=k+\mathcal{V}\left[\frac{1}{\sqrt{2}}\left(|x_1\rangle+|x_2\rangle\right)\right]$$

Remembering that we can choose k to be anything we want, we decide to be sneaky and set $k=-(x_1+x_2)$ which gives us:

$$\mathcal{V}\left[\frac{1}{\sqrt{2}}\left(|x_1-(x_1+x_2)\rangle+|x_2-(x_1+x_2)\rangle\right)\right]=-(x_1+x_2)+\mathcal{V}\left[\frac{1}{\sqrt{2}}\left(|x_1\rangle+|x_2\rangle\right)\right]$$

The left-hand side simplifies to:

$$\mathcal{V}\left[\frac{1}{\sqrt{2}}\left(|-x_2\rangle+|-x_1\rangle\right)\right]=-(x_1+x_2)+\mathcal{V}\left[\frac{1}{\sqrt{2}}\left(|x_1\rangle+|x_2\rangle\right)\right]$$

This is now an ideal opportunity to apply Rule 1, transforming the left-hand side:

$$-\mathcal{V}\left[\frac{1}{\sqrt{2}}\left(|x_2\rangle+|x_1\rangle\right)\right]=-(x_1+x_2)+\mathcal{V}\left[\frac{1}{\sqrt{2}}\left(|x_1\rangle+|x_2\rangle\right)\right]$$

Now we can gather like terms together:

$$-2\mathcal{V}\left[\frac{1}{\sqrt{2}}\left(|x_2\rangle+|x_1\rangle\right)\right]=-(x_1+x_2)$$

Or:

$$\mathcal{V}\left[\frac{1}{\sqrt{2}}\left(|x_2\rangle+|x_1\rangle\right)\right]=\frac{1}{2}(x_1+x_2)$$

which completes what we set out to show. The value of a game played using $|\varphi\rangle$ is the same as the expectation value from conventional quantum theory.

This is the key result in Deutsch's paper. Using it, he was able to show that the value of a game played on a more general state $|\Psi\rangle=\sum a_i|x_i\rangle$ is:

$$\mathcal{V}\left[|\Psi\rangle\right]=\mathcal{V}\left[\sum a_i|x_i\rangle\right]=\sum|a_i|^2x_i$$

As the value of a game played on an eigenstate is $\mathcal{V}\left[|x_i\rangle\right]=x_i$, then:

$$\mathcal{V}\left[|\Psi\rangle\right]=\sum|a_i|^2x_i=\sum|a_i|^2\mathcal{V}\left[|x_i\rangle\right]$$

So, the value of the game played on $|\Psi\rangle$ is the weighted sum of the values of individual games played on eigenstates $|x_i\rangle$ with $|a_i|^2$ being the weighting factor. In other words, it's the expectation value of a measurement on $|\Psi\rangle$.

Deutsch's argument is that adding some assumptions from decision theory into a quantum theory without the Born rule allows him to recreate a role for the square of the amplitude.

Since its initial publication, Deutsch's proof has been criticized and revised by others. For example, if you read the original paper, it's not clear how much the work depends on the Many Worlds interpretation. In essence, using such a value measure will only work if you can be sure that all outcomes will happen. After all, if we are in a probabilistic world with only one outcome happening, the value of that outcome changes when we know that it has happened and none of the others have. Strictly then, a value such as the one that Deutsch uses can only be consistently defined in the context of the Many Worlds interpretation.

Deutsch's proof has also been accused of being circular. After all, when we write a state as $|\varphi\rangle = \frac{1}{\sqrt{2}}\left(|x_1\rangle + |x_2\rangle\right)$, are we not already importing probability into the proof via the $\frac{1}{\sqrt{2}}$ normalization term? This criticism could be countered by regarding the mathematical formalism as an abstract set of rules that are not connected with the 'real world' until the formalism runs into an experimental situation.

The debate about Deutsch's work is likely to continue for some time yet.

28.6.2 State Reduction

This one is easy: there is no such thing. In a conventional interpretation, state collapse is needed to explain why a state covering a wide range of possibilities converts into a situation in which there is only one observed result. In the Many Worlds view the combined state exists and describes the whole universe, the different measuring results playing out in the branching worlds contained within that universe. We may see only one result in our branch, but the others are equally real to the observers occupying parallel branches.

28.6.3 Entanglement

> In the case of observation of one system of a pair of spatially separated, correlated systems, nothing happens to the remote system to make any of its states more "real" than the rest. It had no independent states to begin with, but a number of states occurring in a superposition with corresponding states for the other (near) system. Observation of the near system simply correlates the observer to this system, a purely local process - but a process which also entails automatic correlation with the remote system. Each state of the remote system still exists with the same amplitude in a superposition, but now a superposition for which element contains, in addition to a remote system state and correlated near system state, an observer state which describes an observer who perceives the state of the near system. From the present viewpoint all elements of this superposition are equally "real." Only the observer state has changed, so as to become correlated with the state of the near system and hence naturally with that of the remote system also. The mouse does not affect the universe - only the mouse is affected.[23]

The surprise introduction of a mouse into Everett's deliberations reflects Einstein's address[24] to the Palmer Laboratory, when he apparently expressed his criticism of quantum theory by commenting that a mouse could not bring about a drastic change to the universe simply by looking at it. More importantly, in this quote, Everett neatly dissolves the EPR paradox in the context of the relative state approach, and with it the non-local correlations of an entangled state.

28.6.4 Measurement

Much of the mystery surrounding the process of measurement is removed when we cast out state reduction. In each branch, a measurement reports a single result. There is no need to have a separate theory for measuring equipment or rely on a classical world to break the spread of entanglement. Everything is described perfectly well by the Schrödinger equation (or the $\hat{U}(t)$ operator if you prefer).

28.6.5 Bell's Inequality and the K-S Theorem

Once again, much of the mystery surrounding these constraints on a hidden variable theory, and hence the illumination of quantum reality, is dissolved when you adopt a Many Worlds approach.

There is no state collapse bringing about correlations between distant measurements. Those correlations are baked into the wave function and the process of observation simply correlates the states of the observer with the states of the combined entangled system. Equally, the constraint placed on physical variables, that they cannot be unambiguously assigned values prior to measurement, is part of the flow of the wave function and the branching realities that it describes. The one physical variable has multiple values at the same time, played out in different branches.

28.7 CRITICISMS OF THE MANY WORLDS VIEW

The Many Worlds interpretation is probably the most radical revision to our view of reality ever suggested in the name of quantum physics. As DeWitt himself puts it:

> I still recall vividly the shock I experienced on first encountering this multiworld concept. The idea of 10^{100} slightly imperfect copies of oneself all constantly spitting into further copies, which ultimately become unrecognizable, is not easy to reconcile with common sense. Here is schizophrenia with a vengeance.[25]

The interpretation raises all sorts of philosophical problems about individuality (which one of my future possible selves will I become? Does it make any sense to ask that question?). Supporters claim that they can all be trumped by saying that each possible me will have a complete memory record, including the me that I am now. In a sense, it is a meaningless question, as all of them think that they're me anyway.

There is, perhaps, a bigger criticism that can be levelled. Is our confidence in quantum theory great enough to support such a philosophical leap? After all, we know that quantum theory, or even quantum field theory, is not the last word in physical theory due to various issues that arise when you try to develop a quantum theory of gravity (to be discussed later in chapter 32). Such a move is crucial to cosmologists, so even if the interpretation is popular in that domain, it should at best be placed in the 'pending' category while we await a fully consistent quantum gravity to work with.

On the other hand, Hilary Greaves has put the case for the defence quite elegantly:

> To the charge of ontological extravagance it can be countered that, firstly, such 'extravagance' is preferable to the theoretical extravagance and inelegance required to eliminate other branches when our best formalism predicts their existence [i.e. having to use the collapse of state to eliminate other branches]; secondly, this ontological extravagance is anyway not too damaging when the extra entities are of the same kind as those already admitted to existence. Complaints that the Everettian picture is simply unbelievable may be well taken, but as far as philosophy goes, we can reply only by borrowing David Lewis's memorable remark: "I cannot refute an incredulous stare."[26]
> [my addition]

From a practical point of view, however, there is another problem with the Many Worlds interpretation, which can be illustrated with a simple example involving the entangled spin states of two electrons. If we focus on the vertical spin component, we will have to write the state as $|\varphi\rangle = \frac{1}{\sqrt{2}}\left(|U\rangle|D\rangle - |D\rangle|U\rangle\right)$. In the Many Worlds interpretation, this state doesn't describe a mixture of different possibilities, one of which actualizes when we take a measurement. Instead, it tells us about a universal state $|\varphi\rangle$ that branches into two parallel worlds. In one world the spin states of the two particles are $|U\rangle|D\rangle$ and in the other $|D\rangle|U\rangle$.

The problem is that we can write $|\varphi\rangle$ in a different way if we choose to focus on some other spin component such as the (L, R) combination. That would lead to a state $|\varphi\rangle = \frac{1}{\sqrt{2}}\left(|L\rangle|R\rangle - |R\rangle|L\rangle\right)$ and so two *differently* branched worlds.

This is a simple example of a wider issue.

The same state $|\Psi\rangle = \sum_i \sum_k g_{ik} |s_i\rangle |m_k\rangle$ can be expanded over any eigenstates of S and M that we happen to fancy. In which case, we appear to have several different descriptions of the same universe to apply. We need to find some *preferred basis* that correctly describes the universe and all the branching worlds within it.

This is where *decoherence* plays a part.

We first came across decoherence in Section 25.4, in the context of quantum systems becoming entangled with measuring devices, which in turn become entangled with the environment (the rest of the world). The number of variables involved with 'the rest of the world' is clearly too many to keep track of, especially as their values could be changing on a relatively rapid timescale. Such changes don't have much influence on the *macroscopic* picture (who cares if that one molecule of oxygen in the corner of the room moved a bit…?), but they can have significant *microscopic* effects.

To try and cope with this, we trace over the environmental variables in the density operator. Interestingly, when the density operator is represented in matrix form, the decoherence effect of the environment tends to diagonalize the matrix (the off-diagonal terms tend to average to zero).

This could be deeply significant for the Many Worlds interpretation as diagonalizing the density operator effectively picks out a basis. In Chapter 25, we showed how a diagonal density matrix was formed from the sum of density matrices of eigenstates. Perhaps the solution to the preferred basis problem in the Many Worlds interpretation is that the environmental entanglement picks out a preferred basis for the worlds to form along.

This might have an added benefit.

In the original relative state formulation, the existence of separate orthogonal worlds was an important consequence of having orthogonal relative states. In turn, this elevates the category of interaction that we call measurement into a significant role. Without measurements, the world does not develop into distinct branches. However, environmental decoherence could replace measurement as the root cause of branching. In practical terms, the different branches would slowly lose their ability to interfere with one another as their differing entanglements scrambled the phase relationships between them (remember that we need fixed phase differences to see interference). Orthogonality would emerge from the decoherence. In this model, the different worlds would gradually develop and 'diverge' from one another as decoherence took hold.

I have tried to illustrate the difference in Figure 28.1. With measurement as the cause of branching, the world divides into distinct branch points at 'kinks'. If we take a vertical slice through Figure 28.1 A, effectively looking at the world from God's perspective at a given time, then we can easily count the number of branches that exist. However, if branching is a gradual emergence due to

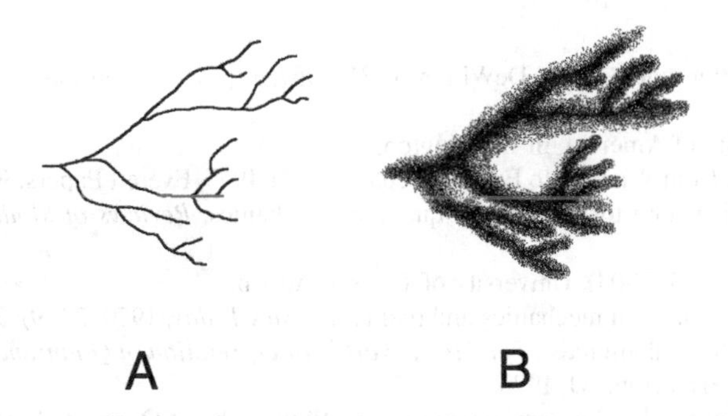

FIGURE 28.1 Contrasting ways in which separate world branches emerge over time. In both diagrams, time is advancing along the horizontal direction. In A, measurement interactions produce definite kinks in the evolution of the world, hence separate branches can be clearly defined. In B decoherence steadily develops the different branches but makes it harder to tell when a branch has emerged sufficiently to be counted as distinct.

decoherence (Figure 28.1b), then things on any vertical slice are rather fuzzier. It is much harder to define the moment at which different branches have separated sufficiently to be counted as distinct.

This is not necessarily problematic. For one, Wallace[27] has argued that the different worlds are not part of the *fundamental ontology*, but instead are an aspect of *emergent ontology* and hence no more an issue than any other approximation that takes place in science.

Another ongoing aspect of the debate is the role of probability. Some contend that the whole decoherence approach is predicated on probability, which supporters of Many Worlds contend can be shown to emerge from the theory (or at least an equivalent to probability can be established). Others feel that decoherence hinges on the ability to take approximations rather than fundamentally relying on probability.

28.8 TIME THOUGHTS

One mark of a good idea in theoretical physics is the cross-pollination that it generates. If the idea opens new perspectives on matters and enables work to push off in a different, and sometimes unthought of direction, then it is probably a good idea.

Some physicists, inspired by the idea of parallel universes, have started serious theoretical work on time travel.

One of the reasons that the community has been so reluctant to accept the notion of time travel is the facility with which you could set up paradoxes. Imagine you build a time machine and with it travel back in time to before you were conceived. For some reason, you decide to murder your father. As a result, you were never born and so could not invent the time machine. So, you did not travel back in time and your father lived to produce a son who invented the time machine....

Although science fiction writers can step around this argument, or even completely ignore it, the problem can be recast in terms of the interactions between particles. Imagine that a particle enters a wormhole and emerges in the past. Inconveniently, the emerging particle then goes on to strike the original particle, diverting its path so that it does not enter the wormhole. In which case, the particle does not go back in time, and hence does not divert the path so that it does go back in time...etc. Phrased in these terms, physicists take the issues more seriously.

This is one reason why time travel has not, until quite recently, been given any degree of theoretical credence. However, the suggestion now is that a journey backwards in time would inevitably move you into one of the parallel worlds, not the one that you occupied at the start of the journey. That would leave you free to exterminate any one of 'your' ancestors, as they would not be part of your timeline.

NOTES

1. Excerpt from a letter sent to Bryce DeWitt, May 31, 1957, https://www.pbs.org/wgbh/nova/manyworlds/orig-02.html.
2. Catholic University of America, in Washington, DC.
3. Handwritten note from Wheeler to Everett, September 21 1955, Everett Papers, Box 1, Folder 5.
4. H. Everett, Relative state formulation of quantum mechanics, *Reviews of Modern Physics*, 1957, 29: 454–462.
5. Bryce S. DeWitt (1923–2004), University of Texas at Austin.
6. Bryce S. DeWitt , Quantum mechanics and reality, *Physics Today*, 1970, 23 (9): 30–35.
7. B.S. DeWitt and N. Graham (eds), *The Many-Worlds Interpretation of Quantum Mechanics*, Princeton University Press, Princeton, NJ, 1973.
8. Letter from Everett to Max Jammer, quoted in The Philosophy of Quantum Mechanics, John Wiley & Sons, 1974.
9. See note 6.
10. See note 8.
11. See note 7.

12 The consistent histories approach in the next chapter also tackles this issue.
13 The poll was conducted by L. David Raub and the results quoted in Tipler, Frank (1994). The Physics of Immortality, Bantam Doubleday Dell Publishing Group pp. 170–171.
14 D. Wallace, Magdalen College Oxford and member of the Philosophy of Physics group at Oxford University.
15 D. Wallace, Everettian rationality: defending Deutsch's approach to probability in the Everett interpretation, *Studies in History and Philosophy of Modern Physics*, 34 (3): 415–438 available online at http://arxiv.org/abs/quant-ph/0303050v2.
16 See note 20.
17 S. Saunders, Time, quantum mechanics, and probability, *Synthese*, 1998, 114: 373–404.
18 L. Vaidman, On schizophrenic experiences of the neutron or why we should believe in the many-worlds interpretation of quantum theory. *International Studies in Philosophy of Science*, 1998, 12: 245–261. Available online at http://www/arxiv.org/abs/quant-ph/9609006.
19 D. Deutsch, Quantum theory of probability and decisions, *Proceedings of the Royal Society of London*, February 1999.
20 As per 20, also Wallace, David (2002). "Quantum Probability and Decision Theory, Revisited". arXiv:quant-ph/0211104, Wallace, David (2003). "Quantum Probability from Subjective Likelihood: Improving on Deutsch's proof of the probability rule". arXiv:quant-ph/0312157 and Wallace, David (2009). "A formal proof of the Born rule from decision-theoretic assumptions". arXiv:0906.2718 [quant-ph].
21 S. Saunders, . Derivation of the Born rule from operational assumptions,. *Proceedings of the Royal Society A*, 2004, 460 (2046): 1771–1788, and S. Saunders, What is Probability?. Quo Vadis Quantum Mechanics? The Frontiers Collection. pp. 209–238. arXiv:quant-ph/0412194.
22 Generally, the value of an outcome is measured in terms of its utility, which is a more general concept than simple financial reward. Interestingly, von Neumann had a large influence on decision theory through his book Theory of Games and Economic Behavior (von Neumann and Morgenstern) published in 1944.
23 See note 14.
24 See note 14, as an address to the Palmer Physical Laboratory, Princeton. Spring, 1954.
25 See note 14.
26 H. Greaves, Understanding Deutsch's probability in a deterministic multiverse, Studies in the History and Philosophy of Modern Physics, September 2004.
27 D. Wallace, Everett and structure, *Studies in History and Philosophy of Science*, 2003, 34 (1): 87–105. arXiv:quant-ph/0107144.

29 Assorted Alternatives

29.1 BEING IN TWO MINDS ABOUT SOMETHING...

One great virtue of the Many Worlds interpretation is the way that it purports to solve the measurement problem without altering the mathematical form of quantum theory.

Strictly speaking, the measurement problem is *dissolved* rather than solved, as the Many Worlds approach sets out to demonstrate how continuous evolution of the wave function, coupled with decoherence and the ontologically bold step of introducing branching worlds, is consistent with our experimental and daily experience.

Of course, certain issues remain. As we have seen, it is difficult to definine a probability across a branching set of worlds, and there is the preferred basis problem. Supporters of Many Worlds are working on solutions to these problems and hanging on by their ontological fingernails in asserting branching worlds to be no more ridiculous than state collapse.

Some physicists, however, have taken the argument in a different direction by putting the mind of the observer into the centre of things.

Let's see how their argument develops in the context of a standard measurement problem.

We start with a state:

$$|\Psi\rangle = \left(a_1|U\rangle + a_2\langle D|\right)|\text{P=R}\rangle_{\text{Device}}|\text{B=R}\rangle_{\text{Carolyn}}$$

describing a system made up of an electron in an arbitrary spin configuration, a device ready to measure that spin with a pointer set to the 'ready' position (P=R), and an observer's brain with its particles in some state indicating that Carolyn is ready (B=R) to look at the device once the measurement is made. If things evolve according to standard quantum processes, we end up with a state:

$$|\Psi'\rangle = a_1|U\rangle|\text{P=U}\rangle_{\text{Device}}|\text{B=U}\rangle_{\text{Carolyn}} + a_2|D\rangle|\text{P=D}\rangle_{\text{Device}}|\text{B=D}\rangle_{\text{Carolyn}}$$

According to the Many Worlds view, this is the start of the branching that leads, via decoherence, into practically orthogonal noninterfering worlds within the quantum universe.

Physicists David Albert and Barry Loewer[1] believe this state to be a correct *physical* description of the universe, but *not a complete description of reality*. For that, we need to consider *nonphysical minds* belonging to observers. In this view, the physical material of the world evolves in a perfectly deterministic way, to give us states such as the one quoted earlier, but minds are probabilistic things, which is why we have the *illusion* of state collapse. The probability issue associated with the Many Worlds interpretation is addressed by moving the probabilistic aspect out of the material world into non-physical minds.

Albert and Loewer propose that each brain, described by a basis set of brain states $\{|B_i\rangle\}$, is associated with an *infinite number of nonphysical minds* which change their mental states in a random manner, but with relative probabilities determined by the Born rule. Thus, the minds become the ensemble needed for an expression of probability. This also removes the need for a branching infinity of worlds. There is just one world, perceived by an infinite number of minds for each physical brain.

To express this mathematically, we need to add to our state $|\Psi'\rangle$ an indication of the mind states involved. At the start of the process, each one of Carolyn's minds, from the infinite set available to her, is associated with the same 'ready' brain state $|\text{B=R}\rangle_{\text{Carolyn}}$. After the measurement, a fraction m of those minds (where $m = |a_1|^2$) are associated with the brain state $|\text{B=U}\rangle_{\text{Carolyn}}$, and a fraction n

DOI: 10.1201/9781003225997-33

(where $n = |a_2|^2$) are coupled to the brain state $|\mathrm{B{=}D}\rangle_{\mathrm{Carolyn}}$. For any one of the minds, it is impossible to determine which final state they will end up with; the process is truly random. We then have:

$$|\Psi'(m,\,n)\rangle = a_1|U\rangle|\mathrm{P{=}U}\rangle_{\mathrm{Device}}|\mathrm{B}(m){=}\mathrm{U}\rangle_{\mathrm{Carolyn}} + a_2|D\rangle|\mathrm{P{=}D}\rangle_{\mathrm{Device}}|\mathrm{B}(n){=}\mathrm{D}\rangle_{\mathrm{Carolyn}}$$

As each of these minds will have a complete memory record of events up to the point of measurement, [2] they will all be entitled to think of themselves as Carolyn. Interesting things start to happen, however, when one observer asks another observer questions about what has been observed, as we will see in the next section.

The theory requires a disconnection between brain and mind states. A brain, according to Albert and Loewer, can be in a quantum superposition of states, but a mind can't, otherwise, we would literally be in two minds about something.

29.1.1 Mindless Hulks...

An infinite number of nonphysical minds hardly seems a better situation than an infinite number of parallel worlds. As far as ontological mudslinging is concerned, both ideas seem capable of collecting their fair share. Unfortunately, if the many minds view is to work, an infinite number of minds for *each* brain is needed.

Let's presume for a second that the theory can get away with only one mind linked to each brain, a mind that is capable of existing in any one of a collection of different states but only one at a time, i.e., not in any quantum superposition.

Before the measurement, the mind is associated with the brain state $|\mathrm{B{=}R}\rangle_{\mathrm{Carolyn}}$. After the measurement, the mind is *either* connected with $|\mathrm{B{=}U}\rangle_{\mathrm{Carolyn}}$ *or* $|\mathrm{B{=}D}\rangle_{\mathrm{Carolyn}}$ with probabilities determined by the Born rule. The final state is then *either*:

$$|\Psi'(\mathrm{C{=}U})\rangle = a_1|U\rangle|\mathrm{P{=}U}\rangle_{\mathrm{Device}}|\mathrm{B}(1){=}\mathrm{U}\rangle_{\mathrm{Carolyn}} + a_2|D\rangle|\mathrm{P{=}D}\rangle_{\mathrm{Device}}|\mathrm{B}(0){=}\mathrm{D}\rangle_{\mathrm{Carolyn}}$$

with probability $|a_1|^2$ or:

$$|\Psi'(\mathrm{C{=}D})\rangle = a_1|U\rangle|\mathrm{P{=}U}\rangle_{\mathrm{Device}}|\mathrm{B}(0){=}\mathrm{U}\rangle_{\mathrm{Carolyn}} + a_2|D\rangle|\mathrm{P{=}D}\rangle_{\mathrm{Device}}|\mathrm{B}(1){=}\mathrm{D}\rangle_{\mathrm{Carolyn}}$$

with probability $|a_2|^2$. Here $B(1)$ is the brain state with a mind associated with it, and $B(0)$ the brain state without an associated mind. Carolyn's brain is in a quantum superposition of states, but her single mind is associated with only one of those states.

The problem comes when we introduce a second observer into the mix. The final quantum state is then (excluding minds for the moment):

$$|\Psi''\rangle = a_1|U\rangle|\mathrm{P{=}U}\rangle_{\mathrm{Device}}|\mathrm{B{=}U}\rangle_{\mathrm{Carolyn}}|\mathrm{B{=}U}\rangle_{\mathrm{Toby}} + a_2|D\rangle|\mathrm{P{=}D}\rangle_{\mathrm{Device}}|\mathrm{B{=}D}\rangle_{\mathrm{Carolyn}}|\mathrm{B{=}D}\rangle_{\mathrm{Toby}}$$

If we factor in Carolyn's mind, we have:

$$\begin{aligned}|\Psi''(\mathrm{C{=}U})\rangle &= a_1|U\rangle|\mathrm{P{=}U}\rangle_{\mathrm{Device}}|\mathrm{B}(1){=}\mathrm{U}\rangle_{\mathrm{Carolyn}}|\mathrm{B{=}U}\rangle_{\mathrm{Toby}}\\ &+ a_2|D\rangle|\mathrm{P{=}D}\rangle_{\mathrm{Device}}|\mathrm{B}(0){=}\mathrm{D}\rangle_{\mathrm{Carolyn}}|\mathrm{B{=}D}\rangle_{\mathrm{Toby}}\end{aligned}$$

or:

$$\begin{aligned}|\Psi''(\mathrm{C{=}D})\rangle &= a_1|U\rangle|\mathrm{P{=}U}\rangle_{\mathrm{Device}}|\mathrm{B}(0){=}\mathrm{U}\rangle_{\mathrm{Carolyn}}|\mathrm{B{=}U}\rangle_{\mathrm{Toby}}\\ &+ a_2|D\rangle|\mathrm{P{=}D}\rangle_{\mathrm{Device}}|\mathrm{B}(1){=}\mathrm{D}\rangle_{\mathrm{Carolyn}}|\mathrm{B{=}D}\rangle_{\mathrm{Toby}}\end{aligned}$$

Adding in Toby's mind, this extends to:

$$\begin{aligned} |\Psi''(\text{C=U, T=U})\rangle = a_1|U\rangle|\text{P=U}\rangle_{\text{Device}}|\text{B}(1)\text{=U}\rangle_{\text{Carolyn}}|\text{B}(1)\text{=U}\rangle_{\text{Toby}} \\ +a_2|D\rangle|\text{P=D}\rangle_{\text{Device}}|\text{B}(0)\text{=D}\rangle_{\text{Carolyn}}|\text{B}(0)\text{=D}\rangle_{\text{Toby}} \end{aligned} \quad (1)$$

$$\begin{aligned} |\Psi''(\text{C=U, T=D})\rangle = a_1|U\rangle|\text{P=U}\rangle_{\text{Device}}|\text{B}(1)\text{=U}\rangle_{\text{Carolyn}}|\text{B}(0)\text{=U}\rangle_{\text{Toby}} \\ +a_2|D\rangle|\text{P=D}\rangle_{\text{Carolyn}}|\text{B}(0)\text{=D}\rangle_{\text{Carolyn}}|\text{B}(1)\text{=D}\rangle_{\text{Toby}} \end{aligned} \quad (2)$$

$$\begin{aligned} |\Psi''(\text{C=D, T=D})\rangle = a_1|U\rangle|\text{P=U}\rangle_{\text{Device}}|\text{B}(0)\text{=U}\rangle_{\text{Carolyn}}|\text{B}(0)\text{=U}\rangle_{\text{Toby}} \\ +a_2|D\rangle|\text{P=D}\rangle_{\text{Carolyn}}|\text{B}(1)\text{=D}\rangle_{\text{Carolyn}}|\text{B}(1)\text{=D}\rangle_{\text{Toby}} \end{aligned} \quad (3)$$

$$\begin{aligned} |\Psi''(\text{C=D, T=U})\rangle = a_1|U\rangle|\text{P=U}\rangle_{\text{Device}}|\text{B}(0)\text{=UP}\rangle_{\text{Carolyn}}|\text{B}(1)\text{=U}\rangle_{\text{Toby}} \\ +a_2|D\rangle|\text{P=D}\rangle_{\text{Device}}|\text{B}(1)\text{=D}\rangle_{\text{Carolyn}}|\text{B}(0)\text{=D}\rangle_{\text{Toby}} \end{aligned} \quad (4)$$

As Toby and Carolyn are blessed with only one mind each, either of which can flip into either possible state, there must be a chance that Toby's mind will be in the state that sees result UP whereas Carolyn's mind sees result DOWN.

In case (2) Toby's mind registers the result DOWN as it is correlated with that brain state. If Toby's brain were to ask Carolyn what she saw, then his *brain* (sans mind) would receive a reply of UP from Carolyn's *brain/mind*, but his *brain/mind* would receive DOWN from Carolyn's *brain* (sans her mind). In the first instance, he would be a *mindless hulk*[3] receiving information from Carolyn's mind. In the second instance, his mind would be talking to a mindless hulk. Only in cases (1) and (3) would genuine mind-to-mind communication happen.

If we back-track, the mindless hulk scenario is a direct result of our initial assumption; that there is only one mind for each of our characters. We can avoid mindlessness if we follow Albert and Loewer and introduce an infinite number of minds coupled to each brain. In which case, direct mind-to-mind communication is always happening as every brain state has more than one associated mind.

29.1.2 The Advantages of Having More Than One Mind

Not many physicists are comfortable with the notion of a mind that is distinct from brain states and nonphysical, not to mention an infinite number of them. The theory though does have some advantages.

Quantum probabilities are associated with the nonphysical mind, rather than some physics going on in the universe.

State collapse doesn't happen, but we're protected from observing strange quantum superpositions as each mind observes only one branch of the quantum state.

The measurement problem is dissolved as the quantum state evolves according to $\hat{U}(t)$ but the mind states evolve randomly.

Entangled states are quantum superpositions and the entanglement is 'broken' by the association with mind states.

The cost, however, is that we have placed some of the most intriguing features of quantum theory outside the realm of physics, or at least the physics of the material world.

The many minds interpretation could be seen as an example of a hidden variable theory. Our typical amplitude expansion $|\psi\rangle = \sum_i a_i |i\rangle$ was set up to model the observation of different possible measurement outcomes with relative probabilities $|a_i|^2$. As a result, we have ended up with state collapse and quantum probabilities potentially being physical features of the world 'out there.' The many minds view regards the amplitudes as governing the random way in which *minds* flip from state to state, determining the branches of evolving states that we observe. We are looking out from our minds into the world, and so the amplitudes are not part of the world as such, which from the point of view of physics makes them hidden variables.

The many minds view is not without its supporters, both among physicists and philosophers, but they are in a minority in the community.

29.2 OBJECTIVE COLLAPSE

As we have repeatedly stressed, the Schrödinger equation, which governs the way amplitudes, or wave functions if you prefer, evolve in time, is *linear*. The time development operator, $\hat{U}(t)$, is based on the Hamiltonian that goes into the Schrodinger equation, so behaves in the same manner. Consequently, any state of the form $|\psi\rangle = \sum_i a_i |i\rangle$, i.e., a mixed state, remains mixed during its time development. The contrast between the evolution of such states and the outcome of measurements, where a specific value of a physical variable is extracted[4], leads to the measurement problem and the various ways in which this has been tackled. Objective collapse theories seek to replace the state collapse *postulate* with a distinct *physical process*. The Schrödinger equation, and hence $\hat{U}(t)$, is regarded as *approximately* true. Some extra physics needs to be added in, which has a negligible effect in the microscopic domain but becomes progressively more significant as the mass and complexity of the system increases. Having said that, we would still expect any new terms in the equations to have some microscopic impact and be potentially detectable by suitably sensitive experiments.

To collapse a wave function, the extra physics must be *non-linear* and have some *stochastic* (randomizing) element. There are several different objective collapse theories on the table, but I am going to focus on one which happens to be my favourite but also illustrates aspects of the general manner in which these adaptations of quantum theory work.

29.2.1 THE PENROSE INTERPRETATION

Essentially, this is an (expert level) speculation regarding the relationship between quantum theory and Einstein's theory of gravity. In a quantum gravity theory, quantum probabilities are presumed to govern the manifestation of different space-time geometries. Penrose relates this to the measurement problem by suggesting that quantum states remain in superposition until the difference between their respective space-time geometries becomes unsustainable.

An evolving quantum state would contain components, or branches if you prefer, governing the manifestation of the same particles at slightly different space-time positions. As space-time geometries are governed by mass distributions[5], each branch develops its own, increasingly distinct, space-time. According to Penrose, this is sustainable until the 'one-graviton' level is reached, which is related to the natural mass scale known as the *Planck mass*[6].

The units that we use to measure physical properties are based on human choices, so in a universal sense, they are somewhat arbitrary. The values of fundamental physical constants depend on the choices that we make when units are set up. However, if we combine the physical constants in adroit ways, the result should relate to a 'natural' scale within the universe. In essence, our colleagues researching physical science on the planet Zog would, almost certainly, not be using kilograms to measure mass, but they would agree on the natural mass scale:

$$M_P = \sqrt{\frac{\hbar c}{G}} \sim 2.2 \times 10^{-8} \text{ kg}$$

with G being the universal gravitational constant. We arrive at this result by noting the following:

$$c \to LT^{-1} \qquad \hbar \to L^2MT^{-1} \qquad G \to L^3M^{-1}T^{-2}$$

hence

$$M_P \to \sqrt{\frac{(L^2MT^{-1})(LT^{-1})}{(L^3M^{-1}T^{-2})}} = \sqrt{M^2} = M$$

There is also the *Planck length*, l_P, which we will come across again in our brief discussion of quantum gravity in Chapter 32:

$$l_P = \sqrt{\frac{\hbar G}{c^3}} \sim 1.6 \times 10^{-35} \text{ m}$$

Although it is difficult to develop a rigorous collapse theory of this nature without a fully refined quantum gravity, Penrose has proposals appropriate for the Newtonian limit. The energy difference between two configurations is calculated from:

$$\Delta E = \frac{1}{G}\int \{g_A(\boldsymbol{r}) - g_B(\boldsymbol{r})\}^2 dV$$

In this formula, $g_A(\boldsymbol{r})$ and $g_B(\boldsymbol{r})$ are the gravitational field strengths at location $\boldsymbol{r} = (x, y, z)$ within the two branches A and B and $dV = dxdydz$. The fields are, in turn, derived from the potential. For example:

$$g_A(\boldsymbol{r}) = -\left(\frac{d\Phi_A}{dx} + \frac{d\Phi_A}{dy} + \frac{d\Phi_A}{dz}\right)$$

The potential, in turn, is obtained by solving a modified version of Poisson's equation[7]:

$$\frac{d^2\Phi_A}{dx^2} + \frac{d^2\Phi_A}{dy^2} + \frac{d^2\Phi_A}{dz^2} = 4\pi Gm|\Psi_A|^2$$

where we see that the source of the field is a probabilistically based mass distribution $m|\Psi_A|^2$. The wave functions come from the *Schrödinger-Newton equation*:

$$-\frac{\hbar^2}{2m}\left(\frac{d^2\Psi(\boldsymbol{r},t)}{dx^2} + \frac{d^2\Psi(\boldsymbol{r},t)}{dy^2} + \frac{d^2\Psi(\boldsymbol{r},t)}{dz^2}\right) + V(\boldsymbol{r},t)\Psi(\boldsymbol{r},t) + m\Phi\Psi(\boldsymbol{r},t) = i\hbar\frac{d\Psi(\boldsymbol{r},t)}{dt}$$

As the extra term here compared to the base Schrödinger equation is $m\Phi\Psi(\boldsymbol{r},t)$, with Φ depending on Ψ, there is a non-linearity introduced. Collapse would take place in one of the stationary state solutions of this equation, over a time scale found using the energy-time inequality from Section 14.3, $\Delta E \times \delta t \geq \hbar/2$.

In summary, Penrose accepts that the wave function is a mathematical description of some objective aspect of reality. As a consequence, matter can exist at more than one space-time location, but

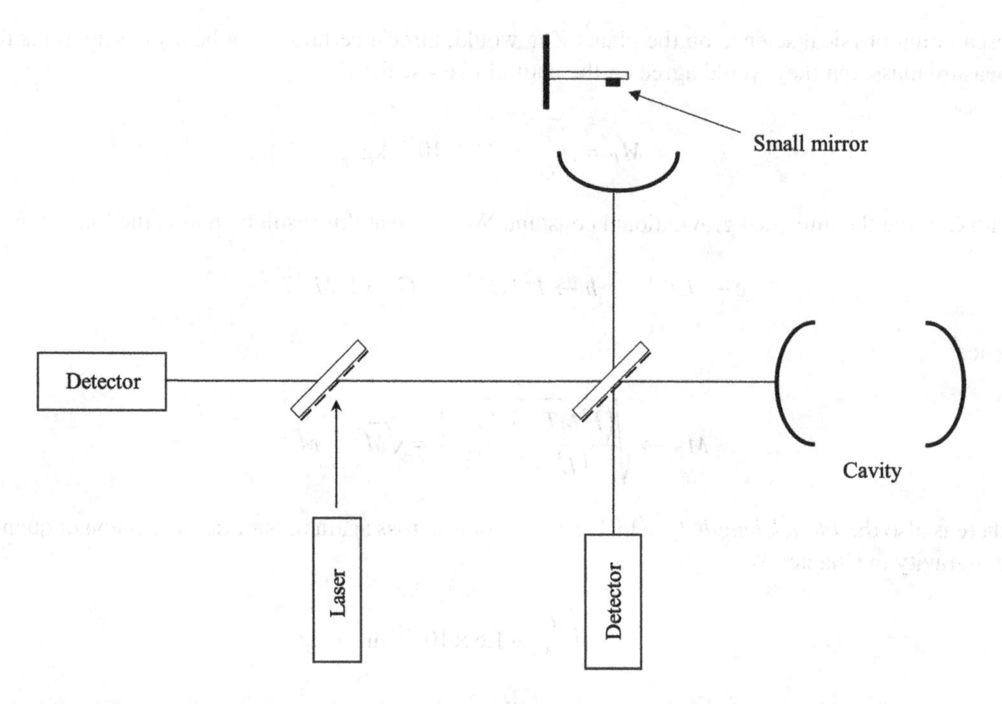

FIGURE 29.1 An arrangement to test objective collapse using a macroscopic superposition of states for a small mirror.

as the possibilities diverge, increasing amounts of energy are required to sustain the differences. At some scale, quite possibly the Planck mass, the natural tendency to fall into the lowest energy state takes over and the quantum state collapses into a steady-state solution of the S-N equation. For microscopic objects, such as electrons, it is possible that the different branches can co-exist for considerable periods of time, possibly thousands of years, but macroscopic objects would collapse in a time scale that would be too small to measure. Penrose and colleagues[8] have proposed an experiment to test his theory by deploying photons and mirrors in an arrangement such as that shown in Figure 29.1.

The basic set up is rather like a Mach-Zehnder interferometer. A single photon entering the experiment is in a superposition of states after the beam splitter. One state, $|A\rangle$, has the photon travelling the horizontal arm to a reflecting cavity built from two concave mirrors. The photon remains in this cavity reflecting back and forth before it 'leaks' out again towards the beam splitter. With state $|B\rangle$ the photon travels the vertical path before passing into another cavity. However, in this case, the second mirror is very small and mounted on a cantilever arm which allows it to move. In normal circumstances, this mirror will vibrate with a small amplitude about its equilibrium position. This is state $|M_e\rangle$. Under the pressure of continually reflecting the photon, the equilibrium position is displaced backwards; state $|M_d\rangle$, by a distance sufficient to ensure $\langle M_d|M_e\rangle \sim 0$. So, the initial state:

$$|\psi\rangle = \frac{1}{\sqrt{2}}\big(|A\rangle + |B\rangle\big)|M_e\rangle$$

evolves into:

$$|\psi'\rangle = \frac{1}{\sqrt{2}}\big(|A\rangle|M_e\rangle + |B\rangle|M_d\rangle\big)$$

placing the mirror into an entangled superposition of macroscopically distinguishable states. If the mirror can be isolated from the environment sufficiently well to prevent or delay environmental decoherence, then sustained single photon interference will occur at the second beam splitter. However, should the state collapse, this effectively measures the photon's path, and there will be no interference.

The constraints on the experiment are extreme, but potentially within the span of current technology. The team propose a mirror $10 \times 10 \times 10$ μm, photon wavelength ~ 630 nm with 5 cm cavities. Such a mirror would contain ~ 10^{14} atoms, which makes this a test of quantum theory in an utterly different regime (see Section 27.6.1). To avoid decoherence, the temperature will need to be maintained below 3 mK and the whole experiment must take place in a high vacuum.

Given the potentially ground-breaking impact of an experiment of this sort, hopefully, we will see results at some point in the near future.

NOTES

1 D. Albert and B. Loewer, Interpreting the Many Worlds interpretation, *Synthese,* 1988, 77: 195–213.
2 Or rather, up to the point at which the mind perceives the results of the measurement.
3 A term introduced by Albert and Loewer.
4 Subject to experimental uncertainty, etc.
5 At least in part. One of the delightful novelties in General Relativity, compared to Newtonian gravity, was the discovery that gravity could itself cause space-time curvature. In a crude sense, the energy in a gravitational field can warp the space-time around it, just as matter can. This led to the prediction of gravitational waves, which were confirmed experimentally in 2016, leading to the 2017 Nobel Prize for Rainer Weiss, Kip Thorne and Barry Barish.
6 Planck was the first to work out a combination of units for length, mass, time and temperature that led to natural scales in this fashion. These days we use slight variations on Planck's suggestions, as it is fashionable to employ $\hbar$ rather than h.
7 This being, in essence, a re-arrangement of the Newtonian equation for the gravitational field.
8 W. Marshall et al., Towards Quantum Superpositions of a Mirror, https://arxiv.org/abs/quant-ph/0210001v1.

30 Consistent Histories

It may be true, as some people say, that everything is in Bohr, but this has been a matter for hermeneuticsm, with the endless disputes any scripture will lead to. It may also happen that he guessed the right answers, but the pedagogical means and the necessary technical details were not yet available to him. Science cannot, however, proceed by quotations, however elevated the source. It proceeds by elucidation, so that feats of genius can become ordinary learning for beginners.

Roland Omnés[1]

In my (single?) mind, one of the most important issues a quantum interpretation must address is the emergence of our 'classical' world from the underlying quantum business. The *consistent histories* approach was initially proposed by Robert Griffiths[2] in 1984. Roland Omnés[3] then developed a more mathematically rigorous version based on the rules of classical probability and logic. A few years later, Murray Gell-Mann[4] and Jim Hartle[5] presented the related *decoherent histories*[6] viewpoint, with a particular emphasis on defining a quantum theory of *closed systems*, i.e., without the need to divide the world into a system and an observer/measuring device. The whole universe is an especially important example of a closed system.

The aim is not to replace traditional quantum theory, but to develop a clear and consistent expression of the fundamentals. It is sometimes viewed as 'Copenhagen done right'.

In this approach,[7] every system has a set of possible *histories*, i.e., a time sequence of values for physical properties, which can extend into the future. The Born rule is taken as a key axiom, and hence quantum probabilities are regarded as fundamental. Consequently, proponents accept that quantum theory can, at best, calculate a set of relative probabilities for each possible history. It is not able to specify which particular history may happen.

Measurements are treated as an example of a broader class of physical processes. Traditional interpretive difficulties, such as Schrödinger's cat, the Kochen-Specker theorem, single-particle interference etc., are dissolved by regarding them as either illustrative of the fundamental differences between classical and quantum physics, rather than a challenge to the quantum view, or the result of making illegal logical moves within the formalism. In the limit of large systems and strong decoherence, the laws of classical physics provide a good approximation to the practical outcome of the quantum business at work. In which case, there is no need to draw a line across the world (à la Bohr/Heisenberg) and have classical physics on one side and quantum physics on the other. Such a view is specifically rejected within consistent histories.

30.1 FRAMEWORKS

The consistent histories approach makes significant use of the projection operators (Section 12.2.4) $\{\hat{P}_i^A\}$ of a physical variable $\hat{A}$, which has eigenvalues $\{a_i\}$. In which case:

$$\sum_i \hat{P}_i^A = \hat{I} \quad \hat{P}_i^A \hat{P}_j^A = \delta_{ij}\hat{P}_j^A \quad \hat{A} = \sum_i a_i \hat{P}_i^A$$

The relationship $\sum_i \hat{P}_i^A = \hat{I}$ is known as a 'projective decomposition of unity', or simply a *projective decomposition*.[8] The most *fine-grained* projective decomposition using $\{\hat{P}_i^A\}$ is $\sum_i \hat{P}_i^A = \hat{I}$, but other

DOI: 10.1201/9781003225997-34

more *coarse-grained* decompositions are possible. Let's say that there are five projection operators for a given variable, so that $\sum_{i=1}^{5} \hat{P}_i^A = \hat{I}$.

We can also define projectors such as:

$$\left(\hat{P}^A\right)' = \hat{P}_1^A + \hat{P}_2^A \quad \left(\hat{P}^A\right)'' = \hat{P}_4^A + \hat{P}_5^A$$

from the set, giving another decomposition:

$$\left(\hat{P}^A\right)' + \hat{P}_3^A + \left(\hat{P}^A\right)'' = \hat{I}$$

If we view the projection operators as mathematical embodiments of a question or proposition, then $\hat{P}_1^A$ = 'are you in state $|1\rangle$?' and $\left(\hat{P}^A\right)'$ = 'are you in a superposition of states $|1\rangle$ and $|2\rangle$?'.

To see how this works, consider a spin-1 particle with the basis:

$$|+\rangle = \begin{pmatrix} 1 \\ 0 \\ 0 \end{pmatrix} \quad |0\rangle = \begin{pmatrix} 0 \\ 1 \\ 0 \end{pmatrix} \quad |-\rangle = \begin{pmatrix} 0 \\ 0 \\ 1 \end{pmatrix}$$

for spin components on the z-axis. Using tensor multiplication, we can construct the appropriate projection operators:

$$\hat{P}_+^Z = |+\rangle\langle+| = \begin{pmatrix} 1 \\ 0 \\ 0 \end{pmatrix} \begin{pmatrix} 1 & 0 & 0 \end{pmatrix}$$

$$= \left(\begin{pmatrix} 1 \\ 0 \\ 0 \end{pmatrix} \times 1 \quad \begin{pmatrix} 1 \\ 0 \\ 0 \end{pmatrix} \times 0 \quad \begin{pmatrix} 1 \\ 0 \\ 0 \end{pmatrix} \times 0 \right) = \begin{pmatrix} 1 & 0 & 0 \\ 0 & 0 & 0 \\ 0 & 0 & 0 \end{pmatrix}$$

$$\hat{P}_0^Z = |0\rangle\langle 0| = \begin{pmatrix} 0 \\ 1 \\ 0 \end{pmatrix} \begin{pmatrix} 0 & 1 & 0 \end{pmatrix}$$

$$= \left(\begin{pmatrix} 0 \\ 1 \\ 0 \end{pmatrix} \times 0 \quad \begin{pmatrix} 0 \\ 1 \\ 0 \end{pmatrix} \times 1 \quad \begin{pmatrix} 0 \\ 0 \\ 1 \end{pmatrix} \times 0 \right) = \begin{pmatrix} 0 & 0 & 0 \\ 0 & 1 & 0 \\ 0 & 0 & 0 \end{pmatrix}$$

$$\hat{P}_-^Z = |-\rangle\langle-| = \begin{pmatrix} 0 \\ 0 \\ 1 \end{pmatrix} \begin{pmatrix} 0 & 0 & 1 \end{pmatrix}$$

$$= \left(\begin{pmatrix} 0 \\ 0 \\ 1 \end{pmatrix} \times 0 \quad \begin{pmatrix} 0 \\ 0 \\ 1 \end{pmatrix} \times 0 \quad \begin{pmatrix} 0 \\ 0 \\ 1 \end{pmatrix} \times 1 \right) = \begin{pmatrix} 0 & 0 & 0 \\ 0 & 0 & 0 \\ 0 & 0 & 1 \end{pmatrix}$$

From here, straightforward matrix manipulation confirms the various properties of these operators. For example:

$$\hat{P}_{+}^{Z}|+\rangle = \begin{pmatrix} 1 & 0 & 0 \\ 0 & 0 & 0 \\ 0 & 0 & 0 \end{pmatrix}\begin{pmatrix} 1 \\ 0 \\ 0 \end{pmatrix} = \begin{pmatrix} 1 \\ 0 \\ 0 \end{pmatrix} = (+1)\times\begin{pmatrix} 1 \\ 0 \\ 0 \end{pmatrix}$$

$$\hat{P}_{+}^{Z}|-\rangle = \begin{pmatrix} 1 & 0 & 0 \\ 0 & 0 & 0 \\ 0 & 0 & 0 \end{pmatrix}\begin{pmatrix} 0 \\ 0 \\ 1 \end{pmatrix} = \begin{pmatrix} 0 \\ 0 \\ 0 \end{pmatrix} = 0\times\begin{pmatrix} 0 \\ 0 \\ 1 \end{pmatrix}$$

shows that the eigenvalues are +1, 0. Visual inspection indicates that $\hat{P}_{+}^{Z} + \hat{P}_{0}^{Z} + \hat{P}_{-}^{Z} = \hat{I}$.

As there are only three operators in play, the number of coarse-grained decompositions is limited, but we can build:

$$\hat{P}_{A}^{Z} = \hat{P}_{+}^{Z} + \hat{P}_{-}^{Z} = \begin{pmatrix} 1 & 0 & 0 \\ 0 & 0 & 0 \\ 0 & 0 & 1 \end{pmatrix}$$

so that $\hat{P}_{A}^{Z} + \hat{P}_{0}^{Z} = \hat{I}$. If we then take the normalized state $|\varphi\rangle = \frac{1}{\sqrt{a^2+b^2}}\begin{pmatrix} a \\ 0 \\ b \end{pmatrix}$ and ask the question, via $\hat{P}_{A}^{Z}$, 'are you in a superposition of $|+\rangle$ and $|-\rangle$?':

$$\hat{P}_{A}^{Z}|\varphi\rangle = \frac{1}{\sqrt{a^2+b^2}}\begin{pmatrix} 1 & 0 & 0 \\ 0 & 0 & 0 \\ 0 & 0 & 1 \end{pmatrix}\begin{pmatrix} a \\ 0 \\ b \end{pmatrix} = \frac{1}{\sqrt{a^2+b^2}}\begin{pmatrix} a \\ 0 \\ b \end{pmatrix} = (+1)|\varphi\rangle$$

i.e., answer 'yes'. If we ask the same question of state $|0\rangle$:

$$\hat{P}_{A}^{Z}|0\rangle = \begin{pmatrix} 1 & 0 & 0 \\ 0 & 0 & 0 \\ 0 & 0 & 1 \end{pmatrix}\begin{pmatrix} 0 \\ 1 \\ 0 \end{pmatrix} = \begin{pmatrix} 0 \\ 0 \\ 0 \end{pmatrix} = (0)|\varphi\rangle$$

we get 'no'. We can ask state $|0\rangle$ the question 'are you in state $|0\rangle$?' as follows:

$$\hat{P}_{0}^{Z}|0\rangle = \begin{pmatrix} 0 & 0 & 0 \\ 0 & 1 & 0 \\ 0 & 0 & 0 \end{pmatrix}\begin{pmatrix} 0 \\ 1 \\ 0 \end{pmatrix} = \begin{pmatrix} 0 \\ 1 \\ 0 \end{pmatrix} = (+1)|0\rangle$$

However:

$$\hat{P}_{0}^{Z}|\varphi\rangle = \frac{1}{\sqrt{a^2+b^2}}\begin{pmatrix} 0 & 0 & 0 \\ 0 & 1 & 0 \\ 0 & 0 & 0 \end{pmatrix}\begin{pmatrix} a \\ 0 \\ b \end{pmatrix} = \frac{1}{\sqrt{a^2+b^2}}\begin{pmatrix} 0 \\ 0 \\ 0 \end{pmatrix} = (0)|\varphi\rangle$$

Notice that within this decomposition, we cannot ask 'are you in state $|+\rangle$?' nor 'are you in state $|-\rangle$?' *as those projectors are not part of this set*. Importantly, these questions are *not* equivalent to 'are you in the superposition?', as being in a superposition is not the same as being in a 'sum' of the individual states. This is quantum physics, not classical physics.

Within the context of a particular physical variable, the various eigenvalues $\{a_i\}$ define a sample space (Section 26.2.1). Equivalently, we can view the fine-grained projective decomposition $\sum_i \hat{P}_i^A = \hat{I}$ as a quantum sample space. The related quantum event algebra $\{\hat{\varepsilon}_k^A\}$ is formed from all the projectors constructed via:

$$\hat{\varepsilon}_k^A = \sum_i \pi_i \hat{P}_i^A$$

where $\pi_i = 0$ or 1. This is equivalent to the choice sequence used to define a compound event back in Section 26.2.1. Once again, with a list of n *elementary projection operators* in the sample space, there will be 2^n *compound projectors* in the event algebra. This list will include $\hat{I}$, with $\pi_i = 1$ for all i, and $\hat{\emptyset}$, built from $\pi_i = 0$ for all i.

As the $\hat{P}_i^A$ commute with each other, it follows that all possible compound projectors commute as well. That being the case, the elements of the event algebra will follow Boolean logic. For example, if M is the proposition encoded into compound projector $\hat{\varepsilon}_M^A$, and N is represented by $\hat{\varepsilon}_N^A$, then $(M\ AND\ N)$ is the product $\hat{\varepsilon}_M^A \hat{\varepsilon}_N^A$, a link that only makes sense if $\hat{\varepsilon}_M^A \hat{\varepsilon}_N^A = \hat{\varepsilon}_N^A \hat{\varepsilon}_M^A$. Equally, the combination $(M\ OR\ N)$ is $\hat{\varepsilon}_M^A + \hat{\varepsilon}_N^A - \hat{\varepsilon}_M^A \hat{\varepsilon}_N^A$, (Figure 30.1) which again is only unambiguous if the two operators commute (in Boolean logic $\hat{\varepsilon}_M^A$ OR $\hat{\varepsilon}_M^A$ means that $\hat{\varepsilon}_M^A$ is true, or $\hat{\varepsilon}_N^A$ is true or $\hat{\varepsilon}_M^A$ and $\hat{\varepsilon}_N^A$ are simultaneously true).

If we then assign probabilities to the sample space and hence the event algebra, those probabilities will follow the normal classical rules.

In the consistent histories terminology, each projective decomposition defines a *framework*, which is, in essence, a context for discussing the system. As Griffith describes it:

> In quantum mechanics, each Boolean event algebra constitutes what is in effect a "language" out of which one can construct a quantum description of some physical system, and a fundamental rule of quantum theory is that a description (which may, but need not be couched in terms of probabilities) referring to a single system at a single time must be constructed using a single Boolean algebra, a single "language". (This is a particular case of a more general "single-framework rule...[9]

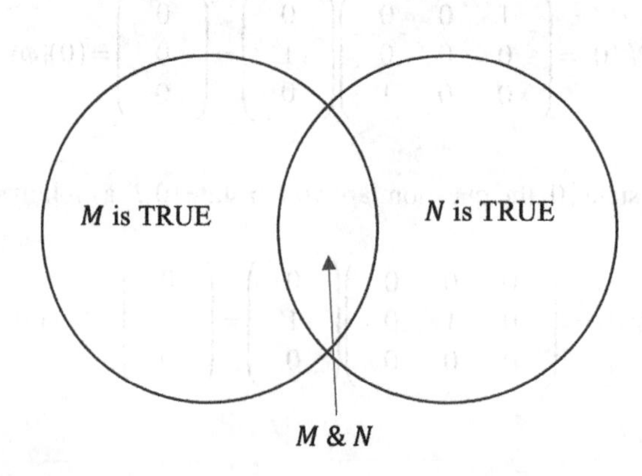

FIGURE 30.1 The proposition $(M\ OR\ N)$. The left-hand circle represents the space of events containing M. The right-hand circle is the space of events containing N. Hence, the central overlapping segment is the region where M and N are both true. The sum of both circles is not quite the set for which M is true OR N is true or both are true, as this counts the overlapping region twice. Hence, in quantum mechanical terms, $(\hat{M}\ OR\ \hat{N}) = \hat{M} + \hat{N} - \hat{M}\hat{N}$, which can only be sensibly defined if the operators commute.

Importantly, within a given framework, *every compatible observable has a definite value*, albeit one that is probabilistically determined. A measurement then reveals which value is the case in each situation.

A framework does not have to be constrained by dealing with a single physical variable and associated set of projectors. You can bring another variable, *B*, into the game with its $\{\hat{P}_i^B\}$, provided they commute with the original set, i.e., $\hat{P}_i^B\hat{P}_j^A = \hat{P}_j^A\hat{P}_i^B$ for all *i* and *j*. The two frameworks are then *compatible* with each other.

A crucial and distinctive feature of the consistent histories approach is the *single framework rule*, which provides guidance for quantum reasoning. Attempting to work with two or more incompatible frameworks in the same quantum description is going to lead to trouble. In essence, any reasoning based on combining inferences from incompatible frameworks will be suspect, *as the sample spaces of the projectors do not overlap.*

30.2 QUANTUM REASONING

According to the supporters of the consistent histories approach, most of our confusions regarding quantum theory arise from violations of the single framework rule. To illustrate one example, consider a system with two possible states $|\psi_A\rangle$ and $|\psi_D\rangle$. We then have $\hat{P}_A = |\psi_A\rangle\langle\psi_A|$ and $\hat{P}_D = |\psi_D\rangle\langle\psi_D|$ with $\hat{P}_A + \hat{P}_D = \hat{I}$. The sample space is $\{\hat{P}_A, \hat{P}_D\}$. However, we also have another possible projective decomposition, $\hat{P}_S$ and $\hat{I} - \hat{P}_S$ where $\hat{P}_S = |\psi_S\rangle\langle\psi_S|$ and $|\psi_S\rangle = \langle\psi_A| + |\psi_D\rangle$. The projector $\hat{P}_S$ asks the question 'are you in a superposition of $|\psi_A\rangle$ and $|\psi_D\rangle$?' to which the answer is either 'yes' or 'no'. Now the sample space is $\{\hat{P}_S, \hat{I} - \hat{P}_S\}$ *which does not contain either* $\hat{P}_A$ *or* $\hat{P}_D$. In this framework, it is invalid (or *meaningless*) to ask, 'are you in state $|\psi_A\rangle$?' or 'are you in state $|\psi_D\rangle$?' as those states are not part of the sample space.

30.2.1 Moggies and Sample Spaces

In the parable of Schrödinger's cat, while we are working in the superposition framework, we are not allowed to ask if the cat is alive, $|\psi_A\rangle$ or dead, $|\psi_D\rangle$. To do so would run roughshod over the single framework rule. Griffith comments:

> To be sure, this does not prevent one from asking whether $|\psi_S\rangle$ by itself has some intuitive physical meaning. What the preceding discussion shows is that whatever that meaning may be, it cannot possibly have anything to do with whether the cat is dead or alive, as these properties will be incompatible with $\hat{P}_S$. Indeed, it is probably the case that the very concept of a "cat" (small furry animal, etc.) cannot be meaningfully formulated in a way which is compatible with $\hat{P}_S$.[10]

30.2.2 Meaningless Statements

As another example, consider the idea that a spin ½ particle has both $S_z = +\frac{1}{2}$ AND $S_x = +\frac{1}{2}$ at the same time. We have already argued that this is not possible by considering sequences of SG experiments (Chapter 2) and from the lack of commutation between $\hat{S}_z$ and $\hat{S}_x$, (Chapter 10), but it is interesting to look at it from a logical perspective.

The Hilbert space for representing a spin ½ particle is the collection of all column vectors, or kets if you prefer, of the form $\begin{pmatrix} a \\ b \end{pmatrix}$. To match the proposition $S_z = +\frac{1}{2}$ AND $S_x = +\frac{1}{2}$, there would have to be a sub-space of at least one ket, $|\phi\rangle = A\begin{pmatrix} a \\ b \end{pmatrix}$, that was an eigenstate of the projector $\hat{S}_z\hat{S}_x$. We already know that this cannot happen as the spin operators do not commute.

Another way of drawing the same conclusion is to note that any ket $|\phi\rangle = \frac{1}{\sqrt{a^2+b^2}}\begin{pmatrix} a \\ b \end{pmatrix}$ corresponds to spin UP along some angle, ϑ. Using Eqs. (3.1) and (3.2):

$$\left|S_\vartheta = +\frac{1}{2}\right\rangle = \cos(\vartheta/2)|U\rangle + \sin(\vartheta/2)|D\rangle$$

$$\cos(\vartheta/2) = a/\sqrt{a^2+b^2} \quad \sin(\vartheta/2) = b/\sqrt{a^2+b^2}$$

and the orthogonal ket, $|\varphi^\perp\rangle = \frac{1}{\sqrt{a^2+b^2}}\begin{pmatrix} b \\ -a \end{pmatrix}$ corresponds to spin DOWN:

$$\left|S_\vartheta = -\frac{1}{2}\right\rangle = \sin(\vartheta/2)|U\rangle - \cos(\vartheta/2)|D\rangle$$

Hence there are no remaining kets with $a, b > 0$ that could describe a situation where two components have spin UP at the same time.

Perhaps instead the *whole Hilbert space* represents this proposition?

This can't be true, as some kets in the Hilbert space must represent $S_z = -\frac{1}{2}$, for example.

Perhaps the sub-space $\varnothing$ does the trick, i.e., the ket $\begin{pmatrix} 0 \\ 0 \end{pmatrix}$. If $\varnothing$ were the correct sub-space, then $S_z = +\frac{1}{2}$ AND $S_x = +\frac{1}{2}$ would always have to be false. However, we could have argued the whole thing so far using $S_z = +\frac{1}{2}$ AND $S_x = -\frac{1}{2}$, so we conclude that this must be false. So, if both are always false, then the combination $\left(S_z = +\frac{1}{2} \text{ AND } S_x = +\frac{1}{2}\right)$ OR $\left(S_z = +\frac{1}{2} \text{ AND } S_x = -\frac{1}{2}\right)$ must always be false as well. Logically, we can show that:

$$\left(S_z = +\frac{1}{2} \text{ AND } S_x = +\frac{1}{2}\right) \text{ OR } \left(S_z = +\frac{1}{2} \text{ AND } S_x = -\frac{1}{2}\right) = \left(S_z = +\frac{1}{2}\right) \text{AND} \left(S_x = +\frac{1}{2} \text{ OR } S_x = -\frac{1}{2}\right)$$

Now, if the right-hand side is always false, this can't be because $S_z = +\frac{1}{2}$ is always false, as we know that particles can have this value. So, $S_x = +\frac{1}{2}$ OR $S_x = -\frac{1}{2}$ must always be false. However, we also have examples of both of those, so this must be wrong as well.

This is a mess, so we are forced to conclude that $\varnothing$ is not able to represent our proposition $S_z = +\frac{1}{2}$ AND $S_x = +\frac{1}{2}$ either.

The point of the argument is not to demonstrate that $S_z = +\frac{1}{2}$ AND $S_x = +\frac{1}{2}$ is *false*, we are showing that it is *meaningless*. There is no ket or sub-space of the Hilbert space which could possibly correspond to that proposition. The question (projector) 'does $S_x = 1$' is **false** for (e.g.) $\begin{pmatrix} 1 \\ 0 \end{pmatrix}$ and **true** for $\frac{1}{\sqrt{2}}\begin{pmatrix} 1 \\ 1 \end{pmatrix}$. In this sense, true statements and false statements have kets. The statement $S_z = +\frac{1}{2}$ AND $S_x = +\frac{1}{2}$ is ketless, hence meaningless, *which is not the same as being false.*

The moral of the story is that the z and x spin components belong to different frameworks and trying to combine them leads to meaningless results.

This has implications for the measurement problem. It is impossible to measure S_z and S_x for a spin ½ particle at the same time, as such a combination is meaningless and hence *does not exist*. It is tempting to see this as a modern take on complementarity.

30.2.3 Contextuality

Viewed from within the consistent histories approach, the Kochen-Specker theorem is also an issue to do with frameworks. We simply find a projector which takes part in two or more incompatible decompositions. If we then assume that it has the same value in each decomposition, and hence each framework, we will end up with a contradiction, as only one projection can be true at any one time. This, in a few words, is the proof of the Kochen-Specker theorem, the result being a direct indication that you can only use one or more *compatible frameworks* at a time.

30.2.4 Non-Locality

Whenever you have a logically correct proof of a theorem that is contradicted by some other well-established result, you are forced to question the assumptions needed by the argument. Bell-type inequalities are violated by experimental results, so there must be something 'un-natural' about the assumptions behind their derivation. The quantum predictions match the experimental data, so implicit aspects of quantum theory are in line with nature, but contrary to the Bell assumptions. Most pundits point to the non-locality of the postulated hidden variables as being the issue. However, Griffith concludes that the problem lies with aspects of probability theory that are deployed in Bell-type proofs. These sorts of approaches will work in quantum theory, but only within a single framework. As Bell's argument specifically spans more than one incompatible framework, this will lead, in Griffiths' view, to 'mistaken', i.e., not quantum compatible, conclusions. There are non-local aspects of quantum theory, but Griffith does not see the violation of Bell's inequality as being an indication of the correct way in which this non-locality appears:

> Although quantum and classical mechanics use many of the same words, such as "energy" and "momentum" and "position", the concepts are not exactly the same. Thus when discussing "location" or "locality" of things in the real (quantum) world, one needs to employ well-defined quantum concepts that make sense in terms of the quantum Hilbert space, and not simply import ideas from classical physics… in the quantum domain one must be willing to allow quantum mechanics itself to suggest a suitable concept of locality, along with whatever limitations are needed to make it agree with the mathematical structure of the theory.[11]

30.3 HISTORIES

The history projection operator (HPO) approach, as suggested by Chris Isham,[12] starts with the state of a system, specified by a density matrix $\hat{D}$, and the physics governing the evolution of that system, encoded in the appropriate Hamiltonian, $\hat{H}$ and hence also the time evolution operator $\hat{U}(t)$. It then deploys the exhaustive and exclusive set of possibilities at any time, t_1, represented by a set of projection operators $\left\{ {}_{t_1}\hat{P}_i^a \right\}$ constructed from the eigenstates $\left\{ |i\rangle \right\}$ of observable property a and subject to:

$$\sum_i {}_{t_1}\hat{P}_i^a = 1\,(\text{exhaustive}) \quad {}_{t_1}\hat{P}_i^a\,{}_{t_1}\hat{P}_j^a = \delta_{ij}\,{}_{t_1}\hat{P}_i^a\ (\text{exclusive})$$

A *homogeneous history* is a time-ordered sequence of *events*[13]:

$$\hat{H} = {}_{t_1}\hat{P}_i^a \odot {}_{t_2}\hat{P}_j^b \odot {}_{t_3}\hat{P}_k^c \cdots \tag{30.1}$$

where ${}_{t_1}\hat{P}_i^a$ is the projection operator for eigenstate i of observable property a at time t_1, ${}_{t_2}\hat{P}_j^b$ is the projection operator for eigenstate j of observable property b at time t_2 and $t_2 > t_1$ etc. We should read this homogeneous history as "property a had value i at time t_1 and then property b had value j at time t_2 etc.". Of course, it is not *necessary* that a, b, c be different.

The symbol '$\odot$' used in this expression is a variation on the tensor product, $\otimes$, and is commonly deployed to indicate a time ordering to the sequence of products; $\otimes$ denoting a product of terms at the same time. The mathematical operation is the same in either case.

Note that $\hat{H}$ is a projector, as $\left(\hat{H}\right)^2 = \hat{H}\hat{H} = \hat{H}$, a result which follows from the properties of the individual projectors.

{**Technical note**: each projector in the sequence acts in its own Hilbert space. Hence, a history projector, $\hat{H}$, is acting in a product of Hilbert spaces $\mathcal{H} = \mathcal{H}^a \odot \mathcal{H}^b \odot \mathcal{H}^c \cdots$ The difference is important. If we have $\hat{P}_+^z$ for $S_z = 1$ and $\hat{P}_+^x$ for $S_x = 1$, the projection $\hat{P}_+^z\hat{P}_+^x$ in the *same* Hilbert space is very different to $\hat{P}_+^z\hat{P}_+^x$ acting on *separate spaces*. In the former case, they do not commute, in the latter case they do.}

If all the projectors embedded with the history relate to pure states, ${}_{t_1}\hat{P}_i^a = |i\rangle_a \langle i|_a$, then the construct is comparatively fine-grained. More coarse-grained histories contain some projectors like $\hat{P}_A^Z$ from earlier, for example. In either case, the history remains coarse-grained in one regard: the use of a finite (albeit possibly very large) number of distinct time points, $(t_1, t_2, \ldots, t_n)$, along its *temporal support*.

Constructing a history as a sequence of distinct time points, like beads on the timeline, tells us nothing about the system between these times. Griffith uses a parable reminiscent of one told by Heisenberg regarding quantum paths, streetlamps, and a walk he took in Faelled Park (Section 19.2.3):

> Imagine being outdoors on a dark night during a thunder storm. Each time the lightning flashes you can see the world around you. Between flashes, you cannot tell what is going on. To be sure, if we are curious about what is going on at intermediate times in a quantum history of the form (Eq. 30.1), we can refine the history ... by writing the projector as a sum of history projectors which include non-trivial information about the intermediate times, and then compute probabilities for these different possibilities.[14]

30.3.1 Combining Histories

It would be intolerably restrictive if the only possible histories were homogeneous. Fortunately, histories can be manipulated and combined in a variety of ways. For example, given two histories $\hat{H}_1 = \hat{P}^1 \odot \hat{P}^2 \odot \hat{P}^3 \cdots, \hat{H}_2 = \hat{Q}^1 \odot \hat{Q}^2 \odot \hat{Q}^3 \cdots$, (dropping the time references for convenience as the presence of the $\odot$ removes any ambiguity) we can form the joint history $\hat{H}_1$ AND $\hat{H}_1$ as $\hat{H}_1 \wedge \hat{H}_2 = \hat{H}_1\hat{H}_2$, provided that the two histories commute.[15] This is a clear extension of the requirement in Section 30.1 for individual projectors. Indeed, if the projectors in the two histories are defined on the same time sequence, then:

$$\hat{H}_1 \wedge \hat{H}_2 = \hat{P}^1\hat{Q}^1 \odot \hat{P}^2\hat{Q}^2 \odot \hat{P}^3\hat{Q}^3 \cdots$$

It is also possible to combine histories with a logical OR:

$$\hat{H}_1 \vee \hat{H}_2 = \hat{H}_1 + \hat{H}_2 - \hat{H}_1\hat{H}_2 \tag{30.2}$$

Here we have an example of a *compound* or *inhomogeneous history*, which cannot be written as a product of projectors.

Finally, the negation of $\hat{H}$, the history 'H did not occur', would be $\underline{\hat{H}} = \hat{I} - \hat{H}$ {where $\hat{I}$ is the product of the separate identity operators on each Hilbert space...}.

For $\hat{H}$ to not occur, it is only necessary that *one* of the events in the series fails to happen. This has interesting consequences. If we had a very simple history with two events, $\hat{H} = \hat{P}^1 \odot \hat{P}^2$, then:

$$\underline{\hat{H}} = \underline{\hat{P}}^1 \odot \hat{P}^2 + \hat{P}^1 \odot \underline{\hat{P}}^2 + \underline{\hat{P}}^1 \odot \underline{\hat{P}}^2 = \left(\hat{I} - \hat{P}^1\right) \odot \hat{P}^2 + \hat{P}^1 \odot \left(I - \hat{P}^2\right) + \left(\hat{I} - \hat{P}^1\right) \odot \left(I - \hat{P}^2\right)$$

A term such as $\left(\hat{I} - \hat{P}^1\right)$ being a sum over the projectors $k \neq 1$, from the appropriate set.

30.3.2 Probabilities

To start, let's consider a simple history containing only two time points, t_0 and t_1. We have an initial state of $|\psi_0\rangle$ at t_0 and at t_1 the system is in one of the orthonormal states $\{|k\rangle\}$. We then have a collection of histories of the form:

$$\hat{H}_k = |\psi_0\rangle\langle\psi_0| \odot |k\rangle\langle k|$$

If we add in the history $\hat{H}_0 = \left(\hat{I} - |\psi_0\rangle\langle\psi_0|\right) \odot \hat{I}$, we have a complete set:

$$\sum_k \left(|\psi_0\rangle|\psi_0\rangle\langle\psi_0| + \hat{I} - |\psi_0\rangle\langle\psi_0|\right) \odot |k\rangle\langle k| = \hat{I} \odot \hat{I}$$

This gives us a sample space with an event algebra, and we are off to the probabilistic races.

Using the Born rule (which is an assumption of the histories approach), we can calculate the probability of history $\hat{H}_k$:

$$\Pr\left(\hat{H}_k\right) = \left|\langle k|\hat{U}(1,\ 0)|\psi_0\rangle\right|^2$$

Note that must advance $|\psi_0\rangle$ through time using $\hat{U}(1,\ 0) = \hat{U}(t_1,\ t_0)$ before taking the product with $\langle k|$. (It will be useful going forward to specify the start and end times of the time evolution operator, $\hat{U}(t_1,\ t_0)$, whereas up to now, we have always (passively) assumed that $t_0 = 0$, hence $\hat{U}(t_1,\ 0) = \hat{U}(t_1)$.)

Now:

$$\Pr\left(\hat{H}_k\right) = \left|\langle k|\hat{U}(1,\ 0)|\psi_0\rangle\right|^2 = \langle\psi_0|\hat{U}^\dagger(1,\ 0)|k\rangle\langle k|\hat{U}(1,\ 0)|\psi_0\rangle$$

If we write $\hat{A} = \hat{U}^\dagger(1,\ 0)|k\rangle\langle k|\hat{U}(1,\ 0)$, then:

$$\Pr\left(\hat{H}_k\right) = \left\langle\psi_0|\hat{A}|\psi_0\right\rangle$$

and I may cunningly deploy Eq. (25.3):

$$\Pr\left(\hat{H}_k\right) = \mathrm{Tr}\left(|\psi_0\rangle\langle\psi_0|A\right) = \mathrm{Tr}\left(|\psi_0\rangle\langle\psi_0|\hat{U}^\dagger(1,\ 0)|k\rangle\langle k|\hat{U}(1,\ 0)\right)$$

It is now helpful to write $\hat{P}^0 = |\psi_0\rangle\langle\psi_0|$ and $\hat{P}^k = |k\rangle\langle k|$ so that:

$$\Pr\left(\hat{H}_k\right) = \mathrm{Tr}\left(\hat{P}^0\hat{U}^\dagger(1, 0)\hat{P}^k\hat{U}(1, 0)\right)$$

Next, we amuse ourselves with the properties of projection operators. As $\hat{P}^2 = \hat{P}$

$$\Pr\left(\hat{H}_k\right) = \mathrm{Tr}\left(\hat{P}^0\hat{P}^0\hat{U}^\dagger(1, 0)\hat{P}^k\hat{P}^k\hat{U}(1, 0)\right) = \mathrm{Tr}\left(\hat{P}^0\hat{U}^\dagger(1, 0)\hat{P}^k\hat{P}^k\hat{U}(1, 0)\hat{P}^0\right)$$

using the cyclic rotation of terms inside a trace.

Defining a *history chain operator* by:

$$\hat{\mathcal{H}} = \hat{P}^k\hat{U}(1, 0)\hat{P}^0 \quad \hat{\mathcal{H}}^\dagger = \hat{P}^0\hat{U}^\dagger(1, 0)\hat{P}^k$$

gives us:

$$\Pr\left(\hat{H}_k\right) = \mathrm{Tr}\left(\hat{\mathcal{H}}^\dagger\hat{\mathcal{H}}\right)$$

While this has been quite a lot of work to calculate something simple and familiar, we can now proceed to generalize to more intricate histories.

Given a history:

$$\hat{H} = {}_{t_1}\hat{P}_i^a \odot {}_{t_2}\hat{P}_j^b \odot {}_{t_3}\hat{P}_k^c \cdots$$

We can build a chain operator:

$$\hat{\mathcal{H}} = \cdots\hat{P}_k^c\hat{U}(3, 2)\hat{P}_j^b\hat{U}(2, 1)\hat{P}_i^a\hat{U}(1, 0)\hat{P}_0$$

{note that within the chain, there is only one Hilbert space.}

with its partner:

$$\hat{\mathcal{H}}^\dagger = \hat{P}_0 U^\dagger(1, 0)\hat{P}_i^a U^\dagger(2, 1)\hat{P}_j^b U^\dagger(3, 2)\hat{P}_k^c\cdots$$

and so:

$$\Pr\left(\hat{H}\right) = \mathrm{Tr}\left(\hat{\mathcal{H}}^\dagger\hat{\mathcal{H}}\right) \tag{30.3}$$

Some people prefer the aesthetic of writing:

$$\Pr\left(\hat{H}\right) = \mathrm{Tr}\left(\hat{\mathcal{H}}^\dagger\hat{D}_0\hat{\mathcal{H}}\right)$$

as it makes the dependence on the initial state explicit. The two are equivalent:

$$\Pr\left(\hat{H}_k\right) = \mathrm{Tr}\left(\hat{\mathcal{H}}^\dagger\hat{\mathcal{H}}\right) = \mathrm{Tr}\left(\hat{\mathcal{H}}\hat{\mathcal{H}}^\dagger\right) = \mathrm{Tr}\left(\cdots\hat{U}(1, 0)\hat{P}_0\hat{P}_0\hat{U}^\dagger(1, 0)\cdots\right)$$

using, once again, $\hat{P}^2 = \hat{P}$, we spot that $\hat{P}_0\hat{P}_0 = \hat{P}_0^2\hat{P}_0 = \hat{P}_0\hat{P}_0\hat{P}_0 = \hat{P}_0|\psi_0\rangle\langle\psi_0|\hat{P}_0 = \hat{P}_0\hat{D}_0\hat{P}_0$. Therefore:

$$\Pr\left(\hat{H}_k\right) = \mathrm{Tr}\left(\hat{\mathcal{H}}^\dagger\hat{D}_0\hat{\mathcal{H}}\right)$$

We can specify distinct chains by introducing an index $\{\alpha,\beta,\gamma\ldots\}$, where in the specific chain:

$$\hat{\mathcal{H}}_\alpha = \cdots \hat{P}_k^c \hat{U}(3,\, 2)\hat{P}_j^b \hat{U}(2,\, 1)\hat{P}_i^a \hat{U}(1,\, 0)\hat{P}_0$$

$$\alpha = \left(a_i,\, b_j,\, c_k,\, \ldots.\right)$$

in the same order as the temporal support. The chain $\hat{\mathcal{H}}_\beta$ would have a different selection of eigenvalues, but from the same projector sets and using the same temporal support. This gives us a family of chains $\left\{\hat{\mathcal{H}}_\alpha\right\}$, which forms the quantum event space for the histories. The event algebra is the set of all possible combinations of these histories.

To be sure that the numbers we calculate from Eq. (30.3) sensibly represent probabilities, three things must be true:

1. the values must always be positive; $\Pr\left(\hat{H}\right) = \mathrm{Tr}\left(\hat{\mathcal{H}}^\dagger \hat{\mathcal{H}}\right) \geq 0$;
2. the values from a set of histories must add to 1; $\sum_\alpha \Pr\left(\hat{H}_\alpha\right) = 1$;
3. for any two mutually exclusive histories, the probabilities add; $\Pr\left(\hat{H}_\alpha \vee \hat{H}_\beta\right) = \Pr\left(\hat{H}_\alpha\right)$ $\Pr\left(\hat{H}_\alpha\right)$.

It is easy to show that Eq. (30.3) naturally ensures (1) and (2) are correct. For (3) to be the case, the histories must be *orthogonal*, $\hat{H}_\alpha \hat{H}_\beta = \hat{H}_\beta \hat{H}_\alpha = 0$ as (Eq. 30.2) $\hat{H}_\alpha \vee \hat{H}_\beta = \hat{H}_\alpha + \hat{H}_\beta - \hat{H}_\alpha \hat{H}_\beta$. In essence, we are eliminating the possibility that the two histories quantum interfere to ensure that the histories within the set have sensible probabilities.

(In the case of a framework, the viability of the quantum event space was guaranteed by the orthogonality of the projectors within the set.)

The *decoherence functional* is defined for any pair of chain operators:

$$DF(\alpha,\beta) = \mathrm{Tr}\left(\hat{\mathcal{H}}_\alpha^\dagger \hat{\mathcal{H}}_\beta\right)$$

Any histories $\left(\hat{H}_\alpha, \hat{H}_\beta\right)$ are orthogonal if the decoherence functional of their chains $DF(\alpha,\beta) = 0$.

If for *any* (α,β) from $\left\{\hat{\mathcal{H}}_\alpha\right\}$, $DF(\alpha,\beta) \approx 0$ we have a *weakly consistent* set of histories and the probabilities derived for them and from them will approximately obey the axioms and rules of probability (i.e., (3) is approximately true).

If for any (α,β) from $\left\{\hat{\mathcal{H}}_\alpha\right\}$, $DF(\alpha,\beta) = 0$, the collection is *strongly consistent.*

As will be evident from the name of the functional, the impact of environmental decoherence on chains is thought to progressively guarantee their orthogonality and tighten the extent to which the rules of probability apply.

30.3.3 Consistent Histories

If we have two families of histories $\left\{\hat{\mathcal{H}}_\alpha\right\}$ and $\left\{\hat{\mathcal{H}}_\beta\right\}$ each of which is at least weakly consistent, then, like frameworks, we must be careful about combining them. The issue is, at least in part, related to the use of the term *incompatible.* Within a family, two histories $\hat{\mathcal{H}}_{\alpha 1}$ and $\hat{\mathcal{H}}_{\alpha 2}$ can be incompatible if the truth of one requires the falsity of the other. This is incompatibility in the sense of *exclusive*; they can't both be true. However, a history from one set is incompatible with another as their combination is *meaningless* (i.e., neither true nor false), as per the discussion in 30.2.2. Once again, the issue hinges on the event algebras of the two families not overlapping.

One might even suggest that the incompatibility or otherwise of families is what Bohr was groping towards with his idea of complementarity.

30.3.4 Histories and Mach-Zehnder

Using a somewhat simplified set of states compared to the full analysis of the Mach-Zehnder interferometer in Chapter 1, (Figure 30.2) we can outline how the histories approach tackles this iconic experiment.

The photon arriving at the first beam splitter is assumed to be in state $|a\rangle$ at t_0. Conventional quantum mechanical would stipulate the time evolution of $|a\rangle$ to a time after the photon has reached the beam splitter, t_1:

$$\hat{U}(1,\ 0)|a\rangle = \frac{1}{\sqrt{2}}\big(|c\rangle + |d\rangle\big)$$

By the time the photon has arrived at the detectors, t_2, we have:

$$\hat{U}(2,\ 1)|c\rangle = \frac{1}{\sqrt{2}}\big(-|e\rangle + |f\rangle\big) \quad \hat{U}(2,\ 1)|d\rangle = \frac{1}{\sqrt{2}}\big(|d\rangle + |f\rangle\big)$$

ensuring the phase difference required to prevent any photons from arriving at Detector Y:

$$\hat{U}(2,\ 1)\hat{U}(1,\ 0)|a\rangle = \frac{1}{\sqrt{2}}\left\{\frac{1}{\sqrt{2}}\big(-|e\rangle + |f\rangle\big) + \frac{1}{\sqrt{2}}\big(|e\rangle + |f\rangle\big)\right\} = |f\rangle$$

As a result $\left|\langle a|f\rangle\right|^2 = 1,\ \left|\langle a|e\rangle\right|^2 = 0$.

Approaching this as if the photon arriving at Detector X either travelled the top path, and hence via $|d\rangle$, or the bottom path, $|c\rangle$, there are two possible histories:

$$\hat{H}_1 = |a\rangle\langle a| \odot |c\rangle\langle c| \odot |f\rangle\langle f| \quad \hat{H}_2 = |a\rangle\langle a| \odot |d\rangle\langle d| \odot |f\rangle\langle f|$$

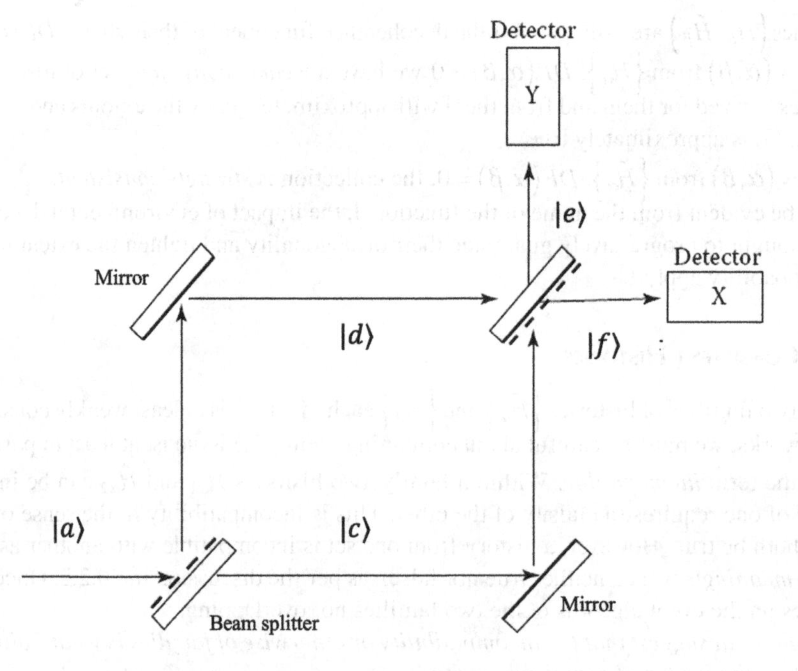

FIGURE 30.2 The Mach-Zehnder interferometer with states for the consistent histories analysis.

Our chains then take the form:

$$\hat{\mathcal{H}}_1 = |f\rangle\langle f|\hat{U}(2,\, 1)|c\rangle\langle c|\hat{U}(1,\, 0)|a\rangle\langle a| \quad \hat{\mathcal{H}}_2 = |f\rangle\langle f|\hat{U}(2,\, 1)|d\rangle\langle d|\hat{U}(1,\, 0)|a\rangle\langle a|$$

After the first time-step we have:

$$\hat{\mathcal{H}}_1 = |f\rangle\langle f|\hat{U}(2,\, 1)|c\rangle\langle c|\frac{1}{\sqrt{2}}\big(|c\rangle + |d\rangle\big)\langle a| = |f\rangle\langle f|\hat{U}(2,\, 1)\frac{1}{\sqrt{2}}|c\rangle\langle a|$$

$$\hat{\mathcal{H}}_2 = |f\rangle\langle f|\hat{U}(2,\, 1)|d\rangle\langle d|\frac{1}{\sqrt{2}}\big(|c\rangle + |d\rangle\big) = |f\rangle\langle f|\hat{U}(t_2,\, t_1)\frac{1}{\sqrt{2}}|d\rangle\langle a|$$

Advancing to t_2:

$$\hat{\mathcal{H}}_1 = \frac{1}{2}|f\rangle\langle f|\big(-|e\rangle + |f\rangle\big)\langle a| = \frac{1}{2}|f\rangle\langle a|$$

$$\hat{\mathcal{H}}_2 = \frac{1}{2}|f\rangle\langle f|\big(|e\rangle + |f\rangle\big)\langle a| = \frac{1}{2}|f\rangle\langle a|$$

The decoherence functional is then:

$$DF(1,2) = \text{Tr}\left(\hat{\mathcal{H}}_1^\dagger \hat{\mathcal{H}}_2\right) = \text{Tr}\left(\frac{1}{4}|a\rangle\langle f|f\rangle\langle a|\right) = \frac{1}{4}\text{Tr}\big(|a\rangle\langle a|\big) = \frac{1}{4} \neq 0$$

Hence the two histories are inconsistent and cannot be combined into the same framework.

Note that these two histories are *exclusive* (if one happens the other can't), but *inconsistent.* Exclusivity does not guarantee consistency.

Straight away, we are limited in the sorts of questions that we can legitimately ask. For example, the combination $\left(\hat{\mathcal{H}}_1 \wedge \hat{\mathcal{H}}_2\right)$ would be a history where the photon travelled along both c and d on its way to the detector. Such a query can only be posed within a consistent set, and so is illegitimate according to the rules of consistent histories. Equally, $\left(\hat{\mathcal{H}}_1 \vee \hat{\mathcal{H}}_2\right) \wedge \underline{\left(\hat{\mathcal{H}}_1 \wedge \hat{\mathcal{H}}_2\right)}$ is 'the photon went along c or d but not both', which is also outlawed as a proposition.

If we introduce further histories, $\hat{H}_3 = |a\rangle\langle a| \odot |c\rangle\langle c| \odot |e\rangle\langle e|$ and $\hat{H}_4 = |a\rangle\langle a| \odot |d\rangle\langle d| \odot |e\rangle\langle e|$ then $\hat{\mathcal{H}}_3 = -\frac{1}{2}|e\rangle\langle a|$ and $\hat{\mathcal{H}}_4 = \frac{1}{2}|e\rangle\langle a|$ respectively. Using them in our decoherence functional shows us that:

$$DF(3,4) = \text{Tr}\left(\hat{\mathcal{H}}_3^\dagger \hat{\mathcal{H}}_4\right) = \text{Tr}\left(\frac{1}{4}|a\rangle\langle e|e\rangle\langle a|\right) = \frac{1}{4}$$

$$DF(1,3) = \text{Tr}\left(\hat{\mathcal{H}}_1^\dagger \hat{\mathcal{H}}_3\right) = -\text{Tr}\left(\frac{1}{4}|a\rangle\langle f|e\rangle\langle a|\right) = 0$$

$$DF(2,4) = \text{Tr}\left(\hat{\mathcal{H}}_2^\dagger \hat{\mathcal{H}}_4\right) = \text{Tr}\left(\frac{1}{4}|a\rangle\langle f|e\rangle\langle a|\right) = 0$$

So, we can make the consistent sets $\{\mathcal{H}_1,\, \mathcal{H}_3\}$ and $\{\mathcal{H}_2,\, \mathcal{H}_4\}$. Other combinations are inconsistent. In abbreviated form, these sets are:

$$\left\{\hat{\mathcal{H}}_1,\, \hat{\mathcal{H}}_3\right\} = \left\{\hat{P}_a\hat{P}_c\hat{P}_f,\, \hat{P}_a\hat{P}_c\hat{P}_e\right\}$$

$$\left\{\hat{\mathcal{H}}_2,\, \hat{\mathcal{H}}_4\right\} = \left\{\hat{P}_a\hat{P}_d\hat{P}_f,\, \hat{P}_a\hat{P}_d\hat{P}_e\right\}$$

Now we can see that 'did the photon travel along c on the way to f?' and 'did the photon travel along c on the way to e?' are consistent (part of the same set), so we can legitimately build the history $\left(\hat{\mathcal{H}}_1 \vee \hat{\mathcal{H}}_3\right)$. Pinning our histories so that there is a definite specification of either c or d has the divided the possibilities into two consistent groups which are inconsistent with each other.

Note that the probability $\Pr\left(\hat{H}_1\right) = \text{Tr}\left(\mathcal{H}_1^\dagger \mathcal{H}_1\right) = \frac{1}{4}$ appears to contradict our earlier suggestion that $\left|\langle a|f\rangle\right|^2 = 1$. However, we are answering two different questions. The histories calculation for $\hat{\mathcal{H}}_1$ determines the probability that the photon arrives at f having travelled via c. For this to happen, the 50:50 chance at the first beam splitter must have selected c and the 50:50 chance at the second beam splitter must have selected f. Our initial consideration did not specify a path. If we construct a more coarse-grained history:

$$\hat{H}_{MZ1} = |a\rangle\langle a| \odot \left(|c\rangle\langle c| + |d\rangle\langle d|\right) \odot |f\rangle\langle f|$$

$$\begin{aligned}\hat{\mathcal{H}}_{MZ1} &= |f\rangle\langle f|\hat{U}(2,\ 1)\left(|c\rangle\langle c| + |d\rangle\langle d|\right)\hat{U}(1,\ 0)|a\rangle\langle a| \\ &= |f\rangle\langle f|\hat{U}(2,\ 1)\hat{I}\hat{U}(1,\ 0)|a\rangle\langle a| \\ &= |f\rangle\langle f||a\rangle\langle a|\end{aligned}$$

Then the probability is:

$$\Pr\left(\hat{H}_{MZ1}\right) = \text{Tr}\left(\mathcal{H}_{MZ1}^\dagger \hat{\mathcal{H}}_{MZ1}\right) = \text{Tr}\left(|a\rangle\langle a|f\rangle\langle f|f\rangle\langle f|a\rangle\langle a|\right) = \left|\langle f|a\rangle\right|^2 \text{Tr}\left(|a\rangle\langle a|\right) = \left|\langle f|a\rangle\right|^2$$

The alternative history would be:

$$\hat{H}_{MZ2} = |a\rangle\langle a| \odot \left(|c\rangle\langle c| + |d\rangle\langle d|\right) \odot |e\rangle\langle e|$$

$$\hat{\mathcal{H}}_{MZ2} = \left(|e\rangle\langle e|\right)\hat{U}(2,\ 1)\hat{I}\hat{U}(1,\ 0)|a\rangle\langle a| = |e\rangle\langle e|a\rangle\langle a|$$

making the probability:

$$\Pr\left(\hat{H}_{MZ2}\right) = \text{Tr}\left(\mathcal{H}_{MZ2}^\dagger \hat{\mathcal{H}}_{MZ2}\right) = \text{Tr}\left(|a\rangle\langle a|e\rangle\langle e|e\rangle\langle e|a\rangle\langle a|\right) = \left|\langle e|a\rangle\right|^2 \text{Tr}\left(|a\rangle\langle a|\right) = \left|\langle e|a\rangle\right|^2$$

agreeing with the standard analysis.

As a final point, $\hat{H}_{MZ1}$ and $\hat{H}_{MZ2}$ are consistent with each other:

$$\text{DF}\left(\hat{H}_{MZ1}, \hat{H}_{MZ2}\right) = \text{Tr}\left(\mathcal{H}_{MZ1}^\dagger \hat{\mathcal{H}}_{MZ2}\right) = \text{Tr}\left(|a\rangle\langle a|f\rangle\langle f|e\rangle\langle e|a\rangle\langle a|\right) = 0$$

but not with other previous ones:

$$\text{DF}\left(\hat{H}_{MZ1}, \hat{H}_1\right) = \text{Tr}\left(\mathcal{H}_{MZ1}^\dagger \hat{\mathcal{H}}_{MZ2}\right) = \frac{1}{2}\text{Tr}\left(|a\rangle\langle a|f\rangle\langle f|f\rangle\langle a|\right) \neq 0$$

showing that an experiment designed to track the path of the photons cannot be described in the same framework as one that does not track paths.

Another way of making the same point is to spot that:

$$\Pr\left(\hat{\mathcal{H}}_{MZ1}\right) \neq \Pr\left(\hat{\mathcal{H}}_1\right) + \Pr\left(\hat{\mathcal{H}}_2\right)$$

which, prima-face, does not follow the rules of probability. However, this is understandable within the histories formalism as we are attempting to add probabilities derived from different frameworks. Or, to put it another different way, there is no sample space that spans these events.

The extension of this approach to other experiments, such as the double slit, is evident. A pair of histories constructed to chart an electron's progress from source via one of the slits to the screen/detector would be inconsistent with each other. A collection of coarse-grained histories that included both slits, or rather, did not specify a slit, but which picked out different detection points, would show interference. The coarse graining is, in essence, the histories version of Important Rule 3.

These sorts of discussions go some way to showing why consistent histories have gathered the 'Copenhagen done right' description.

30.3.5 Measurement

> "collapse" is something which takes place in the theorist's notebook, rather than the experimentalist's laboratory.[16]
>
> In this formulation, it is the internal consistency of probability sum rules that determines the sets of alternatives of the closed system for which probabilities are predicted rather than any external notion of 'measurement'.[17]

Consider Figure 30.3, which shows a Mach-Zehnder interferometer with the second beam splitter removed.

In this configuration, we have a rudimentary measurement situation, with detectors X and Y providing the instrumentation. Let's imagine that the quantum state of detector Y is $|Y\rangle$ in its 'ready' mode and $|\boldsymbol{Y}\rangle$ when it has detected a photon. Similar states apply to detector Z. In this case, the initial quantum state of the combined photon/detectors arrangement (a closed system) is $|\Psi\rangle = |a\rangle|Y\rangle\langle X|$. After the beam splitter, we have:

$$|\Psi'\rangle = U(t_1, t_0)|\Psi\rangle = \frac{1}{\sqrt{2}}\left(|c\rangle + |d\rangle\right)|Y\rangle|X\rangle$$

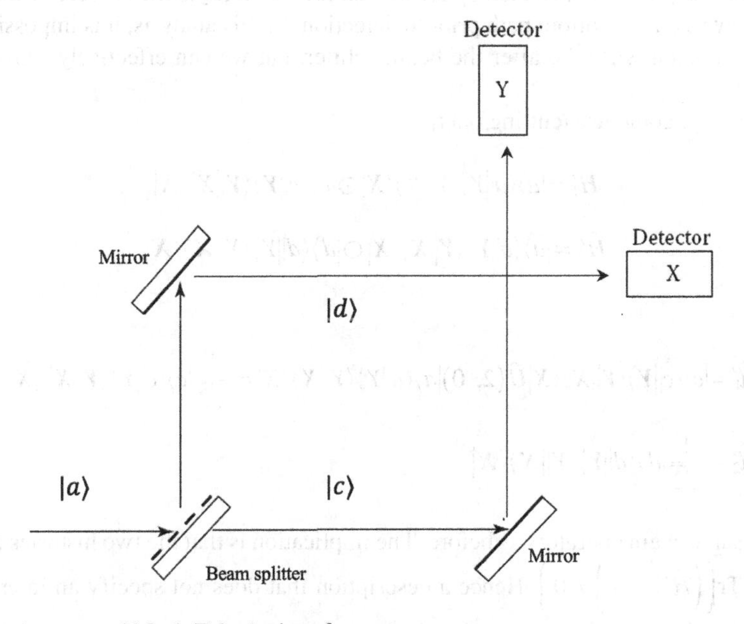

FIGURE 30.3 A truncated Mach-Zehnder interferometer.

After detection:

$$|\Psi''\rangle = U(t_2, t_1)|\Psi'\rangle = \frac{1}{\sqrt{2}}\left(|c\rangle|\mathbf{Y}\rangle|X\rangle + |d\rangle|Y\rangle|\mathbf{X}\rangle\right)$$

which is a standard entangled state. Conventionally, we now invoke state collapse into one of the alternatives, leading to our measurement result.

If the particle is detected at X, we know the quantum state to be $|d\rangle|Y\rangle|\mathbf{X}\rangle$, *but we cannot deduce from this that the photon necessarily travelled along the top path*. Following the Copenhagen view, we would either say that there was no device detecting the path, so we have no right to say anything about it, or we would point to the state at t_1, $\frac{1}{\sqrt{2}}\left(|c\rangle + |d\rangle\right)|Y\rangle|X\rangle$, which is a quantum superposition of states.

If we analyse the same scenario from the histories perspective, we would construct two histories:

$$\hat{H}_1 = |a\rangle\langle a||Y\rangle\langle Y||X\rangle\langle X| \odot |c\rangle\langle c||Y\rangle\langle Y||X\rangle\langle X| \odot |c\rangle\langle c||\mathbf{Y}\rangle\langle \mathbf{Y}||X\rangle\langle X|$$

$$\hat{H}_2 = |a\rangle\langle a||Y\rangle\langle Y||X\rangle\langle X| \odot |d\rangle\langle d||Y\rangle\langle Y||X\rangle\langle X| \odot |d\rangle\langle d||Y\rangle\langle Y||\mathbf{X}\rangle\langle \mathbf{X}|$$

The corresponding chain operators are:

$$\hat{\mathcal{H}}_1 = |c\rangle\langle c||\mathbf{Y}\rangle\langle \mathbf{Y}||X\rangle\langle X|\hat{U}(2, 1)|c\rangle\langle c||Y\rangle\langle Y||X\rangle\langle X|\hat{U}(1, 0)|a\rangle\langle a||Y\rangle\langle Y||X\rangle\langle X| = \frac{1}{\sqrt{2}}|c\rangle\langle c||\mathbf{Y}\rangle\langle \mathbf{Y}||X\rangle\langle X|$$

$$\hat{\mathcal{H}}_2 = \frac{1}{\sqrt{2}}|d\rangle\langle d||Y\rangle\langle Y||\mathbf{X}\rangle\langle \mathbf{X}|$$

with probability:

$$\Pr\left(\hat{\mathcal{H}}_1\right) = \Pr\left(\hat{\mathcal{H}}_2\right) = 1/2$$

If we know that the photon is detected by X, we can infer that $\hat{H}_2$ is the correct history, and hence that the photon was on the bottom path prior to detection. In this analysis, it is impossible to predict which path the photon will take after the beam splitter, but we can effectively *retrodict* its paths from the result.

Alternatively, we could use (cutting out t_1):

$$\hat{H}_1' = |a\rangle\langle a||Y\rangle\langle Y||X\rangle\langle X| \odot |c\rangle\langle c||\mathbf{Y}\rangle\langle \mathbf{Y}||X\rangle\langle X|$$

$$\hat{H}_2' = |a\rangle\langle a||Y\rangle\langle Y||X\rangle\langle X| \odot |d\rangle\langle d||Y\rangle\langle Y||\mathbf{X}\rangle\langle \mathbf{X}|$$

with:

$$\hat{\mathcal{H}}_1' = |c\rangle\langle c||\mathbf{Y}\rangle\langle \mathbf{Y}||X\rangle\langle X|\hat{U}(2, 0)|a\rangle\langle a||Y\rangle\langle Y||X\rangle\langle X| = \frac{1}{\sqrt{2}}|c\rangle\langle c||\mathbf{Y}\rangle\langle \mathbf{Y}||X\rangle\langle X|$$

$$\hat{\mathcal{H}}_2' = \frac{1}{\sqrt{2}}|d\rangle\langle d||Y\rangle\langle Y||\mathbf{X}\rangle\langle \mathbf{X}|$$

These are the same chain operators as before. The implication is that the two histories $\hat{H}_1$ and $\hat{H}_1'$ are inconsistent $\left(\mathrm{Tr}\left(\left(\hat{\mathcal{H}}_1'\right)^{\dagger}\hat{\mathcal{H}}_1\right) \neq 0\right)$. Hence a description that does not specify an intermediate path cannot be used at the same time as one that does.

Consistent historians claim to have solved the measurement problem by:

1. asserting that measurements reveal the pre-measurement values of physical variables (Section 30.1);
2. regarding state collapse as a calculational tool to help the theorist calculate the probability of each possible value;
3. confirming that quantum theory has an inherent randomness active in the physical value presented pre-measurement by the particle, not in the process of measurement itself.

State collapse is then no more mysterious than selecting components of the state expansion to use in further calculations.

However, this is not the full story of measurement, as it does not address the contextuality inherent in quantum measurements. Proponents address that via *relative realism*, which we will take up in Section 30.4.2.

30.3.6 Decoherence and the Classical World

From the perspective of consistent histories, the emergence of classical physics from the quantum level is linked to the suppression of interference, so that the normal rules of probability apply.

Back in Section 25.4, we introduced the idea that environmental decoherence could diagonalize the density matrix describing a quantum system interacting with its environment. To illustrate how this process works, consider a quantum system described by some basis $\{|i\rangle\}$ which can be either discrete or continuous, and an environment with states $|\psi\rangle$. The process of decoherence then evolves the combined system in the following way:

$$\left(\sum_i a_i|i\rangle\right)\otimes|\psi\rangle \rightarrow \sum_i a_i|i\rangle\otimes|\psi(i)\rangle$$

i.e., the environmental states become correlated with the values $\{i\}$ and furthermore $\langle\psi(i)|\psi(j)\rangle \approx 0$. In a simple example, the different $|i\rangle$ might be distinct wave packets indicating possible paths of a particle passing through an SG field. As they emerge, $\langle i|j\rangle \sim 0$ and we expect that each packet will imprint on the environment in distinct ways, leading to $\langle\psi(i)|\psi(j)\rangle \approx 0$. Perhaps the environment contains some measurement needle which points to different parts of a scale depending on the path of the emerging particle. In which case, the (very complicated) wave packets $|\psi(i)\rangle$ describing the centre of mass position of the pointer will not overlap.

Given a density operator $\hat{D}(t)$ for a system at time t, the probability that the system will have value I of $\{i\}$ is $\Pr(I, t) = \langle I|\hat{D}(t)|I\rangle$ Eq. (25.5). In this case, the probability that the system will display value J of $\{i\}$ at a later time t' is:

$$\Pr(J,t') = \langle J|\hat{D}(t')|J\rangle = \sum_I \Pr\langle J|I\rangle\langle I|\hat{D}(t)|I\rangle \tag{30.4}$$

provided the laws of classical probability apply (with $\Pr(J|I)$ being the *conditional probability* of J given I). In which case, we can see that the density operator at time t' is linearly related to that at time t. Generally, this will not be the case and so Eq. (30.4) will not apply. In particular, $\langle J|\hat{D}(t')|J\rangle$ will depend on off-diagonal terms in the density matrix. This is where the preferred basis comes in. If we rotate the density matrix into a representation using the preferred basis, it will be approximately diagonal, with the off-diagonal terms very close to being zero.

However, this still leaves a very important question open: how do we know which basis will work in a given situation?

The study of decoherence revolves around constructing various models to see if model-independent conclusions can be drawn. A suitable scenario might be an atom interacting with an electromagnetic field, or an oscillator (such as those considered inside a black body) within an ensemble of others, or a particle of dust in space interacting with the radiation bath of the cosmic microwave background. In these cases, the interaction between the system and the environment picks out a preferred set of states and suppresses interference between them (by scrambling the phases). For example, with the atom placed in an EM field, the preferred states turn out to be the stationary states of the energy levels. Intuitively, the environment is continually 'measuring' the system and recording the values associated with the preferred basis. The states that are honoured in this way turn out to be those which are most stable under the interaction with the environment.

As the interaction potentials at work between system and environment are common functions of position, the preferred bases are often position-dependent as well. In the oscillator example, the environment picks out a preferred basis which is an approximate joint eigenstate of position and momentum.

The extent to which these position states are localized is often very small.[18] For example, a speck of dust ~ 10^{-7} m floating in the air will have interference suppressed between sections of its internal structure over a scale ~ 10^{-15} m and in a timescale ~ 10^{-6} s. Incidentally, this example illustrates an important point; with a macroscopically sized system, some of the decohering environment can be *within the object itself*, not just its surroundings.

The pace at which decoherence happens is crucially important. It is not simply that decoherence diagonalizes the density matrix, it does it at a rate which is *faster than the evolution of the system*. Hence, *probabilistic predictions of the dynamics become possible* (Eq. 30.3 works). This is where we see the link back to consistent histories. If we wish to chart the evolution of some system variable by specifying possible histories, those histories must be orthogonal for the probabilities to work as we expect classically. Proponents believe that decoherence will select out bases for which this is true and hence allow the emergence of a classically proximate picture:

> In constructing a quasiclassical description decoherence plays a useful role in the sense that it removes the effects of quantum interference which might otherwise render a quasiclassical family inconsistent, preventing the application of the Born rule and its generalizations.[19]
>
> An important mechanism of decoherence is the dissipation of phase coherence between branches into variables not followed by the coarse graining. Consider by way of example, a dust grain in a superposition of two positions deep in interstellar space [my ref[20]]. In our universe, about 10^{11} cosmic background photons scatter from the dust grain each second. The two positions of the grain become correlated with different, nearly orthogonal states of the photons. Coarse graining that follows only the position of the dust grain at a few times therefore correspond to branch state vectors that are nearly orthogonal and satisfy [the rules of probability][21]
>
> [my additions]

In Section 14.2, we introduced Heisenberg's equation of motion:

$$\frac{d\langle\hat{O}\rangle}{dt} = \frac{i}{\hbar}\left\langle\left[\hat{H},\hat{O}\right]\right\rangle + \left\langle\frac{d\hat{O}}{dt}\right\rangle$$

and showed how it leads to Erenfest's theorem:

$$\langle\hat{p}_x\rangle = m\frac{d\langle\hat{x}\rangle}{dt} \quad \frac{d\langle\hat{p}_x\rangle}{dt} = -\left\langle\frac{dV(x)}{dx}\right\rangle$$

While it was reassuring to see that classical laws can be recovered from quantum theory, at the time we were considering position and momentum operators being applied to single point-like particles. For a macroscopic object, we have the centre of mass coordinates and average over large numbers of molecules for the bulk momentum and energy of the collective. In the case of fluid flow, we routinely coarse-grain the description by averaging over tiny volumes that still contain vast numbers of atoms. In any circumstance, our classical description of macroscopic objects works by coarse-graining and averaging to reduce the myriad physical variables involved in the detail to a few special quantities that can map the motion of the object.

In principle, these collective variables would transpose into very large Hilbert spaces with appropriate projectors. Quite probably a range of quantum projectors would all correspond, within the coarse graining, to a classical property. The change in such property over time could be mapped to a quantum history temporally grained by a set of discrete times, rather than a continual progression. Most likely there will be a range of quantum histories all of which reasonably describe the classical evolution, deploying the appropriate projectors. In the literature, the term *quasi-classical* is used to describe a family of histories of this sort. Decoherence should help to pick out the preferred bases for coarse-grained projectors corresponding to the appropriate macroscopic quantities (then the $\{i\}$ in Eq. (30.4) will be values of one of these variables), in which case the quasi-classical histories should be very nearly orthogonal. Equally, as there will be a range of appropriate projectors, one can always tweak the description by using projectors that help the orthogonality.

From this perspective, it is likely that many different quantum frameworks will equally lead to classical variables, including incompatible frameworks. This is seen as being analogous to the coarse graining that happens in classical physics and the expectation is that the same classical laws will emerge. Some frameworks and histories will not be quasi-classical at all, and prima-face there does not appear to be any quantum principle what would exclude them. There is still work to be done in this area.

One of the biggest obstacles to matching quantum descriptions to our classical experience remains the issue of probability. Classical laws are deterministic in nature, whereas quantum theory has inherent randomness and a probabilistic nature (at least in those interpretations that accept this fundamental aspect). In truth, however, classical experiments are never 'clean' and a system is rarely totally isolated, or isolatable, from the 'noise' inherent in its environment. In such instances, classical predictions tend to invoke classical probabilities, so perhaps there is not such as gap after all.

30.3.7 Histories in Cosmology

Some cosmologists have adopted the consistent histories approach to give them a method of working without having to tackle state collapse nor the somewhat sci-fi overtones of the Many Worlds view. Their goal is to discover the initial quantum state that leads, via a high-probability history, to a universe somewhat like the one we observe.

This project must deal with Einstein's theory of relativity and especially his view that gravity is a distortion in the fabric of space-time. Applying quantum theory to gravity, using the consistent histories framework, produces histories that describe different ways in which space-time might evolve from the Big Bang. Then it's a matter of tailoring the physical description and the initial quantum state so that the high-probability histories contain notable characteristics similar to the ones that we observe in nature. In particular, the emergency of quasi-classical families of histories.

Initial quantum states for the universe must be derived or obtained using criteria outside of quantum theory. Dirac's beauty and elegance play a certain role here. One quantum state candidate that has generated a degree of interest is the Hawking-Hartle[22] 'no-boundary state'. In this proposal, the expanding universe is likened to the base of a shuttlecock, a smoothly curving shape with no precise 'point' at the bottom (Figure 30.4)

Early in the history of the universe, using the Hawking-Hartle model, there was no parameter that we would recognize as time. Instead, a variable called 'lapse' is used which, in a quantum

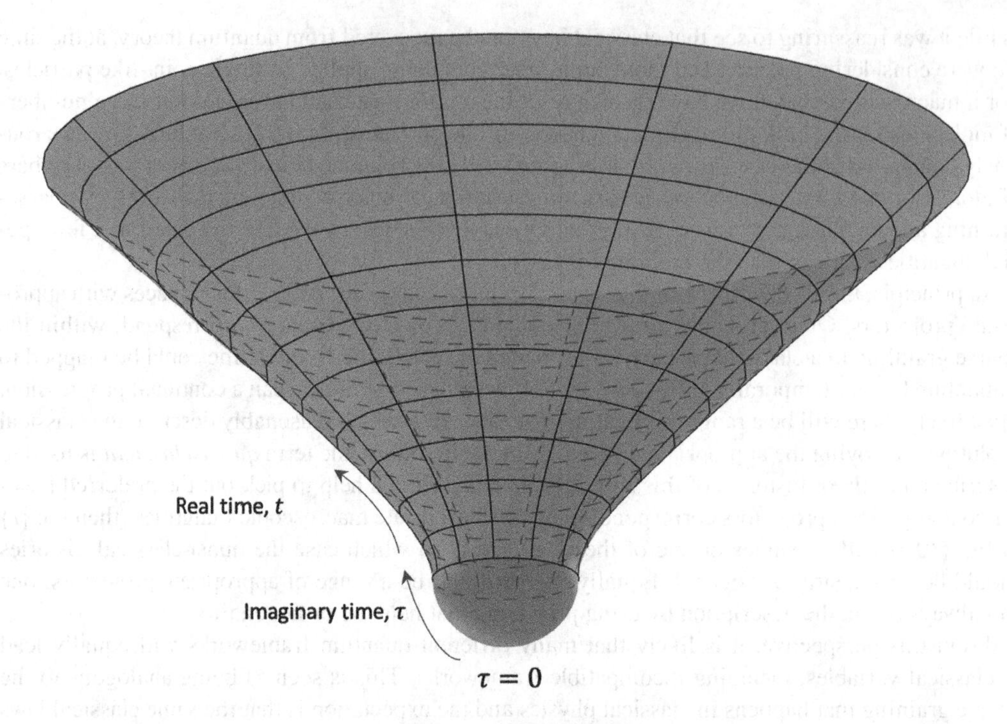

FIGURE 30.4 The Hawking-Hartle no-boundary proposal for the initial quantum state of the universe leads to an evolution shown here. At any 'height' on this diagram, the cross section through the shape represents a 2-d model of the expanding universe. The height gradually transitions from the imaginary time parameter, lapse, into more conventional time.

manner, emerges from the separate spatial dimensions. This emergence gives the universe no definite starting point, nor a history beforehand. More controversially, Hawking and Hartle deploy imaginary values (i.e., $L = ib$) of the lapse parameter in their models, although ordinary real-valued time eventually takes over. Other researchers have explored using real values of lapse, in which case the encouraging outcomes that Hawking and Hartle find are not so evident. In any case, approximations and simplifications have to be made, and despite the bold title of their paper 'wave function of the universe' there is a great deal of crucial detail still to be ironed out.

30.4 ONTOLOGY

Consistent historians accept the fundamental nature of quantum probabilities, rather than trying to explain them in other terms as different approaches set out to do:

> Classical unpredictability arises because one is using a coarse-grained description where some information about the underlying deterministic system has been thrown away, and there is always the possibility, in principle, of a more precise description in which the probabilistic element is absent, or at least the uncertainties reduced to any extent one desires. By contrast, the Born rule or its extension to more complicated situations, [histories], enters quantum theory as an axiom, and does not result from coarse graining a more precise description.[23]
> [my addition]

As to the prospect of quantum theory being superseded by a more deterministic approach, Griffith comments:

> The fact that it was only with great reluctance that physicists abandoned classical determinism in the course of developing a theory capable of explaining experimental results in atomic physics strongly suggests, though it does not prove, that stochastic time development is part of physical reality.[24]

The uncertainty principle is also regarded as a fundamental ontological limitation:

> a quantum particle, in contrast to a classical particle … does not possess a precise position or a precise momentum. In addition, the precision with which either of these quantities can be defined is limited by the Heisenberg uncertainty principle. This does not mean that quantum entities are "fuzzy" and ill-defined, for a ray in the Hilbert space is as precise a specification as a point in phase space. What it does mean is that the classical concepts of position and momentum can only be used in an approximate way when applied to the quantum domain… The limitations on measurements come about because of the nature of quantum reality, and the fact that what does not exist cannot be measured.[25]

So far, this is conventional thinking, reminiscent of the Copenhagen interpretation. However, differences start to develop in discussions of the wave function and *unicity*.

30.4.1 Pre-Probabilities

In the terminology of consistent histories, both the wave function and the density matrix are regarded as *pre-probabilities*, that is *calculational tools* employed to calculate probabilities. As such, they do not directly reflect any aspect of reality:

> …it does not make sense to suppose that at t_1 the system possesses the physical property $|\psi_1\rangle$ Instead, $|\psi_1\rangle$ must be thought of as a mathematical construct suitable for calculating certain probabilities.[26]

The fact that any wave function can be decomposed using more than one basis set, $|\Psi\rangle = \sum_i a_i |\varphi_i\rangle = \sum_j b_j |\phi_j\rangle$ is taken as a signal that we are dealing with a mathematical device. The probability for a system in state $|\Psi(t=0)\rangle$ to manifest a physical property associated with state $|\phi_J(t=1)\rangle$ can be calculated by:

$$\text{Prob}(\phi_J) = \left|\langle \phi_J(t=1)|\hat{U}(1,\ 0)|\Psi(t=0)\rangle\right|^2$$

i.e., advancing $|\Psi\rangle$ forward in time. However, as $|\phi_J(t=1)\rangle = \hat{U}(1,0)|\phi_J(t=0)\rangle$ gives $\langle\phi_J(t=1)| = \langle\phi_J(t=0)|U^\dagger(1,\ 0)$ we can equally write:

$$\text{Prob}(\phi_J) = \left|\langle \phi_J(t=0)|U^\dagger(1,\ 0)\hat{U}(1,\ 0)|\Psi(t=0)\rangle\right|^2 = \left|\langle\phi_J(t=0)|\Psi(t=0)\rangle\right|^2$$

something that is also taken to indicate the calculational status of the wave function.

From this perspective, as mentioned earlier, wave function collapse is simply part of the calculational prescription.

30.4.2 Unicity

In classical physics, it is quite normal to combine multiple different descriptions of an object or system to gain a more complete picture. With a spinning top, for example, we can specify its z-component of angular momentum, L_z, and its x-component, L_x, quite separately and in different contexts, but also combine them with L_y in order to have the complete picture, L. With quantum systems, this is not allowed. We are at liberty to describe a system in many *different*, but *incompatible* ways. In the correct circumstances, more than one incompatible description can be *true*, in the sense of being correctly extracted from the same initial data but using incompatible frameworks. Hence, they cannot be true *at the same time* and so incapable of being combined to give a more complete picture of the system. Again, in contrast to classical physics, the incompatible quantum descriptions are not *exclusive*, either can be true in the correct context:

> In the histories approach probabilities are linked to frameworks, and for this reason the notions of "true" and "false" are also framework dependent. This cannot lead to inconsistencies, a proposition being both true and false, because of the single framework rule. But it is contrary to a deeply rooted faith or intuition, shared by philosophers, physicists, and the proverbial man in the street, that at any point in time there is one and only one state of the universe which is "true", and with which every true statement about the world must be consistent.[27]

Unicity is the name given to the assumption that a complete and exhaustive description of a system can be constructed out of several partial descriptions which are true at the same time. That quantum mechanics is inconsistent with unicity is rooted in the existence of non-commuting operators which are needed to describe physical properties. There is no classical analogue for this. As a result:

> two incompatible quantum frameworks *F* and *G* do not represent mutually-exclusive possibilities in the sense that if the world is correctly described by *F* it cannot be correctly described by *G*, and vice versa. Instead it is best to think of *F* and *G* as means by which one can describe different aspects of the quantum system...Either framework can be employed to answer those questions for which it is appropriate, but the answers given by the two frameworks cannot be combined or compared.[28]

This has led to the consistent histories proposal that reality is a *relative construct*. If we term a collection of consistent appropriately course-grained histories (which hence can have probabilities assigned to them) a *realm*, then different realms can be incompatible yet alternative descriptions of a quantum system. Reality then becomes *realm dependent*. In support of this radical notion, proponents suggest that the history of physics is replete with situations where something that was taken as absolute turns out to be relative. Space and time suggest themselves immediately as one example. However, in relativity, while it is true that different observers in different states of motion will describe events in differing spatiotemporal terms, there is an underlying space-time that ties these relative descriptions together. The claim of realm dependent reality suggests that real is a relative concept without anything underlying to bind the relative reals together. Hence the comparison is not an exact one, and the claim is far more ontologically significant than it would suggest.

30.4.3 Probability (Again...)

Histories can be assigned probabilities provided they form part of a consistent set. Consistent sets arise out of decoherence. In truth then, no set of non-trivial histories is completely and definitively consistent, just approximately so. In which case, do we have any right to regard the various weights ascribed to histories as probabilities? In a nutshell, they will never *fully* obey the rules of probability, hence should not really be regarded as probabilities. The case for the defence would suggest that while any deviation from the rules is below a level of approximation that can be determined experimentally, then all is well.

However, some see this as contrary to the stated assumption that quantum reality is inherently probabilistic. At the very least, it raises familiar issues regarding the nature of probability, indeed the exact sense in which a wave function can be a pre-probability.

30.4.4 Other Issues

Some have criticized the consistent histories approach because it lacks predictive power. Given a set of initial conditions, different realms suggest different futures for the same system. The formalism allows us to calculate the different probabilities, but not to say which one will happen. In truth, this criticism seems a little harsh as we are used to accepting the impossibility of predicted (non-eigenstate) measurements.

The consistent histories claim to have dissolved the measurement problem is, however, also open to question. The discussion of 30.3.5 omitted issues to do with frameworks. As different frameworks are equally viable descriptions of quantum systems, what criteria do we use for selecting the appropriate framework to use in the context of a given measurement?

> in the x framework any state has two disjoint possibilities or properties $\left[\hat{P}_x^+\right]$ and $\left[\hat{P}_x^-\right]$, so for a system prepared in the state $|z^+\rangle$ a measurement of S_x, when viewed in the x framework, will reveal one or the other of these properties, each with probability ½. In the z framework, on the other hand, I am unable to interpret an S_x measurement.[29]

If we build a new apparatus, and describe it in purely quantum terms, with an initial state, Hamiltonian etc., and if there are a set of equally valid frameworks applicable, then how do we know, *from within quantum theory itself*, what the possible measurement outcomes are going to be? The answer would appear to be, as is implicit in the quote above, that we need our practical experience of what happens to be able to select the appropriate framework.

> without a selection of a framework, no predictions are possible, and the selection or identification of the appropriate framework relies on extraneous notions (i.e., notions not specified in the Hamiltonian or in the initial state of the system) and not identified by the general rules of the CH approach[30]

This looks like Copenhagenism, using a different form of words. One might even go on to argue that if only one framework is appropriate for a given measurement instance, then shouldn't the formalism dispense with the others? Equally, what ontological status do the inappropriate frameworks have? Does relative realism work *in practice*?

The single framework rule has also been questioned. Asserting the meaningless of a state such as $|\psi_A\rangle + |\psi_D\rangle$ is one thing, but one might hope that a properly formulated theory would prevent meaningless states from having a mathematical representation.

There are other objections to the coherence of the relative reality notion. If reality is framework dependent, then (for example) how could one go about determining unambiguously the initial state of a system? In particular, the initial state of the universe. In part, this is why the selection of the initial state has to be made using extra conditions from outside quantum theory (Section 30.3.7), for any closed system. However, the selection must have some experimental verification, and it is unclear how one might do this in a framework dependent reality. In a nutshell, how do you determine an absolute in a relative world?

NOTES

1 R. Omnès, Consistent interpretations of quantum mechanics, *Rev. Mod. Phys.*, 1992, 64 (1992).

2 Robert B. Griffiths (1937-), Otto Stern University professor of physics at Carnegie Mellon University, R. B. Griffiths, *J. Stat. Phys.*, 36, 219 (1984).

3 Roland Omnès (1931-) professor emeritus of theoretical physics in the Faculté des Sciences at Orsay, at the Université Paris-Sud XI, R. Omnes, *J. Stat. Phys.*, 53, 893 (1988); 53, 933 (1988); 53, 957 (1988).

4 Murray Gell-Mann (1929–2019), winner of the 1969 Nobel Prize in physics for his work on the quark model.

5 James B. Hartle (1939-) emeritus professor of physics at the University of California, Santa Barbara.

6 M. Gell-Mann and J. B. Hartle, Quantum mechanics in the light of quantum cosmology. In W. Zurek, ed., *Complexity, Entropy, and the Physics of Information*, Addison-Wesley, Reading (1990).

7 In this development, I am following Griffiths' outline and formalism from R. B. Griffiths, *Consistent Quantum Theory*, Cambridge University Press (2002).

8 Note that it is not the summation that is the decomposition, although the terminology suggests that, it is the set of projectors that go into the summation that form the decomposition.

9 R. B. Griffiths, *Consistent Quantum Theory*, Chapter 5, Cambridge University Press, Canbridge (2002).

10 As per 10, Chapter 9, where I have replaced the symbols used in that text with the relevant ones used in the preceding paragraph of this work.

11 R. B. Griffiths, 2011, "Quantum locality", *Foundations of Physics*, 41, 705–733 (2011).

12 C. J. Isham. Quantum logic and the histories approach to quantum theory, *J. Mathematical Phys.*, 35, 2157–2185 (1994), gr-qc/9308006.
13 Different authors use different conventions to the sequence of operators – time increasing left to right, or right to left. I am following Griffiths' choice.
14 See note 11, Chapter 9.
15 The histories can also be combined if they are orthogonal (see the later discussion on the decoherence functional). Orthogonal histories necessarily commute, even if not all of their projection operators commute.
16 See note 11, Chapter 18.
17 J. B. Hartle, The reduction of the state vector and limitations on measurement in the quantum mechanics of closed systems. In B.-L. Hu and T.A. Jacobson, eds., *Directions in Relativity*, vol. 2, Cambridge University Press, Cambridge (1993).
18 https://plato.stanford.edu/entries/qm-decoherence/#EssDec.
19 See note 21.
20 E. Joos and H. D. Zeh, The emergence of classical properties through interaction with the environment, *Zeit. Phys. B*, 59, 223 (1985).
21 J. B. Hartle, The quasiclassical realms of this quantum universe, https://arxiv.org/pdf/0806.3776.pdf.
22 J. B. Hartle and S. W. Hawking, Wave function of the Universe, *Phys. Rev. D*, 28, 2960 – Published (15 December 1983).
23 See note 11, Chapter 27.
24 See note 11, Chapter 27.
25 See note 11, Chapter 27.
26 See note 11, Chapter 9.
27 R. B. Griffith, *The Consistent Histories Approach to Quantum Mechanics*, Stanford Encyclopaedia of Philosophy: https://plato.stanford.edu/entries/qm-consistent-histories/#toc
28 See note 11, Chapter 27.
29 P. C. Hohenberg, An introduction to consistent quantum theory, *Rev. Mod. Phys.*, 82 (2010).
30 E. Okon and D. Sudarsky, Measurements according to Consistent Quantum Theory, (2013), arXiv, 1309.0792.

31 The Ontological Interpretation

The usual interpretation of the quantum theory is self-consistent, but it involves an assumption that cannot be tested experimentally, viz., that the most complete possible specification of an individual system is in terms of a wave function that determines only probable results of actual measurement processes. The only way of investigating the truth of this assumption is by trying to find some other interpretation of the quantum theory in terms of at present "hidden" variables, which in principle determine the precise behaviour of an individual system, but which are in practice averaged over in measurements of the types that can now be carried out.

David Bohm[1]

31.1 PHYSICS AND PHILOSOPHY

David Bohm is another fascinating and intriguing figure in the history of quantum theory. As a student, I discovered his textbook on quantum mechanics[2] and the refreshingly large number of words, compared with equations, that it contained. As I was just as interested in the meaning of quantum theory as the mathematical machinery, Bohm's book became a frequent reference (apparently Einstein was also a fan). It was only much later that I found out how Bohm was dissatisfied with the conventional understanding of quantum theory as he had presented it and that part of his motivation for writing the book was to explore the orthodox interpretation (broadly Copenhagen), so that he would have a better basis from which to critique things.

Bohm's career started in America where he did his PhD at the University of California at Berkley, working with J. Robert Oppenheimer, the so-called father of the atomic bomb. In 1947 Bohm moved to Princeton, where he encountered Einstein. In 1949 Oppenheimer came under suspicion by the notorious McCarthy committee on un-American activities, and Bohm was called to testify. When he refused, Princeton told him not to set foot on campus again. Bohm was arrested and put on trial, although he was acquitted in the end. As a result of this, Bohm moved from America, first to Brazil and then to Israel. He ultimately came to England in 1957, starting to work at Bristol and then moving to a professorship at Birkbeck College, London, in 1961, where he remained for the rest of his life. Bohm died of a heart attack in 1992.

Although David Bohm made important contributions to general physics (especially in the study of plasmas) and quantum theory, his work spread into brain research, creativity, and cognition. Linking these areas together was his interest in philosophy. In 1959 he was recommended a book *The First and Last Freedom*[3] by the Indian philosopher J. Krishnamurti.[4] Bohm was struck by the way in which his own ideas about the nature of reality, developed out of quantum theory, meshed with Krishnamurti's mystical philosophy. As Bohm himself wrote in an introduction to Krishnamurti's work:

> What particularly aroused my interest was his deep insight into the question of the observer and the observed. This question has long been close to the centre of my own work, as a theoretical physicist, who was primarily interested in the meaning of the quantum theory. In this theory, for the first time in the development of physics, the notion that these two cannot be separated has been put forth as necessary for the understanding of the fundamental laws of matter in general.[5]

Bohm contacted Krishnamurti, and the two were close friends for over 25 years.

As Bohm's thinking developed, he placed an increasing emphasis on the notions of *implicate and explicate orders*. I can't do any justice to the complexity and subtlety of these ideas in a short

DOI: 10.1201/9781003225997-35

passage like this, but in essence, Bohm viewed the visible world around us with all its apparently separate objects, structures, and events as the explicate result of an implicate underlying process or flow. An object, such as an elementary particle, is hence seen as a stable subprocess in the underlying business going on. One of the Bohm's favourite analogies compared the implicate and explicate orders with the flow of a stream:

> On this stream, one may see an ever-changing pattern of vortices, ripples, waves, splashes, etc., which evidently have no independent existence as such. Rather, they are abstracted from the flowing movement, arising and vanishing in the total process of the flow. Such transitory subsistence as may be possessed by these abstracted forms implies only a relative independence or autonomy of behaviour, rather than absolutely independent existence as ultimate substances.[6]

31.2 WAVE AND PARTICLE

One way of summing up Bohm's approach to quantum theory is to replace the wave/particle idea of wave *or* particle, with the declaration wave *and* particle.

De Broglie had originally suggested that some form of (ontologically) real wave influencing the motion of particles might explain the wave aspect of quantum physics. Bohm independently developed his ideas into a full theory, capable of dealing with multiple particles and, with his collaborator, Basil Hiley,[7] into a relativistic version comparable to quantum field theory.

The term 'ontological interpretation' is sometimes applied to Bohm's approach (as is the somewhat less elegant *Bohmian Mechanics*[8]), as he set out to tackle the ontological problems of conventional quantum theory. As Hiley has written:

> it is well known that Heisenberg favoured the use of potentialities. What is well less known is that Bohm also proposed that the wave function should be thought of in terms of potentialities. Bohm argued that the potentialities were latent in the particle and that they could only be brought out more fully through interaction with a classical measuring apparatus. This of course is essentially the conventional view, so why did Bohm bother to make alternative proposals? It was the complete absence of any account of the actual that troubled him. In the quantum formalism nothing seemed to happen unless and until there was an interaction with a measuring apparatus. There was no actualisation until some form of instrument was triggered. Surely something triggered the instrument? Why was the measuring instrument so different? Isn't it just another collection of physical processes governed by the same laws of physics?[9]

In truth, Bohm's approach might better be viewed as a different way of doing quantum theory, rather than a different interpretation. His version of quantum theory takes waves and particles equally seriously and treats them both as ontologically real – hence the phrase 'wave and particle.'

31.2.1 Bohm's Version of the Schrödinger Equation

From a mathematical perspective, Bohm started from the standard quantum wave function $\Psi(x, t)$, but written in the form $\Psi(x, t) = Re^{iS/\hbar}$, where $R(x, t)$ and $S(x, t)$ are both real-valued functions. Inserting this into the Schrödinger equation[10]:

$$-\frac{\hbar^2}{2m}\frac{d^2\Psi(x,t)}{dx^2} + V(x,t)\Psi(x,t) = i\hbar\frac{d\Psi(x,t)}{dt}$$

gives:

$$-\frac{\hbar^2}{2m}\frac{d^2\left(Re^{iS/\hbar}\right)}{dx^2} + V(x,t)Re^{iS/\hbar} = i\hbar\frac{d\left(Re^{iS/\hbar}\right)}{dt}$$

What follows is a sequence of mathematical re-arrangements, which can be skipped over to get to the final results, Eqs. (31.1) and (31.2), if you do not wish to inspect the details right now:

$$-\frac{\hbar^2}{2m}\frac{d}{dx}\left(R\frac{d\left(e^{iS/\hbar}\right)}{dx}+e^{iS/\hbar}\frac{dR}{dx}\right)+VRe^{iS/\hbar}=i\hbar\left(R\frac{d\left(e^{iS/\hbar}\right)}{dt}+e^{iS/\hbar}\frac{dR}{dt}\right)$$

Carrying out the first stage of the differentials:

$$-\frac{\hbar^2}{2m}\frac{d}{dx}\left(\frac{iRe^{iS/\hbar}}{\hbar}\frac{dS}{dx}+e^{iS/\hbar}\frac{dR}{dx}\right)+VRe^{iS/\hbar}=i\hbar\left(\frac{iRe^{iS/\hbar}}{\hbar}\frac{dS}{dt}+e^{iS/\hbar}\frac{dR}{dt}\right)$$

and now the second layer of differentials:

$$-\frac{\hbar^2}{2m}\left(\frac{iRe^{iS/\hbar}}{\hbar}\frac{d^2S}{dx^2}+\frac{d}{dx}\left(\frac{iRe^{iS/\hbar}}{\hbar}\right)\frac{dS}{dx}+\frac{d\left(e^{iS/\hbar}\right)}{dx}\frac{dR}{dx}+e^{iS/\hbar}\frac{d^2R}{dx^2}\right)+VRe^{iS/\hbar}$$

$$=i\hbar\left(\frac{iRe^{iS/\hbar}}{\hbar}\frac{dS}{dt}+e^{iS/\hbar}\frac{dR}{dt}\right)$$

Continuing the consequences of the nested differentials:

$$-\frac{\hbar^2}{2m}\left(\frac{iRe^{iS/\hbar}}{\hbar}\frac{d^2S}{dx^2}+\frac{ie^{iS/\hbar}}{\hbar}\frac{dR}{dx}\frac{dS}{dx}+\frac{iR}{\hbar}\frac{d\left(e^{iS/\hbar}\right)}{dx}\frac{dS}{dx}+\frac{ie^{iS/\hbar}}{\hbar}\frac{dS}{dx}\frac{dR}{dx}+e^{iS/\hbar}\frac{d^2R}{dx^2}\right)+VRe^{iS/\hbar}$$

$$=i\hbar\left(\frac{iRe^{iS/\hbar}}{\hbar}\frac{dS}{dt}+e^{iS/\hbar}\frac{dR}{dt}\right)$$

$$-\frac{\hbar^2}{2m}\left(\frac{iRe^{iS/\hbar}}{\hbar}\frac{d^2S}{dx^2}+2\frac{ie^{iS/\hbar}}{\hbar}\frac{dR}{dx}\frac{dS}{dx}-\frac{Re^{iS/\hbar}}{\hbar^2}\frac{dS}{dx}\frac{dS}{dx}+e^{iS/\hbar}\frac{d^2R}{dx^2}\right)+VRe^{iS/\hbar}$$

$$=i\hbar\left(\frac{iRe^{iS/\hbar}}{\hbar}\frac{dS}{dt}+e^{iS/\hbar}\frac{dR}{dt}\right).$$

and gathering terms:

$$-\frac{\hbar^2}{2m}\frac{\cancel{Re^{iS/\hbar}}}{\hbar}\left(i\frac{d^2S}{dx^2}+\frac{2i}{R}\frac{dR}{dx}\frac{dS}{dx}-\frac{1}{\hbar}\frac{dS}{dx}\frac{dS}{dx}+\frac{\hbar}{R}\frac{d^2R}{dx^2}\right)+V\cancel{Re^{iS/\hbar}}=i\hbar\cancel{Re^{iS/\hbar}}\left(\frac{i}{\hbar}\frac{dS}{dt}+\frac{1}{R}\frac{dR}{dt}\right)$$

$$-\frac{\hbar}{2m}\left(i\frac{d^2S}{dx^2}+\frac{2i}{R}\frac{dR}{dx}\frac{dS}{dx}-\frac{1}{\hbar}\left(\frac{dS}{dx}\right)^2+\frac{\hbar}{R}\frac{d^2R}{dx^2}\right)+V=-\frac{dS}{dt}+\frac{i\hbar}{R}\frac{dR}{dt}$$

Now we can separate the terms that depend on i from those that do not, as these must be independent equations:

$$\frac{\hbar}{2m}\left(-\frac{1}{\hbar}\left(\frac{dS}{dx}\right)^2+\frac{\hbar}{R}\frac{d^2R}{dx^2}\right)-V=\frac{dS}{dt} \qquad (31.1)$$

$$-\frac{1}{2m}\left(i\frac{d^2S}{dx^2}+\frac{2i}{R}\frac{dR}{dx}\frac{dS}{dx}\right)=\frac{i}{R}\frac{dR}{dt} \tag{31.2}$$

On the face of it we don't seem to have gained much by turning one equation into two, but each of these equations has something interesting to tell us.

Conservation of Probability

Taking Eq. (31.2) first, we start by making the (not entirely obvious) move of turning the equation into one using R^2 rather than R by writing:

$$\frac{1}{R}\frac{dR}{dx}=\frac{1}{2R^2}\frac{d(R^2)}{dx} \qquad \frac{1}{R}\frac{dR}{dt}=\frac{1}{2R^2}\frac{d(R^2)}{dt}$$

and substituting (cancelling out the factors of i in the process):

$$-\frac{1}{2m}\left(\frac{d^2S}{dx^2}+\frac{1}{R^2}\frac{d(R^2)}{dx}\frac{dS}{dx}\right)=\frac{1}{2R^2}\frac{d(R^2)}{dt}$$

Now, letting $\mathcal{P}=R^2$:

$$-\frac{1}{2m}\left(\frac{d^2S}{dx^2}+\frac{1}{\mathcal{P}}\frac{d\mathcal{P}}{dx}\frac{dS}{dx}\right)=\frac{1}{2\mathcal{P}}\frac{d\mathcal{P}}{dt}$$

Cross multiplying and simplifying we get:

$$-\frac{1}{m}\left(\mathcal{P}\frac{d^2S}{dx^2}+\frac{d\mathcal{P}}{dx}\frac{dS}{dx}\right)=\frac{d\mathcal{P}}{dt}$$

$$-\frac{1}{m}\frac{d}{dx}\left(\mathcal{P}\frac{dS}{dx}\right)=\frac{d\mathcal{P}}{dt}$$

which is known as the *conservation of probability equation*. Given that we started with $\Psi(x,t)=Re^{iS/\hbar}$, it is clear that $|\Psi|^2=R^2=\mathcal{P}$, which is the *probability density* (the probability of finding a particle in a small volume of space). The right-hand side, $\frac{d\mathcal{P}}{dt}$, is then the rate at which this probability density varies with time. The only way for $\mathcal{P}$ to change is for it to 'leak out' of the volume into the surrounding space. If we now define the particle's momentum as $p=\frac{dS}{dx}$, the velocity becomes $v=\frac{1}{m}\frac{dS}{dx}$. Our equation is then:

$$-\frac{d}{dx}(\mathcal{P}v)=\frac{d\mathcal{P}}{dt}$$

Hence the rate of change of probability with time is equal to the spatial gradient of probability times velocity (the *probability current*).

Two things to note: firstly, Eq. (31.2) does not contain any factors of $\hbar$, so, we would not expect its effects to be limited to the quantum scale set by that value. Secondly, our definition of $p=\frac{dS}{dx}$ is entirely consistent with the earlier use $\hat{p}=-i\hbar\frac{d}{dx}$, which in this case would give $\hat{p}\psi=-i\hbar\frac{d}{dx}\left(Re^{iS/\hbar}\right)=(-i\hbar)\left(\frac{i}{\hbar}\right)\frac{dS}{dx}Re^{iS/\hbar}=\frac{dS}{dx}\psi$.

Conservation of Energy

Returning to Eq. (31.1), and making the further identification $E = -\frac{dS}{dt}$, we now get:

$$\frac{\hbar}{2m}\left(-\frac{1}{\hbar}p^2 + \frac{\hbar}{R}\frac{d^2R}{dx^2}\right) - V = -E$$

or:

$$\frac{p^2}{2m} - \frac{\hbar^2}{2mR}\frac{d^2R}{dx^2} + V = E$$

which looks like an energy balance equation if we define:

$$Q = -\frac{\hbar^2}{2mR}\frac{d^2R}{dx^2}$$

to be the *quantum potential energy*. In a typical double-slit experiment, $Q < 10^{-4}$ eV against electron kinetic energies ~ 45 keV.

Note that as Q is proportional to $\hbar^2$, it will vanish in the classical limit, which can be obtained by (hypothetically) taking $\hbar \rightarrow 0$. Indeed, we could turn this around and characterize the classical level of reality as being *the domain in which Q is negligible.*

In Hiley's words:

> Thus equation (31.1) can be regarded as a generalised expression for the conservation of energy provided we regard Q as a new form of potential energy which is negligible in the classical world and is apparent only in quantum systems. This energy has traditionally been called the quantum potential energy. It should not be thought as the source of some mysterious new force to be put into the Newtonian equations of motion.[11]

So, we have by this analysis revealed an interesting and remarkable fact. By writing $\Psi(x, t) = Re^{iS/\hbar}$, we have decomposed the Schrödinger equation into two coupled equations: one for the conservation of probability and one for an extended form of energy conservation. We have not added anything in by carrying out this process, simply revealed some hidden truths of the formalism, if we can, at least temporarily, accept the role of Q as a form of energy.

31.2.2 The Quantum Potential Energy

In the previous quote, Hiley is at pains to stress that the quantum potential energy should not be associated with some form of force acting on particles. This is something new.

In classical physics, generic fields of force, are simply related to potential energies:

$$F = -\frac{dU}{dx} \text{ or } U = -\int F\,dx$$

where $U = U(q_A, q_B, r)$ is the potential energy as a function of the appropriate 'charge', q, (which in the case of gravity is mass) and separation, r. The formalism can be extended by introducing a suitable *field*, $\mathcal{F}$, and *potential*, $\mathcal{V}$ in the following manner:

$$F = F_{AB} = q_A\mathcal{F}(q_B, r) = F_{BA} = q_B\mathcal{F}(q_A, r)$$

$$U = U_{AB} = q_A\mathcal{V}(q_B, r) = U_{BA} = q_B\mathcal{V}(q_A, r)$$

$$\mathcal{F} = -\frac{d\mathcal{V}}{dx} \text{ or } \mathcal{V} = -\int \mathcal{F}\,dx$$

In which case, one of the charges can be arbitrarily identified as the source of both the field and potential. This approach also makes it evident that the force between the two charges naturally follows Newton's third law, which for an electrical interaction, reads:

> If charge A exerts a force on charge B, via an electrical field, then charge B exerts an equal and opposite force on A via its electrical field.

From the general relationship between force and potential energy:

$$F = -\frac{dU}{dx}$$

we can see that the magnitude of the force is dependent on the magnitude of the potential energy. If we scale the potential energy by taking $U' = aU$ with constant a, then the resulting force is given by:

$$F' = -\frac{dU'}{dx} = \frac{d}{dx}(aU) = U\cancel{\frac{da}{dx}} + a\frac{dU}{dx} = a\frac{dU}{dx}$$

striking out the term that is zero. Hence the force scales with the potential energy.

We can now construct an appropriate definition of a 'mechanical' force, in the sense of an ordinary push/pull type of force, whether mediated by a field or not. A mechanical force is characterized by:

1. there is an evident generalized 'charge' that can be considered as the source/sink of the field;
2. the force between charges obeys Newton's third law of motion;
3. the magnitude of the force scales with the magnitude of the potential energy.

The quantum potential energy is a fundamentally different sort of animal to the potential energy associated with a mechanical force.

Firstly, the quantum potential energy is determined by:

$$Q = -\frac{\hbar^2}{2mR}\frac{d^2R}{dx^2}$$

where R is extracted from the wave function $\Psi(x, t) = Re^{iS/\hbar}$, so there is no evident 'charge' to identify as the source. This is strike one against the idea that the quantum potential energy gives rise to a mechanical force.

As the quantum potential energy has no source, any particle affected by it has nothing to 'push back' against. If there were some sort of force associated with the quantum potential energy, then it would violate Newton's third law. This is strike two.

Finally, the presumed quantum mechanical force, F_Q, would relate to Q via:

$$F_Q = -\frac{dQ}{dx} = -\frac{d}{dx}\left(-\frac{\hbar^2}{2mR}\frac{d^2R}{dx^2}\right)$$

$$= \frac{\hbar^2}{2m}\left(-\frac{1}{R^2}\frac{d^2R}{dx^2} + \frac{1}{R}\frac{d^3R}{dx^3}\right)$$

$$= \frac{\hbar^2}{2mR}\left(\frac{d^3R}{dx^3} - \frac{1}{R}\frac{d^2R}{dx^2}\right)$$

If we scale R according to $R' = aR$ then:

$$F'_Q = \frac{\hbar^2}{2maR}\left(\frac{d^3(aR)}{dx^3} - \frac{1}{aR}\frac{d^2(aR)}{dx^2}\right)$$

$$= \frac{\hbar^2}{2maR}\left(a\frac{d^3R}{dx^3} - \frac{a}{aR}\frac{d^2R}{dx^2}\right)$$

$$= \frac{\hbar^2}{2mR}\left(\frac{d^3R}{dx^3} - \frac{1}{R}\frac{d^2R}{dx^2}\right)$$

$$= F_Q$$

The hypothetical quantum force does not scale according to the magnitude of R. Strike three.

This last point has an even deeper significance.

If we take the value of Q and see how it scales with R:

$$Q' = -\frac{\hbar^2}{2mR'}\frac{d^2R'}{dx^2} = -\frac{\hbar^2}{2maR}\frac{d^2(aR)}{dx^2} = -\frac{\hbar^2}{2maR}\frac{ad^2(R)}{dx^2} = Q$$

it is clear that Q is independent of the magnitude of ψ. Hence, Q can have a large effect on proceedings, even if the magnitude of the wave function is small.

It's quite normal for a field to be influenced by its local environment. We can manipulate the electrical fields surrounding a wire by accelerating electrons along the wire's length. The resulting disturbance in the field radiates out from the wire in all directions. In fact, it's a radio wave. However, the strength of this disturbance drops with distance as the energy spreads out over an increasing volume of space. Therefore, the further away you are from the transmitter, the harder it is to pick up the signal from a given station.

The quantum potential energy does not necessarily suffer from this decline with distance. Altering the environment will have a direct influence on the wave function, and hence R, but as the quantum potential energy does not depend on the magnitude of R *the effect of the change can be felt everywhere*. This is directly encoding non-locality into the quantum potential.

Bohm himself did not think of his quantum potential energy giving rise to a force. The analogy he used was that of a ship being guided by its radar: the information about the obstacles and other ships located nearby helping to inform the course being set. The quantum potential energy, derived from the wave function, carries information about the entire experimental arrangement and influences the possible paths the particle can take. Think of the quantum potential energy as defining a set of contours through the experiment that guide how the particle moves, in addition to the classical forces at work.

In any given situation, classical forces act on particles to influence the paths they follow. These forces are inserted into the equivalent quantum description via the potential energy V in the Schrödinger equation. However, in Bohm's approach the quantum potential energy Q also has a very important say in what the particle gets up to. As Q derives from the wave function, which is in turn influenced by the classical potential energy V, Q carries 'information' about the experimental context, which helps to explain how the same particle can behave in radically different ways in different experiments:

> … while the Newtonian potential drives the particle along the trajectory, the quantum potential organises the form of the trajectories in response to the experimental conditions.[12]

Viewed in this way, the quantum potential energy is the 'wave' part of the declaration wave *and* particle.

It is illuminating to see how Q can be related to the probability density, $\mathcal{P} = R^2$:

$$
\begin{aligned}
Q &= -\frac{\hbar^2}{2mR}\frac{d^2R}{dx^2} = -\frac{\hbar^2}{2m\sqrt{\mathcal{P}}}\frac{d^2\left(\sqrt{\mathcal{P}}\right)}{dx^2} \\
&= -\frac{\hbar^2}{2m\sqrt{\mathcal{P}}}\frac{d}{dx}\left(\frac{d\left(\sqrt{\mathcal{P}}\right)}{dx}\right) = -\frac{\hbar^2}{2m\sqrt{\mathcal{P}}}\frac{d}{dx}\left(\frac{1}{2\sqrt{\mathcal{P}}}\frac{d\mathcal{P}}{dx}\right) \\
&= -\frac{\hbar^2}{2m\sqrt{\mathcal{P}}}\left[-\frac{1}{4\left(\sqrt{\mathcal{P}}\right)^3}\frac{d\mathcal{P}}{dx} + \frac{1}{2\sqrt{\mathcal{P}}}\frac{d^2\mathcal{P}}{dx^2}\right] \\
&= -\frac{\hbar^2}{4m}\left[\frac{1}{\mathcal{P}}\frac{d^2\mathcal{P}}{dx^2} - \frac{1}{2\mathcal{P}^2}\frac{d\mathcal{P}}{dx}\right]
\end{aligned}
$$

which re-enforces the notion that the quantum potential energy is encoding information about the configuration of the experiment.

31.3 PROBABILITY

So far everything about Bohm's theory is perfectly deterministic and calculable. The quantum potential energy can be found for any experimental arrangement, and the classical forces acting on a particle can be defined and calculated. So, all we need to know, to predict the exact path followed by each particle through an experiment, is the starting positions and momenta of each particle. And there lies the snag.

The action of the quantum potential ensures that the particle's path is precisely determined by its initial state, *something that we could never measure accurately enough to give an exact prediction.* All we can do is obtain from our experimental measurements, subject to the uncertainty principle, some possible set of initial values that will then predict a set of likely paths, rather than one actual path.

Bohm's particles have fixed values for all the physical variables associated with their properties. We may not *know* some of the values, but they are ontologically real and 'out there.' The uncertainty principle enters the measurement arena, but not due to limitations imposed on eigenfunctions of conjugate variables:

> in the model we are considering, the role played by the quantum potential itself is such as to make it impossible to reduce the uncertainty in the initial conditions below that given by quantum mechanics in any given measurement.[13]
>
> The standard approach merely assumes a particle cannot possess well-defined values of these variables simultaneously, whereas Bohm assumes the particle can have simultaneous x and p, but we do not know what the values are.
>
> So how does the uncertainty principle enter? Here we follow Wheeler and assume that measurement does not merely reveal those values present, the measuring instrument becomes active in the measuring process, producing the appropriate eigenvalue of the operator that is being measured. In this participatory act, the complementary variables are changed in an un-controllable way. Thus the statistical element of a quantum process remains.[14]

This is how quantum probability enters Bohm's account, as a subtle version of a classical probability: a cover for our ignorance of the initial conditions.

Bohm's original paper, published the year after his textbook, was entitled *A Suggested Interpretation of the Quantum Theory in Terms of "Hidden" Variables I,* which suggests that, at least initially, he thought the approach was an example of a hidden variable theory. Indeed, when

Bell read Bohm's original paper and noted the non-locality built into the approach, he was inspired to think more about non-locality in quantum theory and ultimately develop his inequalities.

However, according to Hiley, they both somewhat regretted Bohm's original reference to hidden variables. In this case, the 'hidden variables' are the initial positions and momenta of the particles, not new physics. The variables are not hiding from us so much as playing hard to get.

31.4 QUANTUM POTENTIAL ENERGY IN ACTION

One of the interesting side-effects of exploring Bohm's approach to quantum theory is the light that it casts on how different the quantum world is from our classical expectations. Clearly, then, it pays to see how the quantum potential energy applies in a range of 'classic' experiments.

31.4.1 Quantum Potential Energy and the Double Slit Experiment

> There is no longer a mystery as to how a single particle passing through one slit "knows" the other slit is open. This information is carried by the quantum potential so that we no longer have a conceptual difficulty in understanding the results obtained in very low intensity interference experiments.[15]

The quantum potential is responsible for the *apparent* interference effect in the double-slit experiment. Particles are steered to a point on the far screen having passed through one slit or the other. In each individual case, we can't be sure which slit the particle will pass through (as we can't measure its starting parameters precisely enough), but there is no suggestion that a particle passes through both slits. Nor is it a waveform.

Figure 31.1 shows the spatial distribution of Q for a specific double slit experiment.[16] The quantum potential energy has the greatest influence where its distribution *changes steeply*. For that reason, you're more likely to find particles in a region where the spatial distribution of potential energy is *flat*. If a particle approached the vicinity of one of the thin, steep valleys, it will be accelerated transversely across the valley towards a plateau region. Where the valleys line up with the screen, dark spots will result. The light bands are at the ends of the plateau regions.

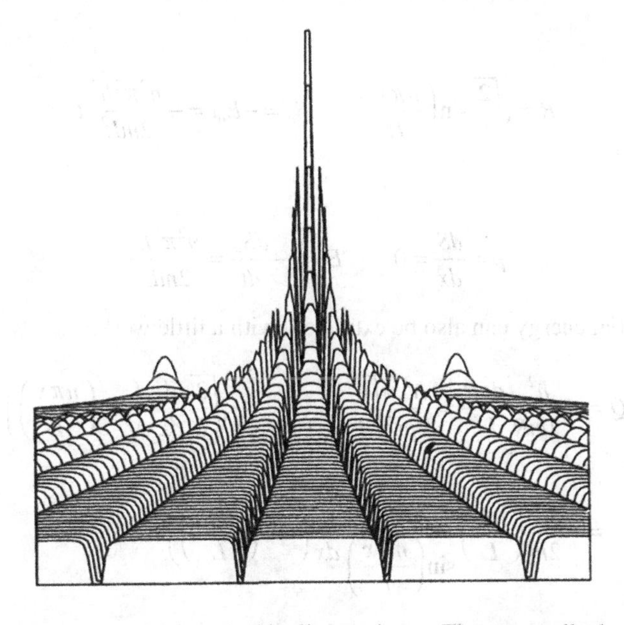

FIGURE 31.1 The quantum potential for a double slit experiment. The two smaller bumps towards the back of the distribution are quantum potential peaks where the slits are located; we are looking back towards eh slits from the far screen. (Reproduced with permission from Philippidis, C., Dewdney, C., Hiley, B.J, Nvovo Cimento, S2B, 15-28, 1979).

A detailed analysis, which is unfortunately far too complex to go into here, shows that the quantum potential depends on the momentum of the particles as well as the width of the slits and their separation. So, any attempt to isolate an individual path for the particle, say by blocking off parts of the beam so that we single out a restricted range of starting points for the electrons, will in turn alter the quantum potential energy, defeating the object.

Equally, blocking off one of the slits alters Q right throughout the experiment, ensuring that the interference pattern is replaced by a distribution of electrons arriving opposite the open slit. The contextuality of the quantum physics is guaranteed by the structure of the quantum potential energy. Bluntly, any change in the experimental setup will have an influence on the particles, no matter how far away that change takes place:

> Thus altering the experimental arrangement can radically change the outcome of the experiment. This feature of quantum mechanics was continually emphasized by Bohr when he talked about "the impossibility of subdividing quantum phenomena". The quantum potential actually provides a clear expression of this inseparability. For example, the quantum potential calculated [for the double-slit experiment] shows that the properties of the particle (such as mass and velocity) cannot be isolated from those of the apparatus (such as width and separation of the slits). In other words, the observed system and the observing apparatus are linked in an essential and irreducible way.[16]
>
> *[my emphasis]*

The quantum potential energy embodies both the contextuality and the nonlocal nature of quantum theory (see further in Section 31.4.4).

31.4.2 Quantum Potential Energy and the Particle in a Box

In Chapter 22, we derived the wave function for a particle trapped in a box with rigid walls:

$$\Psi(x,t)=\sqrt{\frac{2}{L}}\sin\left(\frac{n\pi x}{L}\right)e^{-\frac{E_n t}{\hbar}}$$

so it follows that:

$$R=\sqrt{\frac{2}{L}}\sin\left(\frac{n\pi x}{L}\right) \qquad S_n=-E_n t=-\frac{n^2\pi^2\hbar^2}{2mL^2}t$$

Furthermore:

$$p=\frac{dS}{dx}=0 \qquad E_n=-\frac{dS_n}{dt}=\frac{n^2\pi^2\hbar^2}{2mL^2}$$

The quantum potential energy can also be extracted, with a little work:

$$Q=-\frac{\hbar^2}{2mR}\frac{d^2R}{dx^2}=\frac{\hbar^2}{2m}\sqrt{\frac{L}{2}}\frac{1}{\sin\left(\frac{n\pi x}{L}\right)}\sqrt{\frac{2}{L}}\frac{d^2}{dx^2}\left(\sin\left(\frac{n\pi x}{L}\right)\right)$$

$$=-\frac{\hbar^2}{2m}\left(\frac{n\pi}{L}\right)\frac{1}{\sin\left(\frac{n\pi x}{L}\right)}\frac{d}{dx}\left(\cos\left(\frac{n\pi x}{L}\right)\right)$$

$$=\frac{\hbar^2}{2m}\left(\frac{n\pi}{L}\right)^2\left[\frac{\sin\left(\frac{n\pi x}{L}\right)}{\sin\left(\frac{n\pi x}{L}\right)}\right]=\frac{n^2\pi^2\hbar^2}{2mL^2}=E_n$$

So, we have a rather curious state of affairs: $p = \frac{dS}{dx} = 0$, so the particle is not moving and the quantum potential energy, $Q = E_n$. As the wave function is a superposition of the ψ_+ and ψ_- momentum eigenstates, we might expect that $p = 0$ which in turn means that there is no kinetic energy involved. With no forces or classical potential energies acting with the box, the only form of energy left to the particle is the quantum potential energy, hence we really should not be surprised that $Q = E_n$.

One way to measure momentum would be to remove one of the walls and allow the particle to escape. Then we can see how long it takes for particle to cover a given distance. Classically we would be certain that removing the wall has no effect on the particle,[17] hence the emerging momentum must be the momentum that it had inside.

Quantum mechanically, though, this is never going to work. Removing the walls *changes the wave function,* and hence the quantum potential energy. A free particle has:

$$\psi = Ae^{\frac{i}{\hbar}(px-Et)} \quad R = A \quad S = px - Et$$

$$p = \frac{dS}{dx} \quad E = -\frac{dS}{dt} \quad Q = 0$$

the quantum potential energy having been converted into kinetic energy.

31.4.3 Spin

The three *Euler angles* (α, β, γ) denote a rotation about the *z*-axis of angle α (yaw), followed by a rotation β about the *y*-axis (pitch) and finally γ about the *x*-axis (roll). They can be used to specify how to rotate a co-ordinate system to come into line with a particular orientation in space. Hence, they can indicate a spin axis if we return (tentatively) to a conventional rotating top model of quantum spin. We can then write the wave function in the form:

$$\Psi(x, t, \alpha, \beta, \gamma) = Re^{i\gamma/2}\begin{pmatrix} \cos(\alpha/2)e^{i\beta/2} \\ i\sin(\alpha/2)e^{-i\beta/2} \end{pmatrix}$$

and develop an appropriate Hamiltonian using the magnetic moment of the rotating particle. If we do this, we can then extract a version of the quantum potential:

$$Q_T = -\frac{\hbar^2}{2mR}\frac{d^2R}{dx^2} - \frac{\hbar^2}{8m}\left[\left(\frac{d\alpha}{dx}\right)^2 + \sin^2\alpha\left(\frac{d\beta}{dx}\right)^2\right]$$

although it evidently needs to be in three dimensions to work properly. The first part is the familiar spatial quantum potential and the second is a new term relating to the spin orientation. Under the influence of this term, the Bohm trajectories of the particles passing through an SG magnet diverge into our familiar UP and DOWN paths. At the same time, in a short distance after the particles exit the magnetic field region, the orientation of the spin axis rotates to be UP relative to the magnetic axis for the particles following the UP path and down for those following the DOWN path. Note that in this model, there is a smooth evolution of the spin axis taking place, not a sharp change as suggested by the collapse of state.

31.4.4 Entanglement

Suppose that we have a wave function $\Psi = R(x_1, x_2, t)e^{\frac{i}{\hbar}S(x_1, x_2, t)}$ describing a general entangled state involving two particles with positions x_1, x_2 respectively. In this case, the quantum potential energy would be:

$$Q(x_1, x_2, t) = -\frac{\hbar^2}{2mR(x_1, x_2, t)}\left[\frac{d^2R(x_1, x_2, t)}{dx_1^2} + \frac{d^2R(x_1, x_2, t)}{dx_2^2}\right]$$

The intriguing aspect of this version is that it is *non-local*; Q depends on the position of both particles *at the same time*. Alongside the non-locality inherent in this potential energy, we also have the established independence of field intensity, which does not limit the range at which it can act. The non-local features disappear if the wave function can be decomposed into a product $\Psi = \psi_1(x_1, t)\psi_2(x_2, t)$.

Although Q can evolve with time, nothing is flowing from one particle to another and the overall structure (pattern) of Q is determined by the global experimental conditions. Indeed, this non-local aspect of the quantum potential energy is at the heart of Bohm's approach to the EPR paradox. Picking up on Bohr's reply to the EPR paper, there is:

> no question of a mechanical disturbance of the system under investigation during the last critical stage of the measuring procedure. But even at this stage there is essentially the question of an influence on the very conditions which define the possible types of predictions regarding the future behavior of the system.[18]

Hiley adds:

> ...in our view also there is no mechanical force because the quantum potential does not operate mechanically. The meaning of the word 'influence' used by Bohr is unclear, but it can be understood in terms of active information, which, as we have explained, contains information about the overall experimental conditions and it is these conditions that are responsible for the correlations. Furthermore since we are dealing with a global effect there is no direct energy transfer involved.[19]

The notion of *active information* will be explored further in Section 31.5.

31.5 INFORMATION AND WAVE FUNCTION COLLAPSE

The term 'information' carries various overtones, depending on the context. When humans discuss matters, 'information' is generally considered to be data that carries *meaning*, whereas in physics, information is defined in terms of the probability that a certain character from a lexicon will appear, irrespective of any meaningful content. Physicists like to remove any messy subjective references, if possible.

However, the central stem of the word is *forma*, which primarily refers to a visible shape or outward appearance. In that case, *informare* indicates the process of bringing a certain shape into being. Hiley uses the term *information* in this sense; especially when referring to the *objective process* by which dynamics is constrained by the environment.

He further distinguishes between *active information*, *passive information* and *inactive information*.[20] To see the distinction, we can apply these ideas to the Mach-Zehnder interferometer. As per our earlier discussions, in the Bohm approach photons have real paths guided by the quantum potential energy, but the paths are not predictable due to the uncertainty in the initial conditions.

The quantum potential energy exists throughout the experimental arrangement, and so is carrying and distributing information about the complete setup. If a photon is travelling through the top path, then the information along that path is *active* and informing the photon's motion. Information also exists along the bottom path, but it is *passive* as there is no photon, in this case, for it to act on, assuming that we are running the experiment in a low-intensity mode.

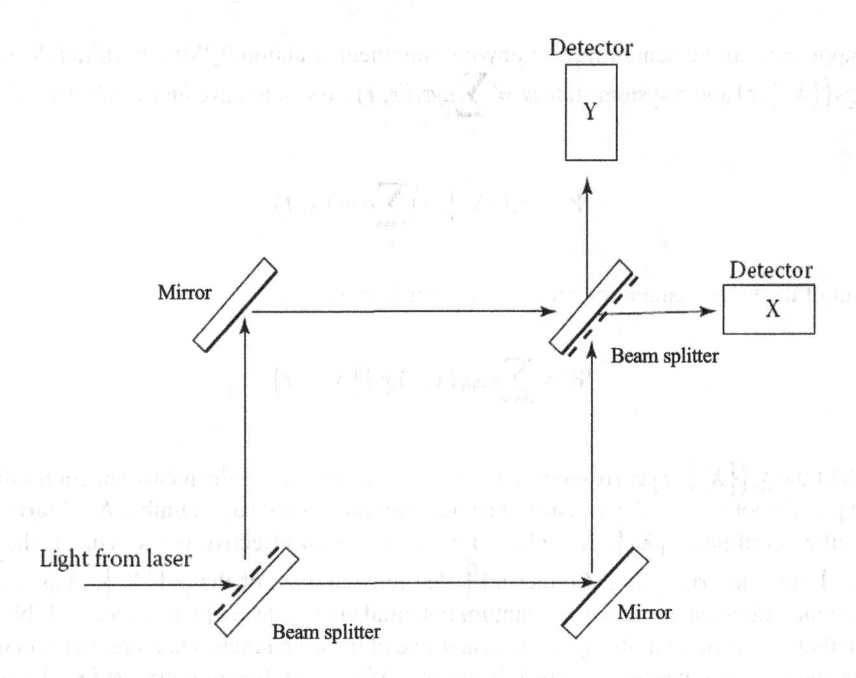

FIGURE 31.2 The Mach-Zehnder interferometer, yet again.

If we now introduce some form of detector on the top path which does not divert or 'capture' the photon, merely registers its presence as it passes by, then the wave function arriving at the final beam splitter is:

$$\Psi = a\psi_{\text{top}}\chi_{\text{hit}} + b\psi_{\text{bottom}}\chi_{\text{miss}}$$

where χ_{hit} is the wave function of the device having registered a photon passing and χ_{miss} the device when the photon has not been observed, as it travelled the other route. We assume that the detector is 100% efficient so that it always sees the photon if it happens to be on the top path. Even if the impact of the device on the photon is gentle, so that it hardly disturbs the photon's wave function, the process of detecting its passage still destroys the interference, as we shall see.

For the apparatus to successfully record the result of the measurement, the two wave functions χ_{hit} and χ_{miss} must not 'overlap'. The measurement record would not be sustained if there was an overlap. Either the result would be ambiguous (with the device registering something between hit and miss) or the particles in the apparatus described by χ_{hit}, for example, might spontaneously transition to χ_{miss}.

To see what happens to the interference, we need to evaluate the quantum potential energy at the second beam splitter. This is derived from Ψ with the coordinates of the photon at the beam splitter along with, in principle, the coordinates of the particles within the device. Assuming that we are in a 'hit' situation, these coordinates will correspond to the macroscopic configuration of the device registering the passage of the photon. If we put the values of those coordinates into χ_{miss}, it will come out to be zero, as the two states do not overlap. Hence the second term in Ψ will not impact on Q. Any information in ψ_{bottom} will be rendered inert, or inactive. A similar argument would follow if the photon had traversed the bottom path, in which case the coordinates of the apparatus' particles would make $\chi_{\text{hit}} = 0$. Either way, the interference is destroyed. The very presence of the detection device removes the interference, even if the device does not observe the passage of a photon.

This approach can be generalized to any measurement situation.[21] With an initial device ready state of $\chi_0\left(\left\{X_j\right\}, t\right)$ and a system state $\psi = \sum_i a_i\varphi_i(x, t)$ we would have an overall wave function:

$$\Psi = \chi_0\left(\left\{X_j\right\}, t\right)\sum_i a_i\varphi_i(x, t)$$

As a result of the measurement interaction, this would evolve into:

$$\Psi' = \sum_i a_i\varphi_i(x, t)\chi_i\left(\left\{X_j\right\}, t\right)$$

with each of the $\chi_i\left(\left\{X_j\right\}, t\right)$ corresponding to the device registering the measurement result i. Here we are explicitly noting that the measuring device contains a very large number N of particles with their specific coordinates $\left\{X_j\right\}$. As before, for this to be an effective measurement, the various $\chi_i\left(\left\{X_j\right\}, t\right)$ must be orthogonal. Bohm and Hiley refer to each of the $\chi_i\left(\left\{X_j\right\}, t\right)$ as a *channel.* During the measurement process, the quantum potential energy develops a structure of 'bifurcation points' so that the particles in the apparatus enter one of these channels and thereafter remain there. From that stage on, the measured particle behaves as if its wave function were $\varphi_i(x, t)$, even if the various φ_i have an overlap. The orthogonality of the apparatus states ensures that only the appropriate channel continues to influence Q. The information in the other channels has become inactive. For any of this information to become active again, either the particles in the apparatus would have to spontaneously re-distribute themselves and enter a different χ_i, which would be fantastically unlikely, or human intervention would be needed to shepherd the particles into place. This would be equivalent to developing some system for getting 'a kettle of water placed on ice to boil'.

So it follows that everything will from this point on take place as if the wave function had "collapsed" to the actual result, without the need for any such collapse ever to occur.[22]

As you will have probably noticed, Bohm and Hiley's channels correspond to Everett's evolving branches. However, the ontological interpretations are strikingly different. In the Many Worlds view, the differing branches are simultaneously and equally real, whereas Bohm's inactive channels remain as unmanifested potentialities, with steadily declining likelihoods as more and more irreversible process (like measurement) pile up along the manifest branch. Bohm and Hiley make the comparison to a human decision process, where different possibilities are weighed up before action is taken:

> Before the decision is made, each of these possibilities constitutes a kind of information. This may be displayed virtually in imagination as the sort of activities that would follow if we decided on one of these possibilities. Immediately after we make such a decision, there is still the possibility of altering it. However, as we engage in more and more activities that are consequent on this decision, we will find it harder and harder to change it. For we are increasingly caught up in its irreversible consequences and sooner or later we would have to say that the decision can no longer be altered. Until that moment, the information in the other possibilities was still potentially active, but from that point on such information is permanently inactive.[23]

Another striking aspect of this approach is its general similarity to the notion of *potentia* espoused by Heisenberg. Indeed, Hiley has pointed out this comparison and suggested that the quantum potential energy is a physical manifestation of *potentia.* As a hint to their deeper philosophical approach, which would, regrettably, take far more pages to explore than we can afford, we have the authors' comment:

> …the information content in the wave function is quite generally in a non-manifest (implicate) order (generally multi-dimensional), while the particles are in the ordinary (explicate) order of space time.[24]

One of the biggest challenges faced by those who would like to view amplitudes and wave functions as ontologically real is their multi-dimensionality.[25] A two-particle amplitude, whether entangled or not, requires six spatial coordinates to specify and so cannot occupy real space. No such dimensional restrictions apply to the realm of the *implicate order* – a layer of reality below manifest space time.

31.6 DEEPER WATERS

> As I have said above and will repeat again, quantum phenomena require us to think in a radical new way, a way in which we will have to ultimately give up both the notion of particles and fields. Nevertheless, as we explained in our book, Bohm and Hiley (1993),[26] by adopting the simplifying assumption of a particle with a well-defined position and momentum, we can obtain a consistent interpretation of the quantum formalism that does not contain many of the perplexing paradoxes that one faces daily in the standard approach.[27]

If you are happy to go along with the unconventional quantum potential energy, then the notion of an ontologically real and classically proximate particle being guided by quantum rails is an attractive one. However, the view cannot be sustained. In the history of physics, the quantum revolution started with Planck, Einstein, and Bohr – the architects of the so-called 'old quantum theory'. In hindsight, this looks like an attempt to shore up classical physics by introducing a few quantum ideas. Not that this is a criticism. You can only do the work that is possible at the time. With the developments of Schrödinger, Heisenberg, Jordan, Born, Dirac etc. the 'new old quantum theory' matured[28] and placed physics on a purely quantum footing. This is the arena for Bohm's ideas as outlined so far. The quantum potential energy, as we have discussed it, emerges from an analysis of the single-particle Schrödinger equation.

We have seen the form of Q for a general entangled state, but even here (and as will be developed more fully in Chapter 32), we are really pasting together single-particle states rather than developing a fully multiparticle theory. The quantum potential energy can be refined to take account of spin (Section 31.4.3), but a full understanding of the quantum nature of spin only arose when Dirac produced his relativistic equation (Section 20.1). If the ontological interpretation is to be sustainable, it must extend into the relativistic regime and ultimately into quantum field theory, which as we will see in the next chapter, really underlies the whole of quantum reality.

Of course, Bohm and his co-workers were aware of this. On the nature of the particle model being an approximation, Hiley comments:

> I want to make it absolutely clear here that I am not concluding from these arguments that the quantum particle actually is a 'small lump of substance' changing mechanically as the process develops in time. Something much subtler is involved. Nevertheless as I have remarked above, we can think of this process as being represented as a particle to within a certain approximation without arriving at a contradiction. When we go deeper and use field theory, a more complex process is revealed and clearly the simple particle picture must be modified in some radically new way.[29]

Work has been done on extending the quantum potential energy formalism into the field case, albeit with more success for bosonic fields rather than fermion fields.[30] The development re-enforces the idea of the particle being an excitation of the underlying field and the quantum potential energy arising out the field as part of a self-organizing process.

While an approach for fermions is possible, the anti-symmetry inherent to their nature presents challenges requiring some ad-hoc adaptations. Nevertheless, Bohm Hiley and Kaloyerou somewhat reasonably point out:

> … the quantization of the gravitation field implies that space and time would lose their customary meaning… in black holes and near the "big bang" with which the universe is said to have originated. Moreover, near singularities of this type, there could be no atoms or other particles or any other structure of the kinds that we know. Therefore, there could be no measuring instruments and this means that

> the usual interpretation will have nothing to talk about in these domains. Even the causal interpretation could find no foothold if there is no clear significance for space-time. For these reasons we feel that a serious interpretation of the relativistic domain would best be deferred until we obtain a more consistent theory. For the present, we should regard the relativistic quantum theory mainly as a set of algorithms from which we can derive a large number of useful and interesting results.[31]

31.7 REACTIONS TO BOHM'S THEORY

> But in 1952 I saw the impossible done. It was in papers by David Bohm. Bohm showed explicitly how parameters could indeed be introduced, into non-relativistic wave mechanics, with the help of which the indeterministic description could be transformed into a deterministic one. More importantly, in my opinion, the subjectivity of the orthodox version, the necessary reference to the 'observer,' could be eliminated.
>
> Moreover, the essential idea was one that had been advanced already by de Broglie in 1927, in his "pilot wave" picture.
>
> But why then had Born not told me of this 'pilot wave'? If only to point out what was wrong with it? Why did von Neumann not consider it? More extraordinarily, why did people go on producing 'impossibility' proofs, after 1952, and as recently as 1978? Why is the pilot wave picture ignored in text books? Should it not be taught, not as the only way, but as an antidote to the prevailing complacency? To show us that vagueness, subjectivity, and indeterminism, are not forced on us by experimental facts, but by deliberate theoretical choice?
>
> *J S Bell*[32]

> The quantum potential suggests a radical change in our conceptual outlook and provides two new interesting possibilities that could have a direct bearing on the subsequent development of the theory. Firstly it provides a fresh perspective on the microworld by giving clear intuitive representations of physical processes without the need for the ambiguous relation between the individual and the wave function…
>
> Secondly the quantum potential offers a clearer insight into the quantum interconnectedness or "quantum wholeness" that Bohr saw as the essential new feature of quantum phenomena.
>
> *B J Hiley*[33]

The general reaction from the physics community to Bohm's approach has not been positive, although it is sometimes difficult to see where the fundamental problem is.

Often the criticism centres around the apparent contrivance of the theory – the quantum potential energy is a device introduced to guarantee that Bohm's theory gives the same result as quantum mechanics, without any justification or derivation behind it. Heisenberg felt that the approach required[34] "some strange quantum potentials introduced *ad hoc* by Bohm".

It strikes me that this criticism is somewhat unfair. After all, the quantum potential appears in Eq. (31.1):

$$\frac{1}{2m}\left(\frac{dS}{dx}\right)^2 - \frac{\hbar^2}{2mR}\frac{d^2R}{dx^2} + V = -\frac{dS}{dt}$$

and if we take $\hbar \to 0$, this becomes:

$$\frac{1}{2m}\left(\frac{dS}{dx}\right)^2 + V = -\frac{dS}{dt}$$

which is known as the Hamilton-Jacobi equation and is widely used in classical physics. Indeed, the classical quantity S appearing in this equation is known as the *action*, and $p = \frac{dS}{dx}$, $E = -\frac{dS}{dt}$

are standard classical formulations. In which case, Hiley argues quite reasonably, it seems natural to suppose that $Q = -\frac{\hbar^2}{2mR}\frac{d^2R}{dx^2}$ is a form of quantum energy that is not apparent at the classical scale. What tips the balance between taking the view that this is a contrivance rather than a brilliant insight is the extent to which you are willing to adapt to the non-classical properties of this energy (e.g., that it is not related to a mechanical force).

However, science is based on gut feelings and insights as much as rational argument, and the majority opinion is that Bohm's approach just doesn't smell right.[35]

That reaction makes it very tough for Bohm's supporters to argue against their critics. Despite that, there have been some prominent physicists who have signed up for the general approach. Basil Hiley continues to work in the field and John Bell was a vocal supporter in his day.

At the very least, Bohm's theory serves to illuminate some of the more interesting aspects of quantum theory (its contextuality and nonlocal nature) and demonstrates what is at stake from an ontological point of view. It also rather nicely draws a flexible boundary between classical and quantum behaviours. Systems will behave classically if the quantum potential energy is small compared with the classical energies at work. Generally, that happens for large-scale objects.

Arguably, Bohm's approach clarifies a lot of Bohr's philosophical thinking.

NOTES

1. D. Bohm, A suggested interpretation of the quantum theory in terms of "Hidden" variables I, *Physical Review*, 1952, **85**: 166.
2. D. Bohm, *Quantum Theory*, Prentice Hall, 1951, ISBN-10: 0137478739.
3. J. Krishnamurti, *The First and Last Freedom*, Victor Gollancz; REPRINT edition, 1958.
4. J. Krishnamurti (1895–1986), Philosopher and writer.
5. Available at http://www.kinfonet.org/Biography/bohm_intro.htm.
6. D. Bohm, *Wholeness and the Implicate Order*, Routledge & Kegan Paul, London, 1980.
7. B.J. Hiley, 1935 - , emeritus professor of physics, Birkbeck College, London.
8. A term that Bohm himself wouldn't have liked as he didn't regard his approach as *mechanical* in the traditional sense, indeed he wasn't fond of the term *quantum mechanics* either.
9. B.J. Hiley, From the Heisenberg Picture to Bohm: A New Perspective on Active Information and Its Relation to Shannon Information, *Proceedings of the Conference on Quantum Theory: Reconsideration of Foundations*, A. Khrennikov (ed.), 141–162, Växjö University Press, Sweden, 2002.
10. We are working in one spatial dimension here to make the calculations a little more straightforward.
11. See note 4.
12. B.J. Hiley, *Active Information and Teleportation, Epistemological and Experimental Perspectives on Quantum Physics*, D. Greenberger et al. (eds.), 113–126, Kluwer, Netherlands, 1999.
13. C. Philippidis et al., Quantum interference and the quantum potential, *Nuovo Cimento*, 1979, **52B**: 15–28.
14. B.J. Hiley, *Some Personal Reflections on Quantum Non-locality and the Contributions of John Bell*, https://arxiv.org/pdf/1412.0594.pdf.
15. As per 13.
16. As per 15.
17. See note 20.
18. Provided we do not carry out a "rapid unscheduled dismantling" of the wall, i.e., blow it up.....
19. N. Bohr, *Phys. Rev.*, 1935, **48**: 696–702.
20. See note 15.
21. See note 12.
22. D. Bohm, B.J. Hiley, P.N. Kaloyerou, An ontological basis for the quantum theory, *Physics Reports (Review Section of Physics Letters)*, 1987, **144**(6): 321–348, North Holland, Amsterdam.
23. See note 22.
24. See note 22.
25. See note 22.
26. As well as their complex-valued nature...

27 D. Bohm, B.J. Hiley, *The Undivided Universe: An Ontological Interpretation of Quantum Theory*, Routledge, London, 1993.
28 See note 12.
29 Or old new quantum theory, depending on your view as to the commuting properties of 'new' and 'old'.
30 See note 12.
31 See, for example, D. Bohm, B.J. Hiley, P.N. Kaloyerou, A causal interpretation of quantum fields, *Physics Reports (Review Section of Physics Letters)*, 1987, **144**(6): 349–375, North Holland, Amsterdam.
32 See note 31.
33 From *On the Impossible Pilot Wave*, in John S. Bell on the Foundations of Quantum Mechanics, World Scientific, 2001.
34 See note 15.
35 W. Heisenberg, *Physics and Philosophy: The Revolution in Modern Science*, George Allen and Unwin, London, 1959.
36 "The start need not be rational, as long as the end is": many scientific discoveries have come about through inspired (skilled?) guesswork or insight into the right answer, followed up after the fact by constructed rational argument.

32 Quantum Field Theory

32.1 WHY ARE WE DOING THIS?

Quantum field theory is the ultimate expression of quantum mechanics. Its formalism provides the framework for all modern fundamental theories about the nature of matter. However, it involves difficult and intricate mathematics, which is why physics students don't generally come across it until postgraduate level, and then only if their subject specialism requires the insights that it provides.

Compared to the wealth of detailed analysis that's been applied to the 'meaning' of regular quantum theory, there has been rather less work done on the philosophy of the quantum field. There may be many several reasons for this. Some feel that quantum field theory adds little, philosophically, to what we can glean from standard quantum mechanics. There is also the distinct possibility that the advanced mathematics involved has put people off tackling this subject from a philosophical perspective.

It doesn't have to be that way, though. There is an approach to quantum field theory that can get us some distance into the formalism without running into any tricky mathematics, while at the same time illuminating the basic conceptual insights. This is the path shown in Merzbacher's classic text on quantum theory[1], and subsequently developed by Princeton philosopher Paul Teller[2].

32.2 TAKING IDENTICAL PARTICLES SERIOUSLY

In ordinary quantum theory, observable properties are represented by Hermitian operators, with eigenstates that form a basis for the expansion of a system's states. So, let's imagine that a particle can be described by the eigenstates $\{|i\rangle\}$ of some observable operator $\hat{O}$, corresponding to one of the particle's physical properties.

Now let's add another particle into the mix. Combining two particles may very well alter the *values* of certain properties (e.g., position), but we'll assume that the pairing does not change their *ability* to have those properties. The same basis set will continue to describe each particle perfectly adequately.

We need to construct a basis for the combined system of both particles. The obvious staring point is to collect products of single-particle states such as:

$$|1\rangle_A |1\rangle_B, |1\rangle_A |2\rangle_B, |2\rangle_A |2\rangle_B, \dots$$

where I am using A and B to refer to the two particles.

From this point of view, the states $|1\rangle_A |2\rangle_B$ and $|2\rangle_A |1\rangle_B$ are different, as in the first case particle A is in state 1 and in the second case particle A is in state 2, and vice versa for B.

From our discussion in Chapter 8, we know that such distinctions can't be maintained when we're faced with quantum identical particles (e.g., two photons or two electrons). If we're dealing with a pair of bosons, then only states that are *symmetrical* in the particle's labels are allowed; states such as:

$$|1\rangle_A |1\rangle_B \qquad |2\rangle_A |2\rangle_B \qquad \frac{1}{\sqrt{2}}\left(|1\rangle_A |2\rangle_B + |2\rangle_A |1\rangle_B\right)$$

However, with two identical fermions combined states must be *antisymmetric:*

$$\frac{1}{\sqrt{2}}\left(|1\rangle_A |2\rangle_B - |2\rangle_A |2\rangle_B\right)$$

All we've done here is take 'single-particle quantum theory' and paste bits of it together to describe multiparticle systems. It's a process that works, but it has some significant shortcomings:

1. There are many ways of combining the single-particle basis states, some of which are neither symmetric nor antisymmetric. For example, with the state $|1\rangle_A |2\rangle_B$, switching

DOI: 10.1201/9781003225997-36

the particles produces $|1\rangle_B|2\rangle_A = |2\rangle_A|1\rangle_B$: a completely different state. Experimentally, such states don't exist for fermions or bosons. It is surely reasonable to expect that a properly functioning theory would prevent us from forming such states in the first place.

It's often the case that physical theories produce solutions that must be eliminated as being 'not physical.' For example, Dirac struggled with his negative energy states. However, this situation is rather different. Something as fundamental as the distinction between bosons and fermions ought to be more deeply engrained in the theory. What we have done, gluing single-particle quantum states together to form multiparticle descriptions, generates *an excess of structure.* We are building general combinations and then crossing off the ones that nature doesn't allow us to use.

Put like that, it doesn't sound like the best way of concocting a theory.

2. Both symmetric and antisymmetric states get rather complicated when more than two particles are involved. An antisymmetric combination of three particles with three possible states looks like this:

$$|1\rangle_A|2\rangle_B|3\rangle_C + |2\rangle_A|3\rangle_B|1\rangle_C + |3\rangle_A|1\rangle_B|2\rangle_C - |3\rangle_A|2\rangle_B|1\rangle_C - |2\rangle_A|1\rangle_B|3\rangle_C - |1\rangle_A|3\rangle_B|2\rangle_C$$

and it only gets worse with more particles and more states to choose from. Of course, we shouldn't be afraid of complication when it is necessary, but it's always worth trying to simplify. Perhaps being forced to use such fiddly combinations of states is another signal that our approach is fundamentally flawed.

3. There is a more serious issue as well. When we paste together single-particle states and cross off the ones that don't work, *we can't really claim that we are taking the indistinguishability of identical particles seriously.* After all, we're setting up these combinations of states by giving the particles *labels* such as *A*, *B*, and so on, *as if they could be separated or sorted.* According to what we already know about identical particles, this whole approach should be called into question.

Addressing this third point is the central concern of this chapter.

32.2.1 Particle Labels

The idea of giving a particle a name or a label seems so basic and natural that it's hard to see what harm it can do. However, on several occasions, quantum reality has shown how simple-sounding assumptions can lead us astray. Perhaps we need to think a little more carefully about what we're doing when we say: "that is particle *A*".

There are two possibilities. Either we mean something as straightforward as "look there is a particle in state 1, let's call it *A*", or we have some way of recognizing particle *A irrespective of what state it happens to be in.*

The first option boils down to naming the *state* rather than the *particle*, which isn't much use, as we'll see in a moment.

The second option doesn't present a problem in classical physics, as you can always find some property of a particle that can be used as an identifier, irrespective of the state it's in. In our discussion of classical physics in Chapter 3, we made a distinction between *system-* and *state properties.* In quantum physics, system properties allow us to tell the difference between an electron (negatively charged, mass of 9.11×10^{-31} kg, etc.) and a photon (no charge, no mass etc.), but they don't allow us to distinguish one electron, for example, from another. All of them have exactly the same mass and charge. In macroscopic physics, we can rely on small differences in the system properties to differentiate one example from another. No two snooker balls have precisely the same size, mass, and colour, but the differences are small enough that we can still tell that they're all snooker balls. We don't have the same luxury in the quantum world.

Quantum identical particles behave in measurably different ways to particles that can be distinguished. If we walk up to a system of two particles with one in state $|1\rangle$ and the other in state $|2\rangle$, then we can call the first one A and the second B, if we like. When we come back to look again sometime later, we may still find that there is one in state $|1\rangle$ and the other in state $|2\rangle$, *but we can't be sure that* $|1\rangle$, *is still* A. We can still call it A, but there's no way to know that it's the same particle as the one we labelled as A earlier.

No system property can be used to identify a particle independent of its state. Noting that there is a particle in state $|1\rangle$ is all there is to say on the matter.

Unfortunately, there's no state property, such as position, that we can use to keep track of which one's which either. We have discussed several experiments that demonstrate how particles don't have traceable tracks or paths between one measurement and the next.

So, in conclusion, there is *no plausible basis for labelling a particle in a manner that 'sticks' to it permanently.*

32.2.2 Substance Abuse

Now, from a philosophical point of view, this is rather big stuff. Our whole manner of speech (X has the property Y) naturally guides us to imagine some *fundamental substance* on which properties can, in a sense, be attached. It encourages us to visualize taking a particle and removing its properties one by one until we're left with a featureless 'thing' devoid of properties.

Philosophers have been debating the correctness of such arguments, but now it seems that experimental science has come along and shown that, at the quantum level at least, the objects we study *have no substance to them, independent of their properties.*

Let's say that we have two particles with substance, but with different properties. Now imagine removing the properties of both particles, to leave the featureless substance, and then replacing the properties again, *but the other way round* (giving A the properties that B used to have and vice versa). If the two particles are identical bosons, we know that the state we end up with must be the same as the one we started with (symmetrical under exchange). In which case, what's the point of the labels A and B? As far as the physics of the situation is concerned, we've made no difference, so is the substance real?

The argument is even more striking for two identical fermions, as in their case nature clearly forbids them to be in the same state. But, if we're stripping the properties from them one by one, it's entirely possible that at some point *their remaining properties might add up to them being in the same state* (all the differences having been removed). What stops us from doing that?

Equally, take a fully featured electron alongside a lump of featureless substance. We start gluing properties to the lump, but we're not allowed to make this lump into a new electron in the same state as the one we already have. Does the lump vanish if we manage to do this by mistake? What suddenly prevents us from applying the final property which completes the transformation of the lump into an electron in the same state?

This isn't a purely philosophical debate, as it impinges directly on physics. If there is no featureless substance, then how can we even *in principle* apply labels to particles; and if we can't do that, then *why set up our formalism as if we can*?

I take these arguments to be conclusive, or at the very least to strongly motivate us in looking for a way of doing quantum theory that does without particle labels in any form.

Labels cause trouble.

32.3 STATES IN QUANTUM FIELD THEORY

We start again with the idea that a particle has a set of observable properties represented by operators with bases $\{|i\rangle\}$. Let's list these single-particle states in some order, $(|1\rangle, |2\rangle, |3\rangle, \ldots, |k\rangle, \ldots)$. Now, if we have more than one particle, and we assume that the same set of states is available to both, perhaps we can include all the information we need by saying "there is one particle in state $|1\rangle$ and one in state $|3\rangle$," and go no further than that.

We could represent this situation by writing (1, 0, 1,..., 0, ...), to show how many particles there are in each of the states in our list. We can introduce a new notation in which the ket $|1, 0, 1, \ldots, 0, \ldots\rangle$, is the state with one particle in state $|1\rangle$ and the other in state $|3\rangle$. Equally, a state with 11 particles in state $|k\rangle$ would be $|0, 0, 0, \ldots, 11, \ldots\rangle$, and we see how the idea develops.

Note, and this is very important, that the state $|0, 0, 0, \ldots, 11, \ldots\rangle$ is *not* the same as $11 \times |0, 0, 0, \ldots, 1, \ldots\rangle$. The former has 11 particles in state $|k\rangle$ and the latter looks like some amplitude to be in state $|0, 0, 0, \ldots, 1, \ldots\rangle$.

This formalism symbolically expresses how many particles there are in each of the possible single-particle states, and by inference how many particles there are in total, but not which particle is which. We have resisted the temptation to label them in any way.

There is a rather nice analogy for this. Imagine going into a bank and handing over a pile of coins to be paid into your account. Once the credit has entered your account electronically, *the individual existence of these pound coins has vanished.* We can always find out how much money there is in the account, but we don't do this by counting coins. The total can be *aggregated,* but the individual lumps of currency in your account can't be given labels to help them be counted ("there's one, there's another—keep still while I count you").

Here is another way of thinking about this.

Imagine that you were given a sack containing some marbles and told to figure out how many marbles were inside without opening the sack. Short of trying to feel the marbles through the material, the best way of doing this would be to weigh the sack. Provided you had an idea of how much a single marble weighed, you could estimate the *aggregate total* of marbles present, but you wouldn't have *counted* them (in the normal sense of the word). One might say that the sack had certain properties that depended on the aggregate total of marbles, but not on the ability to point to individuals within the sack. The same would be true in a bank account, where the bottom line is the aggregate number of units of currency in the account. In short, if we can't label particles then we should not try and 'count' them, if we use the word 'count' to indicate a process whereby individuals are isolated, identified and totalled.

When we write a ket such as $|2, 3, 1, \ldots, 7, \ldots\rangle$ we are listing the aggregate total of objects in each state.

32.3.1 Fock States

As it's going to be important to distinguish between classical particles and quantum objects that can be aggregated but not counted, I propose to follow a convention that is relatively common in quantum field theory and refrain from talking about *particles* and call them *quanta* instead.

Quanta are objects that can be aggregated, but not counted. By extension, quanta are objects that have properties, but not substance.

A symbol such as $|1, 0, 1, \ldots, 0, \ldots\rangle$ is a *Fock ket,* after the Russian physicist Vladimir Aleksandrovich Fock, [3] who was instrumental in developing this way of representing quanta. The states represented by such kets are *Fock states,* which are a different sort of animal to the standard quantum states that we've been dealing with up to now. Although I claimed earlier that $|1, 0, 1, \ldots, 0, \ldots\rangle$ was *the same* as $|1\rangle$, strictly speaking, all we can say is that the two representations contain the same quantum information.[4] We must avoid thinking that $|1, 0, 1, \ldots, 0, \ldots\rangle = \frac{1}{\sqrt{2}}\left(|1\rangle|3\rangle + |3\rangle|1\rangle\right)$ (for the boson case). The state $\frac{1}{\sqrt{2}}\left(|1\rangle|3\rangle + |3\rangle|1\rangle\right)$ is built by pasting together single-particle states, where the particles have been given labels, and then worrying about the overall state having to be symmetric. In other words, *it is talking about countable objects, not quanta.*

What we're hoping is that $|1, 0, 1, \ldots, 0, \ldots\rangle$ will play the same *role* in the theory as $\frac{1}{\sqrt{2}}\left(|1\rangle|3\rangle + |3\rangle|1\rangle\right)$, but without the baggage that comes with labels.

However, the Fock ket formalism gives us something extra as well. The set of all possible Fock kets for given observable *forms a basis for describing the states of a multiquanta system*, which must also include states such as

$$|\Psi\rangle = A|1,\ 0,\ 1{,}1,\ 3,\ \ldots,\ n,\ \ldots\rangle + B|1,\ 1,\ 2,\ 1{,}4,\ \ldots,\ n,\ \ldots\rangle + \cdots$$

In a state like this, we can't be sure exactly how many quanta there are in total. Such states are very important, from both physical and interpretative points of view, as we will see later.

32.3.2 The Vacuum

As each Fock ket is supposed to represent a specific number of quanta in each state, there must be a unique ket is called the *vacuum*, $|0\rangle$, with no quanta in any state:

$$|0\rangle = |0,\ 0,\ 0,\ 0,\ 0, \ldots,\ 0, \ldots\rangle$$

32.3.3 Up and Down We Go...

To develop the formalism further, we define *raising and lowering operators*[5] $\hat{b}_k^+$ and $\hat{b}_k^-$ which can be applied to each state $|k\rangle$:

$$\hat{b}_k^+|\ldots,\ n_k,\ \ldots\rangle = \alpha_k(n_k)|\ldots,\ n_k+1,\ \ldots\rangle$$

$$\hat{b}_k^-|\ldots,\ n_k,\ \ldots\rangle = \beta_k(n_k)|\ldots,\ n_k-1, \ldots\rangle.$$

In these expressions, I have only shown the number of quanta in the state concerned and relegated any other quanta present to three dots.

The raising operator, $\hat{b}_k^+$, increases the number of quanta in state $|k\rangle$ by one. I have put a constant $\alpha_k(n_k)$ in front, acknowledging that this factor may depend on the number of quanta. We will need to figure out what that should be later.

Conversely, the lowering operator, $\hat{b}_k^-$, removes a single quantum from state $|k\rangle$ every time that it is applied (with a constant $\beta_k(n_k)$), the one exception being $\hat{b}_k^-|\ldots,\ 0,\ \ldots\rangle = 0$.

Any Fock ket can be constructed by applying raising operators to the vacuum, for example:

$$|2,\ 1,\ 3,\ \ldots\rangle = \left(\hat{b}_1^+\hat{b}_1^+\right)\left(\hat{b}_2^+\right)\left(\hat{b}_3^+\hat{b}_3^+\hat{b}_3^+\right)\ldots|0\rangle$$

(Note that I have not worried about normalizing the state in this construction.)

32.3.4 Change of Basis

In single-particle quantum theory, we have the option of switching from one basis, $\{|i\rangle\}$, to another, $\{|I\rangle\}$, using:

$$|K\rangle = \sum_i \langle i|K|i\rangle\rangle \qquad \langle L| = \sum_j \langle L|j\rangle\langle j| \tag{32.1}$$

$$|i\rangle = \sum_K \langle K|i\rangle|K\rangle \qquad \langle j| = \sum_L \langle j|L\rangle\langle L| \tag{32.2}$$

In the Fock ket formalism, we would write Eq. 32.1 as:

$$\hat{B}_K^+|0\rangle = \sum_i \langle i|K\rangle \hat{b}_i^+|0\rangle = \left(\sum_i \langle i|K\rangle \hat{b}_i^+\right)|0\rangle$$

$$\langle 0|\hat{B}_L^+ = \sum_j \langle j|L\rangle\langle 0|\hat{b}_j^+ = \langle 0|\left(\sum_j \langle j|L\rangle \hat{b}_j^+\right)$$

which suggests that the transformations applied directly to the operators are:

$$\hat{B}_K^+ = \sum_i \langle i|K\rangle \hat{b}_i^+ \qquad \hat{B}_L^- = \sum_j \langle L|j\rangle \hat{b}_j^- \tag{32.3}$$

$$\hat{b}_i^+ = \sum_K \langle K|i\rangle \hat{B}_K^+ \qquad \hat{b}_j^- = \sum_L \langle j|L\rangle \hat{B}_L^- \tag{32.4}$$

A consequential result that will prove useful later is:

$$\sum_i \hat{b}_i^+\hat{b}_i^- = \sum_i \left(\sum_K \langle K|i\rangle \hat{B}_K^+\right)\left(\sum_L \langle i|L\rangle \hat{B}_L^-\right)$$

$$= \sum_K\sum_L \left(\sum_i \langle K|i\rangle\langle i|L\rangle\right)\hat{B}_K^+\hat{B}_L^- = \sum_K\sum_L \langle K|L\rangle \hat{B}_K^+\hat{B}_L^- = \sum_K \hat{B}_K^+\hat{B}_K^-$$

Also note that $\sum_i \hat{b}_i^-\hat{b}_i^+ = \sum_K \hat{B}_K^-\hat{B}_K^+$.

32.3.5 Orderly Matters

Applying raising operators $\hat{b}_i^+$ and $\hat{b}_j^+$ in any sequence must produce the same *physical state*, but as we know from our previous dealings, this does not mean the same *mathematical state*. It's all about phase. In which case, we ought to write:

$$\hat{b}_i^+\hat{b}_j^+|0\rangle = \lambda\hat{b}_j^+\hat{b}_i^+|0\rangle \qquad \left(\hat{b}_i^+\hat{b}_j^+ - \lambda\hat{b}_j^+\hat{b}_i^+\right)|0\rangle = 0$$

If we now use Eq. 32.4 to rotate this expression into a different basis:

$$\left(\left\{\sum_K \langle K|i\rangle \hat{B}_K^+\right\}\left\{\sum_M \langle M|j\rangle \hat{B}_M^+\right\} - \lambda\left\{\sum_M \langle M|j\rangle \hat{B}_M^+\right\}\left\{\sum_K \langle K|i\rangle \hat{B}_K^+\right\}\right)|0\rangle = 0$$

$$\left(\sum_M\sum_K \langle K|i\rangle\langle M|j\rangle \hat{B}_K^+\hat{B}_M^+ - \lambda\langle M|j\rangle\langle K|i\rangle \hat{B}_M^+\hat{B}_K^+\right)|0\rangle = 0$$

$$\left(\sum_M\sum_K \langle K|i\rangle\langle M|j\rangle\left(\hat{B}_K^+\hat{B}_M^+ - \lambda\hat{B}_M^+\hat{B}_K^+\right)\right)|0\rangle = 0$$

the vacuum being invariant under such rotations.

The only constraint on the complex numbers $\langle K|i\rangle$ and $\langle M|j\rangle$ is that $\sum_M \langle j|M\rangle\langle M|i\rangle = \delta_{ij}$, so if this relationship is to work for any values of K and M, the portion bounded by the inner brackets must be zero, $\hat{B}_K^+\hat{B}_M^+ - \lambda\hat{B}_M^+\hat{B}_K^+ = 0$. As our initial order was arbitrary, $\hat{B}_M^+\hat{B}_K^+ - \lambda\hat{B}_K^+\hat{B}_M^+ = 0$. If we insert one inside the other:

$$\hat{B}_K^+\hat{B}_M^+ - \lambda\left(\lambda\hat{B}_K^+\hat{B}_M^+\right) = \hat{B}_K^+\hat{B}_M^+\left(1-\lambda^2\right) = 0$$

Clearly $\lambda = \pm 1$, resulting in two possibilities:

$$\hat{b}_i^+\hat{b}_j^+ - \hat{b}_j^+\hat{b}_i^+ = 0 \qquad\qquad \hat{b}_i^+\hat{b}_j^+ + \hat{b}_j^+\hat{b}_i^+ = 0$$

Remembering the definition of the commutator, $[A, B] = AB - BA$, and introducing by analogy the *anticommutator*, $\{A, B\} = AB + BA$, we have:

$$\left[\hat{b}_i^+, \hat{b}_j^+\right] = 0 \qquad\qquad \left\{\hat{f}_i^+, \hat{f}_j^+\right\} = 0$$

where I now distinguish between operators $\hat{b}$ and $\hat{f}$ depending on their commutation/anticommutation. So, if our raising and lowering operators are to generate the same physical state no matter what order they are applied, then these operators must fall into one of two distinct classes: those that commute or those that anticommute.

32.3.6 Fermions and Bosons

The whole point of the Fock ket formalism has been to replace the notion of 'particle,' with all that it implies as far as substance and particle labels, with 'quanta.' Consequently, while exploring the essential differences between fermions and bosons and the antisymmetry/symmetry of their states, we can't transfer the old idea of switching two particles to switching two quanta, as quanta are not distinguishable. However, we can think of this in terms of building up a ket from the vacuum. As we have seen, developing this point has brought us to two possibilities. Taking the commuting operators first:

$$|\ldots, 1, \ldots, 1, \ldots\rangle = \left(\hat{b}_k^+\hat{b}_j^+\right)|0\rangle = \left(\hat{b}_j^+\hat{b}_k^+\right)|0\rangle$$

Evidently the order in which we apply the operators is symmetric under exchange. For the anticommuting operators:

$$|\ldots, 1, \ldots, 1, \ldots\rangle = \left(\hat{f}_k^+\hat{f}_j^+\right)|0\rangle = -\left(\hat{f}_j^+\hat{f}_k^+\right)|0\rangle$$

i.e., they are anti-symmetric under exchange of the order of application. This strongly suggests that our $\hat{b}$ operators apply to *bosons* and the $\hat{f}$ operators to *fermions*. To confirm this, consider $j = k$, in which case $\left(\hat{f}_k^+\hat{f}_k^+\right)|0\rangle = -\left(\hat{f}_k^+\hat{f}_k^+\right)|0\rangle$, which can only happen if $\hat{f}_k^+\hat{f}_k^+ = 0$. From this, it follows that:

$$\hat{f}_k^+|\ldots, 1, \ldots\rangle = \hat{f}_k^+\left(\hat{f}_k^+|\ldots 0 \ldots\rangle\right) = \hat{f}_k^+\hat{f}_k^+|\ldots 0 \ldots\rangle = 0$$

which is clearly an expression of the Pauli exclusion principle, which we know is characteristic of fermions.

It's now possible to tabulate a variety of different relationships between operators:

Boson Operators	Fermion Operators
$\left[\hat{b}_j^+,\hat{b}_k^+\right]=0$	$\left\{\hat{f}_j^+,\hat{f}_k^+\right\}=0$
$\left[\hat{b}_j^-,\hat{b}_k^-\right]=0$	$\left\{\hat{f}_j^-,\hat{f}_k^-\right\}=0$
$\left[\hat{b}_j^-,\hat{b}_k^+\right]=\delta_{jk}\hat{I}$	$\left\{\hat{f}_j^-,\hat{f}_k^+\right\}=\delta_{jk}\hat{I}$

but I won't put you through the tedium of proving each one.

32.3.7 The Number Is Up

The next step is to define a number operator, $\hat{N}_i$, which returns the aggregate of quanta in state $|i\rangle$, $\hat{N}_i|\ldots, n, \ldots\rangle = n|\ldots, n, \ldots\rangle$, with the aggregate of all quanta being $\hat{N} = \sum_i \hat{N}_i$.

If we rotate this into another basis, then we expect the total count of quanta to be the same. Hence $\hat{N}$ must be invariant under such a transformation. So, the most general form for $\hat{N}_i$ is:

$$\hat{N}_i = X\hat{b}_i^+\hat{b}_i^- + Y\hat{b}_i^-\hat{b}_i^+ + Z\hat{I}$$

as we demonstrated in Section 32.3.4 that $\sum_i \hat{b}_i^+\hat{b}_i^-$ and $\sum_i \hat{b}_i^-\hat{b}_i^+$ were invariant under basis change. The first step in determining the values (X, Y, Z) is to realize that $\hat{N}|0\rangle = 0$, so:

$$X\hat{b}_i^+\hat{b}_i^-|0\rangle + Y\hat{b}_i^-\hat{b}_i^+|0\rangle + Z\hat{I}|0\rangle = 0 + Y + Z = 0$$

allowing us to write:

$$\hat{N} = X\hat{b}_i^+\hat{b}_i^- + Y\left(\hat{b}_i^-\hat{b}_i^+ - \hat{I}\right)$$

Now, for bosons we have $\left[\hat{b}_j^-,\hat{b}_k^+\right] = \delta_{jk}$ so $\hat{b}_i^-\hat{b}_i^+ = 1 - \hat{b}_i^+\hat{b}_i^-$ and therefore:

$$\hat{N}_i^b = (X+Y)\hat{b}_i^+\hat{b}_i^-$$

For fermions, $\left\{\hat{f}_j^-,\hat{f}_k^+\right\} = \delta_{jk}\hat{I}$ and so:

$$\hat{N}_i^f = (X-Y)\hat{f}_i^+\hat{f}_i^-$$

Finally, as $\hat{N}_i^b|\ldots,1,\ldots\rangle = 1|\ldots,1,\ldots\rangle$, $X+Y=1$ for bosons, and as $\hat{N}_i^f|\ldots,1,\ldots\rangle = 1|\ldots,1,\ldots\rangle$, $(X-Y)=1$ for fermions. So:

> **NUMBER OPERATORS**
>
> $\hat{N}_i^b = \hat{b}_i^+\hat{b}_i^-$ and $\hat{N}_i^f = \hat{f}_i^+\hat{f}_i^-$.

32.3.8 Normalization

It is high time that we figured out the values of the constants in:

$$\hat{b}_k^+|\ldots, n_k, \ldots\rangle = \alpha_k(n_k)|\ldots, n_k + 1, \ldots\rangle$$

$$\hat{b}_k^-|\ldots, n_k, \ldots\rangle = \beta_k(n_k)|\ldots, n_k - 1, \ldots\rangle.$$

In section 9.3, we concluded that the amplitude for creating a boson in each state is increased by a factor $\sqrt{m+1}$ when there are already m identical bosons present. However, when a boson is absorbed out of a state with m present, the amplitude for that absorption is proportional to $\sqrt{m}$.

In our current situation, we have raising and lowering operators that do precisely this: take us from states of m bosons to $(m+1)$ or $(m-1)$. Hence, I suggest, without formal proof, that we accept the values $\alpha_k(m) = \sqrt{m+1}$ and $\beta_k(m) = \sqrt{m}$, so we have:

$$\hat{b}_k^+|\ldots, m, \ldots\rangle = \sqrt{m+1}|\ldots, m+1, \ldots\rangle$$

$$\hat{b}_k^-|\ldots, m, \ldots\rangle = \sqrt{m}|\ldots, m-1, \ldots\rangle$$

As confirmation, we apply these operators to a state:

$$\hat{b}_k^+\hat{b}_k^-|\ldots, m, \ldots\rangle = \sqrt{m}\hat{b}_k^+|\ldots, m-1, \ldots\rangle$$

$$= \sqrt{m}\sqrt{(m-1)+1}|\ldots, m, \ldots\rangle = m|\ldots, m, \ldots\rangle$$

Exactly as we would hope, given that we showed in the previous section that $\hat{N}_k = \hat{b}_k^+\hat{b}_k^-$.

We can now address the issue of normalization, which I declined to discuss earlier. Examining a boson state built up from the vacuum:

$$|2, 1, 3, \ldots, n_K\rangle = \left(\hat{b}_1^+\hat{b}_1^+\right)\left(\hat{b}_2^+\right)\left(\hat{b}_3^+\hat{b}_3^+\hat{b}_3^+\right)\ldots\left(\hat{b}_K^+\right)^{n_K}|0\rangle$$

and looking at the state $|K\rangle$, we can parse it like so:

$$\left(\hat{b}_K^+\right)^{n_K-1}\hat{b}_K^+|0\rangle = \left(\hat{b}_K^+\right)^{n_K-1}\sqrt{1}|1\rangle = \left(\hat{b}_K^+\right)^{n_K-2}\sqrt{1}\sqrt{2}|2\rangle = \left(\hat{b}_K^+\right)^{n_K-3}\sqrt{1}\sqrt{2}\sqrt{3}|3\rangle = \cdots = \sqrt{n_K!}|\ldots n_K\rangle$$

This representation can be streamlined by using the symbolism $\prod_j \hat{b}_j^+ = \hat{b}_1^+\hat{b}_2^+\hat{b}_3^+\ldots\hat{b}_j^+$. In which case a general normalized boson Fock state is:

$$|n_1, n_2, n_3, \ldots n_j \ldots, n_K\rangle = \prod_{j=1}^{K}\left(\hat{b}_j^+\right)^{n_j}\frac{1}{\sqrt{n_j!}}|0\rangle$$

In one sense, normalization is easier for fermion kets, but in another sense a bit trickier. The *magnitude* of the normalization constant is straightforward, as the occupation number, n_k, for any state

$|k\rangle$ is either 0 or 1. However, selecting the correct *sign* is more of an issue. In building our boson ket from the vacuum, the order in which we apply the individual creation operators is not significant, as they all commute with each other, $\left[\hat{b}_j^+,\hat{b}_k^+\right]=0$. We cannot be so laissez faire building up fermion kets, as their creation operators are anticommuting, $\left\{\hat{f}_j^+,\hat{f}_k^+\right\}=0$ hence $\hat{f}_j^+\hat{f}_k^+=-\hat{f}_k^+\hat{f}_j^+$.

The issue is tackled by first adopting a convention: that a fermion ket is built from the vacuum *starting with the highest values of k in* $\{|k\rangle\}$, and then *working down in sequence*, thus:

$$|1,0,1,\ldots,1\rangle=\hat{f}_1^+\hat{f}_3^+\ldots\hat{f}_K^+|0\rangle$$

where $1\le k\le K$. The subsequent action of another raising operator on the ket:

$$\hat{f}_i^+|1,0,1,\ldots,n_i,\ldots,1\rangle=\hat{f}_i^+\left\{\hat{f}_1^+\hat{f}_3^+\ldots\hat{f}_K^+|0\rangle\right\}$$

is determined by anticommuting the $\hat{f}_i^+$ through the chain, picking up a factor of -1 every time, until it gets to the right place. Hence:

$$\hat{f}_i^+|1,0,1,\ldots,n_i,\ldots,1\rangle=0 \qquad \text{if } n_i=1$$

$$\hat{f}_i^+|1,0,1,\ldots,n_i,\ldots,1\rangle=(-1)^m|1,0,1,\ldots,n_i,+1\ldots,1\rangle \qquad \text{if } n_i=0$$

where m is the number of *occupied* states, $n_k=1$, up to i in the sequence, i.e:

$$m=\sum_{k=1}^{i-1}n_k$$

A similar argument can be constructed for the lowering operators.

32.3.9 Round and Round We Go...

As $\hat{b}_k^+|\ldots,m,\ldots\rangle=\sqrt{m+1}|\ldots,m+1,\ldots\rangle$, then:

$$\langle\ldots,m+1,\ldots|\hat{b}_k^+|\ldots,m,\ldots\rangle=\sqrt{m+1}\langle\ldots,m+1,\ldots|\ldots,m+1,\ldots\rangle=\sqrt{m+1}$$

Here, $\hat{b}_k^+$ could be viewed as a raising operator acting on the ket (as before), or as a *lowering operator acting on the bra*:

$$\langle\ldots,m+1,\ldots|\hat{b}_k^+|m\rangle=\sqrt{m+1}\langle m|m\rangle=\sqrt{m+1}$$

as before. Note that:

$$\langle\ldots,m+1,\ldots|\hat{b}_k^+=\sqrt{m+1}\langle\ldots,\mathrm{m},\ldots|$$

is equivalent to:

$$\langle\ldots,n,\ldots|\hat{b}_k^+=\sqrt{n}\langle\ldots,\mathrm{n}-1,\ldots|$$

As we have already established $\hat{b}_k^-|..., m, ...\rangle = \sqrt{m}|..., m-1, ...\rangle$ it follows that:

$$\langle ..., m, ...|\left(\hat{b}_k^-\right)^\dagger = \sqrt{m}\langle ..., \text{m}-1, ...|$$

and hence $\left(\hat{b}_k^-\right)^\dagger = \hat{b}_k^+$.

A similar relationship holds for the lowering operator; it lowers when acting on a ket and raises when acting on a bra. The same is true for the fermion raising and lowering operators.

32.3.10 Multiparticle Operators Representing Observables

We started by listing a set of eigenstates, $\{|i\rangle\}$, for some observable operator, $\hat{O}$, in the form $(|1\rangle, |2\rangle, |3\rangle, ..., |k\rangle, ...)$. Next, we generalized that into a Fock ket $|1, 2, 3, ..., k, ...\rangle$. Now we need to see how we can represent the original operator in this new regime. If this operator $\hat{O}$ has a collection of eigenvalues $\{o_i\}$ associated with the eigenstates $\{|i\rangle\}$, then we can surely write:

$$\hat{O}_M = \sum_i o_i \hat{N}_i = \sum_i \langle i|O|i\rangle \hat{N}_i$$

as the multiparticle version of the operator.

Kets such as $|1, 2, 3, ..., k, ...\rangle$ are natural eigenkets of $\hat{O}_M$, representing definite numbers of quanta in the system:

$$\hat{O}_M|1, 2, 3, ..., k, ...\rangle = \sum_i o_i \hat{N}_i|1, 2, 3, ..., k, ...\rangle$$

$$= (o_1 + 2o_2 + 3o_3 + \cdots + ko_k + \cdots)|1, 2, 3, ..., k, ...\rangle$$

Combinations such as:

$$|\Psi\rangle = A|1, 0, 1,1, 3, ..., n, ...\rangle + B|1, 1, 2, 1,4, ..., n, ...\rangle +$$

are not eigenkets of the number operator for any single-particle state, and hence represent a situation where there is no definite number of quanta present. This intriguing sounding situation is something that needs careful physical interpretation, which we will come to shortly.

Any single-particle observable represented by an operator that commutes with $\hat{O}$ will have the same eigenstates, and so the multiquanta version of that operator will share the same eigenkets. In that physical situation, we have a definite number of quanta displaying properties that are eigenvalues of the two observables. In contrast, any two non-commuting operators can't share the same Fock eigenkets. So, an eigenket of one operator can only be expanded over a series of eigenkets of the other operator. This produces an interesting physical situation: a definite number of quanta as far as one observable is concerned, but an indefinite number of quanta from the point of view of the other observable.

Everything so far is based on the observable having discrete values. Of course, many physical properties are continuous; position and momentum are two examples. Clearly, we must extend our scheme to deal with continuous properties and their operators as well, but we're not quite ready to deal with such situations yet.

In the meantime, it is informative to see what happens to our operator under a change of basis.

Our first step is to play a small trick, by converting the operator using $\langle i|\hat{O}|j\rangle = \delta_{ij}o_i$, after all:

$$\sum_i\sum_j \langle i|\hat{O}|j\rangle \hat{b}_i^+\hat{b}_j^- = \sum_i\sum_j \delta_{ij}o_i\hat{b}_i^+\hat{b}_j^- = \sum_i o_i\hat{b}_i^+\hat{b}_i^- = \sum_i o_i\hat{N}_i^b = \hat{O}_M$$

Now using our transformation equations, Eq. 32.3/32.4:

$$\hat{O}_M = \sum_i\sum_j \langle i|\hat{O}|j\rangle \left(\sum_K \langle K|i\rangle \hat{B}_K^+\right)\left(\sum_L \langle j|L\rangle \hat{B}_L^-\right)$$

$$= \sum_i\sum_j\sum_K\sum_L \langle K|i\rangle\langle i|\hat{O}|j\rangle\langle j|L\rangle \hat{B}_K^+\hat{B}_L^-$$

$$= \sum_K\sum_L \langle K|\hat{O}|L\rangle \hat{B}_K^+\hat{B}_L^-$$

What we can see here is a sum of matrix elements, each one of which has the form $\langle K|\hat{O}|L\rangle \hat{B}_K^+\hat{B}_L^-$. A natural way of interpreting these expressions is to say that they remove one quantum from the L state and create another in the K state under the influence of the transition matrix element $\langle K|\hat{O}|L\rangle$.

Although this gives us a glimpse of how quantum field theory deals with particle interactions, it's dangerous to take such an interpretation too literally. There are good reasons for hesitating before thinking that $\hat{B}_K^+$ and $\hat{B}_L^-$ *factually* create and destroy quanta.

32.4 BASIS FOR PROGRESS

The previous section dealt with a change of basis from one set of eigenstates into another, but there is still something missing. As I mentioned earlier, some physical properties have a continuous set of eigenstates and eigenvalues. We must see how to deal with these.

In fact, the extension is relatively simple.[6] The formula for changing basis, Eq. 32.3:

$$\hat{B}_K^+ = \sum_i \langle i|K\rangle \hat{b}_i^+ \qquad \hat{B}_L^- = \sum_j \langle L|j\rangle \hat{b}_j^-$$

just needs to be adapted to deal with the continuous situation.

Let's start with momentum and its eigenstates. For the Fock ket representation to work, we imagine that the *continuous* range of momentum available is *quantized* into discrete values that are very close to one another. It's a bit like having a particle in a box with the momentum values, $(p_x)_m = m\pi\hbar/L$ but choosing a very big box (the universe?) so that the difference between $(p_x)_m$ and $(p_x)_{m+1}$ is very small. We then define raising and lowering operators for the momentum states, $\hat{p}_m^+$ and $\hat{p}_m^-$. These operators will be subject to the same rules as before. For example, you still can't create two fermions in the same state. A momentum basis Fock ket P would the be:

$$P = |n_1, n_2, n_3, \ldots, n_K\rangle$$

where n_1 quanta had momentum p_1, n_2 quanta had momentum p_2 etc.

That's not the end of the story, however. We could equally decide that we wanted to work in a position representation where the eigenstates corresponded to position measurements, having divided the universe up into localized regions so we can use a Fock ket again. The raising and lowering operators connected with a position basis can be found by converting the momentum operators using the standard rules:

$$\hat{\Psi}^{+}(x_i) = \sum_m \langle x_i | p_m \rangle \hat{p}_m^{+}$$

$$\hat{\Psi}^{-}(x_i) = \sum_m \langle x_i | p_m \rangle \hat{p}_m^{-}$$

The operators $\hat{\Psi}^{+}(x_i)$ and $\hat{\Psi}^{-}(x_i)$ are responsible for creating or destroying a quantum localized within a small region of position x_i, respectively. Although I have generally called similar operators raising and lowering operators in this chapter, there is a widespread tendency to refer to them as *creation* and *annihilation* operators, which I will adopt from now on. These names are not entirely helpful as they suggest that the creation operator is literally responsible for a quantum appearing at a point x out of the vacuum (or vanishing if it is an annihilation operator). From an ontological viewpoint it's doubtful whether such an interpretation can be carried through, but more of that later.

In Chapter 7 we wrote down the momentum eigenstate:

$$|\psi(x_i,t)\rangle = |p_m\rangle = A\sum_i \exp\left[\frac{i}{\hbar}(p_m x_i - Et)\right]|x_i\rangle$$

using a discrete position basis $\{|x_i\rangle\}$ and the symbolism appropriate to this chapter. From here we see that $\langle x_I | p_m \rangle = \exp\left[\frac{i}{\hbar}(p_m x_I - Et)\right]$ which makes our creation and annihilation operators:

$$\hat{\Psi}^{+}(x_i, t) = \sum_m A e^{\frac{i}{\hbar}(p_m x_i - Et)} \hat{p}_m^{+}$$

$$\hat{\Psi}^{-}(x_i, t) = \sum_m A e^{\frac{i}{\hbar}(-p_m x_i + Et)} \hat{p}_m^{-}$$

The operators $\hat{\Psi}^{+}(x_i, t)$ and $\hat{\Psi}^{-}(x_i, t)$ are defined for a specified (x_i, t), so there are (close to) an *infinite number of such operators spread over space and time.* They are sometimes referred to as *field operators.*

In classical physics the idea of a field, such as the gravitational or electric field, is based on assigning a quantity, representing a physical variable, to every space–time point in a region. The quantum field is sometimes called an *operator-valued field:* in the spatial representation an operator, such as $\hat{\Psi}^{+}(x_i, t)$, is associated with every space–time point.

To be honest, the idea of an operator-valued field is rather dubious philosophically and doesn't work well as a quantum analogy for a classical field, but it's an approach that is sometimes taken in textbooks. The route we have followed in developing quantum field theory has arrived at the creation and annihilation operators without any mention of fields on the way. So, we are left puzzling as to why it is called quantum *field* theory in the first place.

32.4.1 So Why Is It Called Quantum Field Theory?

In fact, the reason is probably mostly historical. The first steps into quantum field theory were taken when Dirac quantized the electromagnetic field by treating it as a collection of imaginary oscillators and applying the quantum rules which had been worked out for the more mechanical oscillations we find in atoms and solids. As a result, he produced operators rather like $\hat{\Psi}^{+}(x_i, t)$ and $\hat{\Psi}^{-}(x_i, t)$ which acted to create and annihilate photons (quanta). After that success, the same technique was applied to the Schrödinger wave function $\Psi(x, t)$ by treating that as a field. Then came the field theoretic version of the Dirac wavefunction, which is compatible with relativity. In each case, the trick was to imagine that the wave functions were fields spread out through space and time fed by imaginary oscillators, which became the quanta.

In the Dirac case, this turned out to be a masterstroke. At last, the negative energy solutions that Dirac had struggled with found their correct context. Once the Dirac wave function was 'second quantised' and the corresponding field operators produced, creation and annihilation operators for negative energy states appeared. However, with a neat twist, a creation operator for a negative energy Dirac quantum could be turned into *an annihilation operator for a positive energy antiquantum* (see Chapter 21). Equally, annihilation operators for negative energy states turned into creation operators for positive energy antiquanta. Knitting all this together required the use of Fock kets that had separate states for quanta and antiquanta. Hence, the multiquanta version of the single-particle Dirac state automatically included quanta and antiquanta, an amazing theoretical triumph. The two field operators are, understandably, slightly more intricate:

$$\hat{\Phi}_{\text{Dirac}} \approx \sum_m U e^{\frac{i}{\hbar}(p_m x_i - Et)} \hat{f}_m^- + V e^{\frac{i}{\hbar}(-p_m x_i + Et)} \hat{F}_m^+$$

$$\hat{\Phi}^*_{\text{Dirac}} \approx \sum_m U e^{\frac{i}{\hbar}(-p_m x_i + Et)} \hat{f}_m^+ + V e^{\frac{i}{\hbar}(p_m x_i - Et)} \hat{F}_m^-$$

where $\hat{f}_m^{\pm}$ are the creation and annihilation operators for quanta, $\hat{F}_m^{\pm}$ the creation and annihilation operators for antiquanta. U and V are functions that take account of the electron's spin. Quite a few additional details have been left out of these expressions, hence the use of the $\approx$ sign.

All of which still leaves open the question of the name.

When we refer to 'quantum field theory' we appear to be drawing an analogy with a classical field. If we have a gravitational field in a region of space, we're entitled to say that each point in space at a given time has a value of a specific physical quantity, the gravitational field strength, measured according to an appropriate unit. As we know, physical properties are related to operators in quantum theory, so at first sight, we can think in terms of a *field of operators*. However, that connection doesn't stand up to detailed analysis.

Quantum operators don't represent specific values of a physical quantity. That job falls to their eigenvalues. It's closer to think that operators represent the process of measurement, or at least the whole collection of possible values $\{o_i\}$, rather than one specific value, o_I. Of course, if the operator acts on a state that is not one of its eigenstates, then the best comparison we can make with measurement is between the *expectation value* of the operator and the *average result* of a series of measurements. In neither case do we see anything that would support much comparison with a classical field.

On balance, the best approach is not to worry too much about a direct theoretical link or analogy between quantum field theory and classical fields. Just get used to the name and move on. Quantum field theory transcends the notion of classical fields in the same manner that quanta have transcended particles. The two issues are closely related, which means that it is high time that we tackled wave–particle duality.

32.4.2 Wave-Particle Duality

Quantum field theory is often advertized as containing the resolution to the problem of wave–particle duality that is so characteristic of conventional quantum theory.

This is only partially true. The paradox of wave–particle duality is a *conceptual one*: how is it that an object can take on the characteristics of a particle (localized hard lumps, like cricket balls, that simply change their position with time) in one instance and those of a wave (a spread-out flappy sort of thing that varies in both space and time) in another?

To resolve this, we need to take a closer look at the classical ideas of *wave* and *particle* and break them down into their component parts. Then we can see which parts can be put back together in a manner consistent with quantum physics.

A classical particle is localized in space, has a definite trajectory, can be distinguished from another similar particle, and has fixed values of quantities such as mass and electrical charge, along with the presumption of some substance behind those properties. Quanta can be localized to some degree but do not have trajectories, can't be distinguished from other identical quanta, do have fixed values of properties such as charge and mass, but can't be counted, so we rejected the idea of substance.

Classical waves are evidently not localized, but their most important feature revolves around *superposition*. Bluntly, a wave is capable of existing in more than one configuration (state if you like), and you can add wave configurations together and still produce a possible configuration. For example, take Young's interference experiment. A wave can pass through the experiment with one of the slits open, which is a configuration. A second configuration would have that slit closed and the other one open. The experiment normally runs with both slits open, in which case you get the amplitude of the field at any point in this third configuration by adding together the amplitudes at that point in the other two configurations.

One of the biggest conflicts between wave and particle concepts hinges on this notion of superposition. Wave states can be added, but particle states can't. The classical state of a particle lists all the quantities that make up its properties. We can't give a meaning to adding these states together in the same way we can for a wave. The problem is especially serious if we're dealing with classical states of more than one particle. What can it possibly mean to add a state with 10 particles in it to a state with 20 particles?

States in quantum field theory don't have this problem. If we think of the state in Heisenberg's terms, it doesn't list a set of values of properties so much as a collection of potentia. A quantum field state with a definite number of quanta is simply telling us about the potentia for that number of quanta actualizing. If we add two such states together, we are combining alternative sets of possible actualizations, governed by the amplitudes involved in the sum. So, quantum field states can be added together, which give them a crucial feature carried over from classical waves.

It also helps that we have shed the idea of substance. A classical particle is localized as its substance is contained in a little lump. The substance of a classical wave (or more precisely the substance that is waving) is spread out over an extended region, making it very hard to square the idea of wave and particle together.

The quantum case is rather different. If the potentia in a state shows that a collection of properties will be localized in a restricted region of space, what we observe takes on particle-like characteristics. If the state's potentia are spread out over an extended region, a more wave-like aspect is observed.

There are some properties, such as electrical charge for an electron, which only come in exact values characteristic of the particle. Classically we think of these as being examples of a property pasted onto a substance. In a multiquanta state, we don't necessarily observe the separate quanta, but we can measure the aggregated total charge. Once again, we don't have to rely on a substance being there, just potential or actualized properties clustered together.

So, a good way of thinking about the quantum field is to allocate potentia for properties to be actualized at every point in space and time.

Importantly, each distinct type of quantum, electrons, photons, quarks, and so on, gets its own quantum field. That's not a different field for each quantum; it's one field for all quanta of the same species, e.g., the electron field, the photon field etc.

We can put a little more gloss on this by looking at the action of the field operators.

When we use the creation operator $\hat{\Psi}^+(x, t)$ to act on the vacuum:

$$\hat{\Psi}^+(x; t)|0\rangle = \sum_m A e^{\frac{i}{\hbar}(p_m x - Et)} \hat{p}_m^+|0\rangle$$

we produce a set of potentia for a quantum to be localized within a region about x. However, when you look at the expansion over momentum creation operators, it's tempting to think of it in terms of creating an infinite number of quanta (there is no limit in principle to m) of momentum $\hat{p}_m^+$. What it's really saying is that localizing the potentia for a quantum to a given position can't be done without allowing the momentum potentia to spread over a broad range of momenta. It's the quantum field theory version of the uncertainty principle. The individual $\hat{p}_m^+\hat{p}_n^+$ don't *literally* create separate quanta out of the vacuum. They tell us how the state is constructed in terms of potentia to actualize specific values of momentum.

We can of course turn things around and write:

$$\hat{p}_m^+|0\rangle = \sum_{x_i} A e^{\frac{i}{\hbar}(-p_m x_i + Et)} \hat{\Psi}^+(x_i, t)$$

$$\hat{p}_m^-|0\rangle = \sum_{x_i} A e^{\frac{i}{\hbar}(p_m x_i - Et)} \hat{\Psi}^-(x_i, t)$$

showing us how creating a quantum of fixed momentum p_m prevents us from giving it any localized position.

Another way to look at it is to realize that the state $\hat{\Psi}^+(x, t)|0\rangle$ is not an eigenstate of the number operator $\hat{N}_{p_m} = \hat{p}_m^+\hat{p}_m^-$, which counts the number of quanta in momentum state p_m. The quantum localized in space is *equivalent to* (not the same as) an indefinite number of quanta of fixed momentum.

32.5 INTERACTIONS IN QUANTUM FIELD THEORY

So far, our entire development has focused on states of noninteracting quanta. In other words, we have been describing a rather boring world. In reality, quanta are interacting with one another all the time. Electrons, for example, emit and absorb photons when atoms change energy levels.

In classical physics, interactions between particles can be considered by either an examination of forces or from the potential energies involved.

In quantum theory, it is convenient to work with potential energies. We have already seen how the Schrödinger equation includes a potential energy operator, $\hat{V}(x,t) = V(x,t)\hat{I}$:

$$-\frac{\hbar^2}{2m}\frac{d^2\Psi(x,t)}{dx^2} + V(x,t)\Psi(x,t) = i\hbar\frac{d\Psi(x,t)}{dt}$$

although we have never written down an explicit formula for such an operator in a given situation. Quantum field theory carries over the idea that interactions should be considered via potential energies, which means that we must construct a typical quantum-to-quantum potential energy operator to work in the multiquanta case.

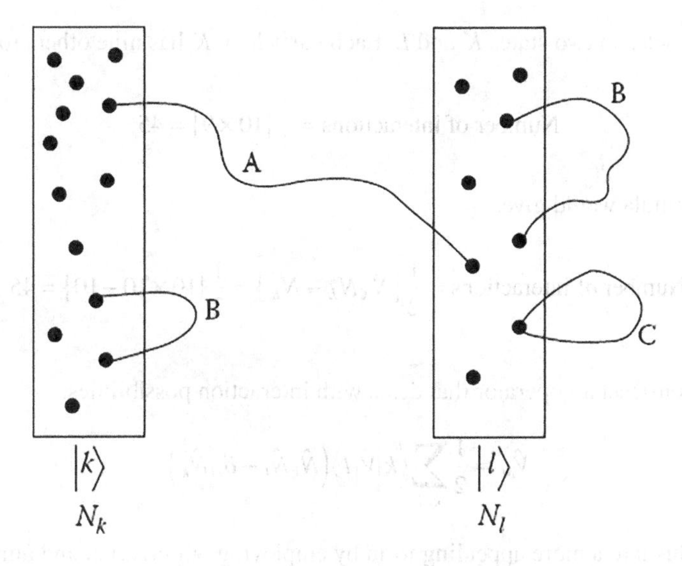

FIGURE 32.1 Possible interactions involving particles in two states.

32.5.1 Interaction Operators

Building an operator to represent the interaction between two quanta is slightly tricky. Let's start with the idea that we have two collections of quanta in states $|k\rangle$ and $|l\rangle$ that are interacting.

There are three possibilities that we must account for (Figure 32.1). Either a quantum in one state can interact with another in a different state (A), or a quantum in one state can interact with another in the same state (B), or the quantum interacts with itself (C). If there are N_k quanta in state $|k\rangle$ and N_l in state $|l\rangle$, then the number of possible interactions between them is:

$$\frac{1}{2}\sum_{k \neq l} N_k N_l$$

as for each of the N_k there are N_l possible quanta to pair up with. The 1/2 in front of this expression takes care of the 'double counting' that takes place: quantum 1 in state $|k\rangle$ interacting with quantum 2 in state $|l\rangle$ is the same as quantum 2 in state $|l\rangle$ interacting with quantum 1 in state $|k\rangle$, both of which would be counted in the sum as it stands.

However, this doesn't account for the possibility of quanta interacting with others in the same state. We can include these if we allow the sum to cover $k = l$ as well. The snag with this is that we want to *exclude* a quantum interacting with itself.

Self-interactions are a recurring theoretical issue. Although there is no reason in principle why a particle (or quantum) should not interact with itself via some form of potential energy, this idea does not mix well with point (or size less) objects. You tend to get infinite answers to the calculations. It's generally best to exclude that possibility.

There is a neat mathematical way of constructing a sum that covers all the possible interactions, but at the same time excludes the self-interactions:

$$\text{Number of interactions} = \frac{1}{2}\sum_{k,\,l} N_k N_l - \delta_{kl} N_k$$

Try it with 10 particles in two states K and L. Each particle in K has nine others to interact with:

$$\text{Number of interactions} = \frac{1}{2}\{10 \times 9\} = 45$$

The preceding formula would give:

$$\text{Number of interactions} = \frac{1}{2}\{N_K N_L - N_K\} = \frac{1}{2}\{10 \times 10 - 10\} = 45$$

so it works.

Now we can construct an operator that deals with interaction possibilities:

$$\hat{V}_M = \frac{1}{2}\sum_{k,\,l} \langle k|\hat{V}|l\rangle \left(\hat{N}_k \hat{N}_l - \delta_{kl}\hat{N}_k\right)$$

We can convert this into a more appealing form by employing our creation and annihilation operators. After all (using a to indicate an operator that could be either fermion or boson):

$$\hat{N}_k \hat{N}_l - \delta_{kl}\hat{N}_k = \hat{a}_k^+ \hat{a}_l^- \hat{a}_l^+ \hat{a}_k^-$$

which is not too difficult to demonstrate if you treat the boson and fermion situations separately and use the various commutation and anticommutation relationships from Section 32.3.6.

Our operator for the interaction potential has become:

$$\hat{V}_M = \frac{1}{2}\sum_{k,\,l} \langle k|\hat{V}|l\rangle \hat{a}_k^+ \hat{a}_l^- \hat{a}_l^+ \hat{a}_k^-$$

but even this is not its final form, as we must apply the change of basis rules to get the most general representation of the operator:

$$\begin{aligned}\hat{V}_M &= \frac{1}{2}\sum_{k,\,l} \langle k|\hat{V}|l\rangle \left(\sum_q \langle q|k\rangle \hat{a}_q^+\right)\left(\sum_r \langle l|r\rangle \hat{a}_r^-\right)\left(\sum_s \langle s|l\rangle \hat{a}_s^+\right)\left(\sum_t \langle k|t\rangle \hat{a}_t^-\right)\\ &= \frac{1}{2}\sum_{k,\,l} \langle k|\hat{V}|l\rangle \sum_{q,\,r,\,s,\,t} \langle q|k\rangle\langle l|r\rangle\langle s|l\rangle\langle k|t\rangle \hat{a}_q^+ \hat{a}_r^- \hat{a}_s^+ \hat{a}_t^-\end{aligned}$$

Now, if we extract:

$$\sum_{k,\,l} \langle q|k\rangle\langle k|t\rangle\langle k|\hat{V}|l\rangle\langle s|l\rangle\langle l|r\rangle = \langle q,\, t|\hat{V}|r,\, s\rangle$$

and re-insert:

$$\hat{V}_M = \frac{1}{2}\sum_{q,\,r,\,s,\,t} \langle q,\, t|\hat{V}|r,\, s\rangle \hat{a}_q^+ \hat{a}_t^- \hat{a}_s^+ \hat{a}_r^-$$

With the operator displayed in this form, we are encouraged to think in very pictorial terms, as in Figure 31.2, where the jagged cloud represents the matrix element responsible for scattering the

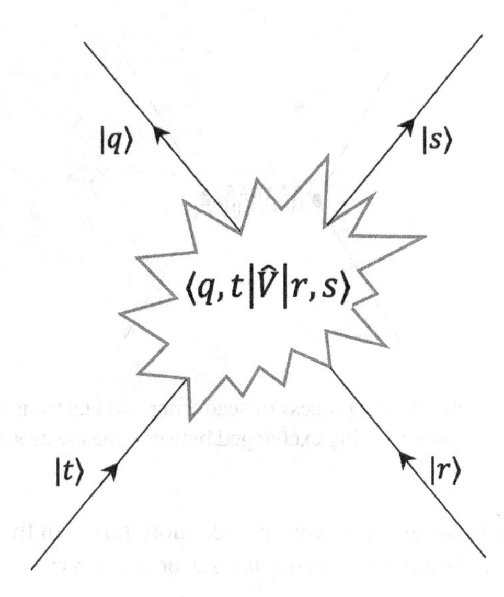

FIGURE 32.2 The multiquanta interaction operator can be viewed in terms of a figure like this. The operator is responsible for scattering a quantum from state $|t\rangle$ into state $|q\rangle$ while at the same time another is scattered from $|r\rangle$ to $|s\rangle$.

incoming particles into their final states. Here we are inching toward deploying *Feynman diagrams*, but we need one more piece before we can close the story.

32.5.2 Interaction Potentials

Although have an interaction operator, we do not have a specific form for the potential energy operator $\hat{V}$ to put in the middle of our matrix element.

The search for appropriate operators was a central theoretical concern through the 1970s.

Physicists had demonstrated that four *fundamental forces* exist in nature, i.e., forces that can't be explained in terms of other forces. For example, friction is pretty fundamental to our daily lives[7], but is not a *fundamental force* in the physical sense as it is due to the electromagnetic interactions between the atoms in materials in contact.

Gravity and electromagnetism are genuinely fundamental forces, as are the so-called *weak* and *strong* forces that act over very short ranges between fundamental particles (which are represented as quanta in quantum field theory). Each of these forces must be modelled in a quantum field theory using an appropriate interaction potential. Gravity has turned out to be something of a headache (see Section 32.7), but great progress has been made on the other three.

We are now bordering on areas where our theoretical and mathematical understanding doesn't allow us to tread, fascinating subjects though these are. Hence, from this point, our discussion becomes rather qualitative.

The various options for these operators are constrained by physical laws, the requirements of relativity, for example, but physicists have had to use a degree of theoretical imagination to come up with good prospects to be checked by experiment. In the case of the electromagnetic interaction between two electrons, a suitable combination turns out to be $\hat{\Phi}^*_{\text{Dirac}}\left(e\hat{\phi}\right)\hat{\Phi}_{\text{Dirac}}$ ($\hat{\Phi}^*_{\text{Dirac}}$ is something more than a simple complex conjugate, creation and annihilation operators have swapped places), and $\hat{\phi}$ is the boson field of photons associated with the electromagnetic interaction. The electron charge, e, acts as a 'coupling' between the fields.

Given an element of this type, quantum field theorists can calculate the energy of the interaction as a series of mathematical terms. Taking a few terms in the series gives an approximation to the

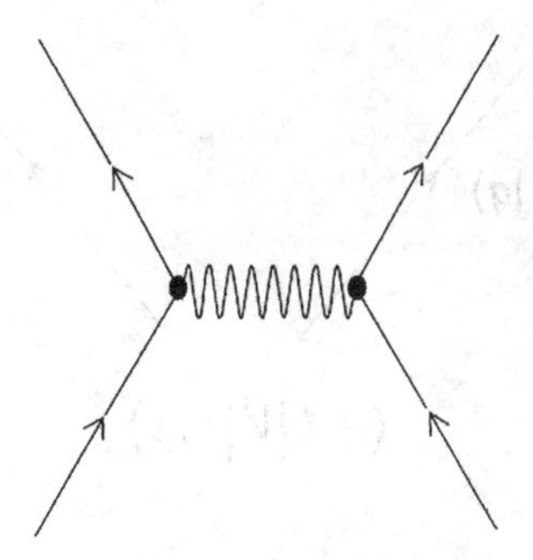

FIGURE 32.3 A Feynman diagram for the process of scattering two electrons via an electromagnetic interaction. The wavy line represents a photon being exchanged between the electrons.

actual answer that increases in accuracy as you include more terms in the final sum[8]. It is also possible to calculate something called the *scattering matrix* or *S matrix*:

$$S = \left\langle \text{final field configuration} \middle| \hat{S} \middle| \text{initial field configuration} \right\rangle$$

which can be used to obtain the probability of a given interaction occurring.

In essence, we are filling in the spiky cloud at the centre of Figure 32.2, but not with one single diagram. Instead, we use a sum of possibilities represented by Feynman diagrams.

These diagrams can cause a certain degree of confusion, especially if they are interpreted too literally. In Figure 32.3 I have drawn the standard Feynman diagram for scattering one electron off another via an electromagnetic interaction. Compare this with Figure 32.2, where I drew a diagram to pictorially capture the interaction operator and its effect on quanta. We should look at the Feynman diagram in a similar manner and 'see' how various creation and annihilation operators for electrons and photons have contributed to building the overall picture. That's not the end of the story though[9]. Each diagram represents a term in a series that would have to be evaluated in its entirety to get a completely accurate answer. There are other terms, or diagrams, that need to be included, such as Figure 32.4.

Fortunately, not every diagram is equally important in the sum. The more complicated they are, the smaller the effect of adding them in. In practice, including just a few diagrams can produce remarkably accurate predictions[10].

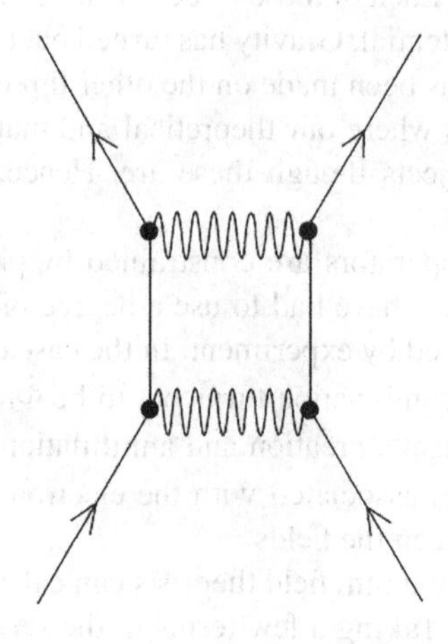

FIGURE 32.4 Another diagram for electron–electron scattering.

The interpretation tends to go wrong when people think of these diagrams as being *literally* true, rather than illustrative of a mathematical term in an approximation series. Physicists often don't help this problem by talking about 'exchange forces' and how the electromagnetic force comes about due to 'an exchange of photons', as if electrons were throwing photons back and forth.

A more useful and accurate way of thinking about these diagrams, and the whole quantum field theory method for dealing with interactions, is to go back to Important Rule 3, clause 2b which states: when two or more alternatives are possible, the *amplitudes add* if the *alternatives cannot be distinguished.* When two particles interact, we only observe their states on the way in and their states on the way out. Each Feynman diagram is then an indistinguishable route from one state to another, and so the mathematical terms that they represent must be summed in a series. No one diagram is a literally true representation of what is happening on its own.

32.6 VACUUM FLUCTUATIONS

As we can set up creation and annihilation operators for any set of basis states, we can also define a number operator to go alongside. However, not all number operators are equal. One, in particular, is of fundamental importance.

If we go back to the Schrödinger equation and apply it to the free particle case, so there is no potential energy involved, the Hamiltonian operator is $\hat{H} = \frac{1}{2M}\left(-i\hbar\frac{d}{dx}\right)\left(-i\hbar\frac{d}{dx}\right)$. (Note that I have used M for mass to save it being confused with a summation index.)

In standard quantum theory, the expectation value of this Hamiltonian for a state $|\Psi\rangle$ would be $\langle\hat{H}\rangle = \langle\Psi|\hat{H}|\Psi\rangle = \int \Psi^* \hat{H}\Psi\, dx$. If we carry this over into field theory, by replacing the wave function Ψ with field operators:

$$\langle\hat{H}\rangle = \int \hat{\Psi}^+ \hat{H}\hat{\Psi}^-\, dx$$

making the expectation value an operator.

Now things get a bit messy. If we substitute in the expansions of the field operators over momentum eigenstates:

$$\langle\hat{H}\rangle = \int\left(\sum_m Ae^{\frac{i}{\hbar}(p_m x - Et)}\hat{p}_m^+\right)\hat{H}\left(\sum_n Ae^{\frac{i}{\hbar}(-p_n x + Et)}\hat{p}_n^-\right)dx$$

$$= \frac{1}{2M}\int\left(\sum_m Ae^{\frac{i}{\hbar}(p_m x - Et)}\hat{p}_m^+\right)\left(-i\hbar\frac{d}{dx}\right)\left(-i\hbar\frac{d}{dx}\right)\left(\sum_n Ae^{\frac{i}{\hbar}(-p_n x + Et)}\hat{p}_n^-\right)dx$$

$$= \frac{-\hbar^2}{2M}\int\left(\sum_m Ae^{\frac{i}{\hbar}(p_m x - Et)}\hat{p}_m^+\right)\left(-\frac{d}{dx}\right)\left(\sum_n A\left(\frac{i}{\hbar}p_n\right)e^{\frac{i}{\hbar}(-p_n x + Et)}\hat{p}_n^-\right)dx$$

$$= \frac{-\hbar^2}{2M}\int\left(\sum_m Ae^{\frac{i}{\hbar}(p_m x - Et)}\hat{p}_m^+\right)\left(\sum_n A\left(-\frac{1}{\hbar^2}p_n^2\right)e^{\frac{i}{\hbar}(-p_n x + Et)}\hat{p}_n^-\right)dx$$

$$= \frac{A^2}{2M}\int\sum_{m,\,n} p_n^2 e^{\frac{i}{\hbar}((p_m - p_n)x)}\hat{p}_m^+\hat{p}_n^-\, dx$$

$$= \frac{A^2}{2M}\sum_{m,\,n} p_n^2\left(\int e^{\frac{i}{\hbar}((p_m - p_n)x)}\, dx\right)\hat{p}_m^+\hat{p}_n^-$$

Taking a deep breath, we note that the integral in the bracket is the Dirac delta function, which vanishes for $m \neq n$. Hence our result is:

$$\langle \hat{H} \rangle = \frac{A^2}{2M} \sum_n p_n^2 \hat{p}_n^+ \hat{p}_n^- = \frac{1}{2M} \sum_n p_n^2 \hat{N}_n$$

given that $A^2 = 1$, if I have got the normalization right.

As $p_n^2/2M$ is the kinetic energy of a quantum in momentum state $|p_n\rangle$, the Hamiltonian reduces to adding up the kinetic energy of each momentum state times the number of quanta in that state. That shouldn't be a surprise, as we are dealing with a collection of free quanta. However, the relationship between the momentum number operator and the energy is sufficiently important that it is generally taken to be *the* number operator and the number of quanta in momentum eigenstates to be the fundamentally important aggregation.

32.6.1 Fields and Numbers

In standard quantum theory, we know that two operators must commute if their associated physical variables are to have definite values at the same time. This carries over into quantum field theory and presents us with a problem. The field operators $\hat{\Psi}^+$ and $\hat{\Psi}^-$, for example, don't commute with momentum state number operators. Mathematically you can show this by playing with the commutation relationships, but physically this is nothing new as we have already seen how $\hat{\Psi}^+$ requires an indefinite aggregate of momentum quanta.

Things become interesting when you start applying this logic to the vacuum state $|0\rangle$. After all, a vacuum state still contains a definite number of quanta, *it just happens that the number is zero.*

Any operator that doesn't commute with the momentum number operator will have an expectation value in the vacuum that is not zero. That's still going to be true, even if we expand that operator over momentum creation and annihilation operators. Really this is nothing different to our discussion in Section 32.4; it's just more striking in the context of the vacuum state.

We can expect Feynman diagrams for processes such as that shown in Figure 32.5, which is an example of a *vacuum fluctuation.*

Once again, it's very dangerous to take such diagrams literally, i.e., that electrons and positrons are popping in and out of existence. Sometimes the energy-time inequality is used (badly, see Section 14.3.2) to justify this by suggesting that energy is borrowed to create the electron and positron and then paid back over a timescale too short for us to directly observe them. The electron and positron are sometimes said to be *virtual particles,* which is near to the mark but possibly not the most helpful expression.

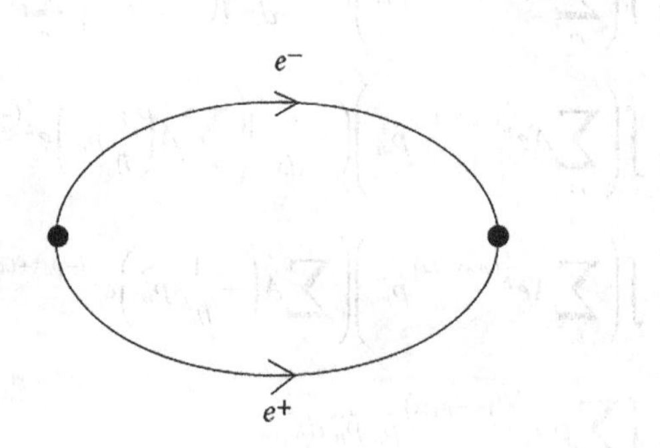

FIGURE 32.5 A Feynman diagram for a typical vacuum fluctuation. In this figure, an electron and a positron have apparently popped into existence out of the vacuum and disappeared again shortly after. Many other such figures, including those with different particles, exist.

Perhaps a better way of thinking about vacuum fluctuations is to use the notion of potentia again. A vacuum state is still a state with the potential to manifest quanta, it just happens that no actual quanta have been actualized. Expectation values look at the averages of properties that would actualize in the right situation so that *specific examples of the properties don't have to be there.* Once again, this is easier to accept once the notion of substance has been removed.

Remarkably, these vacuum fluctuations can have measurable effects when something else interacts with one of the fluctuations that, like a measurement, snaps it into reality. In an extreme case, Stephen Hawking, in a landmark paper[11], noted that vacuum fluctuations in the vicinity of a black hole could be scattered into reality by interacting with the gravitational field. The effect, observed from a distance, would be *Hawking radiation*, which causes the black hole to leak energy to its surroundings and as a result steadily decline in mass, completely against expectations.

In more prosaic instances, vacuum fluctuations are responsible for triggering the movement between atomic energy levels, giving rise to the emission of photons.

32.7 QUANTUM GRAVITY

As mentioned on a few occasions, reconciling quantum mechanics with Einstein's theory of gravity, General Relativity, is still proving problematic. As just one illustration of the problems involved, consider time again. As we have stressed, there is no 'time operator' in quantum theory. Instead of being observable, time is a parameter that marks the progress of a system's evolution. In General Relativity time is an integral aspect of dynamics and interacts with the mass/energy of a system. Here is a fundamental clash of perspectives.

Furthermore, in a successful quantum gravity, one presumes that the probabilistic nature of quantum theory would be retained, leading, via the field quantization of space-time geometry, to a theory in which different geometries would be available to the universe on some probabilistic scale. In which case, we need some way of understanding how this superposition of geometries collapses into a single instance[12]. If we wish to apply a quantum gravity theory to the Big Bang, we are faced with a measurement problem of (literally) universal proportions.

As if that were not enough there is another problem. Within a black hole, General Relativity predicts that matter is crushed to a point of no size, the *singularity*, (see Section 15.3.2) as no degeneracy pressure can resist the gravity involved. Presumably, quantum effects must come into play to avoid the physical nonsense of infinite density. However, when applied to the Big Bang General Relativity indicates that the entire universe should have started with a singularity, raising the stakes in figuring out how such structures can exist.

There is a viable low-energy approximation, called the *effective field theory of gravitation*. It turns out that a bosonic field theory can be constructed based on a linearized weak-field approximation to the full equations of General Relativity. This is the *graviton field*, with *gravitons* being the quanta involved. In which case, given that General Relativity presents the view that gravitational effects are not due to the action of a force, but rather to distortions in space-time, gravitons are quanta of space-time distortion. The problems begin when we start to enquire into their behaviour in the region of a large mass. The mass will distort space-time, but as the gravitons are distortions of space-time, we might think that the mass's effect is mediated by gravitons. In which case, they are moving through distortions caused by other gravitons. Equally, gravitons themselves act as gravitational sources, so gravitons can emit and absorb gravitons. The simple picture in the effective theory, which deals with low-energy cases and a 'flat' background space-time, is not adequate. Especially, if we wish to tackle quantum aspects of the Big Bang.

While tackling quantum cosmology seems a dramatic step, we know that it must be taken at some point. In the early history of the universe, all that existed were quanta of different varieties and the universe itself was microscopic in size, so we know quantum effects had to play a part. This is certainly one motivation for seeking a quantum theory of gravity. Also, one should not discount the maddening frustration inherent in having two marvellously successful theories on the shelf which appear to be incompatible with each other.

Given the clash between the conceptual structures of General Relativity and quantum field theory, and as we are currently lacking new concepts to rescue us, the best thing to do is to treat one theory as primary and develop the other to fit. In broad terms, this corresponds to the approaches taken by the two most popular candidate quantum gravities.

32.7.1 Loop Quantum Gravity (LQG)

Loop quantum gravity is a quantization of space-time starting from General Relativistic principles.

The theory reveals a microstructure to space-time which makes space look like a network with discrete edges. In essence, space is quantized into distinct volumes ~ l_p^3 in size[13]. The surfaces dividing one volume from the next are also quantized ~ l_p^2.

This approach has had some notable successes, not the least of which has been calculating the *Hawking temperature* of a black hole, which leads to Hawking radiation. Applying it to the Big Bang has produced the intriguing suggestion that the initial singularity can be avoided by changes in the quantum structure at high densities. This might permit fundamental constants, including the speed of light, to change, allowing the universe to 'bounce' if it were to collapse back in the future (or it might have already bounced prior to our universe).

At this stage of development, it is unclear if LQG reproduces the General Theory at macroscopic scales, but as an approach, its popularity is starting to rise as some grow frustrated with the alternative, *string theory*.

32.7.2 String Theory

String theory tackles the problem from a different direction, by trying to develop quantum theory to incorporate gravity. The fundamental concept is the suggestion that all quanta are different vibrational modes of exceptionally small, but finite-sized, objects called *strings*. These are 1-dimensional structures ~ a Planck length in size that can close on themselves to form 'vibrating' loops. Each fundamental particle is then a different mode of vibration in the string. Even more extraordinarily, string theory requires six or seven (depending on the variant) new spatial dimensions for its mathematical consistency. These dimensions are assumed to be unobservable at our scales as they are *compactified* (rolled-up) on themselves so that they form closed structures, and hence can't be navigated.

String theory successfully incorporates the fundamental forces into one framework as the graviton appears as one of the string vibrations. However, the movement of strings takes place in a flat space-time. As with LQG, the theory presents formidable mathematical challenges and struggles to produce experimentally testable predictions. This is one reason why some in the community are starting to believe that string theory is a theoretical dead end. Others would, of course, dispute this view.

32.7.3 Prospects

All candidate quantum gravity theories face one grave difficulty, aside from the complexity of the mathematics involved. As we expect their effects to come into play on a mass scale ~ M_P, the Planck mass, it is hard to see how experimental predictions and tests can be made. Penrose's suggested experiment from Section 29.2.1 may serve to illuminate aspects of objective state collapse, but not the details of how quantum gravity operates.

The Planck mass is merely ~ 2.2×10^{-8} kg, but to access the equivalent energy (using $E = mc^2$) in some form of particle experiment requires a stunning level ~ 1.2×10^{19} GeV. Quanta in the very early universe undoubtedly had comparable energies, so one hope is that some 'imprint' of quantum gravity will show through in the evolution of the universe.

We do know that quantum behaviour played a significant role in the Big Bang after the quantum gravity era. One of our primary sources of data regarding the early universe, up to ~300,000 years into history, comes from the *cosmic microwave background*, relic microwave radiation that fills the entire universe. Photons in the evolving universe were constantly interacting with charged quanta, a process that maintained an energy equilibrium. When the universe reached about 300,000 years of age, the typical photon energies dropped to the point where electrons could bind to protons without being ionized again by the photon bombardment. At this point, the universe became 'transparent', and the photons were released from their endless cycle of scattering and interaction. However, the matter density distribution as this happened left an imprint on the radiation. When we observe the microwave background coming to us from different patches of the sky, we see tiny energy differences that reflect matter density variations.

Remarkably, the matter density can be traced to a much earlier epoch when the expansion of the universe was being driven by another quantum field, the *inflaton field*. We know little about this, other than its broad characteristics. Theory suggests that about $\sim 10^{-33}$ seconds into history, the inflaton quanta collapsed into conventional matter. Quantum fluctuations in the inflaton field intensity from point to point resulted in a non-uniform mass density that then carried through to imprint on the cosmic microwave background. Impressively, calculations carried out by Stephen Hawking and others match the expected quantum fluctuations to the energy pattern in the comsic microwave background radiation.

Now that we have a successful technology for detecting gravitational waves[14], another prospect becomes a possibility. Our electromagnetic data cannot predate the cosmic microwave background, as prior to the universe going transparent, the constant photon scattering did not allow any coherent information to be sustained. However, gravity waves are not affected in this way. Presumably, the violent mass/energy churnings in the early moments of the Big Bang set off gravitational waves that would be moving through the universe to this day. Maybe some clue to quantum gravity lies in their detection and characteristics.

As things stand, the choice of theory and its key principles are largely a matter of aesthetics. Without experimental data to guide us, progress has to be made as it can. We should also remember that Dirac found the beauty of equations to be a reliable guide to their physical relevance. However, there is the nagging suspicion that the ultimate solution to squaring quantum theory with General Relativity lies in some new set of concepts, which might have an added bonus in making the correct interpretation of quantum reality clear.

NOTES

1 *Quantum Mechanics,* 2nd Ed., Wiley International Series, 1961.
2 P. Teller, *An Interpretive Introduction to Quantum Field Theory*, Princeton University Press, 1997.
3 Vladimir Aleksandrovich Fock (1898–1974).
4 Mathematicians would say that the two are *isomorphic.*
5 These are sometimes called creation and annihilation operators, a terminology that I will adopt later in the chapter when it makes more sense.
6 Although I am possibly oversimplifying some of the complications to do with normalization.
7 You try moving without using it.
8 There are theoretical complications here. Firstly, in *quantum chromodynamics*, the theory of the strong force between quarks, terms become more significant as they grow in complexity, making the approximation series problematical. Secondly, the series is not convergent, but rather a low-energy approximation.
9 There are additional complications to do with relativity. No specific time ordering, e.g., electron1 emits a photon which is then absorbed by electron 2, should be ascribed to any diagram.
10 In the case where the 'coupling constant' is <1. As per note 9 that is not the case in quantum chromodynamics except at high energy.

11 Hawking, S.W. Particle creation by black holes. *Commun. Math. Phys.* 43, 199–220 (1975). https://doi.org/10.1007/BF02345020.
12 We have, for example, Penrose's suggestion on the table from Section 29.2.1
13 l_p being the Planck length $= \sqrt{\frac{G\hbar}{c^3}} \sim 1.6\times10^{-35}$ m.
14 https://en.wikipedia.org/wiki/LIGO.

33 Personal Conclusions

It is truly surprising how little difference this all makes. Most physicists use quantum mechanics every day in their working lives without needing to worry about the fundamental problem of its interpretation. Being sensible people with very little time to follow up all the ideas and data in their own specialties and not having to worry about this fundamental problem, they do not worry about it. A year or so ago, while Philip Candelas (of the physics department at Texas) and I were waiting for an elevator, our conversation turned to a young theorist who had been quite promising as a graduate student and who had then dropped out of sight. I asked Phil what had interfered with the ex-student's research. Phil shook his head sadly and said, "He tried to understand quantum mechanics".

Steven Weinberg[1]

33.1 POPULAR OPINION

It's difficult to gauge where the majority opinion stands among physicists on the question of interpretation. Partly this is due to there being a range of different perspectives stemming from the specialisms of individuals, from quantum foundations to cosmology.

My hunch is that most professional working physicists (engaged in academic or industrial research of one form or another) subscribe to the Copenhagen interpretation if they worry about interpretations at all. The chances are that they spend their time calculating amplitudes and probabilities and rarely lift their heads from the computer screen to ask what it all means.

Several surveys have been attempted over the years, but the statistics are small. In 2019, 1234 physicists across eight universities were contacted and 149 responded by completing an online questionnaire.[2] Their responses were interesting. The question "What is your favourite interpretation of quantum mechanics" garnered the results shown in Figure 33.1 which tends to confirm the feeling that many physicists "shut up and calculate", otherwise they tend to be Copenhagenists.

Arguably, the general physics population is not a sensible sample space. If you are working in an area that does not require up-to-date knowledge of quantum foundations, then you may not be fully informed with the latest results and their implications. Equally, if your primary research area is quantum cosmology, then your perspective is likely to be very different.

33.2 QUANTUM REALITY

In Hilbert space no one can hear you scream.

Yakir Aharonov[3]

Doubtless, every quantum mechanic has their own view on the distinctive features of the quantum description of the world. In no special order of significance then, here is my list:

- Quanta are fundamental quantum objects which are utterly indistinguishable to an extent that undermines their individual, trackable, existence.
- Quanta come in two broad types, fermions, which are 2π rotationally antisymmetric, and bosons, which are 2π rotationally symmetric. The rotational symmetries have a profound impact on their physical properties.

DOI: 10.1201/9781003225997-37

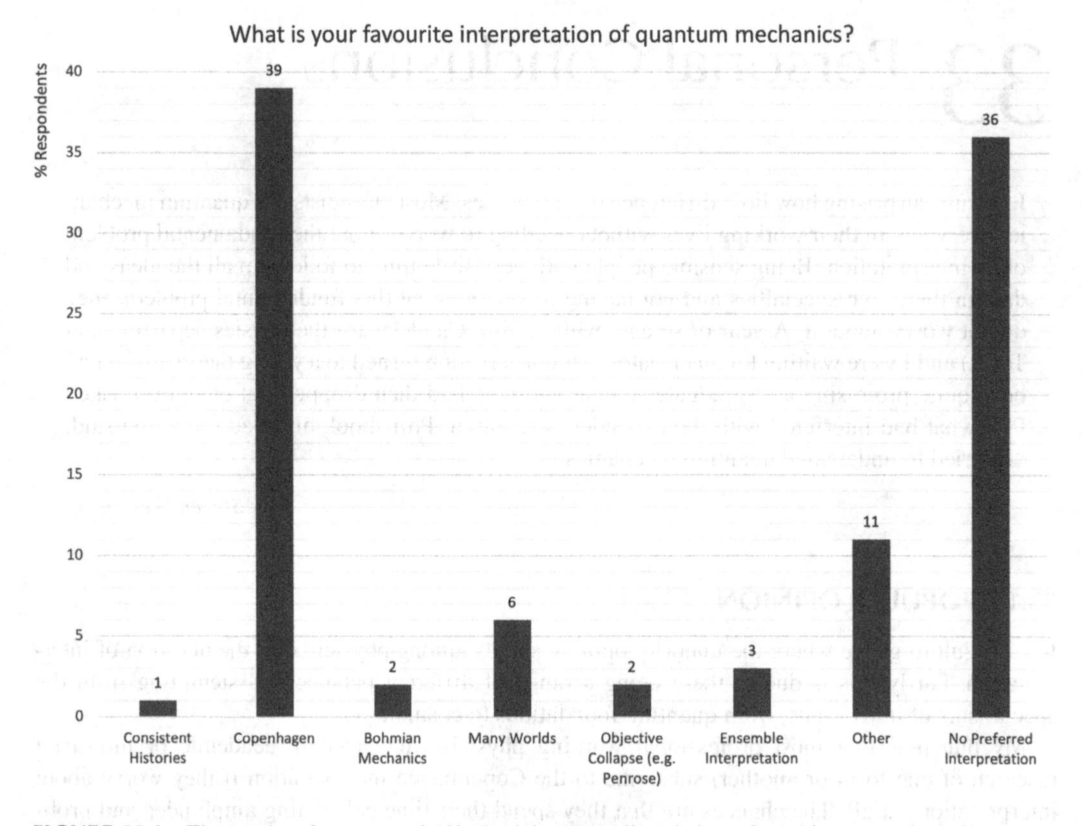

FIGURE 33.1 The results of a survey of 149 physicists regarding their preferred interpretation of quantum mechanics. The Ensemble Interpretation regards the state as representative of a collection of objects, not individuals.

- Spin is a physical property of quanta which is related to their rotational symmetry but has no classical analogue.
- Some physical properties of quantum systems are incompatible and cannot be used in combination to form a more complete description of the system's state.
- Quantum states that are superpositions have no classical analogue, yet are frequently required to describe a quantum system.
- The quantum world exhibits an interconnectedness and contextuality which undermines reductionism.
- Quantum probabilities are a fundamental aspect of reality.

Of course, it is one thing to construct a list of this nature, quite another to determine which characteristics reflect the world and which are simply instrumentally used elements of our description. For that, we need further guidance.

33.2.1 Critical Realism

I stick my personal philosophical flag in a patch of metaphysics which is a version of *critical realism*, an approach that I would summarize as follows:

1. There is an objectively real world open to investigation.
2. However, the nature of the universe is too surprising and subtle to be deduced entirely from philosophical reflection.

3. Consequently, our thinking needs to be guided by experience, observation, and experiment.
4. However, theory and data intertwine in a manner that defies a clean separation between the two. At one level, a complete theoretical understanding of any instrumentation is needed before it can be employed effectively.[4] More significantly, raw data only takes on meaning in the context of a theory.[5]
5. The task then is to devise theories that effectively integrate and explain our experiences.
6. The most effective theories 'reach outside' their initial domain. While they are based on the interpretation of a collection of key data points, within a certain region of understanding, they gather in aspects of other areas and explain them in surprising ways. Equally, they posit new and unexpected features of the world that are then proven to hold true. For example:
   - Quantum theory concisely explains why atoms have ground states, which was one of the primary considerations during its gestation. However, nobody could have expected that it would also explain how white dwarf stars can hold stable against their own gravitational compression.
   - Dirac's relativistic extension to quantum theory encompassed spin and projected the existence of antimatter, taking physics in a totally new direction.
7. The scientific community works by applying peer review, independent verification, and a range of tacit skills, both practical and intellectual, to evaluate, refine and extend our understanding. In a sense, our corporate ideas evolve under the natural selection pressures of the community gaze.
8. *True* and *real* are, within an acceptable level of approximation, synonymous[6]. In other words, we may take the items and aspects of a successful scientific theory to be reflections of ontologically real components of the world.

Many of these points would be accepted by a broad range of philosophical approaches. Critical realism is distinctive in its emphasis on (6), the *critical* part, and on (1) and (2). The process of creating theory is difficult. The trite picture of a theoretician pondering a page of raw numbers and applying sufficient coffee until an idea that explains all the data appears, is drastically oversimplified. As (3) suggests, the data has no meaning outside of the theory, so the theory creation process is one that envisions the data in a context that works to draw threads together into an economic and elegant whole. Any theory that succeeds in doing that must contain elements that convincingly represent reality, (7) and (8).

In a nutshell, proposing the existence of an elementary particle, such as an electron, serves to explain a range of experimental data in a concise and satisfying way (as judged by the community) and also allows us to bring in aspects of physics that were not previously explained nor viewed as part of the context where electrons were normally discussed, then the chances are that something like an electron really exists in nature. It is difficult to believe that our imaginations, on their own, are up to the task of inventing some of the deeply surprisingly aspects of the quantum description of the world, without these aspects being in some sense reflections of what is out there.

So, to my mind, critical realism satisfies our innate scientific instinct to be realist while at the same time providing a criterion for 'real' which does not pre-judge what might turn out to be real.

Therefore, from a critical realist perspective, my thinking is as follows.

33.2.2 Copenhagenism & Consistent Histories

I believe that there is a lot to be gained by grappling with Bohr. While his writing can be obscure, there is no doubt that he thought deeply about the philosophical issues from a front row perspective as quantum theory developed. It is fair to say that the work of consistent historians has done a lot to clarify the probable thrust of Bohr's thinking. Framework incompatibility is a helpful re-interpretation of complementarity, for example. However, while relative realism does capture some distinctive aspects of the quantum world, I do not think we need to go that far.

In many ways, the consistent histories approach reminds me of linguistic philosophy: the view that philosophical problems can be dissolved by an analysis of the correct use of language. Ludwig Wittgenstein is often regarded as one of the key figures in this approach. However, Wittgenstein's focus was not intended to reduce philosophy to a nit-picking debate about linguistic niceties. He was a profound believer in the importance of notions of truth and beauty and that their significance could not be fully captured in language. Consequently, he rejected the extreme views of some schools that purported to follow him, that these 'metaphysical' concepts were literally non-sense.

While I value the clarity that consistent historians have brought to aspects of quantum theory, I believe that they have stepped over at least one important feature.

I do think that Heisenberg was onto something with his (adopted) notion of potentia. I accept the fundamental nature of quantum probabilities but feel that the critical history approach to the wave function, as a pre-probability calculational tool, is underselling a key issue. For me, it is hard to understand how the wave function can be such an effective device for calculating probabilities without reflecting some aspect of reality (critical realism at work). Significantly, complex numbers must be used for amplitudes. In no other physical theory is this a necessity, rather than a calculational convenience. Instead of being a reason to doubt that complex-valued wave functions are representational of some aspect of the world, *I take it as a signal indication that they do.* Consequently, guided by critical realism again, *we must adopt a wider view of reality* that encompasses a layer underneath our explicate (Bohm's words) experience. The multi-dimensional aspect needed to describe more than one quantum (i.e., $\Psi(x_1, x_2, \dots, x_n, t)$ is inconvenient in a realist approach that pre-defines real as 'along classical grounds'. However, we should take the world as we find it, or rather as we are forced by experience to describe it. Hence, I believe this to be another signal that there is an implicate (I like Bohm's terminology) layer to reality which cannot be directly experienced, but that shapes and influences what can. Within this layer, complex numbers have a reality and role similar to that found for real numbers at the explicate, experiential, layer.

33.2.3 Many Worlds and Many Minds

An incredulous stare may not be much of a philosophical argument, but that should not stop us from using one at an appropriate moment.

We are entitled to judge our scientific theories by their consonance with our basic human experience. Of course, when we venture outside direct human experience with phenomena like black holes, sub-atomic particles, and Bose-Einstein condensates, for example, scientific experiments must have the last word. But, when it comes to the Many Worlds interpretation, which tells me that my perspective of being the same person moment by moment is an illusion and that my consistent (if fallible when it comes to car keys) memory of the past is just one branch of many parallel worlds, I begin to feel that it is all too much of an effort.

In Galileo's time, the prevailing view took the heavens as the seat of perfection and hence the Moon and planets were thought to be completely smooth spheres. The view through Galileo's telescope showed the Moon to be pocked marked with craters and in possession of plenty of mountains. In an attempt to counter Galileo's observations, some sceptics suggested that the Moon was coated in a transparent material that filled the craters and rose to the height of the highest mountain. Hence the Moon had an unobservable, smooth surface. Galileo was prepared to allow his opponents the existence of this material, provided they grant him the right to suggest that the Moon was not smooth as it had unobservable mountains of the same stuff.

I acknowledge that there is a certain elegance to the Many Worlds view, but I am not convinced by the approach to probability, which proposes that I should care about future outcomes relative to their $|\psi|^2$ as if I were involved in a game. It all feels like invisible Moon-stuff to me. Furthermore, I don't think that it is necessary.

33.2.4 The Ontological Interpretation

I am puzzled by the reaction to Bohm's approach. At the very least, it serves to bring out some distinctive features of the quantum world. Bell was right to support its study. I am prepared (critically realistically) to accept the non-classical nature of quantum potential energy, which seems to be a stumbling block for many. Problems remain in extending this approach into field theory, which is why I suspect that it might ultimately be missing something, but it demands serious attention.

I appreciate the introduction of concepts such as 'active information' and think that this is a fruitful direction to follow.

33.2.5 Objective Collapse

I tend to agree that state collapse, as an assumption pasted on to quantum theory, is a bit rich. If I am going to accept an extended view of 'real' that incorporates the complex-valued wave function in an implicate layer of the world, then I think it would be perfectly coherent for this to 'collapse' in certain situations. I subscribe to the objective collapse view. The trigger must be something linked to the complexity of a measuring apparatus, compared to the simplicity of a few atoms or particles. The discovery of decoherence is undoubtedly an important step forward, but I think of this again as a clue to the role of the complexity in the macroscopic world having an impact on the microscopic. My sense is that some extension is needed to active information to mediate the contextual impact of the macroscopic on the microscopic. This would be a form of downward causation, taking place within the implicate layer. One must respect the intuition of a physicist of Penrose's caliber,[7] and I think that there is something elegant to his suggestion, but I suspect that information/complexity may be another factor at work, perhaps as well as gravitational collapse. Unless they are aspects of the same thing…

Of course, this is hand-waving, but I think that I can justify that it is at least waving in a fruitful direction.

Having said all of this, and while I applaud the ambition, and some of the results, of trying to apply quantum theory to the creation of the universe, I think something new is needed. I do not believe that a hidden variable theory will 'rescue' us from the weirder parts of quantum theory. I do think that the clash between quantum and relativistic views of the world currently harassing our attempts at a quantum gravity, is a signal that we need some new fundamental concepts. The problem is, I tend to agree with Bohr, that it is struggle to break out of an classical way of thinking. So, where those concepts are coming from, I have no idea. Something is going to take us by surprise. Quite possibly, from an apparently unrelated area.

33.3 CONCLUSIONS

One of my main aims in this book has been to present a sound understanding of quantum theory and its implications for our view of the world. The technical development presented is, in some respects, not far short of first- or second-year undergraduate level. A more formal approach would have explored far more examples and shown how to solve the Schrödinger equation in a variety of circumstances. However, that would not have added anything to the basic conceptual core of the theory.

From a philosophical and interpretational point of view, we have ventured much further than an undergraduate physics course.

Hopefully, we have now reached a vantage point where we can reflect on the impact of quantum theory and perhaps take a view on its importance for our understanding of the world.

Whether you're an instrumentalist or a realist, you are trying to make a map of reality. The only difference is the literal seriousness with which you take the map and the job you want to use it for.

I am a realist: I like to think that what I struggle to understand in physics is the world as it really is out there. As I like mathematics and have an aptitude for science, I find the map that physics creates challenging, interesting, and in the end quite beautiful.

That beauty, however, can be quite beguiling. There is a danger that you come to think of the map physics creates as the only complete map of the world. From time to time, we have talked about some wider philosophical issues such as the nature of free will, morality, and the mind. In my view, physics gives us only one perspective on the subject: an important but a limited vantage point.

The conscious mind is not something that we should ignore, nor should we downplay the importance of mind coming to be in the universe. We have evolved from the dust of the world, and the existence of mind must be exploiting something in the fundamental laws of nature. That perspective has to be acknowledged, but not at the price of an inadequate view of our humanity.

Take free will as an example. No matter what our philosophical position might be on the existence, or not, of free will, we all act as if we are free to determine our own fate, even to the point of which breakfast cereal to buy. We have a basic intuition of freedom. Classical physics is structured in a very deterministic way. Given a sufficiently detailed knowledge of the positions, momenta, and forces for every particle in the universe, classical physics, if it were right, would enable us to predict the future as far ahead as we liked. As a result of this, some philosophers lost nerve and started to doubt their intuitions of free will. On the contrary, and to paraphrase John Polkinghorne's memorable slogan, if physics leaves no room for free will, then so much the worse for physics.

Now that quantum mechanics and chaos theory has swept away this classical view, there is a danger that anything goes in its place. There has been a rise of *quantum hype* – in essence, the use of quantum theory to justify a range of ideas well outside the scope of its application. It is interesting to search for books online with 'Quantum' in their titles. The span of results is far wider than simply physics.

While I have a great deal of sympathy with the intentions behind some of this work, I think that basing any view of the world solely on physics, or any other science, is dangerous and inadequate. I would say the same for any perspective based on a single area of study or experience.

Coming back to free will, I want to defend its existence. I think that quantum theory has made the defense easier, but not because I necessarily think that quantum effects have a role in the brain (they might), but because tackling quantum theory encourages us not to pre-judge what the world might be like and to take it on its own terms.

The real lesson of the quantum revolution is that we were too quick to paint ourselves as mindless mechanistic lumps of matter when the physics seemed to suggest a deterministic world.

Science is an important tool for viewing the world, but it is not the only road to truth.

NOTES

1 S. Weinberg, *Dreams of a Final Theory*, Pantheon Books, New York, 1992, pp. 84–85.

2 S. Sivasundaram & K. H. Nielsen, Surveying the Attitudes of Physicists Concerning Foundational Issues in Quantum Mechanics, arXiv:1612.00676v1.

3 Y. Aharonov (1932 -), Professor of Theoretical Physics and the James J. Farley Professor of Natural Philosophy at Chapman University in California.

4 One challenge that Galileo faced while trying to use his telescopic observations in defence of the Copernican viewpoint is that he lacked a theory of optics to justify the accuracy of the instrument. Referring to its use in observing terrestrial features was no help, as the prevailing opinion held the physics of the heavens to be radically different to that applicable on Earth.

5 Even a relatively innocuous statement like 'the grass is green' presupposes, on critical analysis, a raft of in-built 'theory' identifying the flora concerned, the optics of light reflection and the wavelengths involved.

6 John Polkinghorne was fond of the phrase 'epistemology models ontology'; what we know is an accurate guide to what is real. Indeed, I gather he had a t-shirt to that effect.

7 Especially as I find some of his other arguments to do with AI etc., convincing.

Appendix List of Important Rules

IMPORTANT RULE 1: THE BORN RULE

If

$$|\varphi\rangle = a_1|A\rangle + a_2|B\rangle + a_3|C\rangle + a_4|D\rangle + \cdots$$

then:

$$\text{Prob}\left(|\phi\rangle \rightarrow |A\rangle\right) = a_1^* \times a_1 = |a_1|^2$$

$$\text{Prob}\left(|\phi\rangle \rightarrow |B\rangle\right) = a_2^* \times a_2 = |a_2|^2$$

etc., named after Max Born who first formulated this approach in 1926.

IMPORTANT RULE 2: NORMALIZATION

If:

$$|\varphi\rangle = a_1|A\rangle + a_2|B\rangle + a_3|C\rangle + a_4|D\rangle + \cdots$$

then:

$$|a_1|^2 + |a_2|^2 + |a_3|^2 + |a_4|^2 + \cdots = 1$$

IMPORTANT RULE 3: COMBINING AMPLITUDES

Changes in states (transitions) are governed by amplitudes.

1. When one transition directly follows another, the respective amplitudes are multiplied together.
2. When two or more alternatives are possible:
    a. the probabilities add if the alternatives can be distinguished in the experiment
    b. the amplitudes add if the alternatives cannot be distinguished (and then the probability is the complex square of the total amplitude)

IMPORTANT RULE 4: TRANSITION AMPLITUDES

If a system starts off in state $|\varphi\rangle$ and ends up in state $|\psi\rangle$, then the amplitude for the transition can be calculated by taking the *bra of the final state* and multiplying by the *ket of the initial state*:

$$\text{Amplitude}\left(|\varphi\rangle \rightarrow |\psi\rangle\right) = \langle\psi|\times|\varphi\rangle = \langle\psi|\varphi\rangle$$

IMPORTANT RULE 5: INTERMEDIATE STATES

Any amplitude governing a transition from an initial state to a final state via an intermediate state can be written in the form:

$$\langle \text{final state} | \text{initial state} \rangle = \langle \text{final state} | \text{intermediate state} \rangle \langle \text{intermediate state} | \text{initial state} \rangle$$

If there is more than one indistinguishable intermediate state, we have to sum over all of the transition amplitudes to get the overall amplitude:

$$\langle \text{final state} | \text{initial state} \rangle = \sum_i \left[\langle \text{final state} | i \rangle \langle i | \text{initial state} \rangle \right]$$

QUITE IMPORTANT RULE 6

the amplitude to go from $|i\rangle$ to $|j\rangle$ is the complex conjugate of the amplitude to go from $|j\rangle$ to $|i\rangle$ $\langle j|i\rangle = \langle i|j\rangle^*$

IMPORTANT RULE 7: OPERATORS & THINGS

Every physical variable has an operator $\hat{O}$, associated with it. These operators have eigenstates defined by:

$$\hat{O}|\psi\rangle = a|\psi\rangle$$

where a is the value obtained if you conduct a measurement of that physical variable on a system in the eigenstate $|\psi\rangle$.

The complete set of eigenstates $\{|\psi_i\rangle\}$ for a given operator forms a basis.

The operator associated with a physical variable can be used to calculate the expectation value, $\langle \hat{O} \rangle$, of a series of measurements made on a collection of systems in the same state $|\varphi\rangle$:

$$\langle \hat{O} \rangle = \langle \phi | \hat{O} | \phi \rangle$$

IMPORTANT RULE 8: HEISENBERG'S UNCERTAINTY PRINCIPLE

$$\Delta x \times \Delta p \geq \frac{\hbar}{2}$$

IMPORTANT RULE 9: GENERALIZED UNCERTAINTY

$$\Delta O_1 \times \Delta O_2 \geq \left| \frac{i}{2} \left\langle \left[\hat{O}_2, \hat{O}_1 \right] \right\rangle \right|$$

IMPORTANT RULE 10: SOCIAL BOSONS

The probability of a boson entering a state already occupied by n bosons of the same type is proportional to $(n+1)$

Index